Springer Tracts in Natural Philosophy

Volume 29

Aldo Bressan

Relativistic Theories of Materials

Springer-Verlag
Berlin Heidelberg New York 1978

Aldo Bressan
Institute of Analysis and Mechanics
at the University of Padua, Italy

AMS Subject Classification (1970): 72 B xx, 83 A 05, 83 C xx

ISBN-13:978-3-642-81122-7 e-ISBN-13:978-3-642-81120-3
DOI: 10.1007/978-3-642-81120-3

Library of Congress Cataloging in Publication Data. Bressan, Aldo. Relativistic theories of materials (Springer tracts in natural philosophy; v. 29). Includes bibliographical references and indexes. 1. Relativistic mechanics. I. Title. QA808.5.B7. 530.1'1. 77-2632

Softcover reprint of the hardcover 1st edition 1978

2141/3140-543210

To Anna
in deep gratitude

Preface

The theory of relativity was created in 1905 to solve a problem concerning electromagnetic fields. That solution was reached by means of profound changes in fundamental concepts and ideas that considerably affected the whole of physics. Moreover, when Einstein took gravitation into account, he was forced to develop radical changes also in our space-time concepts (1916).

Relativistic works on heat, thermodynamics, and elasticity appeared as early as 1911. However, general theories having a thermodynamic basis, including heat conduction and constitutive equations, did not appear in general relativity until about 1955 for fluids and appeared only after 1960 for elastic or more general finitely deformed materials. These theories dealt with materials with memory, and in this connection some relativistic versions of the principle of material indifference were considered.

Even more recently, relativistic theories incorporating finite deformations for polarizable and magnetizable materials and those in which couple stresses are considered have been formulated. A broader description of the development of these relativistic topics is contained in § 13.

The purpose of this book is to describe the foundations of the general relativistic theories that include constitutive equations, and to present some applications, mainly to elastic waves, of these theories. This tract is divided into two parts. In the first part only the Eulerian point of view is considered; basic equations of general relativity, other than constitutive equations, are stated in full generality (except for couple stresses which are considered in part 2). Part 1 also thoroughly covers fluids, including constitutive equations.

In the second part, the Lagrangian point of view is introduced and—primarily following Bressan [1963b to 1967b]—some mathematical tools connected with materials coordinates are elucidated; by means of these tools, solids and materials of a very general kind—possibly capable of couple stresses—are dealt with.

An introductory chapter (Chap. 1) reviews the main ideas underlining the origin and development of relativity: among others, the mass-energy equivalence principle, the debated general principle of relativity, its local form, and Fock's privileged harmonic co-ordinates. Because the existence of these co-ordinates, hypothesized by the well-known Russian school headed by V. Fock, has not yet been proved, they will be taken into account only for a few cases of a conceptual nature on constitutive equations (§ 27, § 46).

Various authors view the subject of Chapter 1 from more or less different points of view; the presentation of this author's point of view should prove useful for a better understanding of the theory developed from Chapter 2 onward. However, that theory is discussed independently of Chapter 1 and is complete in itself; all needed axioms are stated and all theorems are fully proved, with the exception of a few marginal theorems that are not applied in the remainder of this work. In order not to obscure the main lines of development of the above theory, the proofs of some theorems of secondary interest for our purposes are given in the appendix.

It is assumed that the reader is familiar with the basic ideas of classical continuum mechanics for finite deformations and also with those of special and general relativity, particularly common tensor calculus. However, for the convenience of the reader, the main definitions and theorems of those subjects are reviewed in Appendix A—without any proofs—using our notations and including extensions to double tensors.

An overview of the theory contained in this tract, chapter by chapter, can be found in § 14.

Although the bibliography contains some works in classical physics, it is not intended to be exhaustive, even for relativity.

I am deeply indebted to Professor Elco, Department of Electrical Engineering at the Carnegie-Mellon University in Pittsburgh, Pa. (USA), for his decisive linguistic help in writing a part of this tract in English and for helpful suggestions, mainly concerning presentation.

I wish to thank Professor C. Truesdell of The Johns Hopkins University for his kind linguistic revision of the whole book. (Only very few additions and changes have been performed since this revision.) I am particularly grateful to him for his invitation to write this tract.

Table of Contents

Chapter 1. Introduction 1

§ 1. On the Beginning of Relativity 1
§ 2. The Space-Time Structure of Special Relativity and First Basic Consequences 2
§ 3. On the Operational Aspect of Physical Concepts 4
§ 4. New Ideas on Mass and Energy, in Contrast with Classical Physics, Accepted on the Basis of Special Relativity Kinematics 5
§ 5. On Forces, Cauchy Equations of Continuous Media, and the First Principle of Thermodynamics in Special Relativity 7
§ 6. On Electromagnetism, Heat Conduction, and Constitutive Equations in Special Relativity 9
§ 7. Gravitation and Relativity 10
§ 8. On the Local Equivalence Principle and the Basic Local Laws of the Electromagnetic Field and Continuous Media, Other than the Poisson Equation, in General Relativity. A Criterion Connecting those Laws with Their Analogues in Classical Physics or Special Relativity . . . 13
§ 9. On the Invariance of Physical Equations and on the Possible Physical Equivalence of the Frames in which these Equations have the Same Form. On a Privileged Absolute Concept of Event Point 16
§ 10. On Harmonic Coordinates and the Existence of General Frames not Physically Equivalent in General Relativity 19
§ 11. Some Distinctive Properties of General Relativity. On the Equivalence of General Frames in General Relativity 20
§ 12. What We Mean by General Theory of Relativity 23
§ 13. On the Development of General Relativity. Inclusion of Elasticity, Electromagnitostriction, Couple Stresses, and Hereditary Phenomena 25
§ 14. Scope and Plan of the Present Tract 27
Footnotes to Chapter 1 28

Part I. Basic Equations of Gravitation, Thermodynamics and Electromagnetism, and Constitutive Equations from the Eulerian Point of View

Chapter 2. Space-Time Kinematics Including Masses 35

§ 15. On the Riemannian Relativistic Space-Time Metric Introduced as a Chronometry. Admissible Frames. Some Possible Axioms for Non-Cosmological Relativity 35
§ 16. On Tensors in Relativistic Space-Time 39
§ 17. On Tensors in S_4 in Connection with a Moving Continuous Body $\mathscr{C}$ or an Ideal Fluid $\mathfrak{F}$. Spatial Projections and Natural Decompositions of Tensors, Spatial Derivatives and Spatial Divergences 40
§ 18. The Spatial Metric $d\overset{\perp}{s}{}^2$. Another Physical Meaning of the Chronometry ds^2. Ordinary Units 44
§ 19. On Some Classical Analogues of Locally Natural Frames from a Physical Point of View 46
§ 20. Material Derivatives, the Spatial Ricci Tensor, and the Relative Rate of Change of Proper Volume Dealt with from the Eulerian Point of View 50
§ 21. On Gravitational Mass and Reference or Conventional Mass. The Continuity Equation 53
§ 22. Angular and Deformation Velocities. Convected and Co-Rotational Fluxes. On $T_{\alpha\beta}{}^{/\beta}$ for $T'''_{[\alpha\beta]}=0$ 56
Footnotes to Chapter 2. 58

Chapter 3. Gravitation and Conservation Equations. Fluids and Elastic Waves 60

§ 23. The Einstein Gravitation Equations and Basic Consequences 60
§ 24. The Case of Interacting Matter Capable of Heat Conduction 61
§ 25. On the Second Principle of Thermodynamics, the Clausius-Duhem Inequality and Fourier's Law. A Relativistic Proof of the Symmetry of the Heat Conduction Coefficient 63
§ 26. On the Paradox of an Infinite Velocity of Heat Propagation from the Classical Point of View and the Relativistic One 69
§ 27. On the Local Spatial Physical Isotropy of S_4 75
§ 28. Free Energy and Relativistic Thermodynamics for Possibly Viscous Fluids. 77
§ 29. On Non-Viscous Fluids in the Presence of Heat Conduction and on Perfect Gases 80
§ 30. Acceleration Waves in Non-Viscous Fluids in the Absence of Electromagnetic Phenomena 81
§ 31. On Elastic Waves in Perfect Gases 84
§ 32. On the Importance of the Thermodynamic Tensor 85
§ 33. Some Historical Remarks on Relativistic Theories of Fluids and Hints at Some Further Results 87
Footnotes to Chapter 3 88

Chapter 4. Electromagnetism from the Eulerian Point of View. Polarizable Fluids . . . 90

§ 34. Introductory Considerations. The Ohm Law and the Relations Between the Electric and Magnetic Fields and the Respective Inductions . . . 90
§ 35. On the Maxwell Equations in Space Time . . . 93
§ 36. On the Electromagnetic Energy Tensor $E_{\alpha\beta}$. Some Requirements for it in the Absence of Polarization. Its Indeterminancy in the Presence of Polarization . . . 97
§ 37. On Some Widely Used Instances of the Electromagnetic Energy Tensor $E_{\alpha\beta}$ and Some Instances of $E^{\alpha\beta}{}_{/\beta}$ . . . 99
§ 38. On Isotropic Functions and Tensors . . . 102
§ 39. Some Uniqueness Properties of the Electromagnetic Energy Tensor $E_{\alpha\beta}$. On Its Arbitrariness in Connection with Heat Conduction . . . 105
§ 40. Some Historical Hints. Basic General Energetic Properties of Minkowski's Tensor and the Instances ${}^5E_{\alpha\beta}$ to ${}^7E_{\alpha\beta}$ of $E_{\alpha\beta}$ . . . 108
§ 41. Some Versions of Poynting's Theorem for Moving Media . . . 112
§ 42. W as the Proper Density of Non-Material Electromagnetic Energy 113
§ 43. On the Equations of Gravitation and Energy Balance in the Presence of Electromagnetic Phenomena . . . 115
Footnotes to Chapter 4 . . . 116

Chapter 5. On Media Capable of Electromagnetic Phenomena from the Eulerian Point of View. Magneto-Elastic Waves in Ideal Conductors 118

§ 44. Introduction . . . 118
§ 45. Black Body and Absolute Temperature in Thermodynamic Equilibrium 119
§ 46. Polarizable Non-Viscous Fluids . . . 121
§ 47. Polarizable Viscous Fluid . . . 125
§ 48. The Cauchy Equations in the Presence of Heat Conduction and an Electromagnetic Field; Preliminaries for Ideal Conductors . . . 128
§ 49. Dynamic Discontinuity Equations for Magneto-Elastic Acceleration Waves in Magnetizable Fluids . . . 129
§ 50. Magneto-Elastic Acceleration Waves in Magnetizable Non-Viscous Fluids . . . 131
Footnotes to Chapter 5 . . . 134

Part II. Materials from the Lagrangian Point of View

Chapter 6. Kinematics and Stresses from the Lagrangian Point of View . . 137

§ 51. Historical Hints at Relativistic Theories of Elastic and More General Materials . . . 137
§ 52. On the Representation of the Motion $\mathcal{M}$ of $\mathcal{C}$ . . . 138
§ 53. Lagrangian Spatial Derivative and Absolute Derivative of a Double Tensor Field with Respect to the Motion $\mathcal{M}$ of $\mathcal{C}$ . . . 141

§ 54. Polar Decomposition of the Position Gradient α^{ρ}_{L} and Principal Axes of Strain 144
§ 55. Fermi Transport 147
§ 56. On the Dilation Coefficients for Line, Volume, and Surface Elements, and the Ratio dC/dC^* 148
§ 57. The Vectors V^*_L and V^L_* for V_ρ Spatial. Expressions of $\grave{\alpha}^{\rho}_{L}$ and $\grave{C}_{LM}$ in Terms of $u_{\rho/\sigma}$ 152
§ 58. New Determination of the General Solution for the Continuity Equation. Connection of $D^c V_\rho$ and $D_c V_\rho$ with $D V^L_*$ and $D V^*_L$ for V_ρ Spatial, and Lagrangian Expression for the Electromagnetic Work $d_3 \lambda$ 153
§ 59. The First and Second Piola-Kirchhoff Stress Tensors $K^{\rho M}$ and Y^{LM}, and Lagrangian Expressions for $dl^{(i)}$ 155
§ 60. Connection Between $X^{\rho\sigma}{}_{/\overset{\perp}{\sigma}}$ and $K^{\rho M}{}_{|M}$ 156
§ 61. On $\overset{\rho}{\alpha}_{LM}$ and the Lagrangian Expression of $\overset{\perp}{g}{}^{\rho}_{\lambda} X^{\lambda\sigma}{}_{/\overset{\perp}{\sigma}}$ 158
§ 62. Explicit Form in Co-Moving Co-Ordinates for Some of the Preceding Lagrangian Formulas 159
Footnotes to Chapter 6 162

Chapter 7. Elasticity, Acceleration Waves, and Variational Principles for Simple Materials 164

§ 63. Foundations of Elasticity 164
§ 64. Some Theorems on Elastic Materials 166
§ 65. On Discontinuity Surfaces in Space-Time 168
§ 66. Dynamic Equations of Elastic Acceleration Waves 172
§ 67. Polarization and Inertial-Mass Quadrics. Acoustic Axes 174
§ 68. Pure Pressure States. Isotropic Elastic Materials. Comparison with the Classical Theory 176
§ 69. A Principle Concerning the Variation of the Metric Tensor of Riemannian Space-Time in the Adiabatic Elastic Case 179
§ 70. Variation of World Lines in the Adiabatic Elastic Case 182
Footnotes to Chapter 7 187

Chapter 8. Piezo-Elasticity and Magnetoelastic Waves from the Lagrangian Point of View 188

§ 71. Introduction 188
§ 72. Foundations of Piezo-Elasticity 188
§ 73. Extension of the Operations $T\ldots \to T^*\ldots$, $T^{\cdots} \to T^{\cdots}_{*}$, D^c and D_c to Tensors of Arbitrary Order 190
§ 74. On Rigid Motions in the Born Sense 192
§ 75. Born Rigidity and Stationary Tensors 194
§ 76. Some Invariance Properties of Ideal Conductors 196
§ 77. Dynamic Equations for Piezo-Elastic Ideal Conductors 197
§ 78. Magneto-Elastic Acceleration Waves in Piezo-Elastic Ideal Conductors 200
Footnotes to Chapter 8 202

Chapter 9. Materials with Memory and Axiomatic Foundations 203

§ 79. Introduction to a Relativistic Theory of Materials with Memory . . . 203

§ 80. Intrinsic Kinematic Histories. Total Geodesic Derivatives 204
§ 81. A Relativistic Version of the Principle of Material (Frame) Indifference 207
§ 82. Some Consequences of the Principle of Material Indifference 212
§ 83. On the Axiomatic Foundations of the Preceding Theory. Primitive Notions and First Axioms 216
§ 84. On Kinematic Axioms and the Notion of Physical Possibility 218
§ 85. Conservation Equations and Maxwell Equations in Our Axiomatic Theory 220
Footnotes to Chapter 9 223

Chapter 10. Couple Stresses and More General Stresses 226

§ 86. Introduction 226
§ 87. Contributions of Couple Stresses to the Expression of $\mathcal{U}_{\alpha\beta}$ and to the Equation of Energy Balance 227
§ 88. The Relativistic Cauchy Equations of Continuous Media in the Case of Couple Stresses 230
§ 89. The Non-Working Part of $m^{\alpha\beta\gamma}$ 233
§ 90. Some Commutation Formulas for Lagrangian Spatial Derivatives . . 233
§ 91. A Useful Expression for $\grave{C}_{LAB}$ 236
§ 92. A Lagrangian Expression for the Work of Stress and Couple Stress in Special or General Relativity 238
§ 93. Elasticity with Couple Stress 239
§ 94. Hints at Non-Viscous Fluids Capable of Couple Stress and at Electromagnetoelasticity with Couple Stress 241
§ 95. Some Preliminary Variational Formulas Related to Second Order Lagrangian Kinematics and the Variation of Space-Time Metric . . . 245
§ 96. A Variational Principle Involving Couple Stress and the Variation of Space-Time Metric 247
§ 97. A Variational Principle Involving the Variation of World Lines in the Presence of Couple Stresses. On Constitutive Equations 249
§ 98. On General Materials of Order $n=2$ in the Adiabatic Case. Variations of $g_{\alpha\beta}$ and World Lines 250
§ 99. Variational Principles for Elastic Materials of any Order $n \geqslant 1$, not Capable of Heat Conduction 253
Footnotes to Chapter 10 257

Appendix A. Double Tensors 258

§A 1. Definition of Double Tensors Related to Two Topological Spaces . . 258
§A 2. Partial Covariant Derivative and Total Covariant Derivative Based on a Mapping 259
§A 3. On Differentiation of Double Tensors, Functions of Double Tensors. Case of Arguments Fulfilling Typical Regular Conditions 261
§A 4. Partial Derivative of any Double Tensor $\hat{T}_{...}^{...}(H_{...}^{...})$ Defined Only for Values of $H_{...}^{...}$ that are Symmetric, Skewsymmetric, Spatial, or Subject to Other Particular Conditions 266

Appendix B. Two Uniqueness Properties of $E_{\alpha\beta}$ 269

Appendix C. On the Divergence of Spatial Vectors in Space-Time 273

Appendix D. On the Lie Derivatives $\mathscr{L}_u, D^c, D_c$ and the Lagrangian Representation $\overset{\perp}{T}_{\dots}^{\dots} \to \overset{*}{T}_{\dots}^{\dots}$. Application to Linear Elasticity 277

References 280

Index 287

Chapter 1

Introduction

§ 1. On the Beginning of Relativity

Scientists were forced to construct a relativistic theory of electromagnetism in order to have a systematic treatment of this subject in agreement with experience. More in particular, when the experimental transformation laws of electric and magnetic fields are taken into account Maxwell's basic equations of electromagnetism are not invariant for Galilean transformations, but are invariant for Lorentz transformations. Moreover Michelson's famous experiment of 1881 can be explained by assuming that light propagation in vacuum is isotropic. This excludes the possibility of determining one privileged inertial space or *absolute* space $S_{(ab)}$.[1)]

The other hypotheses considered to explain Michelson's experiment were soon disproved. Two such hypotheses are the following:

a) The dragging hypothesis, which asserts the existence of a (positive) coefficient $\alpha (\alpha \leqslant 1)$ by which the ether should be dragged along by a moving body (in 1818, on the basis of his elastic ether theory, Fresnel assumed $\alpha = 1 - n^{-2}$ where n is the refraction index; in 1845 Stokes assumed $\alpha = 1$).

b) The ballistic hypothesis, which asserts a (certain) dependence of the speed of a light ray on the velocity of its source.

The hypothesis a) was abandoned on the basis of Fizeau's and Hoek's experiments of 1851 and 1868;[2)] the hypothesis b) was disproved by an analysis of the apparent motions of double stars—cf. Möller [1952].

Lorentz transformations and the above experimental results received a very satisfying physical interpretation by A. Einstein [1905], where upon a new physical theory was born. It stated that inertial spaces are mutually equivalent not only with regard to mechanical phenomena—which was substantially known by Galileo—but also with regard to electromagnetic phenomena. Hence this theory denied the existence of the above absolute space $S_{(ab)}$, and it was called the theory of relativity.

Einstein succeeded in founding the above theory, which explained the preceding theoretical and experimental results, through his criticism of the traditional concepts of simultaneity, past, and future. Therefore the theory of relativity does not differ from the preceding theories merely by improvements or additions of some physical laws. The difference is a fundamental one; it concerns basic concepts and the way of thinking on any physical phenomena. As a consequence most classical concepts and laws required reviewing. In spite of this apparent difficulty, the theory of relativity, or more precisely its first stage of development, special relativity, is now accepted by most physicists as the most convincing (and systematic) way of dealing with electromagnetic phenomena.

In special relativity, space-time S_4 is considered to be pseudo-Euclidean and gravitation is not taken into account. In a second stage of relativity theory, constructed to include gravitation, S_4 is considered to be a Riemannian 4-dimensional space whose structure (metric) depends on phenomena and becomes pseudo-Euclidean in the case of empty space. These are distinguished as special relativity and general relativity.[3)]

§ 2. The Space-Time Structure of Special Relativity and First Basic Consequences

Let us first remark that with relativity—based on Lorentz transformations instead of Galilean transformations—the concept of *event point*, which can be used also in classical physics, became very useful. An event point $\mathscr{E}$ can be determined by means of four real numbers; three of them are spatial co-ordinates —taken e.g. in a Cartesian frame joined to an inertial space, $S_{(i)}$—and the fourth is a time co-ordinate. If one takes the concepts of event point and simultaneity as primitive, then the instants may be introduced as special space-time sections, e.g. by the following definition: *The set of the event-points simultaneous with a same event point is called an instant.*

In classical physics we have one system of instants, the same for all observers connected with the different inertial spaces. Furthermore, any two such observers consider as natural the same systems of time co-ordinates; these systems are determined up to non-singular linear transformations (i.e. changes of time origin and time length units). Instead, the relativistic concept of simultaneity gives us ∞^3 systems of instants, each like the above mentioned one; and they are in a one-to-one correspondence with inertial spaces. Since no inertial space is privileged, the same holds for the above ∞^3 systems of instants.

It is physically meaningful to choose a same time unit for all of the above systems of instants, i.e. for all inertial spaces; furthermore for every inertial space $S_{(i)}$ there is exactly one (natural) time co-ordinate t that: (i) is defined for the instants associated with $S_{(i)}$, (ii) is connected with the above time unit, and (iii) takes the value zero for a given event point $\mathscr{E}_0$ (i.e. for the instant associated with $S_{(i)}$, which constitutes a space-time section through $\mathscr{E}_0$).

There is a Cartesian co-ordinate system (x^1, x^2, x^3) joined to $S_{(i)}$. Then the space-time system (t, x^1, x^2, x^3) is called a Lorentz co-ordinate system or Lorentz frame. Such frames are related to each other by Lorentz transformations, which, in particular afford the mathematical relation between any two of the above natural time co-ordinates; this relation involves also spatial co-ordinates.

Well-known consequences of Lorentz transformations, in contrast with classical kinematics, are time dilation, length contraction, and a non-additive law for velocity composition. This law complies with the condition that, if with respect to an inertial space a signal propagates with the velocity c of light in vacuum, then the same holds for every other inertial space.

Some convincing considerations, based also on the causality principle, show that c is an upper limit for the velocity of every signal, with respect to inertial spaces, and in particular for the mutual relative velocity of any two material particles.

Now let us consider a Lorentz co-ordinate system (t, x^1, x^2, x^3) joined to the inertial space $S_{(i)}$, and any linear element in S_4, with end points (t, x^1, x^2, x^3), $(t+dt, \ldots, x^3+dx^3)$. Then the quadratic form in $dt, \ldots, dx^3$

$$ds^2 = (c^2 - v^2)dt^2\,, \quad \text{where} \quad v^2 = \sum_{s=1}^{3} \left(\frac{dx^s}{dt}\right)^2, \tag{2.1}$$

does not depend on the Lorentz system being considered and constitutes the *Minkowski metric* in S_4.

By a *particle* we mean a very little isotropic body of nonzero rest mass, just as in particle mechanics. By *material point* we understand either a particle or—following Truesdell and Noll [1965]—an ideal zero-dimensional element of a continuous body $\mathscr{C}$, which is determined by three material co-ordinates. Such an element always has zero mass (it generally has a non-zero mass density). Thus a photon, which has zero rest mass but non-zero mass relative to every observer, is not considered as a particle or a material point in this tract.

The set W_M of event points occupied by the material point M is called the *world line* of M. Let M describe the above linear element (t, x^1, x^2, x^3), $(t+dt, \ldots, x^3+dx^3)$, so that this element belongs to W_M. Then v—cf. (2.1)—is the intensive velocity of M with respect to $S_{(i)}$.

It is commonly understood that a material point M can carry along an ideal observer, hence a reference frame. Therefore, the following hypothesis is quite natural: *If the material point M passes through the event point $\mathscr{E}$, there exists an inertial space, $S_{(pr,i)}$, with respect to which M has zero velocity.* Such a space is unique and is called *momentarily proper.*

Now we suppose, as an absurd hypothesis, that the velocity v of M with respect to $S_{(i)}$ equals c. Then a photon could have a world line tangent to W_M at $\mathscr{E}$. Thus the velocity of this photon with respect to the inertial space $S_{(pr,i)}$ would be zero instead of c. We conclude $v \neq c$.

Furthermore, as we said, c is an upper limit on the velocity of every signal, so that $v \leqslant c$. Hence the relation $v < c$ must hold, so that by (1) $ds^2 > 0$ along the elements of the world line of any material point.

Conversely, if along a linear element of S_4 we have $ds^2 > 0$, then it can be described by a particle[4)], which is equivalent to the following condition: For every point $\mathscr{E}$, every vector v_r whose modulus is less than c can be the velocity of some particle at $\mathscr{E}$, with respect to the co-ordinate system (t, x^1, x^2, x^3).[5)]

For $S_{(i)}$ coinciding with $S_{(pr,i)}$ the velocity v of M vanishes momentarily, so that by (1.1) $ds = c\,dt$ holds momentarily. Therefore, setting $\tau = c^{-1}s$, where s is the arc length on the world line W_M of M (relative to a given origin $\mathscr{E}_0$) we momentarily have $dt = d\tau$; hence τ is called the *proper time* of M, of origin $\mathscr{E}_0$.

Some authors identify the proper time of M with the time marked by a clock carried along by M. Other authors accept this identification only when M is not accelerated by some force, because this acceleration might affect the intrinsic motions of different clocks in different ways.

§ 3. On the Operational Aspect of Physical Concepts

In order to consider the operational character of physical concepts, let us take into account the relativistic phenomenon of length contraction. We may consider an infinitesimal material line $d\ell^*$ undergoing a given motion. The length of $d\ell^*$, when its end A^* occupies the event point $\mathscr{E}$, may be measured with what is usually considered to be a "rigid" rod at rest in an inertial space $S_{(i)}$, by marking on this rod the positions simultaneously occupied by the ends of $d\ell^*$ with respect to $S_{(i)}$. According to classical physics, the result of this experimental operation is independent of $S_{(i)}$ and is simply called the *length of* $d\ell^*$. Instead, in (special) relativity the same experimental result depends on $S_{(i)}$ and is called the relative length, precisely, the *length relative to* $S_{(i)}$. When $d\ell^*$—or at least A^*—has zero velocity with respect to $S_{(i)}$, its relative length is called *proper length* and is the largest relative length.

Thus, in special relativity—a second stage in the development of physics—$d\ell^*$ has many different lengths (for A^* in $\mathscr{E}$) which correspond to the above ways of measuring length, connected with different inertial spaces. The length measurements of $d\ell^*$ commonly taken by experimenters are very similar and not even mutually distinguishable. Therefore in classical physics—i.e. in the first stage in the development of physics—one can consider only one length of $d\ell^*$ to be measured in any of the above ways.

Thus, historically, the basilar properties attributed to the concept of length effectively changed with the set of physical operations employed to measure lengths. This is an important effect of the operational character of the concept of length.

Roughly speaking, our concept of a given physical magnitude has an operational character provided we know an experimental procedure to measure the values of this magnitude. This procedure may be an ideal experiment—cf. Synge [1965, § 4, p. 7]—i.e. it may be conceived without regard to practical difficulties.

In case the magnitude in question is length, as appears from the preceding considerations on $d\ell^*$, the fact that the measuring rod is stationary with respect to a particular inertial space $S_{(i)}$ rather than to another constitutes a feature of the experimental procedure to measure the length of $d\ell^*$, which has no importance in connection with crude techniques; however it becomes essential for finer techniques in that it causes us to consider a numerous variety of magnitudes instead of the original one.

Similar considerations, based again on Einstein's criticism, hold for the concept of duration.

The operational character of physical concepts had substantially proved useful also before relativity arose, e.g. in connection with the concept of temperature. This was first based on the thermic dilation property of some fluid, but it depended on that fluid; then it was based on the first and third principles of thermodynamics; more precisely, the *absolute temperature* $T (T>0)$ was introduced.

Relativity's contribution—to realize the importance of operational concepts—is great because Einstein's criticism, on which relativity was based, essentially dealt with the basic concepts of space and time, which were so familiar and unsuspected. Newton intuitively characterized them, not operationally, but in terms of properties—cf. Bridgman [1927], p. 4. These properties were considered to be well known even in everyday life and very reliable, as appears among other things from the fact that E. Kant considered space-time concepts as given "a priori".

Einstein's criticism of such concepts pushed physicists to request every physical concept to be operational, and to abandon intuitive definitions of physical concepts in terms and properties, as was the case with past theories such as Newton's mechanics.[6)]

Let us add that relativity theory—through Einstein's criticism—pushes physicists to reinterpret old physical theories using operational concepts, and especially that such reinterpretations are generally quite possible; of course the degree of approximation with which old theories can describe reality must be taken into account.

To the above intuitive definitions in terms of properties are related the axioms of some modern physical theories where the axiomatic aspect is emphasized; hence, such theories may look abstract. Nonetheless, a suitable interpretation may be possible for them—as well as for old theories—, by which these theories become as close to physical reality as common physical theories.[7)]

§ 4. New Ideas on Mass and Energy, in Contrast with Classical Physics, Accepted on the Basis of Special Relativity Kinematics

Einstein's physical interpretation of Lorentz transformations—by which the transformations of (privileged) time co-ordinates involved also spatial co-ordinates—caused fundamental changes in physicists' ideas on mass and energy. In order to say how this could happen, let us consider, in special relativity, a very little material point M (test particle) having no internal structure (hence a constant internal energy). Furthermore let v be its intensive velocity with respect to an arbitrary inertial space $S_{(i)}$.

Now let us remark, from the theoretical point of view, that, if the mutual translation velocity of two inertial spaces is very small, then the Lorentz transformations connected with them are very similar to Galilean transformations; moreover, as to experiments, in the field concerned by special relativity the phenomena on systems which are moving slowly with respect to the observer are well described by classical physics. Therefore it was natural to assume that in the (momentarily) proper inertial space $S_{(pr,i)}$ [§ 2] the relativistic version of the fundamental equation of dynamics for M is identical with its classical version

$$d(m_0 v_r)/dt = f_r \quad (r=1,2,3), \text{ with } m_0 \text{ constant}. \tag{4.1}$$

However, in classical physics, by Galilean transformations, the acceleration dv_r/dt of M, hence the force f_r, are the same vectors in all inertial spaces. Instead, in special relativity, by Lorentz transformations, the fundamental equation of dynamics for M cannot have the form (4.1) also in $S_{(i)}$ (for $v \neq 0$). The form in a Lorentz frame (t,x^1,x^2,x^3) for this relativistic equation, which appears most natural and most similar to (4.1), is[8)]

$$\frac{d}{dt}(mv_r) = F_r\,, \tag{4.2}$$

where

$$m = \frac{m_0}{\sqrt{1-\dfrac{v^2}{c^2}}}\,, \qquad F_r = \sqrt{1-\frac{v^2}{c^2}}\, f_r\,, \qquad v_r = \frac{dx^r}{dt} \qquad (r=1,2,3)\,. \tag{4.3}$$

Then it is natural to call m_0 *proper mass*, f_r *proper force*, m *relative mass*, mv_r *relative momentum*, and F_r *relative force referred to* $S_{(i)}$.

As is well-known, by (4.2) and (4.3) $c^2\,dm/dt$ equals the power $F_1 v^1 + F_2 v^2 + F_3 v^3$ of the force F_r. According to classical physics that power equals $d\mathscr{I}/dt$ where $\mathscr{I}$ denotes the kinetic energy. Then it is most natural to assume $\mathscr{I} = c^2(m - m_0)$ as the relativistic kinetic energy of M. This agrees with the fact that by (4.3), $c^2 m \approx m_0(c^2 + v^2/2)$ for $v \ll c$.

The above most natural assumption constitutes an equivalence relation between mass and kinetic energy. This relation is in complete agreement with a well known "material" property of electromagnetic energy derived from the classical Maxwell equations. So, more generally, the following relativistic *mass-energy equivalence principle* was stated: *An energy W, of any kind, is physically equivalent to a mass m, with*

$$W = c^2 m\,. \tag{4.4}$$

Nowadays the above principle, in particular with regard to proper mass, has received many experimental confirmations in nuclear physics and, more precisely, in the transformations of some chemical elements.

In classical physics inertial mass and gravitational mass are proportional and hence are identical for suitable choices of units. Therefore in (4.4) one may regard m as gravitational mass. Throughout this tract (4.4) will be understood in this way. This will be useful in dealing with general relativity, for then, on the one hand, gravitational mass occurs in Einstein's gravitation equations, one of which is a substitute for the Poisson equation connecting the mass density with the gravitational potential. On the other hand, a certain kind of inertial mass will be of interest in § 67 in connection with elastic waves, and this inertial mass cannot always be represented by a scalar and hence is not always proportional to gravitational mass.

§ 5. On Forces, Cauchy Equations of Continuous Media, and the First Principle of Thermodynamics in Special Relativity

Let M schematize an element of an elastic three-dimensional continuous body $\mathscr{C}$, i.e. an infinitesimal three-dimensional portion $d\mathscr{C}$ of $\mathscr{C}$. Then the internal energy of $d\mathscr{C}$ generally varies according to the first principle of thermodynamics. Thus the proper mass m_0 of $d\mathscr{C}$ must vary according to the equivalence principle (4.4). For this reason the relativistic fundamental law (4.2) of dynamics for $d\mathscr{C}$ cannot be considered to coincide with its classical analogue (4.1), even if the proper inertial space $S_{(\mathrm{pr,i})}$ is referred to; furthermore this relativistic version is strictly connected with the first principle.

There are also other reasons—due to the relativistic ideas on space time—for changing the classical fundamental equation of dynamics, even when $S_{(\mathrm{pr,i})}$ is referred to: According to classical physics various forces—such as gravitational and Coulomb forces—are at a distance, i.e. they are exerted and received simultaneously. In (special) relativity, to deal with such forces would be troublesome, for simultaneity depends on the inertial space being referred to.

In addition forces at a distance may be considered as signals whose speed is infinite. Hence, in (special) relativity, considering them would be incompatible with the inference, mentioned above, that no signal can travel faster than light [§ 2].

We may conclude that, in relativity, dynamic equations must concern only contact forces and—because of the variability of mass and the equivalence principle (4.4)—they are to be considered in a tight connection with the first principle.

So, through contact forces, we are led to consider the continuous body $\mathscr{C}$ again. Let $X^{rs}(r,s=1,2,3)$ be the classical stress tensor having a pressure character. Then, in case forces at a distance are absent, classical Cauchy equilibrium equations are

$$\sum_{s=1}^{3} \frac{\partial X^{rs}}{\partial x^s} = 0 \quad (r=1,2,3). \tag{5.1}$$

Let us denote mass density by k. Then, taking the continuity equations into account, in the above case the Cauchy dynamic equations can be written like (5.1):

$$\sum_{\beta=0}^{3} \frac{\partial W^{r\beta}}{\partial x^\beta} = 0 \tag{5.2}$$

where

$$W^{r0} = k v^r, \qquad W^{rs} = k v^r v^s + X^{rs} \quad (r,s=1,2,3), \qquad x^0 = t. \tag{5.3}$$

In addition let kw be the proper density of internal energy and q_r the heat flux vector (having the oriented direction of the heat flow). Then, by combining the kinetic energy theorem for continuous media, implied by Cauchy dynamic equations, with the first principle, one proves—cf. Truesdell and Toupin [1960; (285.2), (286.6), (288.9)]—that *in case no forces at a distance occur (and $X^{rs}=X^{sr}$) we have*

$$\sum_{\beta=0}^{3} \frac{\partial W^{0\beta}}{\partial x^{\beta}}=0 \tag{5.4}$$

where

$$W^{00}=k\left(\frac{v^2}{2}+w\right), \quad W^{0r}=k\left(\frac{v^2}{2}+w\right)v^r+q^r+X^{rs}v_s, \quad t=x^0. \tag{5.5}$$

Therefore, in the above case, the Cauchy dynamic equations and the first principle can be expressed, in classical physics, by the vanishing of the space-time divergence of the system $W^{\alpha\beta}(\alpha,\beta=0,\dots,3)$. In accord with a widespread nomenclature (based on $(5.3)_{1,2}$ and $(5.5)_{1,2}$ for $X^{rs}=0$) $W^{\alpha\beta}$ may be called the energy-momentum system. Likewise (5.2) and (5.4) may be referred to as energy-momentum equations.

The interest of this point of view is increased by the possibility—shown by means of the Maxwell electromagnetic tensor—of introducing Coulomb and Lorentz forces as contact forces, at least for negligible (electric and magnetic) polarizations.

Incidentally, in classical physics the conservation equations (5.2) and (5.4) were also extended to any electromagnetic field with any polarizations; of course a privileged inertial space—i.e. the absolute space $S_{(ab)}$ which Michelson tried to determine—was referred to[9)].

Because of the preceding considerations about force, mass, and the coupling of dynamic equations with the first principle, the classical conservation equations (5.2) and (5.4) or particular cases of these have engaged the attention of relativists. Their analogues in special relativity are easy to write, taking the proper inertial space $S_{(pr,i)}$ into account.

Since such analogues—i.e. relativistic (energy momentum) conservation equations—will be explicitly and extensively considered in general cases later [§§ 23, 24, 43, 87] presently we may limit ourselves to pointing out that such equations are very useful in special relativity where there is a space-time metric: Minkowski metric. Moreover the tensor calculus of Ricci and Levi Civita enables us to write easily invariant equations in any Riemannian space—cf. Einstein [1955]—. Consequently the use of the tensor conservation equations is an easy way to obtain invariance under Lorentz space-time transformations and, moreover, under any (regular) space-time transformations. The last kind of invariance, also called world invariance, is of great importance, especially in the second stage of development of relativity; general relativity.

As to world invariance let us incidentally add that it was achieved recently in the classical dynamics of particles and continua[10)], and, in particular, for classical energy-momentum conservation equations—such as (5.2), (5.4)—including electromagnetism.[11)]

§ 6. On Electromagnetism, Heat Conduction, and Constitutive Equations in Special Relativity

Special relativity created so as to solve a fundamental particular problem of electromagnetism: to give Lorentz transformations a physical interpretation. In accordance with this fact, a very compact treatment of electromagnetism was possible in special relativity: The electric and magnetic fields and the corresponding inductions can be expressed by means of two skew-symmetric space-time tensors (of rank 2), such that by taking a Minkowski space-time metric into account, the values of the fields and inductions measured in different inertial spaces are related by the simple rules of tensor calculus. Moreover, in special relativity, electric charge and current are represented by a 4-vector, and the Maxwell equations can be condensed into two tensor space-time equations.

In the development of relativity, heat and temperature were studied very early; however, for a rather long time only adiabatic processes were dealt with in relativistic thermodynamics.[12)] The relativistic energy-momentum conservation equations included heat conduction only after C. Eckart introduced his thermodynamic tensor $Q_{\alpha\beta}$ in 1940. However, the explicit form of $Q_{\alpha\beta}$ has been and still is a continuing matter of discussion, especially in the case of anisotropic materials.

Relativistic heat conduction will be widely dealt with (in Chapter 3) only in the framework of general relativity. Here let us only say that the afore-mentioned difficulties essentially have two main reasons:

First, a classical local equation may have several natural relativizations (based on the consideration of the momentarily proper inertial space $S_{(\mathrm{pr,i})}$) if spatial derivatives occur in it. In particular, the classical divergence of the heat flux vector q_r may be naturally relativized both into a space-time divergence and into a spatial divergence; and the corresponding explicit forms of $Q_{\alpha\beta}$ are different (Chapter 3).[13)]

The other reason—which perhaps is giving more trouble—is the fact that the speed of every signal has c as a relativistic upper limit, whereas the Fourier classical law implies an infinite speed for heat propagation. Moreover, as far as I know, the only classical heat conduction law proposed to avoid the above implication—cf. Cattaneo [1948]—substantially concerns fluids. Several relativistic attempts made thus far to solve the afore-mentioned problem are essentially based on that proposal. Müller [1969] presents a new approach to solve the problem (again for fluids) [§ 33].

Let us add that from the historical point of view the relativistic treatment of heat conduction caused other differences among relativistic theories of thermodynamics. However, for several of these mutually incompatible theories, it is now known that certain more or less physically unsatisfactory theorems can be proved. Thus a certain selection can be made, a selection that may depend on the purpose being considered.

On the whole, in special (and general) relativity the theories of elastic or more general materials were developed more recently than electromagnetism and thermodynamics. It is true, relativistic elasticity was considered by Herglotz [1911] as early as 1911, as is mentioned in the well-known book Möller [1952]. However, until recently most authors on relativistic elastic theories considered only linear theories, and the foundations of these do not appear as complete as the corresponding classical theories. The situation was even worse with relativistic electromagnetostriction.[14)]

In the last decade general theories of mechanical and also thermo-electro-magneto-mechanical constitutive equations have been available in special relativity. These theories have been directly stated in general relativity. Therefore a brief survey of them is put off to § 13, which concerns general relativity. The same holds for the foundations of relativistic theories for materials with memory, non-simple materials[15)], and in particular materials capable of couple stresses.

§ 7. Gravitation and Relativity

Although Maxwell had dealt with electromagnetic actions by means of the so-called Maxwellian tensions, nobody had dealt with gravitation as due to contact forces. Therefore, in special relativity—where S_4 is pseudo-Euclidean—gravitation was not taken into account at all.

To include gravitation in relativity Einstein performed a deep modification of relativistic space-time, briefly hinted at in § 1. Now we wish to consider this modification more in detail, with the aim of considering briefly the physical acceptability of the new theory which so arises, even in the general version to be presented in this tract. However, we do not claim to follow exactly the reasonings which historically led Einstein to construct general relativity. Therefore let us consider a test particle M neither charged nor subject to appreciable gravitation.[16)] In classical physics M moves along a straight line of any inertial space $S_{(i)}$. This holds also in special relativity where, in addition, M describes a geodesic of S_4, along which Minkowski metric is positive—cf. (2.1).

Light rays or photons have the same behavior as M both in classical physics and in special relativity except that the Minkowski metric is zero—and not positive—along the geodesics described by photons in Minkowski space time.

As Einstein proved, it is compatible with actual gravitation experiments to generalize Minkowski pseudo-Euclidean metric ds^2 into a Riemannian and locally Lorentzian space-time metric, which depends on phenomena, becomes pseudo-Euclidean in total absence of matter—so that we may denote it again by ds^2—and in addition fulfills the following conditions:

(a) *Along the world lines of material points* $ds^2 > 0$.

(b) *The possible world lines of uncharged test particles*—cf. footnote 16 in Chapter 1—*are the geodesics of* S_4 *along which* $ds^2 > 0$—cf. footnote 4 in Chapter 1.—

(c) *The possible world lines of photons in empty regions are the geodesics of* S_4 *along which* $ds^2 = 0$.

In case S_4 is empty, conditions (a) to (c) hold according to special relativity. So they appear reasonable also in the presence of gravitational mass. In our theory the metric ds^2 will be characterized by other conditions, substantially following Synge. However, here we consider conditions (a) to (c) by their important physical content.

On the one hand, by taking ideal experiments into account one can prove that in connection with a same physical process $\not p$ of the universe and a given length unit, there exists at most one space-time metric ds^2 which satisfies conditions (a) to (c), up to a constant factor—cf. Levi Civita [1928, p. 85]. This uniqueness of ds^2 contributes essentially to showing the operational character of the metric satisfying the above-mentioned conditions.

However, on the other hand, by the actual state of experimental techniques—not very powerful, at least in connection with purely thermo-mechanical phenomena—there are several space-time metrics, for each of which the validity of conditions (a) to (c) are compatible with all known experimental results. Among these metrics, in connection with a given process $\not p$ of the universe, there are the metrics satisfying either the system $\mathscr{E}_1$ of Einstein's non-cosmological gravitational equations for gravitating and mechanically interacting matter, or anyone of the successive generalizations $\mathscr{E}_2$ to $\mathscr{E}_4$ of $\mathscr{E}_1$, to be considered in this tract in order to deal also with heat conduction [§ 24], general electromagnetic phenomena [§ 43][17] and couple stresses [§ 87] respectively.[18]

In harmony with the preceding considerations is the fact that the explicit form of the system $\mathscr{E}_2$, including heat conduction, has been and still is a matter of discussion.

Also in recent treatises of general relativity the agreement of this theory with ordinary gravitation experiments is explicitly shown only in the purely mechanical case, referred to by the above system $\mathscr{E}_1$. However, from the reasonings commonly used to prove this argument and from the comparison of $\mathscr{E}_1$ with any of its generalizations $\mathscr{E}_2$ to $\mathscr{E}_4$ hinted at above, it is immediately obvious that the same satisfactory conclusion holds also if electromagnetism, thermodynamics, and couple stresses are taken into account.

We shall not explicitly prove the important fact that the Einstein theory of general relativity based on any $\mathscr{E}_i$ of the above hinted systems $\mathscr{E}_1$ to $\mathscr{E}_4$, of increasing generality, is in agreement with gravitational phenomena. Therefore, we are going to summarize in an introductory style a common procedure by which this proof can be done. To this purpose we first remark that the system $\mathscr{E}_i$ constitutes a link between the space-time metric ds^2, on the one hand, and the actual material properties of matter and possibly (for $i > 1$) the actual electromagnetic field on the other hand; furthermore we remember that the following assertion is usually proved as a preliminary:

Let the space-time metric ds^2 differ very little from an everywhere pseudo-Euclidean metric—Lorentzian metric—in the space-time frame (x) *being considered. Moreover, we set*

$$-g_{00}=c^2-2U \quad \left(ds^2=-\sum_{\alpha,\beta=0}^{3} g_{\alpha\beta}(x^0,\dots,x^3)dx^\alpha dx^\beta, \quad g_{00}<0\right) \tag{7.1}$$

and consider the classical motions which correspond to ordinary initial conditions and which are dynamically possible for a free particle subject to the potential U given as a function of the Galilean—i.e. Cartesian inertial—co-ordinates x^1,x^2,x^3 and the natural time $t=c^{-1}x^0$. Then these motions are very close to some geodesics of ds^2 (along which $ds^2>0$).

Let us now consider ordinary physical situations, essentially those holding for the experiments confirming Newton's gravitation theory. Such situations can be satisfactorily described using a static space-time metric ds^2.[19] Therefore, in various treatises of general relativity such a ds^2 is considered in order to realize the compatibility of general relativity with ordinary gravitation experiments, and to the same purpose the purely mechanical theory based on the system $\mathscr{E}_1$ is usually referred to. Then one proves that in a static frame (x)—cf. footnote 19 in Chapter 1—the values of x^1,x^2,x^3 are constant along the world line of any material point. As a consequence, in the purely mechanical case the mass density k is a function of the same co-ordinates:[20]

$$k=f(x^1,x^2,x^3). \tag{7.2}$$

At this point one assumes that the metric ds^2 is very nearly a Lorentzian metric, which is reasonable from the physical point of view in connection with ordinary gravitation experiments, and one proves on the basis of $\mathscr{E}_2$ that the function U of x^1,x^2,x^3, given by $(7.1)_1$ in a static frame, is very nearly the classical gravitational potential corresponding to the mass distribution described by (7.2) when x^1,x^2,x^3 are meant as Galilean co-ordinates. More in detail, U is usually shown to satisfy an equation very similar to the Poisson equation in connection with the mass density (7.2). Therefore, the Poisson equation is considered to be implied by the system $\mathscr{E}_1$.[21]

Hence by the preceding italicized assertion including (7.1), Einstein's theory based on $\mathscr{E}_1$, and in particular the validity of the preceding conditions (a) and (b) for the space-time metric ds^2, are compatible with ordinary gravitation experiments confirming classical physics. The same holds for condition (c) because it does not involve the curvature of S_4 and is true according to special relativity. Likewise, the afore-mentioned compatibility can be proved for the conditions by which ds^2 will be characterized in § 15, substantially following Synge [1960].

In addition the theory based on $\mathscr{E}_i$—which deals with matter in the scheme of continuous bodies—substantially implies that the world lines of uncharged test particles—cf. footnote 16 in Chapter 1—must be some geodesics of S_4 along which $ds^2 > 0$. Then this theory is in agreement with ordinary gravitation experiments.

Gravitation experiments and careful astronomical observations—in particular those of the advancement of the perihelion of Mercury—prove that general relativity describes gravitational phenomena not only at least as well as, but even better than classical physics. Hence general relativity not only reaches the theoretical aim of furnishing a general scheme of natural phenomena, which both includes gravitation and is in accord with the basic relativistic concepts (thereby complying with Michelson's electromagnetic experiment) but it also effectively improves the accuracy of the theory of gravitation itself.

At this point it must be added, however, that the afore-mentioned theoretical aim is very important in itself. Therefore we may not help remarking that general relativity implies that gravitational actions propagate with the speed c, which is a chief result in accord with a basic conclusion concerning the speed of any signal, drawn just from the basic relativistic ideas of special relativity [§ 2].

Usually the considerations hinted at in this section are thoroughly developed only in connection with the system $\mathscr{E}_1$ but it is easy to realize that they hold also for $\mathscr{E}_2$ to $\mathscr{E}_4$.

§ 8. On the Local Equivalence Principle and the Basic Local Laws for the Electromagnetic Field and Continuous Media, Other than the Poisson Equation, in General Relativity. A Criterion Connecting those Laws with Their Analogues in Classical Physics or Special Relativity

On the one hand, within general relativity, we say that the frame—i.e. the co-ordinate system—(x) is *(locally) natural* at the event point $\mathscr{E}$, if it is locally pseudo-Euclidean—cf. (2.1)—and geodesic at $\mathscr{E}$.

If $\mathscr{E}$ belongs to the world line of a material point M, then, up to a spatial rigid displacement and some infinitesimals of order greater than 2 there is only one natural frame $(\bar{x})$ at $\mathscr{E}$ with respect to which M has zero spatial velocity at $\mathscr{E}$.[22] We shall call such a frame *(locally) proper and natural* at $\mathscr{E}$.

On the other hand, within classical physics we say that the space-time frame $(y) = (y^0, y^1, y^2, y^3)$ is *Euclidean* if

(i) y^0 is a natural time co-ordinate, and

(ii) the co-ordinates y^1, y^2, y^3 belong to a Euclidean spatial frame $\mathscr{F}$, i.e. to an orthogonal frame moving rigidly with respect to inertial spaces and as regularly as desired.

The Euclidean frame (y) will be said to be *locally natural* at the event point $\mathscr{E}$ if

(iii) the frame (y)—or the three-dimensional frame $\mathscr{F}$ subordinated by it—is *non-rotating* and *freely falling* at $\mathscr{E}$, i.e. at $\mathscr{E}$ and with respect to inertial spaces the angular velocity of $\mathscr{F}$ vanishes and the acceleration of $\mathscr{F}$ equals the gravitational force per unit mass.

Suppose that the material point M passing through the event point $\mathscr{E}$ has zero velocity with respect to the spatial frame $\mathscr{F}$ at $\mathscr{E}$. Then both $\mathscr{F}$ and the space-time frame (y) are said to be *proper* at $\mathscr{E}$. We shall denote by $\mathscr{F}_{(pr)}$ and $(\bar{y})$ a spatial frame and a space-time frame proper and locally natural at $\mathscr{E}$.

In § 19 the same physical condition will be shown to characterize both the locally natural space-time frames among the locally pseudo-Euclidean frames in general relativity, on the one hand, and the locally natural Euclidean frames among the Euclidean space-time frames in classical physics on the other hand.

Let us also assume that the same space and length units hold for all frames we are using. Then the locally natural Euclidean space-time frames turn out to be the analogues in classical physics for the locally natural frames in general relativity.

In case gravitation is negligible, on the one hand the above classical space-time frames (y) and their corresponding spatial frames $\mathscr{F}$ coincide with inertial frames, at least up to infinitesimals of order greater than 2; on the other hand, in the same case, locally natural frames coincide with Lorentzian frames, up to infinitesimals of order greater than 2 (at $\mathscr{E}$).

In the case of negligible gravitation let us consider the inertial space $S_{(pr,i)}$ which is momentarily proper. We can identify $(\bar{x})$ and $(\bar{y})$ precisely with space-time frames which are joined to $S_{(pr,i)}$ and are pseudo-Euclidean and Euclidean respectively. ($S_{(pr,i)}$ can be considered to be freely falling and non-rotating.)

According to classical physics, in the above frame $\mathscr{F}_{(pr)}$ the gravitational force balances the dragging force and Coriolis force in every case. Hence these forces can be dropped from every dynamic equation referred to $\mathscr{F}_{(pr)}$. Thus, on the one hand, these equations contain in every case only forces admissible in special relativity. On the other hand, according to classical physics, the following *local equivalence principle* is certainly compatible with the results of ordinary mechanical experiments:

In a frame $\mathscr{F}'$ which does not rotate with respect to inertial spaces, local experiments can detect neither gravitational force, nor inertial force, but only their resultant.[23)]

Now let us consider a photon moving in an empty region and passing through $\mathscr{E}$. Then, on the basis of classical electromagnetism—where no connection with gravitation is considered—this photon must have a linear motion with respect to the absolute inertial space $S_{(ab)}$, no matter how gravitational mass is distributed. As a consequence, for $\mathscr{F}'$ locally natural at $\mathscr{E}$, this photon has zero acceleration at $\mathscr{E}$ also with respect to $\mathscr{F}'$ if and only if, the gravitation acceleration vanishes at $\mathscr{E}$. Thus we can decide by means of a local experiment whether the frame joined to the laboratory is locally natural or not.

The preceding example shows that according to classical physics the above equivalence principle is not compatible with experiments of every kind, in particular with electromagnetic experiments. In spite of this, to accept that principle is quite reasonable—and fruitful—because electromagnetic phenomena and, in particular, light propagation imply the presence of electromagnetic energy; hence by the mass-energy equivalence principle (4.4), electromagnetic phenomena and in particular light propagation must be affected by a gravitational field just as matter and mechanical phenomena. (Analogous reasoning holds e.g. for thermodynamic phenomena.) In addition by Michelson experiment the space $S_{(ab)}$ does not exist; hence classical physics has wrong features and they are basilar for the above incompatibility example.

On the basis of the local principle printed above in italics, the tensor form of Maxwell equations, stated in special relativity, is kept completely unaltered in general relativity.[24)] Thus in any locally natural frame (x) Maxwell equations hold in the familiar forms they have in classical absolute frames.

Let us now consider the basic local equations of continuous media, i.e. the first and second Cauchy equations and the energy balance equation (energy-momentum conservation equations [§ 5]) and constitutive equations. On the basis of the above local principle, those equations are accepted in any locally proper and natural frame $(\bar{x})$ within general relativity, with the same mathematical expression they have in a Lorentzian frame within special relativity, at least under the restriction that the second-order derivatives of any tensors of rank $\geqslant 1$, with respect to actual space-time co-ordinates, should be absent—cf. footnote 24 in Chapter 1.

In case couple stresses are absent, the above restriction is met by all of the afore mentioned local equations.[25)] So, by what was said on those equations on considering special relativity [§ 5], at least for vanishing couple stresses *the mathematical forms they possess within general relativity, in any locally proper and natural frame* $(\bar{x})$ *(possibly in the presence of gravitational actions) are very similar to the corresponding forms of the same equations in any classical locally natural frame* (y) *(both in the absence and in the presence of gravitational actions).*

By well-known properties of double derivatives of tensors in Riemannian spaces and the very small values of the space-time Riemannian tensor in general relativity, the preceding italicized similarity holds also for local laws which do not meet the preceding restrictions on second derivatives; hence it holds, in particular, also in the presence of couple stresses.

The criterion hinted at in the title concerns the connection between the versions of the physical laws in general relativity, other than the Poisson equation, and the versions of the same laws in classical physics or special relativity. This criterion, which is widely used for checking the acceptability of any relativistic physical law, is strictly related to the above mass-energy equivalence principle; hence it is natural to consider it now. It consists in the following operations:

(i) *to consider any locally proper and natural frame* $(\bar{x})$, *to set* $x^0 = ct$, *and to express the local law, being studied in general relativity, in the frame* $(t, \bar{x}^1, \bar{x}^2, \bar{x}^3)$;

(ii) *to ascertain whether that expression is equivalent to the vanishing of a power series in* c^{-1} *whose constant term does not vanish on purely mathematical grounds;*

(iii) *to examine whether the vanishing of the above constant term coincides with, or at least is very similar to the corresponding classical law expressed in a Cartesian frame joined to the locally natural and proper (spatial) frame* $\mathscr{F}_{(pr)}$.[26)]

Since c^{-1} is very small, the above criterion is important to make sure whether a local relativistic law is compatible also with non-gravitational experiments, at least as much as the corresponding classical law.

Let us also remark that the operation (iii) is equivalent to considering the limit for $c\to\infty$ of the equation hinted at in (ii) and containing a power series in c^{-1}. Hence, by the above criterion, relativistic local laws are compared with the laws belonging to a classical theory where, e.g., geometrical optics and not Maxwell optics holds.

§ 9. On the Invariance of Physical Equations and on the Possible Physical Equivalence of the Frames in which these Equations have the Same Form. On a Privileged Absolute Concept of Event Point

In connection with given units, in special relativity—as well as in classical physics—there are ∞^{10} privileged frames related to one another by a group of linear transformations. In these frames physical equations have a simple invariant form, where the metric tensor—as well as any other quantity representing pure space-time properties—does not appear.

In harmony with this invariance a "physical principle of relativity" in the sense of Fock [1964] holds; more precisely, the above frames are *physically equivalent* in the following sense: Let (x^α) and (x'^α) be any two of the above privileged frames, and let the physically possible universal process $\mathscr{P}_\phi$ be described in the first frame by the functions $\phi_1, \ldots, \phi_r$. Then the process $\mathscr{P}'_\phi$ described by $\phi_1, \ldots, \phi_r$ in the second frame is also physically possible.

For the moment we assume that a set of privileged frames is given; furthermore every suitably regular solution $\phi_1, \ldots, \phi_r$ of the physical equations, written in any one of those frames, describes a physically possible universal process $\mathscr{P}'_\phi$. (The converse is certainly true.) Then the invariance of physical equations for transformations between any two of the above privileged frames obviously implies the physical equivalence of those frames.

Now let us only assume that (x) and (x') are any two regular frames, so that the space-time metric has the form $(7.1)_{2,3}$. We also consider the physical equations of special relativity in their completely invariant form as given by tensor calculus. These equations are now thought of as including some relations to describe the structure of space time; precisely, the vanishing of Riemann's tensor is included. On the other hand, the sets of functions $\phi_1, \ldots, \phi_r$ and $\phi'_1, \ldots, \phi'_r$, representing the physically possible universal process $\mathscr{P}_\phi(=\mathscr{P}'_{\phi'})$ in the frames (x) and (x') respectively, have to include some fields $\psi_{\alpha\beta}$ and $\psi'_{\alpha\beta}$ to express the components $g_{\alpha\beta}$ and $g'_{\alpha\beta}$ of the metric tensor in the same frames:

$$g_{\alpha\beta}=\psi_{\alpha\beta}(x^\rho), \quad g'_{\alpha\beta}=\psi'_{\alpha\beta}(x'^\rho) \quad (\alpha,\beta=\cdots 0.2\ldots 3). \tag{9.1}$$

Since $\mathscr{P}_\phi$ is physically possible, the sets $\phi_1, \ldots, \phi_r$ and $\phi'_1, \ldots, \phi'_r$ of functions of x^ρ and x'^ρ respectively, satisfy the afore-mentioned invariant equations. So, referring the functions $\phi_1, \ldots, \phi_r$ to the second frame, these functions represent a process $\mathscr{P}'_\phi$ which satisfies all physical equations.

In spite of the above conclusion, in general, $\mathscr{P}'_\phi$ *is not physically possible.* Indeed let $\mathscr{E}$ and $\mathscr{E}+d\mathscr{E}$ [$\mathscr{E}_1$ and $\mathscr{E}_1+d\mathscr{E}_1$] be the event points with co-ordinates x^α and $x^\alpha+dx^\alpha$ respectively in the frame (x) [(x')]. Call ds^2, ds'^2, and $ds_1{}^2$ the values of the space-time metric along the segments $(\mathscr{E},\mathscr{E}+d\mathscr{E})$, $(\mathscr{E}_1,\mathscr{E}_1+d\mathscr{E}_1)$, and $(\mathscr{E}_1,\mathscr{E}_1+d\mathscr{E}_1)$ (again) evaluated in the processes $\mathscr{P}_\phi$, $\mathscr{P}'_\phi$, and $\mathscr{P}_\phi$ respectively. Then $ds^2=ds'^2$. We can identify (x) with a Minkowskian frame and we can set

$$x'^3=2x^3\,, \quad x'^\rho=x^\rho \quad (\rho=0,1,2)\,, \quad \text{hence} \quad g'_{11}=g_{11}\,, \quad g'_{33}=4g_{33}\,. \tag{9.2}$$

Since $\mathscr{P}_\phi$ is physically possible and space-time metric is independent of phenomena, ds^2 and $ds_1{}^2$ are the real values of the space-metric along the segments $(\mathscr{E},\mathscr{E}+d\mathscr{E})$ and $(\mathscr{E}_1,\mathscr{E}_1+d\mathscr{E}_1)$, whereas the analogue holds for ds'^2 only for $dx^3=0$. In particular, for $dx^3\neq 0$ and $dx^0=dx^1=dx^2=0$ we have $ds_1{}^2=g'_{33}(dx^3)^2=4(dx^3)^2=4ds^2=4ds'^2$ $(\neq 0)$. Hence $\mathscr{P}_{\phi'}$ is not physically possible.

Thus, a conclusion emphasized by Fock is reached: Only the invariance of the physical equations written without involving any pure space-time element as an unknown gives rise to a corresponding *physical principle of relativity in the sense of Fock*, i.e. to the mutual physical equivalence of the frames in which the above equations have the same form.

Let us add that from the considerations above it is apparent that the typical regular frames (x) and (x') are not physically equivalent in spite of the tensor invariance of the physical equations being considered, in that the unknown functions $\phi_1, \ldots, \phi_r$ in these equations involve the field $g_{\alpha\beta}$, and therefore their initial values can be considered as arbitrary mathematically but not physically: *the initial value of the distance ds' between the event points $\mathscr{E}_1$ and $\mathscr{E}_1+d\mathscr{E}_1$ cannot be given arbitrarily, unlike the initial positions and velocities of matter elements.*

At this point, especially in view of some future considerations on constitutive equations in general relativity [§§ 27, 46] let us remark that in the preceding reasoning the independence of space time of phenomena is essential; furthermore we considered first that event points are independent of phenomena, and then that the same holds for the space-time metric. In general the former independence does not necessarily imply the latter and can be identified with the possibility of judging according to a determinate natural criterion, whether an event point, $\mathscr{E}$, defined through given phenomena occurring in a universal process $\mathscr{P}$, is the same as the event point $\mathscr{E}_1$ defined through given phenomena occurring in another universal process $\mathscr{P}'$.[27)]

In case one knows the natural criterion mentioned above, we shall say that one has a *natural* (or *privileged*) *absolute concept of event point*. One knows such a criterion in classical physics, special relativity, and general relativity when something such as Fock's conjecture [§ 10] is accepted. When such conjectures are taken out of account—as is done in common textbooks—up to now one has not been able to specify any (physically privileged) absolute concept of event points—cf. Bressan [1974a, § 7].

In order to make the above considerations clearer, we remark that since experimenters can influence the future of reality and not the past, every universal process that experimenters can realize coincides before a spatial section of S_4

with the real universal process $\mathscr{P}_{uR}$. Then it is natural to consider a universal process $\mathscr{P}_u$ as *physically possible* if and only if, for every spatial section S_3 of S_4 the part of $\mathscr{P}_u$ after S_3 can be realized by suitable ideal experimenters working before S_3. (We mean that such experimenters are able to realize a universal process which coincides with $\mathscr{P}_u$ after S_3, and with $\mathscr{P}_{uR}$ before a suitable space-like section, S'_3, of S_4, which in turn precedes S_3).

Therefore to define a natural absolute concept of event point it suffices to determine a natural criterion by which, given two arbitrary universal processes $\mathscr{P}_u$ and $\mathscr{P}'_u$ coinciding before a space-like spatial section, S_3, of S_4, and given two arbitrary event points $\mathscr{E}$ and $\mathscr{E}'_1$ defined within $\mathscr{P}_u$ and $\mathscr{P}'_u$ respectively—cf. the footnote 27 in Chapter 1—, we can decide whether $\mathscr{E}$ and $\mathscr{E}_1$ coincide or not.

Let $\mathscr{P}_u$ and $\mathscr{P}'_u$ coincide before S_3 (and on S_3). Furthermore let $S_4(\mathscr{P}_u)$ $[S_4(\mathscr{P}'_u)]$ be the determination of S_4 corresponding to $\mathscr{P}_u$ $[\mathscr{P}'_u]$. Then (by definition) there is a frame (x) $[(x')]$ in $S_4(\mathscr{P}_u)$ $[S_4(\mathscr{P}'_u)]$ such that (i) the equation of S_3 in (x) $[(x')]$ is $x^0=\phi(x^1,x^2,x^3)$ $[x'^0=\phi(x'_1,x'_2,x'_3)]$ and the metric tensors $g_{\alpha\beta}$ and $g'_{\alpha\beta}$ in $S_4(\mathscr{P}_u)$ and $S_4(\mathscr{P}'_u)$ fulfill the condition

$$g'_{\alpha\beta}(\xi^0,\xi^1,\xi^2,\xi^3)\equiv g_{\alpha\beta}(\xi^0,\xi^1,\xi^2,\xi^3) \quad \text{for} \quad \xi^0\leqslant\phi(\xi^1,\xi^2,\xi^3). \tag{9.3}$$

Now we consider the following existence and uniqueness condition

E. U. Condition. *With every physically possible universal process $\mathscr{P}_u$ a natural class $\Gamma(\mathscr{P}_u)$ of (privileged) frames can be associated in such a way that, for every space-like section S_3 of S_4* (i) *every frame in $\Gamma(\mathscr{P}_u)$ is (uniquely) determined by its part before S_3, and* (ii) *if $\mathscr{P}_u$ and $\mathscr{P}'_u$ coincide before S_3, then the frames in $\Gamma(\mathscr{P}_u)$ coincide, before S_3, with the frames in $\Gamma(\mathscr{P}'_u)$.*

This condition obviously holds (i) in special relativity where $\Gamma(\mathscr{P}_u)$ can be formed with Minkowskian frames[28)], (ii) in the static case of general relativity where $\Gamma(\mathscr{P}_u)$ can be formed with static frames—cf. footnote 19 in Chapter 1 and (iii) even in the general case of general relativity, provided we accept Fock's conjecture [§ 10] which enables us to form $\Gamma(\mathscr{P}_u)$ with Fock's (privileged) harmonic frames [§ 10] or provided we take into account something like this, e.g. following Dirac [1938].

In all of the above cases we obtain an absolute concept of event point as follows. Let (x) belong to $\Gamma(\mathscr{P}_u)$. Then, since $\mathscr{P}_u$ and $\mathscr{P}'_u$ coincide before S_3, the part of frame (x) before S_3 is the part before S_3 of a frame in $\Gamma(\mathscr{P}'_u)$—cf. (ii) in the E. U. Condition— and this frame is unique—cf. (i) in the E. U. Condition—so that it is the frame (x') mentioned in the assertion including (9.3). Incidentally (x) and (x') can be called *corresponding frames*. This relation is an equivalence.

For every event point $\mathscr{E}'$ in $S_4(\mathscr{P}'_u)$ let $\mathscr{E}$ be the event point in $S_4(\mathscr{P}_u)$ whose co-ordinates in the frame $(\bar{x}')$ belonging to $\Gamma(\mathscr{P}'_u)$ are those of $\mathscr{E}$ in the corresponding frame $(\bar{x})$ belonging to $\Gamma(\mathscr{P}_u)$. Then it is obvious that $\mathscr{E}$ is independent of $(\bar{x}')$, i.e. it is determined by $\mathscr{E}'$.

If we identify $\mathscr{P}'_u$ with $\mathscr{P}_{uR}$ and we remember how physically possible universal processes were characterized, we see that we now have an absolute concept of event point.

In the remainder of this section, as well as in most books on relativity Fock's conjecture or something similar—e.g. according to Dirac's ideas—is left out of account. Then no way of constructing the class $\Gamma(\mathscr{P}_u)$ fulfilling the E.U. Condition is available. In order to give a support to this assertion, we again consider $\mathscr{P}_u$, $\mathscr{P}'_u$, S_3, (x), and condition (9.3). Furthermore we assume that (i) formula $(9.3)_2$ has the simple form $\xi^0=0$, that (ii) the co-ordinate lines $x^0=\text{var.}$ $[x'^0=\text{var.}]$ are the geodesics of $S_4(\mathscr{P}_u)$ $[S_4(\mathscr{P}'_u)]$ orthogonal to S_3, i.e. to the hypersurface $x^0=0$ $[x'^0=0]$, and that (iii) x^0 $[x'^0]$ is the arc length on these geodesics, which together with (ii) yields

$$g_{0\rho}\equiv\delta_{0\rho}, \qquad g'_{0\rho}\equiv\delta_{0\rho}. \tag{9.4}$$

We may conclude that the parts of the frames (x) and (x') after S_3 are determined by the common part of them before S_3. This may push us to identify event points of $S_4(\mathscr{P}_u)$ and $S_4(\mathscr{P}'_u)$ with the same co-ordinates in (x) and (x') respectively. However, especially because of the presence of empty regions, S_3 is not determined; furthermore a change of S_3 generally implies a change of the geodesics orthogonal to S_3, hence a change of (x) and (x'). Consequently no absolute concept of event point has been obtained in this way.

Incidentally absolute concepts like the one of event point are studied from a purely logical point of view in Bressan [1972d]—cf. § 84.

§ 10. On Harmonic Coordinates and the Existence of General Frames not Physically Equivalent in General Relativity

In De Donder [1921] and Lancsos [1923] harmonic co-ordinates were introduced and applied in general relativity. In Fock [1939a, b] and [1950] the harmonic frames satisfying certain asymptotic conditions were considered and it was pointed out that, in special relativity, those privileged frames coincide with Lorentzian frames—cf. Fock [1964, §§ 92, 93]; moreover—in the last book on p. 372—Fock substantially says that in general relativity the analogous assertion can be surely proved:

The harmonic frames which fulfill certain reasonable asymptotic conditions[29] *and are connected with given units, are only* ∞^{10}, *furthermore they are related to one another by Lorentz transformations.*

To prove the above assertion by Fock in general relativity in the non-stationary case is still an unsolved problem—cf. Fock [1964, p. 368]. Hence, by motives of rigor, this assertion will be referred to here as *Fock's conjecture.* However, it is the author's opinion that such conjectures deserve much consideration. Therefore in this tract, which is generally independent of Fock's conjecture, the validity of it or any conjecture similar to it is taken for granted within certain considerations involving constitutive equations [§§ 27, 46].

For this tract Fock's conjecture is interesting because it provides the Riemannian space S_4 with ∞^3 inertial spaces, no matter what the actual phenomena in S_4 are.[30] The space-time metric may be assumed to be Lorentzian as in special relativity and to coincide asymptotically with the Riemannian metric; in addition, the metric tensor $g_{\alpha\beta}$ of the Riemannian S_4 may be considered as a tensor potential of the gravitational field. In harmonic co-ordinates the basic equations of general relativity have a slightly modified form; moreover the tensor $g_{\alpha\beta}$ fulfills four conditions.[31]

Let us add that, if Fock's conjecture is proved, and in the way hinted at in Fock [1964, §§ 92, 93] then it will be nearly obvious to prove a certain theorem, very simply related to the above conjecture, which confirms the legitimacy of using a privileged absolute concept of event point—cf. the end of § 9—based on harmonic co-ordinates in general relativity.[32] This concept is interesting e.g. for some remarks on constitutive equations in general relativity [§§ 27, 46]. Furthermore, it is useful in order to show that, in the first place, *also in general relativity the invariance of the physical equations written in harmonic co-ordinates is accompanied by the physical equivalence of the above privileged harmonic frames* and, in the second place, *the analogue does not hold for the same equations written in general co-ordinates according to tensor calculus.*

To realize the above two statements one may consider $g_{\alpha\beta}$ as a tensor gravitational potential and introduce a Lorentzian metric tensor $\mathscr{L}_{\alpha\beta}$ whose components have a canonical form in harmonic co-ordinates, hence they do not appear in the physical equations written in harmonic co-ordinates.

Since, in general relativity, physical equations are invariant in particular in harmonic co-ordinates, the first italicized statement implies a relativistic version of the physical equivalence of inertial spaces, of their physical isotropy and homogeneity, and of the physical homogeneity of the natural time in every inertial space. The second statement, which is negative, can be easily proved in the same way as its analogue for special relativity [§ 9].[33]

§ 11. Some Distinctive Properties of General Relativity. On the Equivalence of General Frames in General Relativity

In the classical theory including electromagnetism, the locally most privileged frames are usually considered to be those locally inertial frames which are momentarily joined to the privileged inertial space $S_{(ab)}$. In all of these frames, physical equations have the same form and contain no elements of space-time such as spatial metric or the local acceleration, angular velocity, and possibly deformation velocity of the reference frame with respect to $S_{(ab)}$. Furthermore the classical form of the physical equations in those frames is the simplest possible.

In the general case the afore-mentioned equations must include gravitational actions[34] and this implies the known non-compliance of the classical theory with the local equivalence principle.

At this point, in order to emphasize the above non-compliance and also in view of later questions connected with the name general relativity [§ 12], it seems to me worthwhile to observe that one can construct a *modified classical theory* including electromagnetism and gravitation which is to general relativity what classical physics is to special relativity. To achieve this, it is enough to assume that (i) there is a privileged inertial space $S_{(ab)}$[35], and (ii) at every event point $\mathscr{E}$ the Maxwell equations should take their well-known simplest form, not necessarily in any (spatial) Cartesian frame joined to $S_{(ab)}$, but in every Cartesian frame $\mathscr{F}_{(ab)}$, locally freely falling and not rotating, which is also locally *absolute* in that its velocity at $\mathscr{E}$ with respect to $S_{(ab)}$ vanishes—cf. (35.19). In the absence of masses $\mathscr{F}_{(ab)}$ can be assumed to be joined to $S_{(ab)}$.

It is clear that the locally most privileged frames in the above theory are the preceding locally absolute frames such as $\mathscr{F}_{(ab)}$. Neither space-time elements nor gravitational actions are contained in the physical equations written in these frames.[36] The same can be said in special or general relativity of the locally natural frames, which are the locally most privileged. The existence in general relativity of such frames, privileged with respect not only to gravitation but also to electromagnetism, is due to the compliance of general relativity with the local principle of equivalence. Incidentally—following Fock [1964]—we used that principle in § 8 just to justify how electromagnetism is dealt with in general relativity.

Let us now remark that, on the one hand, in classical physics including electromagnetism the locally most privileged frames do not depend on phenomena, and in S_4 there exists some space-time frames—by the way harmonic—which are everywhere of the locally most privileged kind. On the other hand, in the suggested modified classical theory of physics and in general relativity the locally most privileged frames—more precisely their class—depend on phenomena; and under typical physical conditions there is no frame in S_4 which is *everywhere* of the locally most privileged kind. Thus the lack of such frames appears to be connected with a partial combination of electromagnetism with the local equivalence principle, no matter whether classical or relativistic kinematics (of space-time) is accepted.

The modified classical theory, historically, was never developed; moreover, it is incompatible with ordinary classical physics. Hence, from the historical point of view, *the lack of any frame which is everywhere of the locally most privileged kind is distinctive of general relativity*. This lack is very important from the practical point of view, and, is due to the world structure according to general relativity; hence it has a physical significance.

Because of the non-existence of frames which are everywhere of the locally most privileged kind, it is often said that in general relativity all frames are equivalent, even if, as it is well known, special classes of frames show considerable advantages. Among them let us consider the frames where the time co-ordinate lines are geodesics, and in addition all harmonic frames.[37]

The afore-mentioned equivalence of general frames, essentially related to the above lack, certainly has a physical significance; however, it is quite different from physical equivalence [§ 9].

Now let us remark that, historically, when the equivalence of general frames was asserted, it was certainly meant in a non-strict sense because it was asserted in spite of the afore-mentioned special classes of frames which were known very early, and because Einstein agreed with Kretschmann [1917], who showed that the requirement of general covariance of physical laws made no assertion about the content of these laws—cf. Anderson [1973, p. 338]. The equivalence of general frames was meant in an even much looser sense when Fock pointed out the existence of his privileged harmonic frames and their important physical significance. Thus now it is a question of personal decision whether or not to continue asserting the equivalence of general frames.

This assertion, meant without any restriction, is certainly objectionable. However, as a support for the favorable attitude towards it, it can be said that the simplifications, borne in general relativity to the physical equations by Fock's privileged harmonic frames, are the same as those due to non-privileged harmonic frames (and the latter frames are many). Hence the afore-mentioned simplifications are by far less relevant than the analogous simplifications borne by the use of the privileged frames—harmonic also—in classical physics and special relativity. In particular, the tensor calculus of Ricci and Levi Civita—considered in Einstein's fundamental paper [1916] as a necessary tool for constructing general relativity and, more in particular, for stating physical equations in space time—is still necessary for practical purposes, and as a matter of fact it is also widely used in dealing with general relativity in privileged harmonic frames—cf. Fock [1964].

Let us now add that tensor calculus is of little use in classical physics. In its space-time version it is very useful in special relativity for dealing with electromagnetism, in connection with inertial frames, which are harmonic. In general relativity—because of the lack of frames which are everywhere of the most privileged kind—it is necessary, practically, to use unspecified and in particular non-harmonic space-time frames, in any physical situations, e.g. even in the absence of electromagnetic phenomena. In connection with unspecified frames tensor calculus is essential. We may conclude that the resulting practical necessity of using tensor calculus is a distinctive property of general relativity.

The formalism of tensor calculus makes the above unspecified frames mutually equivalent in some sense. This, and the lack of frames which are everywhere of the most privileged kind, led to the assertion that a general relativity principle may be asserted in general relativity.

In support of the preceding favorable attitude on the equivalence of arbitrary regular frames, we may also point out that Fock's privileged harmonic frames have, in the historical development of general relativity, a much lesser importance than the analogous privileged frames in classical physics and special relativity. This holds for two reasons: (1) for the last two theories the corresponding privileged frames were known since the beginning of their historical development; (2) in the case of general relativity, which was born in 1916 and rapidly developed, the existence of ∞^{10} privileged harmonic frames was mentioned in 1938—47 and is not yet proved in the general case. Incidentally, the difficulty of finding privileged frames in general relativity is also to be attributed to the world structure according to the same theory; hence it has some physical significance.

Let us further remark that if a finite space-time region, R, is considered, then with respect to experiments localized in R, nowadays all harmonic frames are to be considered as mutually equivalent to the same extent as they were before Fock introduced his privileged harmonic frames through asymptotic conditions (in that the latter frames cannot be determined experimentally within R because of these conditions).

In favor of the attitude against the assertion of a general relativity principle, it may be said that the exceptions to that principle[38)] known before Fock pointed out his privileged frames, were tolerable. But Fock's privileged frames, considered in connection with the whole S_4, determine the inertial spaces in general relativity; so, they are not only special but truly privileged frames. Admitting the existence of privileged frames is, strictly speaking, contradictory to asserting a general relativity principle.

Furthermore, asserting a general relativity principle favors the somehow widespread beliefs that in general relativity all frames of a general kind are physically equivalent in Fock's sense [§ 8] and that the invariance of the laws of general relativity for general transformations is a distinctive property of general relativity or, at least, of relativity. These beliefs are mistaken; in particular the afore-mentioned invariance has no physical content *in itself*. The same invariance can be realized in special relativity; moreover, for the laws of classical physics a world invariant form is now known [§ 5].

It may be concluded that the equivalence of frames of a general kind in general relativity holds to a certain extent and in a certain sense (or, better, in certain senses). In this tract we shall understand the above equivalence simply as an equivalent for the afore-mentioned lack in S_4 of any frame which, at every event point, is locally a frame of the most privileged kind.

It may also be concluded that Fock's privileged frames may induce people to minimize the significance of the general equivalence principle, and this would change the character of that principle from a mostly objective one—see the afore-mentioned lack—towards a rather heuristic character—practical need of using tensor calculus and e.g. locally geodesic frames.

§ 12. What We Mean by General Theory of Relativity

In Fock [1964, pp. 393—395], among some of the critical historical remarks, it is said that the invariance of physical equations for the transformations between general frames in general relativity was confused with the physical equivalence of those frames even by A. Einstein, and that just the validity of physical equivalence is referred to by the term "general theory of relativity", see below. Therefore he replaced this term by "the theory of space, time and gravitation".

From Fock [1964, p. 393] let us quote "The fact that the theory of gravitation, a theory of such amazing depth, beauty and cogency, was not correctly understood by its author, should not surprise us. We should also not be surprised at the gaps in logic, and even errors, which Einstein permitted himself when he derived the basic equations of the theory." V. Fock relates his preceding assertion to the fact —remarked in Einstein's autobiography (Einstein [1955])—that Maxwell did not fully understand the physical meaning of his own theory; only Lorentz established that meaning with full clarity, showing that the electromagnetic field itself is a physical reality, being capable of existence in free space and not requiring any carrier.

From Fock [1964, p. 395] let us further quote "Einstein considered the general covariance of the equations to be a specific peculiarity of the theory of gravitation ... Subsequently it was pointed out to Einstein that general covariance by itself does not express any physical law and, apparently, he agreed with this. But his agreement was rather formal because in fact at the end of his days Einstein connected the requirement of general covariance with the idea of some kind of 'general relativity' and with the equivalence of all frames of reference. He never realized the difference between physical equivalence (or physical relativity) in the sense of corresponding processes in different frames of reference and that formal equivalence which consists in the possibility of using arbitrary co-ordinate systems ... Einstein identified physical relativity with the covariance of differential equations, actually believing that general covariance is an extension of the concepts of the physical relativity that formed the basis of his 1905 theory. This is the origin of Einstein's conviction that there exists a 'General Principle of Relativity'. This conviction evidenced itself in the fact that he called his theory of gravitation 'The General Theory of Relativity' and that in later years he stubbornly adhered to this term ... The confusion of the concepts of physical relativity and formal covariance is particularly clearly shown at one point in Einstein's autobiography where he himself formulates the 'special' theory of relativity (i.e. a theory in which, according to Einstein there is no 'general relativity') in the general covariant form, which, according to Einstein, expresses the idea of 'general relativity'."

Here we shall not attempt to discuss the difficult problem of finding the exact meaning attributed by certain authors to terms such as "relativity principle", "general principle of relativity", "equivalence of all frames", and "general theory of relativity" or the problem of stating to what extent the theses held in Fock [1964, p. 393—396] on this historical subject are true. Fock's work on foundations of general relativity contributes, I think, to making the meaning of relativity clearer. (Of course, there were physicists unaware of Fock's work, yet inclined to emphasize the difference between the mentioned invariance under transformations between general frames and the physical equivalence of those frames.)

Therefore, I conclude that, on the one hand, Fock's work on foundations of relativity is very important and is also useful for some specific considerations to be done in this tract.[39)] On the other hand, I feel rather skeptical about Fock's historical thesis.[40)] Thus, my appreciation for Fock's work does not prevent me from believing that the name "general theory of relativity", given to general relativity by Einstein, is the most appropriate.

I admit that the decision of continuing to assert a general principle of equivalence [§ 11] has certainly contributed to the fact that certain terms such as "general theory of relativity", are understood by people in different ways, and some of these meanings are not physically satisfactory; so, especially in a tract where Fock's privileged frames are used in some admittedly non-essential considerations, it is useful to specify what is meant by these terms.

For this purpose let us first remark that special relativity deserves its name in that in the special case of negligible gravitation it replaces the only inertial space $S_{(ab)}$, admissible in classical electromagnetism, with ∞^3 physically indistinguishable inertial spaces. For a chosen event point $\mathscr{E}$, to $S_{(ab)}$ are joined ∞^3 physically equivalent space-time frames with origin $\mathscr{E}$, while to the ∞^3 inertial spaces of special relativity are joined ∞^6 physically equivalent privileged frames, i.e. Lorentzian frames, with origin $\mathscr{E}$.

In the general case, which includes gravitation, general relativity performs an analogous replacement in a local way. More in particular, let us go back to the modified classical theory of electromagnetism which was mentioned in § 11. In that theory, for every event point $\mathscr{E}$, the Maxwell equations literally hold in every locally natural frame [§ 8] whose velocity with respect to $S_{(ab)}$ vanishes at $\mathscr{E}$. We decide to disregard infinitesimals of order greater than or equal to 2 in the remainder of this section. Thus we may say that the preceding frames, privileged at $\mathscr{E}$ and with origin $\mathscr{E}$, are ∞^3. General relativity replaces these ∞^3 special classical frames, locally privileged and physically equivalent at $\mathscr{E}$, with ∞^6 frames with origin $\mathscr{E}$, locally privileged and physically equivalent at $\mathscr{E}$; these ∞^6 frames are distributed into ∞^3 classes of mutually joined frames.[41)]

On the basis of the preceding point of view "the general theory of relativity" or briefly "general relativity" seems to me a good name.[42)] (The difference between special and general relativity consists in the absence or presence of gravitation.) I also accept interpretations of "general relativity", which refer to the question of equivalence for general frames; more precisely I accept "general relativity" as synonymous with "theory without frames which are everywhere of the locally most privileged kind" or "theory in which dealing with arbitrary frames (by means of tensor calculus) is necessary practically".

§ 13. On the Development of General Relativity. Inclusion of Elasticity, Electromagnitostriction, Couple Stresses, and Hereditary Phenomena

In all stages of the theory of general relativity considered in this tract, the conservation equations are obtained from the corresponding systems $\mathscr{E}_i$ of gravitational equations [§ 7] by taking their divergence. The condition that $\mathscr{E}_i$ should be connected to the first Cauchy equation of continuous media and to the energy balance equation in the above way is important, hence strictly related to the choice of the tensor $A_{\alpha\beta}$ which characterizes the part of $\mathscr{E}_i$ depending only on the metric tensor $g_{\alpha\beta}$—cf. $(7.1)_2$. As is shown in Cartan [1922], in the non-cosmological case[43)] this condition, together with some simplicity assumption on $A_{\alpha\beta}$[44)], implies that $A_{\alpha\beta}$ is determined up to a constant factor.

The electromagnetic field was included into general relativity very early, with the fundamental paper Einstein [1916] on general relativity; this is quite natural because that theory was devised substantially to extend to gravitational phenomena a theory in agreement with Michelson's experiment.

The first theories concerning some exact constitutive equations in general relativity referred almost exclusively to perfect fluids, possibly charged, capable of only adiabatic processes.

Rather general theories including heat conduction, at least within fluids, have been published in general relativity only since 1955.[45] They are mostly based on Eckart [1940b], a fundamental paper on heat conduction in special relativity.

Relativistic theories dealing with materials of general kinds, having a thermodynamic basis and possibly including heat conduction, electromagnetostriction and couple stresses, appeared only after 1960. This subject, which involves general thermoelectromagnitomechanical constitutive equations from the Lagrangean point of view[46], was substantially dealt with at the same time both in special and general relativity. Among other things the relativistic corrections concerning elastic wave propagation were calculated.

Some of the above relativistic theories are based on energymomentum conservation equations or on the Einstein gravitation equations—cf. e.g. Synge [1959] and Rayner [1963] as to purely mechanical linear elasticity; moreover cf. A. Bressan [1963b] to [1967b] as to the general theory of elasticity with a thermodynamic basis. More in particular, some of these works by Bressan are relativizations of the classical theory of elasticity with finite deformations, where the constitutive equations are chiefly derived from the assumption that the material being considered can undergo only reversible processes. Some others of the above works by Bressan are also related to generalizations of the mentioned classical elasticity theory, to electrically and magnetically polarizable materials and to the case of couple stresses.

The other theories on materials of general kinds are mainly based on some variational principles which are equivalent to more or less general versions of the Einstein gravitational equations taken alone or together with the Maxwell equations. The theories constructed from this point of view are due mainly to G. Schöpf [1964a] to [1965b] in the adiabatic case and to a Russian school which considered both adiabatic and irreversible processes—cf. Sedov [1965a, b] and Berdichewski [1966]; see also Bressan [1972b] and Pitteri [1975a, b].

The above subjects of a general kind constitute the main topics of this tract. Part II is devoted to the foundations of a theory able to deal with these subjects, and to the application of the same theory to elastic and magneto-elastic waves.

Our presentation of the afore-mentioned topics is very similar to the above works by A. Bressan, and in particular we shall use the same approach to constitutive equations and in general the same mathematical algorithms. For economy of thought these algorithms will be used also in §§ 69, 70 in dealing with the aforementioned variational principles. This is not the only reason by which our presentation of that subject will show differences from the original papers.

Let us now add, first, that in the last decade considerable advances have been made in the classical theory of materials with memory. Among other things, the set of its principles has been considerably improved. It is interesting to consider some of them in special and especially in general relativity [§ 81].

Moreover, in classical continuum mechanics consistent theories of nonsimple materials—cf. footnote 17 in Chapter 1—began to appear about 1960; in particular couple stresses have been taken into account—cf. Grioli [1960] and Toupin [1962] and [1964]. Now couple stresses are considered also in special or general relativity—cf. Berdichewski [1966], Bressan [1966a], [1972b].

§ 14. Scope and Plan of the Present Tract

As it was said in the Preface (and § 13), the main aim of this tract is to present in a rather autonomous way a relativistic theory where, besides gravitation, mechanics of continuous media, electromagnetism, and thermodynamics, also constitutive equations of general kinds are taken into account. Most of this subject has been treated by scientific literature appeared after 1960.

The theory explicitly presented in this tract is related to Einstein's general theory of relativity for the non-cosmological case—cf. footnote 43 in Chapter 1. From this point of view some more or less general versions of the Einstein gravitational equations are postulated. From these equations, the conservation equations, and also the second Cauchy equation for continuous media (of interest especially in the case of couple stresses) are derived.

If one is willing to consider the analogous theory related to special relativity, it simply suffices to replace the gravitational equations of the preceding theory by the condition that the Riemannian tensor should vanish, and to postulate the conservation equations mentioned above; to postulate the second Cauchy equation of continuous media is optional.

This tract is divided into two parts—cf. the Preface. In the first of them only the Eulerian point of view is taken into account. The mathematical tools required to understand most of this part are those covered in every textbook on general relativity, i.e. the fundamental notions of tensor calculus and Riemannian manifolds.

To deal with electrostriction and magnetostriction in fluids and solids the *co-rotational time flux* and *convected time flux* are useful. These mathematical tools are simple and can be introduced quickly [§§ 22, 58] and in the same form as in a well known treatise by Truesdell and Toupin which mainly concerns classical physics.

In the second part of this tract also the Lagrangean point of view is taken into account to deal with materials of general kinds. In order to provide a theory for these materials, which is similar to the corresponding classical theory, especially in the elastic case, A. Bressan's approach is followed. Thus double tensors and the Lagrangean spatial derivatives [§ 53] are used. The theory of double tensors (of any rank) is also used in the above well known treatise on classical mechanics.[47] However, for the ease of the reader a reduction of that theory to the one of common tensors, as well as the proofs of some theorems on the derivatives of tensors of some special kind, used in this tract, is considered in the appendix.

We deal with elastic waves and magneto-elastic waves in both the first part [§§ 30—31, 49—50] and the second one [§§ 65—68, 77—78]. We treat fluids and solids separately, as far as elastic waves are concerned. However we deal with variational principles for ordinary elastic bodies [§§ 69, 70] and elastic bodies capable of couple stress [§§ 95—97] only in Part II. Since no "natural state" is assumed, fluids can be considered as particular elastic bodies.

The relativistic theory of thermodynamics based on Eckart [1940] is objectionable because no fully satisfactory law of heat conduction is presently available. Many objections have also been raised against electromagnetism for polarizable continuous media [§§ 34, 39]. However, on the basis of some recent works, electromagnetism and in particular the (non-quantistic) relativistic theory of continuous polarizable media presented in this book appears to be satisfactory at least as far as the result of present experiments are concerned—cf. §§ 34, 39 for more detail.

Footnotes to Chapter 1

[1] Before Michelson's experiment most scientists believed that this absolute inertial space should exist and that the mass center of the solar system should move slowly with respect to it.

[2] These experiments disproved Stokes' hypothesis and confirmed Fresnel's according to which α should be a function of n, which implies the existence of many ethers corresponding to different light frequencies.

[3] These names are involved in a big debate—see [§ 12].

[4] Of course, if a motion of a continuous body $\mathscr{C}$ is given in S_4, and the linear element of S_4 being considered belongs to the region $W_{\mathscr{C}}$ of S_4 occupied by $\mathscr{C}$, it is possible to have it described by a particle only after having performed a suitable very little hole in $\mathscr{C}$.

[5] Let us start with the obvious assumption that *for every event point* $\mathscr{E}$, *an isolated particle can have zero velocity at* $\mathscr{E}$ *with respect to some inertial space* S_4.

Then let us take into account the physical indistinguishability of inertial spaces, furthermore Lorentz transformations and the relativistic law for velocity composition. It follows, that *at every event point, every vector of* $S_{(i)}$ *whose modulus is smaller than c can be the velocity of some isolated particle, with respect to* $S_{(i)}$.

Of course, the same holds for particles whose motion is (practically) unaffected by other bodies, e.g. for non-charged particles in case of negligible gravitational actions.

[6] The great contribution of relativity in engaging physicists' attention to operational aspects of physical concepts is emphasized in the well known book Bridgman [1927]. On pp. 4—5 the author emphasizes the non-operational character of old theories such as Newton's mechanics. Among other things he says "... physics, when reduced to concepts of this character, becomes as purely an abstract science, as far removed from reality as the abstract geometry of mathematicians built on postulates".

It must be added that several scientists do not completely agree with Bridgman [1927] and in particular, believe that some statements made there are going too far. Such excess seems to be partially admitted by the author—cf. Bridgman [1949, p. 52].

[7] Sometimes the author of a strictly axiomatic physical theory believes it superfluous to make any reasonable intuitive characterization of the primitive concepts.

[8] In case M has a constant electric charge e but no magnetic dipole, and E_r and H_r are the electric and magnetic fields in $S_{(i)}$ respectively, then $F_1 = e(cE_1 + v_2 H_3 - v_3 H_2), \ldots$. Hence in this most important case the expression of F_r is as simple as the classical expression for f_r; furthermore, it coincides with the latter.

[9] See Truesdell and Toupin [1960, sections 270 to 288, pp. 666 to 697] where, among others, the ideas of Kottler [1922] are followed. According to Truesdell and Toupin [1960, p. 660] these ideas were taken up by a Dutch school and culminated in Van Dantzig [1934a, b] and [1937a, b]. Truesdell and Toupin's theory is based on certain integral conditions which in the absence of forces at a distance are equivalent to conservation equations such as (5.2) and (5.4). These authors also consider the second Cauchy equation.

[10] See, for instance, Friedrich [1928] and Toupin [1957/58].

[11] In Truesdell and Toupin [1960, p. 695] the classical laws for conservation of (charge, magnetic flux) energy and momentum—referred to $S_{(ab)}$—are put into an integral world invariant form. Thence world invariant local energy-momentum conservation equations are derived. R. Toupin did all this, among other things, to allow easier comparison of classical and relativistic theories—cf. Truesdell and Toupin [1960, p. 697]. The particular world invariant form he chose was influenced by his aim to express the mentioned laws in a way independent of the geometry of space-time.

[12] Very interesting results obtained in adiabatic relativistic thermodynamics in the first three or four decades of relativity are considered in the known book Tolman [1949].

[13] However, the use of a spatial divergence in the explicit form of $Q_{\alpha\beta}$ is not compatible with the only natural relativization of Stefan and Boltzmann's laws (in finite terms)—cf. §§ 25, 45.

[14] Cf. de Donder and Dupont [1932—33], [1936—37] on relativistic non-linear elasticity and electromagnetostriction. V. Fock substantially writes (in Zentralblatt), among other things, that [1932] is a purely formal generalization of classical formulas and the considerations developed in this work are purely formal and may scarcely have any physical significance—cf. footnote 2 in Chapter 6.

[15] Let us consider a system y^L of material co-ordinates for a continuum body $\mathscr{C}$; moreover, let the constitutive equations of $\mathscr{C}$ at its "material point" y^L involve the derivatives of the actual co-ordinates with respect to the material co-ordinates, of the orders $1, 2, \ldots, n$. Then the material of $\mathscr{C}$ at y^L is called *at most of order n*. It is called *simple* for $n=1$. For any positive integer n a material at most of order n is called *(effectively) of order n* if it is not at most of order $n-1$.

[16] A particle is a very small and physically isotropic body. Hence its resultant electric and magnetic polarizations must be zero. As a consequence, an uncharged particle is completely unaffected by the electromagnetic field. Furthermore, a test particle usually means a particle not touching other bodies, so that no contact forces are acting on it. We conclude that an uncharged test particle is subject at most to gravitational actions.

[17] $\mathscr{E}_3$ is considered in the fundamental paper Einstein [1916], which does not contemplate heat conduction.

[18] E.g., electromagnetic and thermal phenomena are actually present in the universe and by the Einstein theories based on $\mathscr{E}_2$ or $\mathscr{E}_3$ they influence gravitation. Thus only the Einstein theories based on $\mathscr{E}_3$ or $\mathscr{E}_4$ might be expected to be compatible with actual gravitation experiments. However, the influence is so small that the above compatibility is reached also by the purely mechanical Einstein theory based on $\mathscr{E}_1$.

[19] A space-time metric ds^2 is called *static* if a co-ordinate system (x)—to be called *static*—can be found, such that in it the coefficients $-g_{\alpha\beta}(g_{\alpha\beta}=g_{\beta\alpha})$—cf. $(7.1)_{2,3}$—do not depend on the time co-ordinate x^0, and moreover, $g_{01}=g_{02}=g_{03}=0$.

[20] In the case of heat conduction, or Joule heat, or electrical and magnetic polarization, the preceding conclusions are not rigorously true; however, they are very nearly true.

[21] In Fock [1964, § 55], where harmonic co-ordinates are used, the reasonings hinted at above are generalized to the quasi-static case, again referring to the system $\mathscr{E}_1$. This system may be easily replaced with any one of the systems $\mathscr{E}_2$ to $\mathscr{E}_4$ in the aforementioned generalizations.

[22] We mean that along the world line of M the spatial co-ordinates $\bar{x}^1, \bar{x}^2, \bar{x}^3$ are functions of $\bar{x}^0$, with vanishing derivatives at $\mathscr{E}$.

[23] The principle of material frame-indifference—cf. Noll [1958]—which concerns constitutive equations, is in accord with and closely related to the above equivalence principle.

[24] This is possible also because the tensor form of Maxwell equations does not contain the second-order derivatives $T_{\cdots}^{\cdots}{}_{/\alpha\beta}$ of any tensor $T_{\cdots}^{\cdots}$ of rank >1, with respect to actual space-time co-ordinates. This is of interest, for the derivation indices α, β always commute in special relativity, where S_4 is pseudo-Euclidean, but do not in general relativity.

[25] In the presence of couple stresses, the above restriction on second derivatives is met neither by the energy balance equation, nor by the first Cauchy equation [§§ 87, 88, 93].

[26] The time co-ordinate t introduced in (ii) is related to a time unit of a common size. In (iii) this time unit is supposed to be associated with the frame $\mathscr{F}_{(pr)}$.

[27] For example let us suppose that in each of the universal processes $\mathscr{P}$ and $\mathscr{P}'$ the particles M and M_1 hit one another only once; moreover, let us define $\mathscr{E}$ and $\mathscr{E}_1$ as the event points where M and M_1 hit one another in $\mathscr{P}$ and $\mathscr{P}'$ respectively. The criterion above should decide whether $\mathscr{E}$ and $\mathscr{E}_1$ coincide or not.

[28] In classical physics $\Gamma(\mathscr{P}_u)$ can be formed with Galilean frames.

[29] Among them there is the condition that the metric tensor must be asymptotically Lorentzian, which is very reasonable in the non-cosmological case.

[30] Fock observes that at present we cannot exclude the possibility of characterizing other ∞^{10} privileged frames related to one another by Lorentzian transformations, by using a differential operator different from the Laplacian in S_4—alternatively called Dalembertian—and by using also some reasonable asymptotic conditions. However, he says this is very unlikely.

31) The harmonic conditions in harmonic co-ordinates read

$$\sum_{\beta=0}^{3}\frac{\partial}{\partial x^\beta}(\sqrt{-g}\,g^{\alpha\beta})=0 \quad (\alpha=0,\dots,3) \quad \text{where} \quad g=\det\|g_{\alpha\beta}\| \quad \text{and} \quad g_{\alpha\gamma}g^{\gamma\beta}=\delta_\alpha^{\ \beta} \quad (\alpha,\beta=0,\dots,3).$$

32) In general relativity the privileged absolute concept of event point is as much privileged as the remark made in footnote 30, in Chapter 1 allows. However, it will be of interest in any case.

33) The second italiciced statement was emphasized by Fock, who proved it through his specific conjecture. To prove that statement, it suffices to assume that a reasonable absolute concept of event point is physically definable. Choose the frame (x) harmonic and fulfilling Fock's asymptotic conditions. and let the frame (x') be determined by (9.2). Then by a reasoning similar to that made in special relativity in section 9, we see that (x) and (x') are not physically equivalent.

34) If electromagnetic phenomena are absent, then every cartesian frame $\mathscr{F}$, locally freely falling and not rotating, gives physical equations a form in which gravitational actions are absent, as well as space-time elements. However, in the general case gravitational actions do reappear in the Maxwell equations as dragging forces and these equations have a more complicated form than in a frame momentarily joined to $S_{(ab)}$.

35) The condition that the center of mass of the universe should be at rest in $S_{(ab)}$ would determine $S_{(ab)}$ among the inertial spaces and might be a reasonable assumption.

36) The locally most privileged space-time frames in the afore-mentioned classical theories are obtained by adding any natural time co-ordinate to everyone of the locally most privileged (spatial) frames in the same theories.

37) In special cases stationary or static co-ordinates can also be used and are very useful.

38) I.e. the frames having some special properties.

39) I am referring to the mentioned fact that, using Fock's framework, it is very easy to obtain a privileged absolute concept of event point in general relativity, and this concept is useful in connection with constitutive equations [§ 27] but not at all essential.

40) This scepticism has the following motives. In Einstein's papers I have never found anything compelling me to agree with Fock about how Einstein meant terms such as "general principle of relativity" and "general theory of relativity". Furthermore, on the one hand, as Fock admits, (i) Einstein was aware that special relativity can be put into a (general) covariant form, (ii) he agreed that covariance in itself has no physical content, and (iii) he himself stressed that not every gravitational field can be replaced by acceleration, and that for this to be possible the gravitational field must be uniform—cf. Fock [1964, p. 229]. On the other hand, "general principle of relativity" can be given a reasonable and physically acceptable interpretation, e.g. the one shown in § 11.

41) All the afore mentioned classical and relativistic frames are connected with given units. To leave the units arbitrary would increase the number of privileged frames. However not all of these frames are, locally, physically equivalent.

42) The term "general relativity" may be confusing in that, especially in contraposition to "special relativity", it may seem to denote a theory where general frames are physically equivalent. However, Fock [1964, p. 395] attributes the same meaning to "the general theory of relativity". Thus the latter term is subject as well as the shorter term "general relativity" to bad interpretation. Following a widespread custom, in this tract the latter term is used.

43) The non-cosmological case can be characterized as the one where matter has an "insular" character, and energy vanishes asymptotically, so that in the gravitational equations the terms directly expressing properties of matter or the electromagnetic field must vanish asymptotically.

44) The following simplicity conditions are assumed: $A_{\alpha\beta}$ is a symmetric tensor; it is a function of the components $g_{\alpha\beta}$ of the metric tensor, and the first and second derivatives of these components; it is linear in the second derivatives of $g_{\alpha\beta}$.

45) See e.g. Van Dantzig [1940] where a particular case is dealt with, Pham Mau Quan [1955], Hughes [1961], A. Bressan [1964a] to [1966e], and Müller [1972].

We must mention also the relativistic theory of fluids by Landau and Lifshitz [1959], even though it will not be used in this tract.

46) Following the common usage, we shall speak of Eulerian and Lagrangean points of view, referring to descriptions of the motion of a continuous body made through space-time co-ordinates and material co-ordinates respectively. This usage may cause the reader to think that Lagrange was the first who

used material co-ordinates and that Euler did not use them. Both of these beliefs are wrong as C. Truesdell has shown.

[47)] Here Truesdell and Toupin [1960] is referred to. A presentation of the double tensor theory may be found in Ericksen [1960] which constitutes an appendix to the former work.

Part I

Basic Equations of Gravitation, Thermodynamics and Electromagnetism and Constitutive Equations from the Eulerian Point of View

Chapter 2

Space-Time Kinematics Including Masses

§ 15. On the Riemannian Relativistic Space-Time Metric Introduced as a Chronometry. Admissible Frames. Some Possible Axioms for Non-Cosmological Relativity

An event point may be thought of as the space-time localization of a possible phenomenon whose extension is extremely small both from the spatial and the temporal points of view [§ 2].

It is possible to specify a co-ordinate system for space time, i.e. the set S_4 of all event points, in an operational way—cf. § 3 and Synge [1965, § 4, p. 7]. This means that it is possible to specify some ideal experiment which enables us to assign every event point $\mathscr{E}$ four real numbers $(x^0, \ldots, x^3)$ by means of given one-to-one physical operations. We understand that these operations include both purely physical operations such as measurements, and mathematical operations.

For instance let us consider four observers moving in an arbitrary way. We suppose that each of them is carrying a clock, which may not be very accurate but keeps going. Now let a clear-cut electromagnetic wave start at the event point $\mathscr{E}$ with an initially spherical shape. This wave reaches the $(\alpha+1)$-th of our observers when his clock marks the time $t_\alpha (\alpha=0, \ldots, 3)$. In this way different event points give rise to different quadruples of real numbers. The preceding quadruple $(t_0, \ldots, t_3)$ may be transformed into another quadruple, $(x^0, \ldots, x^3)$, by means of a preassigned suitably smooth one-to-one mathematical transformation.

The system of physical operations just hinted at constitutes a transformation, ϕ, of space-time S_4 into—or onto—the set of quadruples of real numbers. The topology induced on S_4 by ϕ appears to be independent of the above particular observers.

The transformation ϕ may be called a *physically specified frame*—or co-ordinate system—in S_4. The $(\alpha+1)$-th co-ordinate of the event point $\mathscr{E}$ in this frame may be denoted by $x^\alpha = \phi^\alpha(\mathscr{E})$. In accordance with this, sometimes it will be clearer to say "the frame (x)" instead of "the frame ϕ".

When only a frame (x) is presupposed, by the event point x^ρ we mean the event point having the co-ordinates $x^0, \ldots, x^3$ in that frame.

According to the above considerations the existence of the event point $x^0, \ldots, x^3$ is physically meaningful in that in a suitable ideal experiment it is possible to produce four clear-cut electromagnetic waves having an initially spherical shape, to which the quadruple $(x^0, \dots, x^3)$ corresponds through the physical operations defining the frame ϕ.

Now, in order to give S_4 a metric having a physical meaning, we consider a clock C as a small gadget where certain phenomena happen which are periodic with respect to a suitable time co-ordinate. The intrinsic motion of C is not affected by gravitational actions because these actions induce in all parts of C the same acceleration, so that no change in their mutual distances occurs. Some authors take into account the possibility that the intrinsic motion of C may be affected by other external actions. Such effects may be considered to be absent if C is freely moving provided either electric charge or current or else electric or magnetic dipoles are not present in C.

We think of the clocks being considered as particles, as far as their motions with respect to other bodies are concerned. Thus, following substantially Synge [1960, Chap. III, p. 106], we accept the following hypothesis of consistency which is in accordance with the local principle of equivalence and a local principle of physical homogeneity for space time:[1)]

If two clocks describe a same world line ℓ in the absence of non-gravitational external actions, then their intrinsic motions appear periodic with respect to a same time co-ordinate on ℓ, moreover the ratio of their periods is independent of ℓ.[2)]

This hypothesis enables us to speak of the *proper time* along the possible world lines of any uncharged test particle—cf. footnote 16 in Chapter 1—relative to a given time unit. We understand that this unit has been fixed, once and for all; and it need not be mentioned again.

Except when otherwise noted, Greek indices are meant to run from 0 to 3 and Latin indices from 1 to 3. Summation is understood on indices appearing both up and down in the same expression. For example we may write

$$T_\alpha^{\ \alpha} = \sum_{\alpha=0}^{3} T_\alpha^{\ \alpha}, \qquad T_r^{\ r} = \sum_{r=1}^{3} T_r^{\ r}. \tag{15.1}$$

Let x^ρ and $x^\rho + dx^\rho$ be two very near event points on the world line ℓ of an uncharged test particle $\bar{P}$. Let $ds = \psi(x, dx)$ be the time marked on a clock carried along by $\bar{P}$ while it passes through those event points. Such a clock experiences no non-gravitational actions. Then, following substantially Synge [1960, p. 107], we assume that *there are ten scalars $g_{\alpha\beta}$ (with $g_{\alpha\beta} = g_{\beta\alpha}$) which are continuously differentiable functions of x^ρ and fulfill the following three conditions:*

(a) *For every choice of the afore-mentioned event points x^ρ and $x^\rho + dx^\rho$, the time $ds = \psi(x, dx)$ $(ds > 0)$ is given by*

$$ds^2 = -g_{\alpha\beta}\, dx^\alpha dx^\beta \tag{15.2}$$

where $g_{\alpha\beta}$ is calculated at x^ρ.

(b) *The quadratic form $g_{\alpha\beta}\, dx^\alpha dx^\beta$ is of signature $+2$, i.e. for a suitable choice of the frame* (x) *it locally takes the orthogonal—precisely pseudo-Euclidean—form*

$$g_{\alpha\beta} = \delta'_{\alpha\beta} \quad \text{with} \quad \delta'_{\alpha r} = \delta_{\alpha r}, \qquad \delta'_{\alpha 0} = -\delta_{\alpha 0} \tag{15.3}$$

where $\delta_{\alpha\beta}$ is the Kronecker delta.

(c) *For every quadruple δx^α of real numbers, $g_{\alpha\beta}\delta x^\alpha \delta x^\beta$ is negative if and only if, δx^α are the components of a vector tangent at the event point x^ρ to the possible world line of a material point passing through x^ρ.*

We regard the field $g_{\alpha\beta}$ as defined in connection with every frame.[3] Thus it is obviously a covariant tensor field of rank 2.

By the above condition (a), if the quadratic form ds^2—cf. (15.2)—fulfills the conditions (a) to (c), then in harmony with Synge [1960, p. 108] we shall call it a *chronometry* and the tensor $g_{\alpha\beta}$ a *chronometric tensor.*

The initial conditions of a particle, position and velocity, are completely independent of the forces acting on it. Hence by condition (c) every linear element $x^\rho, x^\rho + dx^\rho$ in S_4 along which ds^2 is positive—cf. (15.2)—can be described by an uncharged test particle.

Now we assume that besides $g_{\alpha\beta}$, also $\gamma_{\alpha\beta}$ is a chronometric tensor at the event point $\mathscr{E}$. Furthermore we can assume that (15.3) holds at $\mathscr{E}$. Thus we have, at the event point $\mathscr{E}$,

$$ds^2 = -g_{\alpha\beta}dx^\alpha dx^\beta \equiv (dx^0)^2 - \delta_{hi}dx^h dx^i = -\gamma_{\alpha\beta}dx^\alpha dx^\beta \tag{15.4}$$

for every dx^α by which $ds^2 > 0$. This implies $g_{\alpha\beta} = \gamma_{\alpha\beta}$.[4] Hence there is only one chronometric tensor $g_{\alpha\beta}$ connected with a given time unit.

From now on we shall consider S_4 as a Riemannian or pseudo-Euclidean manifold endowed with the metric ds^2 based on the afore-mentioned tensor $g_{\alpha\beta}$. So, we shall simply call ds^2 the *(space time) metric* and $g_{\alpha\beta}$ the *(space time) metric tensor.*

Now let us consider two event points on a (possible) world line ℓ (of a particle). Then only one of them, $\mathscr{E}'$, *precedes* the other, $\mathscr{E}''$; in other words $\mathscr{E}'$ belongs to the *past* of $\mathscr{E}''$. Then $\mathscr{E}''$ is said to *follow* $\mathscr{E}'$ or to belong to the *future* of $\mathscr{E}'$.

The physically specified frame ϕ considered at the outset of this section can be chosen so that for every x^α and dx^α (i) $ds^2 < 0$ for $dx^0 = 0$ and $dx^r \neq 0$, (ii) $ds^2 > 0$ if $dx^0 \neq 0$ and $dx^r = 0$, and lastly (iii) the co-ordinate x^0 increases toward the future along every possible world line. We call such a frame an *admissible frame* or an *admissible co-ordinate system.*

Remembering how physically specified frames and admissible frames were introduced, we may conclude that the following assertions (A_0) to (A_2) hold:

(A_0) *In every admissible frame the co-ordinate x^0 is a time co-ordinate in that it increases along any world line toward the future.*

(A_1) *The chronometric tensor $g_{\alpha\beta}$ is suitably smooth and fulfills the following inequalities:*

$$g_{00} < 0, \quad g_{11} > 0, \quad \begin{vmatrix} g_{11} & g_{12} \\ g_{21} & g_{22} \end{vmatrix} > 0, \quad \det\|g_{rs}\| > 0, \quad \det\|g_{\alpha\beta}\| < 0. \tag{15.5}$$

Before stating assertion (A_2) we remember that the conditions (15.5) imply the inequalities

$$g_{22}>0, \quad g_{33}>0, \quad \begin{vmatrix} g_{11} & g_{13} \\ g_{31} & g_{33} \end{vmatrix}>0, \quad \begin{vmatrix} g_{22} & g_{23} \\ g_{32} & g_{33} \end{vmatrix}>0, \tag{15.6}$$

so that they are necessary and sufficient for the quadratic form ds^2 to have the signature 2 and more precisely to be both positive for $dx^r=0$ and $dx^0 \neq 0$, and negative for $dx^0=0$ and $dx^r \neq 0$.

(A_2) *The admissible frames are infinitely many. Moreover, in case* (x) *is one of them, the frame* $\bar{x}^\alpha=\bar{\phi}^\alpha(\mathscr{E})$ *is admissible if, and only if, the transformation* $\bar{x}^\alpha=\bar{\phi}^\alpha[\phi^{-1}(x^0,\dots,x^3)]$ *is suitably smooth, preserves the inequalities* (15.5), *and fulfills the condition* $\partial\bar{x}^0/\partial x^0>0$.

Of course the smoothness of the field $g_{\alpha\beta}$ and the transformation $x^\alpha \to \bar{x}^\alpha$ depends on the problem being considered. However we can generally understand that $g_{\alpha\beta}$, $g_{\alpha\beta,\gamma}$ *and* $\bar{\phi}\phi^{-1}$ *should be twice continuously differentiable everywhere, possibly except on some hypersurfaces where the second spatial derivatives have a discontinuity of the first kind.* We shall not be much interested in regularity conditions.

We aim at a relativistic theory of continuous media. Furthermore in the bulk of this theory concepts such as clock, uncharged test particle, and physically specified frame are not used. These concepts, as well as the first two italicized hypotheses in this section have auxiliary offices in that they serve to introduce other concepts such as *admissible frame* and *chronometric tensor*. Hence it seems to me useful to hint at the possibility of considering the last two concepts and *event point* as the only concepts among those introduced in this section, which are primitive in our relativistic theory of continuous media. Likewise the above assertions (A_1) and (A_2) may be taken as axioms on the afore-mentioned primitive concepts, replacing the preceding hypotheses on the above auxiliary concepts. These axioms hold both in special and general relativity.

A line ℓ in S_4 is said to be *time-like*, *space-like*, or *on the null cone* if along every element of ℓ the metric ds^2 is positive, negative, or vanishing respectively.

Incidentally we remark that, on the basis of the afore-mentioned primitive concepts and axioms (A_1) and (A_2), an event point $\mathscr{E}$ can be said to belong to the *past* of the event point $\mathscr{E}'$—and $\mathscr{E}'$ to the *future* of $\mathscr{E}$—if (i) $\mathscr{E}$ and $\mathscr{E}'$ are the ends of a time-like line, and (ii) for some—and hence for every—admissible frame ϕ we have $\phi^0(\mathscr{E})<\phi^0(\mathscr{E}')$. Let us now add that a world line is usually thought of as an unlimited time-like line; moreover a space-like hypersurface is called a *spatial (cross) section* of S_4 if it has no border points.

In this tract only non-cosmological relativity is considered. Then the following assertion (A_3) may be considered as an axiom:

(A_3) *There exists an admissible frame* (x) *such that for every* x^0

$$\lim_{x^r\to\infty} g_{\alpha\beta}(x^0,x^r)=\delta'_{\alpha\beta} \tag{15.7}$$

—cf. $(15.3)_{2,3}$—. Incidentally we also take *material point* and *world line of a given material point* as primitive notions. Then we accept the following axiom:

(A_4) *Every material point* P^* *has a world-line,* ℓ, *and* ℓ *is a time-like line of* S_4. *Furthermore the world lines of distinct material points do not intersect.*

In addition we do not consider the birth or death of any material point. Therefore we accept the following axiom.

(A_5) *Every spatial section of S_4 cuts the world line of every material point, and the set of all these intersections is a non-empty domain in the common topology subordinated on S_4 by admissible frames.*[5)]

For more detail on the axiomatic presentation of the theory developed in this tract see § 83 to § 85.

§ 16. On Tensors in Relativistic Space-Time

Let us use the following notation:

$$f_{,\alpha} = \frac{\partial f}{\partial x^\alpha}, \tag{16.1}$$

where f is any scalar—e.g. a component of a tensor—defined in S_4. Furthermore we may denote by $\|g^{\alpha\beta}\|$ the inverse of the matrix $\|g_{\alpha\beta}\|$—cf. $(15.5)_5$. Now we may introduce Christoffel symbols $\{\alpha\beta,\gamma\}$ and $\left\{ {\gamma \atop \alpha\beta} \right\}$:

$$2\{\alpha\beta,\gamma\} = g_{\gamma\beta,\alpha} + g_{\alpha\gamma,\beta} - g_{\alpha\beta,\gamma}, \qquad \left\{ {\gamma \atop \alpha\beta} \right\} = \{\alpha\beta,\rho\} g^{\rho\gamma}, \tag{16.2}$$

and the gradient, or the covariant derivative, of any tensor $T_{\alpha_1\alpha_2\cdots}{}^{\beta_1\beta_2\cdots}$

$$T_{\alpha_1\alpha_2\cdots}{}^{\beta_1\beta_2\cdots}{}_{/\gamma} = T_{\alpha_1\alpha_2\cdots}{}^{\beta_1\beta_2\cdots}{}_{,\gamma} - \left\{ {\rho \atop \alpha_1\gamma} \right\} T_{\rho\alpha_2\cdots}{}^{\beta_1\beta_2\cdots} \cdots + \left\{ {\beta_1 \atop \rho\gamma} \right\} T_{\alpha_1\alpha_2\cdots}{}^{\rho\beta_2\cdots} \tag{16.3}$$

As is well known, at every event point $\mathscr{E}$ the admissible frame (x) can be so chosen as to be *locally geodesic* at $\mathscr{E}$—i.e. as to have co-ordinate lines with vanishing curvatures at $\mathscr{E}$. This condition is equivalent to any of the following three equalities

$$g_{\alpha\beta.\gamma} = 0, \quad \{\alpha\beta,\gamma\} = 0, \quad \left\{ {\gamma \atop \alpha\beta} \right\} = 0. \tag{16.4}$$

Hence, by (16.3) in locally geodesic co-ordinates, tensor derivatives become partial derivatives.

As is well known—cf. e.g. Finzi-Pastori [1961, p. 265]—it is always possible to choose the frame (x) locally *natural* at $\mathscr{E}$, i.e. both geodesic and orthonormal —cf. $(16.4)_1$, (15.3).

Let $\mathscr{E}_{\alpha\beta\gamma\delta}$ be the completely skewsymmetric scalar system with $\mathscr{E}_{0123} = 1$. We use the versions $\varepsilon_{\alpha\beta\gamma\delta}$ and $\varepsilon^{\alpha\beta\gamma\delta}$ of the Ricci (axial) tensor for which

$$\varepsilon_{\alpha\beta\gamma\delta} = \sqrt{-g}\,\mathscr{E}_{\alpha\beta\gamma\delta} = g\,\varepsilon^{\alpha\beta\gamma\delta} \text{ where } g = \det\|g_{\alpha\beta}\| < 0. \tag{16.5}$$

The symmetric parts of tensors or any scalar systems are denoted by means of round parentheses and the skewsymmetric parts by means of brackets. Thus e.g.,

$$2\,T_{(\alpha\beta}U_{\gamma)}=T_{\alpha\beta}U_{\gamma}+T_{\gamma\beta}U_{\alpha},\qquad 2\,T_{[\alpha\beta]}=T_{\alpha\beta}-T_{\beta\alpha}. \tag{16.6}$$

If $T_{\alpha\beta}$ is a tensor, then, as is usually done, we understand e.g.

$$T_{\alpha}{}^{\beta}=T_{\alpha\rho}g^{\rho\beta},\qquad T_{(\alpha}{}^{\beta)}=T_{(\alpha\rho)}g^{\rho\beta},\qquad T_{[\alpha}{}^{\beta]}=T_{[\alpha\rho]}g^{\rho\beta}. \tag{16.7}$$

We consider the Riemann tensor as defined by

$$R_{\alpha}{}^{\beta}{}_{\gamma\delta}=2\left\{{\beta \atop \alpha[\gamma}\right\}_{,\delta]}+2\left\{{\rho \atop \alpha[\gamma}\right\}\left\{{\beta \atop \rho\delta]}\right\} \tag{16.8}$$

whence, for every vector field T_{α} that is twice continuously differentiable at $\mathscr{E}$, we have at $\mathscr{E}$

$$2\,T_{\alpha/[\beta\gamma]}=T^{\rho}R_{\rho\alpha\beta\gamma}. \tag{16.9}$$

Lastly we define the *gravitational* tensor $A_{\alpha\beta}$ [6] whose divergence vanishes identically:

$$A_{\alpha\beta}=R_{\alpha\rho}{}^{\rho}{}_{\beta}+\tfrac{1}{2}R_{\rho\sigma}{}^{\rho\sigma}g_{\alpha\beta}\quad\text{whence}\quad A_{[\alpha\beta]}=0=A_{\alpha\beta}{}^{/\beta}. \tag{16.10}$$

§ 17. On Tensors in S_4 in Connection with a Moving Continuous Body $\mathscr{C}$ or an Ideal Fluid $\mathfrak{F}$. Spatial Projections and Natural Decompositions of Tensors, Spatial Derivatives and Spatial Divergences

Let $\mathscr{C}$ be a continuous 3-dimensional body. Since we shall not consider irregular motions of $\mathscr{C}$ such as in the case of sliding, fracture or collision with another body, $\mathscr{C}$ can be thought of as a set of material points. The union $W_{\mathscr{C}}$ of the world lines of these points is called the *world tube* occupied by $\mathscr{C}$. We think of $W_{\mathscr{C}}$ as a (4-dimensional) domain in S_4, i.e. a closed subset of S_4, every point of which is an accumulation point for internal points of the same set.

We assume that $\mathscr{C}$ is moving so smoothly that at nearly every event point in $W_{\mathscr{C}}$ there exist the 4-*velocity* u^{α} and the *intrinsic acceleration* A^{α} defined by

$$u^{\alpha}=\frac{Dx^{\alpha}}{Ds},\qquad A^{\alpha}=\frac{Du^{\alpha}}{Ds} \tag{17.1}$$

where D denotes differentiation along the world line W_{P*} of the material point P^* of $\mathscr{C}$, which passes through $\mathscr{E}$. More explicitly and generally, let $x^{\alpha}=x^{\alpha}(s)$ be the equations of W_{P*} where s is an arc length given by (15.2). We recall from tensor calculus that, for every differentiable tensor $T_{\alpha_1\alpha_2\cdots}{}^{\beta_1\beta_2\cdots}$ defined on W_{P*} the absolute derivative along W_{P*} is given by

$$\frac{D}{Ds}T_{\alpha_1\alpha_2\cdots}{}^{\beta_1\beta_2\cdots} = \frac{d}{ds}T_{\alpha_1\alpha_2\cdots}{}^{\beta_1\beta_2\cdots} - u^\gamma \begin{Bmatrix}\rho\\ \gamma\alpha_1\end{Bmatrix} T_{\rho\alpha_2\cdots}{}^{\beta_1\beta_2\cdots} - \cdots + u^\gamma \begin{Bmatrix}\beta_1\\ \gamma\rho\end{Bmatrix} T_{\alpha_1\alpha_2\cdots}{}^{\rho\beta_2\cdots} + \cdots \tag{17.2}$$

By comparing (17.2) with (16.3) in case $T_{\alpha_1\alpha_2\cdots}{}^{\beta_1\beta_2\cdots}$ is defined in a neighborhood of $\mathscr{E}$, and by identifying $T_{\alpha_1\alpha_2\cdots}{}^{\beta_1\beta_2\cdots}$ in (17.2) with the scalar x^α, we respectively deduce

$$\frac{D}{Ds}T_{\cdots}{}^{\cdots} = T_{\cdots}{}^{\cdots}{}_{/\alpha}u^\alpha, \qquad \frac{Dx^\alpha}{Ds} = \frac{dx^\alpha}{ds}. \tag{17.3}$$

The customary formal rules for ordinary derivatives hold also for covariant derivatives and absolute derivatives. As is well known, (17.1) implies

$$A_\alpha = u_{\alpha/\beta}u^\beta, \qquad u^\alpha u_\alpha = -1, \qquad u^\alpha u_{\alpha/\beta} = 0, \qquad u^\alpha A_\alpha = 0. \tag{17.4}$$

Equality $(17.4)_1$ follows from $(17.1)_2$ and $(17.3)_1$, $(17.4)_2$ from $(17.1)_1$, $(17.3)_2$ and (15.2); $(17.4)_2$ implies $(17.4)_3$; $(17.4)_{1,3}$ imply $(17.4)_4$.

We say that the index α of the tensor $T_{\cdots}{}^{\cdots\alpha}$ is *spatial* in case

$$T_{\cdots}{}^{\cdots\alpha}u_\alpha = 0. \tag{17.5}$$

A tensor $T_{\cdots}{}^{\cdots}$ is called *spatial*, whenever all its indices are spatial.

Of course a spatial linear element dx^α ($u_\alpha dx^\alpha = 0$) is also space-like, i.e. with $g_{\alpha\beta}dx^\alpha dx^\beta > 0$, but the converse is not true.

As is customary, we define the spatial projector $\overset{\perp}{g}_{\alpha\beta}$ by

$$\overset{\perp}{g}_{\alpha\beta} = g_{\alpha\beta} + u_\alpha u_\beta \quad \text{whence} \quad \overset{\perp}{g}_{[\alpha\beta]} = 0 = \overset{\perp}{g}_{\alpha\beta}u^\beta, \qquad \overset{\perp}{g}_{\alpha\rho}\overset{\perp}{g}{}^{\rho\beta} = \overset{\perp}{g}_\alpha{}^\beta. \tag{17.6}$$

By $(17.6)_{2,3}$ $\overset{\perp}{g}_{\alpha\beta}$ is a spatial tensor.

For every event point $\mathscr{E}$ in the world tube $W_{\mathscr{C}}$, the frame (x) can be chosen both *locally proper*, i.e. with $u^\alpha = \delta^\alpha{}_0$, and natural at $\mathscr{E}$, so that we have[7)]

$$g_{\alpha\beta} = \delta'_{\alpha\beta}, \qquad \begin{Bmatrix}\gamma\\ \alpha\beta\end{Bmatrix} = 0, \qquad u^\alpha = \delta^\alpha{}_0 \text{ (at } \mathscr{E}). \tag{17.7}$$

In case S_4 is pseudo-Euclidean (special relativity) there is one Lorentz frame —i.e. an everywhere natural frame—which is locally proper at $\mathscr{E}$, up to spatial rigid displacements.

If the frame (x) is locally orthonormal and proper, by $(17.6)_{1,2,3}$ and $(17.7)_{1,3}$ the components of the spatial projector are

$$\overset{\perp}{g}_{rs}=\delta_{rs},\ \overset{\perp}{g}_{0\alpha}=\overset{\perp}{g}_{\alpha 0}=0 \quad (\text{for } g_{\alpha\beta}=\delta'_{\alpha\beta},\ u^{\alpha}=\delta^{\alpha}{}_{0}). \tag{17.8}$$

Let us now incidentally remark that in accord with conditions $(17.7)_{1,3}$ *a covariant or controvariant index α of a tensor $T_{\dots}^{\dots}$ is spatial if and only if all components of this tensor with $\alpha=0$ vanish in locally proper and orthogonal coordinates.*

In accord with the definition $(17.6)_1$ we introduce the following notations

$$T^{\dots}{}_{\dots\overset{\perp}{\alpha}}=T^{\dots}{}_{\dots\rho}\overset{\perp}{g}{}^{\rho}{}_{\alpha},\qquad T_{\dots}{}^{\dots\overset{\perp}{\alpha}}=T_{\dots}{}^{\dots\rho}\overset{\perp}{g}{}_{\rho}{}^{\alpha},\qquad \overset{\perp}{T}_{\alpha\beta\dots}{}^{\gamma\dots}=T_{\overset{\perp}{\alpha}\overset{\perp}{\beta}\dots}{}^{\overset{\perp}{\gamma}\dots}. \tag{17.9}$$

By $(17.6)_1$ the conditions that the index α of the tensor $T^{\dots}{}_{\dots\alpha}$ is spatial and the tensor $T_{\dots}^{\dots}$ is spatial are respectively equivalent to the equalities

$$T^{\dots}{}_{\dots\alpha}=T^{\dots}{}_{\dots\overset{\perp}{\alpha}},\qquad T_{\dots}^{\dots}=\overset{\perp}{T}{}_{\dots}^{\dots}. \tag{17.10}$$

With a view to dealing with derivatives of tensors it is useful to add to conventions (17.9) and (17.3) the following:

$$(T_{\dots}^{\dots})^{\perp}=\overset{\perp}{T}{}_{\dots}^{\dots},\quad \text{hence e.g. } (T_{\rho/\sigma})^{\perp}=\overset{\perp}{g}{}_{\rho}{}^{\alpha}T_{\sigma/\overset{\perp}{\alpha}}\,. \tag{17.9'}$$

We call *natural decomposition of the tensor $T_{\dots\alpha}$ with respect to the index α* the equality

$$T_{\dots\alpha}=T^{(1)}{}_{\dots\alpha}+T^{(0)}{}_{\dots\alpha}\quad \text{where } T^{(1)}{}_{\dots\alpha}u^{\alpha}=0=T^{(0)}{}_{\dots\overset{\perp}{\alpha}}. \tag{17.11}$$

By $(17.11)_{1,3}$ $T_{\dots\overset{\perp}{\alpha}}$ equals the tensor $T^{(1)}{}_{\dots\overset{\perp}{\alpha}}$ which by $(17.11)_2$ equals $T^{(1)}{}_{\dots\alpha}$, hence the two terms into which $T_{\dots\alpha}$ has been decomposed are uniquely determined. Thus by (17.6) equality $(17.11)_1$ may be written

$$T_{\dots\alpha}=T_{\dots\overset{\perp}{\alpha}}-T_{\dots\beta}u^{\beta}u_{\alpha}. \tag{17.12}$$

The above procedure and in particular (17.12) can be applied to decompose the terms $T^{(1)}{}_{\dots\alpha}$ and $T^{(0)}{}_{\dots\alpha}$ with respect to another index β. By so doing a decomposition of $T_{\dots\alpha}$ into four terms is obtained. This procedure can be continued until all indices of $T_{\dots\alpha}$ are exhausted.

Thus 2^r terms are obtained, where r is the rank of our tensor $T_{\dots\alpha}$. These terms afford the (total) *natural decomposition of $T_{\dots\alpha}$*.

We shall use the decomposition above only for $r=1,2$. For $r=2$ it is particularly important, so that, following Cattaneo [1959], in connection with any such tensor $T_{\alpha\beta}$ we introduce the following notations:

$$T=T_{\rho\sigma}u^{\rho}u^{\sigma},\ T'_{\alpha}=-T_{\overset{\perp}{\alpha}\rho}u^{\rho},\ T''_{\beta}=-u^{\rho}T_{\rho\overset{\perp}{\beta}},\ T'''_{\alpha\beta}=T'_{\alpha}u_{\beta}+u_{\alpha}T''_{\beta}\,. \tag{17.13}$$

By conventions $(17.9)_3$ and $(17.13)_{1,2,3}$

$$T_{\alpha\beta}=T u_\alpha u_\beta+T'_\alpha u_\beta+u_\alpha T''_\beta+\overset{+}{T}_{\alpha\beta}, \quad T'_\alpha u^\alpha=0=u^\beta T''_\beta. \tag{17.14}$$

The natural decomposition $(17.14)_1$ of $T_{\alpha\beta}$ is uniquely determined in the sense that, disregarding conventions $(17.9)_3$ and $(17.13)_{1,2,3}$, under the conditions $(17.14)_{2,3}$ and the conditions $u^\alpha \overset{+}{T}_{\alpha\beta}=0=T_{\alpha\beta}u^\beta$ equality $(17.14)_1$ determines T, T'_α and T''_α—cf. $(17.13)_{1,2,3}$—hence also $\overset{+}{T}_{\alpha\beta}$ and $T'''_{\alpha\beta}$.

The tensors $T u_\alpha u_\beta$ and $T'''_{\alpha\beta}$ determined by (17.13) will be called the *temporal and mixed parts* of $T_{\alpha\beta}$ respectively.

By $(17.9)_1$, $(17.6)_1$ and $(17.3)_1$ we have

$$T_{\dots/\overset{+}{\alpha}}=T_{\dots/\rho}\overset{+}{g}{}^\rho{}_\alpha=T_{\dots/\alpha}+u_\alpha \frac{D}{Ds} T_{\dots}. \tag{17.15}$$

We call $T^{\dots\beta}{}_{/\overset{+}{\alpha}}$ *spatial derivative* and $T^{\dots\alpha}{}_{/\overset{+}{\alpha}}$ *spatial divergence* of the tensor $T^{\dots\beta}$. Of course we generally have

$$T_{\dots\overset{+}{\alpha}/\beta}\equiv[T_{\dots\overset{+}{\alpha}}]_{/\beta}\neq \overset{+}{g}_\alpha{}^\rho T_{\dots\rho/\beta}. \tag{17.16}$$

Similar inequalities generally hold for divergences. We now prove the following properties for the spatial divergences of tensors endowed with certain spatial indices:

$$T^{\dots\alpha}{}_{/\alpha}=T^{\dots\alpha}{}_{/\overset{+}{\alpha}}+T^{\dots\alpha}A_\alpha \qquad \textit{for } T^{\dots\alpha}u_\alpha\equiv 0 \tag{17.17}$$

$$T^{\dots\alpha\sigma}{}_{/\sigma}=\overset{+}{g}{}^\alpha{}_\rho T^{\dots\rho\sigma}{}_{/\sigma}+u^\alpha u_{\rho/\sigma}T^{\dots\rho\sigma} \qquad \textit{for } T^{\dots\rho\sigma}u_\rho\equiv 0 \tag{17.18}$$

$$T^{\dots\alpha\sigma}{}_{/\overset{+}{\sigma}}=\overset{+}{g}{}^\alpha{}_\rho T^{\dots\rho\sigma}{}_{/\overset{+}{\sigma}}+u^\alpha u_{\rho/\sigma}T^{\dots\rho\sigma} \qquad \textit{for } T^{\dots}{}_{\rho\sigma}\equiv T^{\dots}{}_{\overset{+}{\rho}\overset{+}{\sigma}}. \tag{17.19}$$

By $(17.17)_2$ and $(17.1)_2$ $u_\alpha DT^{\dots\alpha}/Ds=-T^{\dots\alpha}A_\alpha$. Hence (17.15) implies $(17.17)_1$.

By $(17.15)_1$ and $(17.6)_1$ $T^{\dots\alpha\sigma}{}_{/\sigma}=\overset{+}{g}{}^\alpha{}_\rho T^{\dots\rho\sigma}{}_{/\sigma}-u^\alpha u_\rho T^{\dots\rho\sigma}{}_{/\sigma}$; thence under assumption $(17.18)_2$, we obtain $(17.18)_1$.

Now we assume $(17.19)_2$, i.e. $(17.17)_2$ and $(17.18)_2$, hence $(17.17)_1$ and $(17.18)_1$ follow. By $(17.17)_1$ and $(17.18)_2$ we have the equalities

$$T^{\dots\alpha\sigma}{}_{/\sigma}=T^{\dots\alpha\sigma}{}_{/\overset{+}{\sigma}}+T^{\dots\alpha\sigma}A_\sigma, \quad \overset{+}{g}{}^\alpha{}_\rho T^{\dots\rho\sigma}{}_{/\sigma}=\overset{+}{g}{}^\alpha{}_\rho T^{\dots\rho\sigma}{}_{/\overset{+}{\sigma}}+T^{\dots\alpha\sigma}A_\sigma.$$

The substitution of the left-hand side of these equalities with their respective right-hand sides in $(17.18)_1$ yields $(17.19)_1$.

Among the proved properties (17.17) to (17.19) the first is the most useful. So we remark that in a locally natural and proper frame—cf. (17.7)—property (17.17) is equivalent to

$$T^{\dots r}{}_{/r}=T^{\dots\alpha}{}_{/\alpha}-T^{\dots r}A_r \quad \textit{for } T^{\dots\alpha}u_\alpha=0. \tag{17.17'}$$

As was briefly hinted at in § 8 and will be more thoroughly considered in § 19, in relativistic locally natural frames the local laws of physics have the same form, up to extremely small terms, as in those classical Euclidean frames where the dragging force and the Coriolis force locally balance gravitational forces. Since the A_α, and hence $T^{\cdots\alpha} A_\alpha$, are extremely small in ordinary cases, a priori, a spatial divergence appearing in a classical local law may be relativized both into a spatial divergence and into a space time divergence. Most authors generally prefer the former, others—cf. e.g. Cattaneo [1963]—prefer the latter divergence. In some cases neither of these divergences is used—cf. (24.5).

At last let us remark that, by (17.15) for $T_{\dots} = u_\rho$ and by $(17.4)_4$, we have

$$u_{\rho\overset{\perp}{/\alpha}} = u_{\rho/\alpha} + A_\rho u_\alpha, \qquad u^\alpha{}_{\overset{\perp}{/\alpha}} = u^\alpha{}_{/\alpha}. \tag{17.20}$$

§ 18. The Spatial Metric $d\overset{\perp}{s}{}^2$. Another Physical Meaning of the Chronometry ds^2. Ordinary Units

Now we consider an event point $\mathscr{E} + d\mathscr{E}$ very near $\mathscr{E}$, with co-ordinates $x^\alpha + dx^\alpha$. Let it belong to the world line $W_{P^*+dP^*}$ of the material point $P^* + dP^*$ of $\mathscr{C}$. The event point $\mathscr{E} + d\overset{\perp}{\mathscr{E}}$ having co-ordinates $x^\alpha + d\overset{\perp}{x}{}^\alpha$—cf. (17.9)—may be said to be the event point on $W_{P^*+dP^*}$ *simultaneous* with the event point $\mathscr{E}$ on the world line W_{P^*}, with respect to an observer stationary with respect to P^*. Indeed the frame (x) may be so chosen as to be locally natural and proper at $\mathscr{E}$—cf. (17.7). Then the condition $d\overset{\perp}{x}{}^0 = 0$ holds there. On the one hand, in special relativity the above choice is realized by identifying (x) with the Lorentz frame which is locally proper at $\mathscr{E}$. Then the condition $d\overset{\perp}{x}{}^0 = 0$ characterizes just the simultaneity of $\mathscr{E}$ and $\mathscr{E} + d\overset{\perp}{\mathscr{E}}$ with respect to that frame. On the other hand, locally natural frames are the local analogues of Lorentz frames, for an effectively Riemannian S_4 (general relativity).

Thus it is natural to characterize local simultaneity by means of the local condition $d\overset{\perp}{x}{}^0 = 0$ also in the general case. Of course in this case the concept of simultaneity concerns extremely near event points and refers only to a first approximation.

Incidentally, if the 4-velocity u^α is continuous, then, up to infinitesimals of order greater than 1, $\mathscr{E}$ is the event point on the world line W_{P^*} simultaneous with $\mathscr{E} + d\overset{\perp}{\mathscr{E}}$ not only with respect to an observer stationary with respect to P^* but also with respect to an observer stationary with respect to $P^* + dP^*$. Therefore, since the components of $d\overset{\perp}{\mathscr{E}}$ are $d\overset{\perp}{x}{}^\alpha$, the linear element $\mathscr{E}, \mathscr{E} + d\overset{\perp}{\mathscr{E}}$ can satisfactorily be called the spatial projection of the linear element $\mathscr{E}, \mathscr{E} + d\mathscr{E}$.

We call *spatial metric* the quadratic form

$$d\overset{\perp}{s}{}^2 = \overset{\perp}{g}_{\alpha\beta}\, dx^\alpha dx^\beta. \tag{18.1}$$

As is easy to realize in locally proper and pseudo-Euclidian co-ordinates, this form is positive definite on the spatial linear elements $d\overset{\perp}{\mathscr{E}}$ at $\mathscr{E}$. Furthermore by $(17.6)_4$ the values of $d\overset{\perp}{s}{}^2$ for $d\mathscr{E}$ and $d\overset{\perp}{\mathscr{E}}$ coincide with the value of ds^2 for $d\overset{\perp}{\mathscr{E}}$:

$$\overset{\perp}{g}_{\alpha\beta}\, dx^\alpha dx^\beta = \overset{\perp}{g}_{\alpha\beta}\, d\overset{\perp}{x}{}^\alpha d\overset{\perp}{x}{}^\beta = g_{\alpha\beta}\, d\overset{\perp}{x}{}^\alpha d\overset{\perp}{x}{}^\beta . \tag{18.2}$$

Hence $d\overset{\perp}{s}{}^2$ may be thought of as a metric on a material neighborhood of P^*.

We shall consider only continuous motions—e.g. no collisions fracture or sliding—as we mentioned earlier. Then there is an admissible frame (x) in which the world lines of the material points of $\mathscr{C}$ have equations of the form

$$x^r = \text{const}, \tag{18.3}$$

so that we may say that we are dealing with a *co-moving frame* or *co-moving co-ordinates*.

Conversely, given any admissible frame (x), equations (18.3) may be interpreted as representing the motion of a continuous system, possibly immaterial, such as an *ideal fluid* $\mathfrak{F}$. Then every triple x^r of real numbers characterizes a point of $\mathfrak{F}$ through (18.3).

In classical physics the motion of $\mathfrak{F}$ can be rigid. Then the space joined to the frame (x) is called *Euclidean* and $\mathfrak{F}$ is an equivalent for this space. In the general case, where $\mathfrak{F}$ is deforming, the ideal fluid $\mathfrak{F}$ appears as a generalization of the above Euclidean space. Thus the points of $\mathfrak{F}$ may be regarded as geometrical points. The same may be done in special or general relativity.

As a consequence, the kinematic concepts we introduced in connection with $\mathscr{C}$ at its typical material point P^* may be considered also in connection with $\mathfrak{F}$ at its (geometrical) point P. Consequently in dealing with $\mathfrak{F}$ we consider $\mathscr{E}$ to belong to the world line W_P of P, and $\mathscr{E}+d\mathscr{E}$ and $\mathscr{E}+d\overset{\perp}{\mathscr{E}}$ to belong to the world line W_{P+dP} of the point $P+dP$ of $\mathfrak{F}$.

We now suppose that the linear element $\mathscr{E}, \mathscr{E}+d\mathscr{E}$ of S_4 is described by a moving point $\bar{P}$. Then $\bar{P}$ also describes the linear element $P, P+dP$ of $\mathfrak{F}$, whose spatial length relative to an observer stationary with respect to P, when P passes through $\mathscr{E}$, is $d\overset{\perp}{s}$—cf. (18.1).

Let us remark that, *if* (x) *and* $(\bar{x})$ *are any two locally proper and pseudo-Euclidean frames, then the relation* $x^0 = \bar{x}^0 + \text{const}$—*or* $dx^0 = d\bar{x}^0$—*holds up to infinitesimals of order greater than* 1.

Following C. Cattaneo [1960] we call dx^0 *standard time* relative to $\mathfrak{F}$. It is clear that the ratio $d\overset{\perp}{s}/dx^0$ is determined by the chronometry $d\overset{\perp}{s}{}^2$ and by the motions of $\bar{P}$ and $\mathfrak{F}$, so that the quantity $d\overset{\perp}{s}/dx^0$ has a physical and operational meaning just as ds^2 has. Then the above ratio can be called *standard velocity* of $\bar{P}$ with respect to $\mathfrak{F}$ (or with respect to P).[8]

So far we have considered only the time unit as fundamental in the measurement of ds^2. Furthermore in locally proper and pseudo-Euclidean co-ordinates (17.8) and $(15.4)_{1,2}$ hold, whence

$$ds^2 = (dx^0)^2 - d\overset{\perp}{s}{}^2 \qquad (\text{for } g_{\alpha\beta} = \delta'_{\alpha\beta},\ u^\alpha = \delta^\alpha{}_0). \tag{18.4}$$

Then, as to physical dimensions, the ratio $d\overset{\perp}{s}/dx^0$ is a pure number. For

$$ds=0 \tag{18.5}$$

we have $d\overset{\perp}{s}/dx^0=1$ and this result is independent of the motion of P or $\mathfrak{F}$. Hence if $\bar{P}$ is moving in S_4 along the null cone, so that (18.5) holds along its motion, then with respect to every ideal fluid $\mathfrak{F}$ $\bar{P}$ describes the spatial length $d\overset{\perp}{s}$ in the standard time $dx^0=d\overset{\perp}{s}$ relative to $\mathfrak{F}$.

In many theoretical considerations it is useful to assume as length unit the length described in a time unit by a point, $\bar{P}$, moving with the above invariant speed, which is the speed c of light in vacuo. In particular this is substantially done in considering a locally natural frame—cf. $(17.7)_{1,2}$. Such mutually linked space and time units may be called Römer units. They cannot both be practical because experiment shows that in the C.G.S. units c is very large ($c=3\cdot 10^{10}$).

In relativity the time unit is usually so chosen as to make the Römer length unit a practical one. However we assume that the latter unit turns out to be a centimeter. Since the corresponding Römer time unit is c^{-1} seconds, along the world line of any material point P^* of $\mathscr{C}$ the proper times s and τ, in Römer and C.G.S. units respectively are linked by the relation

$$s=c\tau+\text{const}, \quad \text{whence } ds=c\,d\tau. \tag{18.6}$$

The units being referred to ought to be specified in connection with every magnitude. E.g. the above intrinsic acceleration A_α and absolute derivatives $DT_{\dots}^{\dots}/Ds$ along world lines—cf. $(17.1)_2$, (17.2)—are in Römer units. Thus they may be called Römer magnitudes, while the corresponding magnitudes in ordinary units are $D^2x^\alpha/D\tau^2$ $(=c^{-2}A_\alpha)$ and $DT_{\dots}^{\dots}/D\tau=c\,DT_{\dots}^{\dots}/Ds$. However, the reference to ordinary units is so important in order to compare relativity with classical physics, that in making such comparisons ordinary units are generally referred to without any explicit mention to units. On the other hand, in purely relativistic considerations Römer units are usually understood for the sake of simplicity.

§ 19. On Some Classical Analogues of Locally Natural Frames from a Physical Point of View

The following *geodesic hypothesis* is a basic requirement underlying the construction of general relativity [§ 7] and complying with (15.3).

The possible world lines of uncharged test particles are the geodesics of S_4 along which $ds^2>0$—cf. footnote 16 in Chapter 1.

This hypothesis is substantially implied by the Einstein gravitational equations [§ 23] in that dust-like matter, i.e. matter not subject to any contact or electromagnetic force, moves along time-like geodesics of S_4. By the same hypothesis the following characterization of locally natural frames is possible.

Theorem 19.1. *Let the frame* (x) *be locally pseudo-Euclidean at the event point* $\mathscr{E}$. *Then a necessary and sufficient condition for this frame to be locally natural at* $\mathscr{E}$ *is the following:*

a) *along the possible world lines of an uncharged test particle,* $\bar{P}$, *through* $\mathscr{E}$ *the co-ordinates* x^α *are linear functions of the proper time—marked by a clock carried along by* $\bar{P}$*—up to infinitesimals of order greater than 2.*

Proof. Let us consider any uncharged test particle $\bar{P}$ passing through $\mathscr{E}$. By the geodesic hypothesis, along the world line of $\bar{P}$ we have

$$\frac{d^2 x^\alpha}{ds^2} + \left(\frac{dx^0}{ds}\right)^2 B^\alpha = 0, \tag{19.1}$$

where

$$B^\alpha = \begin{Bmatrix} \alpha \\ rs \end{Bmatrix} v^r v^s + 2\begin{Bmatrix} \alpha \\ 0r \end{Bmatrix} v^r + \begin{Bmatrix} \alpha \\ 00 \end{Bmatrix}, \qquad v^r = \frac{dx^r}{ds}\bigg/\frac{dx^0}{ds}. \tag{19.2}$$

We first assume that the frame (x) is locally natural at $\mathscr{E}$, so that at $\mathscr{E}$ we have $\begin{Bmatrix} \alpha \\ \beta\gamma \end{Bmatrix} = 0$, which by (19.1) yields $d^2 x^\alpha/ds^2 = 0$. Consequently condition (a) above holds.

Conversely we now assume condition (a), whence $d^2 x^\alpha/ds^2 = 0$ at $\mathscr{E}$ along every possible world line of $\bar{P}$ through $\mathscr{E}$. Furthermore every triple (v^1, v^2, v^3) with $(v^1)^2 + (v^2)^2 + (v^3)^2 < 1$ has the expression $(19.2)_2$ in a possible 4-velocity dx^α/ds of $\bar{P}$ at $\mathscr{E}$. Then by (19.1) we have at $\mathscr{E}$

$$B_\alpha = 0 \quad \text{identically in } v^r \tag{19.2'}$$

whence, by $(19.2)_1$ $\begin{Bmatrix} \alpha \\ \beta\gamma \end{Bmatrix} = 0$. Then the locally pseudo-Euclidean frame (x) is also locally natural at $\mathscr{E}$. q.e.d.

Now let us recall that in classical physics a space-time frame, (x), is called *Euclidean* if x^0 coincides with a natural time co-ordinates τ and the spatial co-ordinates x^r are Cartesian (i.e. everywhere orthonormal), so that the motion of the associated ideal fluid $\mathfrak{F}$ is rigid. Let g^r be Newton's gravitational force per unit mass at $\mathscr{E}$; in addition let a^r be the acceleration and $v_{r/s}$ the velocity gradient of $\mathfrak{F}$ with respect to the inertial spaces, at $\mathscr{E}$ $(v_{(r/s)} = 0)$. Then the dynamic equations of an uncharged test particle, $\bar{P}$, in such a frame are

$$\frac{d^2 x^r}{d\tau^2} + \begin{Bmatrix} r \\ hi \end{Bmatrix} \frac{dx^h}{d\tau}\frac{dx^i}{d\tau} = \mathrm{g}^r - a^r - 2v^r{}_{/s}\frac{dx^s}{d\tau} \quad \left(\begin{Bmatrix} r \\ hi \end{Bmatrix} \equiv 0 \equiv v_{(r/s)}\right). \tag{19.3}$$

We say that the classical frame (x) is *natural* at $\mathscr{E}$ if it is freely falling and non-rotating at $\mathscr{E}$ in the sense that the relations $a_r = \mathrm{g}_r$ and $v_{[r/s]} = 0$ hold at $\mathscr{E}$. Obviously this makes sense even disregarding the assumption that (x) is a Euclidean frame. At this point the following classical analogue of Theorem 19.1 can be immediately proved on the basis of (19.3).

Theorem 19.2. *The condition* (a) *considered in Theorem 19.1 is necessary and sufficient for the classical Euclidean frame* (x) *to be natural at* $\mathscr{E}$.

The preceding two theorems show that the same physical condition characterizes both the locally natural frames among the locally pseudo-Euclidean frames, in special or general relativity, and the locally natural (i.e. the locally freely falling and non-rotating) Euclidean space-time frames in classical physics. Thus the latter classical frames are the Euclidean frames analogous (or equivalent) to the relativistic locally natural frames.

This analogy or equivalence is based on an important physical element (the same behaviour of the considered frames with respect to uncharged freely moving test particles). This does not surely imply that the preceding relativistic and classical frames must behave in the same way also with respect to physical phenomena of any kind.

Of course relativistic locally natural frames and classical locally natural Euclidean frames can be equivalent in some sense with respect to any local physical phenomena only if they are connected to the same length and time units. In particular if ordinary units are used, then instead of any relativistic locally natural frame (x^α) the frame (x^r, τ) with $x^0 = c\tau$ has to be referred to.

One may strenghten the analogy established by means of the preceding two theorems by considering frames in classical physics more general than Euclidean frames. For instance one may assume that the frame (x^r, τ) is generally deforming, i.e. the associated ideal fluid $\mathfrak{F}$ is moving in any way, of course as regularly as desired. Then in this frame $\left\{ \begin{matrix} r \\ hi \end{matrix} \right\} \not\equiv 0$ and $v_{(r/s)} \not\equiv 0$; however equation $(19.3)_1$ still holds.

More generally in the preceding frame we may replace τ with any non-natural time co-ordinate x^0, and this replacement does not affect the associated ideal fluid $\mathfrak{F}$. We say that the resulting frame is *locally Euclidean* at $\mathscr{E}$ if (i) the spatial frame (x^1, x^2, x^3) is locally Euclidean, and (ii) when the classical natural time τ is thought of as a function $\hat{\tau}(x^0, \ldots, x^3)$ of the co-ordinates, we have at $\mathscr{E}$

$$\frac{\partial \tau}{\partial x^\alpha} = \delta_{0\alpha} \quad [\tau = \hat{\tau}(x^0, \ldots, x^3)]. \tag{19.4}$$

Then along the motion of any particle through $\mathscr{E}$ we have

$$\frac{d^2\tau}{(dx^0)^2} = \tau_{,\rho\sigma} \frac{dx^\rho}{dx^0} \frac{dx^\sigma}{dx^0} \text{ at } \mathscr{E}. \tag{19.5}$$

In deed we deduce (19.5) from (19.4) and the second of the equalities

$$\frac{d\tau}{dx^0} = \tau_{,\rho} \frac{dx^\rho}{dx^0}, \quad \frac{d^2\tau}{(dx^0)^2} = \tau_{,\rho\sigma} \frac{dx^\rho}{dx^0} \frac{dx^\sigma}{dx^0} + \tau_{,\rho} \frac{d^2 x^\rho}{(dx^0)^2}. \tag{19.5'}$$

By (19.5), *if the general frame* (x) *is locally Euclidean at* $\mathscr{E}$, *then the condition*

$$\tau_{,\alpha\beta}=0 \quad \text{at } \mathscr{E} \tag{19.6}$$

is necessary and sufficient for $d^2\tau/(dx^0)^2$ *to vanish at* $\mathscr{E}$ *along every possible motion of an uncharged test particle P through* $\mathscr{E}$.

Of course the afore-mentioned vanishing is equivalent to x^0 being a linear function of τ, up to infinitesimals of order greater than 2 along every possible motion of $\bar{P}$ through $\mathscr{E}$.

We say that the classical frame (x) is *locally natural* at the event point $\mathscr{E}$, if (i) it is locally Euclidean, freely falling ($a_r = \mathfrak{g}_r$), non-rotating ($v_{(r/s)}=0$), and non-deforming ($v_{(r/s)}=0$) at $\mathscr{E}$, and (iii) it fulfills conditions (19.4) and (19.6).

Theorem 19.3. *Let the frame* (x) *in classical space-time be locally Euclidean at* $\mathscr{E}$. *Then the condition* (a) *considered in Theorem 19.1 is necessary and sufficient for the frame* (x) *to be locally natural at* $\mathscr{E}$.

Proof. We first assume that the frame (x) is locally natural at $\mathscr{E}$. Then (19.6) holds; hence x^0 is a linear function of τ, up to infinitesimals of order greater than 2, along every possible motion of $\bar{P}$ through $\mathscr{E}$.

Equation (19.3) holds for $\bar{P}$ in the frame (x^r, τ), hence also in the frame (x^α), because x^0 does not appear in (19.3).

Then, since $\mathfrak{g}_r = a_r$ and $\left\{ {r \atop hi} \right\} = 0 = v^r{}_{/s}$, we have $d^2 x^r/d\tau^2 = 0$ at $\mathscr{E}$ along every possible motion of $\bar{P}$ through $\mathscr{E}$, i.e. x^r is a linear function of τ up to infinitesimals of order greater than 2 along every possible motion of $\bar{P}$ through $\mathscr{E}$. We conclude that the condition (a) mentioned in Theorem 19.1 holds.

Now we assume conversely condition (a) (in Theorem 19.1). Then the relation $d^2 x^\alpha/d\tau^2 = 0$ can be asserted to hold at $\mathscr{E}$ for every possible motion of $\bar{P}$ through $\mathscr{E}$. This assertion for $\alpha=0$ implies $d^2\tau/(dx^0)^2 \equiv -(d\tau/dx^0)^2\, d^2x^0/d\tau^2 = 0$, which by $(19.5')_2$ and the arbitrariness of dx^r/dx^0 yields (19.6). By equation (19.3) the same assertion for $\alpha=1,2,3$ implies

$$-\frac{d^2x^r}{d\tau^2} \equiv \left\{ {r \atop hi} \right\} \frac{dx^h}{d\tau}\frac{dx^i}{d\tau} + 2v^r{}_{/s}\frac{dx^s}{d\tau} + a^r - \mathfrak{g}^r = 0 \tag{19.7}$$

for every $dx^r/d\tau$. It follows $a_r = \mathfrak{g}_r$ and $v^r{}_{/s} = \left\{ {r \atop hi} \right\} = 0$. We conclude that the locally Euclidean frame (x) is locally natural at $\mathscr{E}$. q.e.d.

As a consequence of Theorems 19.1 and 19.3 a same physical condition characterizes the relativistic locally natural frames among the locally pseudo-Euclidean frames and the classical locally natural frames among the locally Euclidean frames.

The above general locally Euclidean frames are better classical analogues for the locally pseudo-Euclidean frames, than the classical Euclidean frames; and by comparing the above consequence of Theorems 19.1 and 19.3 with its analogue drawn from Theorems 19.1 and 19.2 we conclude that classical locally natural frames are better classical analogues for relativistic locally natural frames than Euclidean locally natural frames.

However the local fundamental equations of classical physics have the same form both in the locally natural frames and in the Euclidean frames which are locally natural. Then, for the sake of simplicity, we may consider only the latter frames in the remainder of this tract.

§ 20. Material Derivatives, the Spatial Ricci Tensor, and the Relative Rate of Change of Proper Volume Dealt with from the Eulerian Point of View

Let (x^r, τ) be a Euclidean space-time frame in classical physics, and $dx^r/d\tau$ the velocity of the continuous body $\mathscr{C}$ at its material point P^*, with respect to this frame; we also assume that the tensor $T_{\dots}^{\dots} = T_{r\dots}^{\;\;s\dots}$ is a function of the coordinates x^r and τ defined on the world tube $W_{\mathscr{C}}$ of $\mathscr{C}$, and we set

$$\dot{T}_{\dots}^{\dots} = \frac{\partial T_{\dots}^{\dots}}{\partial \tau} + T_{\dots}^{\dots}{}_{/r} \frac{dx^r}{d\tau}. \tag{20.1}$$

The tensor $\dot{T}_{\dots}^{\dots}$ is called the *material derivative* of $T_{\dots}^{\dots}$—cf. Truesdell and Toupin [1960, § 72]—*with respect to the frame* (x^r, τ) or (x^r).

If $T_{\dots}^{\dots}$ is a scalar, then $\dot{T}_{\dots}^{\dots}$ is independent of the Euclidean frame being referred to.

If $T_{\dots}^{\dots}$ has some indices, then its components depend on the unit vectors of the frame (x^r), but not on its origin. Consequently the tensor $\dot{T}_{\dots}^{\dots}$ depends on the actual angular velocity of (x^r) but not on the actual acceleration of the same frame at P^*. Thus, in particular, the tensor $\dot{T}_{\dots}^{\dots}$ is the same whether (x) is a Galilean frame or a locally natural Euclidean frame. In these cases $\dot{T}_{\dots}^{\dots}$ is simply called *material derivative*.

If the frame (x) is also locally proper $(dx^r/d\tau = 0)$, then $\dot{T}_{\dots}^{\dots} = \partial T_{\dots}^{\dots}/\partial t$. Locally natural and proper Euclidean frames are the classical Euclidean frames analogous to the relativistic locally natural and proper frames. In the latter frames $\partial T_{\dots}^{\dots}/\partial \tau = c\,\partial T_{\dots}^{\dots}/Ds$—cf. $(17.3)_1$, $(17.7)_{2,3}$ and (18.6). Therefore in relativity it is natural to set

$$\overset{\ast}{T}_{\dots}^{\dots} = \frac{D}{Ds} T_{\dots}^{\dots} = \frac{1}{c} \dot{T}_{\dots}^{\dots} \tag{20.2}$$

and to call $\overset{\ast}{T}_{\dots}^{\dots}$—which is the absolute derivative along the world line of a material point, P^*, of $\mathscr{C}$—the *(Römer) material derivative* of $T_{\dots}^{\dots}$.

Using locally pseudo-Euclidean and proper co-ordinates, it is easy to realize that by (16.5) and convention (17.9), in every frame we have

$$\overset{\perp}{\varepsilon}_{\alpha\beta\gamma\delta}=0, \quad \text{i.e.}\ \varepsilon^{\alpha\beta\gamma\delta}\,T_{\alpha\beta\gamma\delta}=0 \tag{20.3}$$

for every spatial tensor $T_{\alpha\beta\gamma\delta}$.

Then it is natural and it causes no confusion to set—cf. (16.5)—

$$\overset{\perp}{\varepsilon}_{\alpha\beta\gamma}=u^{\rho}\,\varepsilon_{\rho\alpha\beta\gamma} \tag{20.4}$$

whence $\overset{\perp}{\varepsilon}_{\alpha\beta\gamma}=\mathscr{E}_{0\alpha\beta\gamma}$ *for* $g_{\alpha\beta}=\delta'_{\alpha\beta}$ *and* $u^{\alpha}=\delta^{\alpha}{}_{0}$.

By (20.4) $\overset{\perp}{\varepsilon}_{\alpha\beta\gamma}$ is a skewsymmetric spatial tensor of rank 3. We call it the *spatial Ricci tensor.*

Under $(20.4)_{3,4}$ $\overset{\perp}{\varepsilon}_{hrs}$ $(=\overset{\perp}{\varepsilon}^{hrs})$ are the components, in cartesian co-ordinates, of the Ricci tensor in a 3-dimensional Euclidean space. Hence, on the basis of well known properties of this tensor it is easy to prove the following relations

$$\begin{cases}\overset{\perp}{\varepsilon}_{\alpha\beta\rho}\overset{\perp}{\varepsilon}_{\gamma\delta}{}^{\rho}=\overset{\perp}{g}_{\alpha\gamma}\overset{\perp}{g}_{\beta\delta}-\overset{\perp}{g}_{\alpha\delta}\overset{\perp}{g}_{\beta\gamma}\,, \\ \overset{\perp}{\varepsilon}_{\alpha\rho\sigma}\overset{\perp}{\varepsilon}_{\beta}{}^{\rho\sigma}=2\overset{\perp}{g}_{\alpha\beta}, \qquad \overset{\perp}{\varepsilon}_{\alpha\beta\gamma}\overset{\perp}{\varepsilon}^{\alpha\beta\gamma}=6\end{cases} \tag{20.5}$$

using locally natural and proper co-ordinates.

Since by $(17.4)_{3,4}$ the index ρ is spatial in $u_{\rho/\gamma}$ and A_{ρ}, by (20.4) and (20.3) $\overset{\perp}{\varepsilon}_{\alpha\beta\gamma}$ has the following properties:

$$\overset{\perp}{\varepsilon}^{\alpha\beta\gamma}{}_{/\delta}\,T_{\alpha\beta\gamma}=0 \tag{20.6a}$$

for every spatial tensor $T_{\alpha\beta\gamma}$.

By convention (17.9′) the relations (20.6a) and (17.6) easily yield

$$(\overset{\perp}{\varepsilon}^{\alpha\beta\gamma}{}_{/\delta})^{\perp}=0=\left(\frac{D}{Ds}\overset{\perp}{\varepsilon}^{\alpha\beta\gamma}\right)^{\perp}, \qquad (\overset{\perp}{g}_{\alpha\beta/\gamma})^{\perp}=0=\left(\frac{D}{Ds}\overset{\perp}{g}_{\alpha\beta}\right)^{\perp}. \tag{20.6b}$$

We now prove from the Eulerian point of view that the relative rate of change of proper volume has the expression $(20.12)_1$ below—which immediately yields the continuity equation (21.3). In § 58, where the Lagrangian formalism can be used, a quick proof of the same results is presented (and no infinitesimal quantities are used).

For $r=1,2,3$ we consider the material points P^* and $P^*+\delta_{(r)}P^*$ of $\mathscr{C}$, their respective world lines W and W_r, and the event points $\mathscr{E}$ on W and $\mathscr{E}+\delta_{(r)}\mathscr{E}$ on W_r. We denote by x^{α} and $x^{\alpha}+\delta_{(r)}x^{\alpha}$ their respective co-ordinates. Then the event point $\mathscr{E}+\delta_{(r)}\overset{\perp}{\mathscr{E}}$ of co-ordinates $x^{\alpha}+\delta_{(r)}\overset{\perp}{x}_{\alpha}$—cf. (17.9)—belongs to $W_r (r=1,2,3)$.

The proper volume of the infinitesimal spatial parallelepiped $(\mathscr{E},\delta_{(1)}\overset{\perp}{\mathscr{E}},\delta_{(2)}\overset{\perp}{\mathscr{E}},\delta_{(3)}\overset{\perp}{\mathscr{E}})$ is defined by

$$dC=\overset{\perp}{\varepsilon}_{\alpha\beta\gamma}\,\delta_{(1)}x^{\alpha}\,\delta_{(2)}x^{\beta}\,\delta_{(3)}x^{\gamma}=\overset{\perp}{\varepsilon}_{\alpha\beta\gamma}\,\delta_{(1)}\overset{\perp}{x}^{\alpha}\,\delta_{(2)}\overset{\perp}{x}^{\beta}\,\delta_{(3)}\overset{\perp}{x}^{\gamma} \qquad (\delta_{(r)}\overset{\perp}{x}^{\alpha}=\overset{\perp}{g}^{\alpha}{}_{\beta}\,\delta_{(r)}x^{\beta}). \tag{20.7}$$

Let σ be a space-like section of S_4 and $C_{\mathscr{C},\sigma}$ the integral of dC over the intersection of σ with the world tube $W_{\mathscr{C}}$ of $\mathscr{C}$.

In the (always realizable) case that σ is represented by equation $x^0=\text{const}$ in the frame (x) being considered, it is natural to say that $C_{\mathscr{C},\sigma}$ is the *proper volume* of $\mathscr{C}$ at the instant x^0. $C_{\mathscr{C},\sigma}$ is independent of the motion of the ideal fluid $\mathscr{F}$ associated with (x). Of course a particular physical interest exists in the cases where (i) (x) is a Lorentzian frame in special relativity, (ii) (x) is a static frame[9)] in general relativity, which occurs in special physical situations, and (iii) (x) is a privileged harmonic frame, which can always be realized even in general relativity according to Fock's conjecture [§ 10].

Let $d\mathscr{C}$ be the material parallelepiped $P^*, P^*+\delta_{(r)}P^*$ $(r=1,2,3)$. By (20.7) its proper volume $dC=C_{d\mathscr{C},\sigma}$ over the spatial section σ of S_4 through $\mathscr{E}$ does not depend on σ (up to negligible quantities) but only on the event $\mathscr{E}$ occupied by P^*. It can obviously be called the *proper volume* of $d\mathscr{C}$ (for P^*) at $\mathscr{E}$.

Now, in order to calculate the (Römer) relative rate of change $(dC)^{-1}DdC/Ds$ of the proper volume dC we consider the arc-lengths

$$s=\hat{s}(\mathscr{E}') \qquad s_{(r)}=\hat{s}_r(\mathscr{E}'+\delta_{(r)}\overset{\perp}{\mathscr{E}}{}') \tag{20.8}$$

as parameters on the world lines W and W_r respectively. Since $\delta_{(r)}\overset{\perp}{\mathscr{E}}{}'$ is determined by $\mathscr{E}'$ and moreover is infinitesimal, by (20.8) $s_{(r)}$ is a well determined function $\hat{s}_r(s)$ of s; and $d\hat{s}_{(r)}/ds-1=0(1)$, where by $0(n)$ we denote a suitable infinitesimal quantity of order n with respect to the spatial lengths of $\delta_{(r)}\mathscr{E}'$ $(r=1,2,3)$ considered as principal infinitesimals, because

$$ds=-u_\alpha(\mathscr{E})u^\alpha(\mathscr{E}+d\overset{\perp}{\mathscr{E}})ds_{(r)}=(1-u_\alpha u^\alpha{}_{/\beta}d\overset{\perp}{x}{}^\beta)ds_{(r)}\,.$$

The 4-velocity of $P^*+\delta_{(r)}P^*$ at $\mathscr{E}+\delta_{(r)}\overset{\perp}{\mathscr{E}}$ has the expression $D(x^\alpha+\delta_{(r)}\overset{\perp}{x}{}^\alpha)/Ds_{(r)}=u^\alpha+u^\alpha{}_{/\rho}\delta_{(r)}\overset{\perp}{x}{}^\rho+0(2)$ whence

$$\frac{D}{Ds}\delta_{(r)}\overset{\perp}{x}{}^\alpha=\frac{D(x^\alpha+\delta_{(r)}\overset{\perp}{x}{}^\alpha)}{Ds_{(r)}}\frac{ds_r}{ds}-\frac{Dx^\alpha}{Ds}+0(2)=u^\alpha\left(\frac{d\hat{s}_r}{ds}-1\right)+u^\alpha{}_{/\rho}\delta_{(r)}\overset{\perp}{x}{}^\rho\frac{d\hat{s}_r}{ds}+0(2). \tag{20.9}$$

For $T^{\alpha\beta\gamma}=\delta_{(1)}\overset{\perp}{x}{}^\alpha\delta_{(2)}\overset{\perp}{x}{}^\beta\delta_{(3)}\overset{\perp}{x}{}^\gamma$ the tensor $T^{\alpha\beta\gamma}$ is spatial, so that (20.6a) holds. Furthermore $\delta_{(r)}x^\alpha=0(1)$ and $d\hat{s}_r/ds=1+0(1)$. Hence, by $(20.4)_1$, (20.7) and (20.9)

$$\frac{D}{Ds}dC=\overset{\perp}{\varepsilon}_{\alpha\beta\gamma}u^\alpha{}_{/\rho}(\delta_{(1)}\overset{\perp}{x}{}^\rho\delta_{(2)}\overset{\perp}{x}{}^\beta\delta_{(3)}\overset{\perp}{x}{}^\gamma+\delta_{(2)}\overset{\perp}{x}{}^\rho\delta_{(3)}\overset{\perp}{x}{}^\beta\delta_{(1)}\overset{\perp}{x}{}^\gamma+\delta_{(3)}\overset{\perp}{x}{}^\rho\delta_{(1)}\overset{\perp}{x}{}^\beta\delta_{(2)}\overset{\perp}{x}{}^\gamma)+0(4). \tag{20.10}$$

The frame (x) can be so chosen that the vector $\delta_{(r)}\overset{\perp}{x}{}^\alpha$ is locally tangent to the co-ordinate line x^r $(r=1,2,3)$. Then $\delta_{(r)}\overset{\perp}{x}{}^\alpha=0$ for $\alpha\neq r$. As a first consequence, by (20.7) equality (20.10) becomes

$$\frac{D}{D}dC=\overset{\perp}{\varepsilon}_{123}(u^1{}_{/1}+u^2{}_{/2}+u^3{}_{/3})\delta_{(1)}\overset{\perp}{x}{}^1\delta_{(2)}\overset{\perp}{x}{}^2\delta_{(3)}\overset{\perp}{x}{}^3+0(4)=u^r{}_{/r}dC+0(4). \tag{20.11}$$

As a second consequence, by the relations $0 = u_\alpha \delta_{(r)} \overset{\perp}{x}^\alpha$ we locally have $u_r = 0$, so that $(17.4)_3$ becomes $u_0 u^0{}_{/\beta} = 0$, whence $u^0{}_{/0} = 0$. Thence $u^r{}_{/r} = u^\alpha{}_{/\alpha}$. Furthermore by (20.7) $0(4)/dC = 0$. Then by (20.11) and $(17.20)_2$ we have

$$u^\alpha{}_{/\alpha} = \frac{1}{dC} \frac{D}{Ds} dC = u^\alpha{}_{/\overset{\perp}{\alpha}} \tag{20.12}$$

which affords useful expressions for the relative rate of change $DdC/(dCDs)$ of proper volume. The first of them will be used in § 21 to prove that the continuity equation holds for a certain invariant mass. Thus this equation will be completely proved from the Eulerian point of view.

§ 21. On Gravitational Mass and Reference or Conventional Mass. The Continuity Equation

Let $d\mathscr{C}$ be any element of the body $\mathscr{C}$—i.e. any infinitesimal 3-dimensional portion of $\mathscr{C}$—which contains the material point P^*. The (proper) gravitational mass $m_{(g)}$ of $d\mathscr{C}$ can be determined, at least conceptually, by means of experiments such as Cavendish's. We are directly interested in such a mass because we want to consider the Einstein gravitational equations; one of these ten scalar equations is a relativistic analogue of Poisson's equation relating mass density to the gravitational potential [§ 7], and it surely concerns gravitational mass.

In classical physics $m_{(g)}$ equals the inertial mass $m_{(i)}$ of $d\mathscr{C}$; in special relativity $m_{(i)}$ depends on the velocity v_l of P^* with respect to the inertial space $S_{(i)}$ joined to the observer and this dependence is such that for $v_l \ll c$ the increment $dc^2 m_{(i)}$ of $c^2 m_{(i)}$ in every infinitesimal time interval equals the corresponding increment $d\mathscr{I}$ of the kinetic energy of $d\mathscr{C}$ with respect to $S_{(i)}$. As we brought out at length in § 4, this fact and others—partly connected with Maxwell's equations of electromagnetism—induced people to assert an equivalence between mass and energy and, in particular, the quantitative relativistic equivalence principle $W = c^2 m$ —cf. (4.4)—between the energy W and the corresponding mass m. This principle is confirmed by the experiments where mass is transformed into energy, or viceversa, by means of atomic processes which involve a large amount of internal energy.

In the processes considered by the theory of materials the internal energy varies by much smaller amounts, so that, up to now, one was not able to verify experimentally any disappearance of mass in such processes. However, the afore-mentioned equivalence principle (4.4) is strongly expected to hold also for the latter kind of processes, so that $m_{(g)}$ is to be thought of as effectively depending on the instrinsic physical state Σ of $d\mathscr{C}$ and in particular on the thermodynamic state of $d\mathscr{C}$—cf. e.g. Tolman [1949], Eckart [1940b], and Fock [1964].

Roughly speaking, the intrinsic (physical) state $d\Sigma$ of $d\mathscr{C}$ at the instant τ consists of the values taken at time τ in $d\mathscr{C}$ by those physical magnitudes which (i) are measurable inside an ideal very small laboratory L joined to the material point P^* of $d\mathscr{C}$, the motion of L in S_4 being not directly known, and (ii) are relevant for the motion of $\mathscr{C}$ and its energetic exchanges.

Through the constitutive equations of $d\mathscr{C}$ the state $d\Sigma$ is determined by the actual values in $d\mathscr{C}$ of a few physical magnitudes or—in the case of materials with memory—by the history of the same magnitudes in $d\mathscr{C}$, up to the instant τ.

For instance, in case $d\mathscr{C}$ is a common non-viscous fluid (without memory) the state $d\Sigma$ is determined by the actual values of the absolute temperature T and the pressure p of $\mathscr{C}$; in this case T and p determine the proper volume dC of $d\mathscr{C}$. If $d\mathscr{C}$ is a viscous fluid, the spatial velocity gradient $u_{\alpha/\overset{+}{\beta}}$ also contributes to determining the state $d\Sigma$.

According to the preceding considerations, and precisely to the relativistic equivalence principle (4.4), the gravitational mass $m_{(g)}$ of $d\mathscr{C}$ generally varies in that the internal energy varies. Let us remark that the energy of $d\mathscr{C}$ is usually understood to include the kinetic energy $\mathscr{I}$ of $d\mathscr{C}$; however $\mathscr{I}=0$ holds in the Cavendish experiment by means of which $m_{(g)}$ is determined. When this occurs we call $m_{(g)}$ *proper (gravitational) mass* of $d\mathscr{C}$. After setting

$$c^{-2}\rho dC = m_{(g)}\ (\mathscr{I}=0)\,, \tag{21.1}$$

$c^{-2}\rho$ may be called *proper density of (proper) gravitational mass* and ρ proper density of (proper) total internal energy, or of gravitational mass in energy units.

Choose a state, Σ^*, of the body $\mathscr{C}$—containing $d\mathscr{C}$—as a reference state—cf. Bressan [1963c]. Let dC^* be the proper volume and k^*dC^* the proper mass of $d\mathscr{C}$ in the state Σ^*. We call k^*dC^* *proper reference mass* or *conventional mass* of $d\mathscr{C}$.

We think of this mass as invariably joined to $d\mathscr{C}$ in every process and we relate it to the actual proper volume dC of $d\mathscr{C}$ by setting

$$kdC = k^*dC^*, \quad \text{whence} \quad \frac{1}{k}\frac{Dk}{Ds} = -\frac{1}{dC}\frac{DdC}{Ds}\,. \tag{21.2}$$

From $(21.2)_1$ k is the *proper density of conventional mass*, i.e. k has the physical meaning of actual proper density of proper reference mass. Since by $(17.3)_1$ and $(20.2)_1$ $\dot{k} = k_{/\alpha}u^\alpha$, the relations $(20.12)_1$ and $(21.2)_2$ imply the following mutually equivalent forms of the continuity equation for reference proper mass:

$$\frac{1}{k}\dot{k} = -u^\alpha{}_{/\alpha}, \quad (ku^\alpha)_{/\alpha} \equiv \dot{k} + ku^\alpha{}_{/\alpha} = 0 \quad \left(\dot{k} = \frac{Dk}{Ds}\right). \tag{21.3}$$

Let us denote the difference $c^{-2}\rho dC - kdC$ between the gravitational proper masses of $d\mathscr{C}$ in the actual configuration and the reference configuration by $c^{-2}wkdC$. Then

$$\rho = k(c^2 + w)\,, \tag{21.4}$$

where w can be considered as the proper internal energy per unit proper reference mass [§ 23]. We shall briefly call w the *specific internal energy* (of $\mathscr{C}$ at P^*).

Likewise, we shall call *specific* any physical magnitude which, in some sense, is both proper and per unit proper reference mass.

From the mathematical point of view the fact that $(21.2)_1$ implies (21.3) can also be interpreted in this way: Let σ_3^* be a spatial section of the world tube $W_{\mathscr{C}}$ and k^* an arbitrary differentiable function defined on σ_3^*. Let $\mathscr{E}$ be any point in $W_{\mathscr{C}}$. We consider the material point P^* of $\mathscr{C}$ which passes through $\mathscr{E}$, and the intersection $\mathscr{E}^*$ of its world line W_{P^*} with σ_3^*. Let $d\mathscr{C}$ be any element of $\mathscr{C}$ containing P^*. We denote by dC^* and dC the proper volumes of $d\mathscr{C}$ for P^* in $\mathscr{E}^*$ and $\mathscr{E}$ respectively—which volumes are completely determined by $(20.7)_1$. Then relation $(21.2)_1$, where k is evaluated at $\mathscr{E}$ and k^* at $\mathscr{E}^*$, obviously defines the general solution of the continuity equation (21.3) which is a first order partial differential equation (with an obvious uniqueness theorem).

A scalar field k satisfying (21.3), and called e.g. *invariant mass density*, is much used in relativity and is introduced in various ways: In Schöpf [1964a] the density k is introduced as the inverse dC^*/dC of the proper volume ratio dC/dC^* up to a constant factor.

This procedure, where a purely kinematical aspect of classical mass density appears emphasized, obviously is quite similar to our procedure for $k^*=1$. In classical physics mass density may be considered both as a dynamic and a kinematic quantity (the inverse of the specific volume). The respective relativistic analogues $c^{-2}\rho$ and k do not coincide. However, e.g. for a relativistic non-viscous fluid it is equivalent to consider the pressure p either as a function of k and the absolute temperature T, or as a function of ρ and T.

For the classical non-viscous fluid the stress X^{rs} is a function of T and the mass density; moreover classical mass density is never introduced kinematically, e.g. as the inverse of specific volume. Hence several authors are led to the above dynamic view about relativistic fluids (i.e. they consider p as a function of ρ and T).

On the other hand when general classical elastic materials are taken into account, the Lagrangian point of view is considered, the material co-ordinates are introduced, and the stress X^{rs} is thought of as a function of T and the position gradient $\partial x^r/\partial y^L$. A non-viscous fluid may be considered as a particular elastic material for which X^{rs} depends on $\partial x^r/\partial y^L$ through $\det\|\partial x^r/\partial y^L\|$, hence just through the inverse of the specific volume. This leads us naturally to Bressan's and especially Schöpf's ways of introducing the invariant mass density k, which do not involve any additional postulate. A quite different way of introducing the invariant mass density k is to use a postulate for this purpose. This is what Eckart does in connection with non-viscous or viscous fluids within special relativity; he effectively characterizes $k\,dC$ as the number dv of molecules in $d\mathscr{C}$ multiplied by the molecular weight M of $d\mathscr{C}$; moreover Eckart postulates the constancy of $M\,dv$—cf. Eckart [1940b(2)].

In Eckart's theory of fluids within special relativity, the continuity equation $(21.3)_1$ is a consequence of the postulate mentioned above—and the definition $k\,dC=M\,dv$—just as in classical physics the continuity equation follows from the conservation principle for mass.

However, in general relativity, $M dv$ generally is not constant unless the molecular weight M is defined in connection with a particular reference state Σ^* of $\mathscr{C}$. Such a definition of M may induce some authors to consider $M dv$ constant by definition and not by a postulate. Thus, since $M dv = k^* dC^*$, this procedure of introducing k—based on the definition of molecular weight in connection with a particular reference state Σ^* and the definition $k dC = M dv$—substantially coincides with the one considered in Bressan [1963c], i.e. the first considered in this section. Schöpf's and Bressan's ways of introducing k, based on no additional postulate, are in accord with the fact that in relativity the conservation principle for proper mass holds only approximatively.[10)]

On the other hand the fact that Eckart introduced k by postulating the continuity equation $(21.3)_3$ may induce some authors to believe that the same equation is independent of the axioms accepted in every usual relativity theory, so that the validity of $(21.3)_3$ might be denied.[11)] Among other things this shows how the latter way of introducing k is different from the preceding ways.

By the continuity equation (21.3) *for every continuously differentiable scalar field or tensor field* $T_{\dots}^{\dots} = k\mathscr{T}_{\dots}^{\dots}$ *we have*—cf. (17.3), (20.2)—

$$k\dot{\mathscr{T}}_{\dots}^{\dots} = (k\mathscr{T}_{\dots}^{\dots} u^\alpha)_{/\alpha}, \tag{21.5}$$

hence

$$k\frac{D}{Ds}\frac{T_{\dots}^{\dots}}{k} = (T_{\dots}^{\dots} u^\alpha)_{/\alpha} = \dot{T}_{\dots}^{\dots} + T_{\dots}^{\dots} u^\alpha{}_{/\alpha}. \tag{21.6}$$

§ 22. Angular and Deformation Velocities. Convected and Co-Rotational Fluxes. On $T_{\alpha\beta}{}^{/\beta}$ for $T'''_{[\alpha\beta]} = 0$

Let us consider the 4-velocity field u^α of the continuous body $\mathscr{C}$. By $(17.4)_3$ and (17.9) the spatial 4-velocity gradient $u_{\alpha/\overset{\perp}{\beta}}$ is a spatial tensor, hence so are the tensors $c u_{[\alpha/\overset{\perp}{\beta}]}$ and $c u_{(\alpha/\overset{\perp}{\beta})}$. By (16.3), $(17.1)_1$ and (18.6) their spatial components $c u_{[r/s]}$ and $c u_{(r/s)}$ at the event point $\mathscr{E}$ in a locally natural and proper frame (x) turn out to be identical with those of the angular velocity and deformation velocity respectively, referred to a classical Euclidean frame which is freely falling, non-rotating, and momentarily proper. Thus it is natural [§ 19] to call $c u_{[\alpha/\overset{\perp}{\beta}]}$ *(local) angular velocity* and $c u_{(\alpha/\overset{\perp}{\beta})}$ *deformation velocity* at $\mathscr{E}$.

In dealing e.g. with polarizable elastic materials, for many a one spatial vector V_α defined at least along the world line of the material point P^* of $\mathscr{C}$, it is useful to consider the *contravariant convected (time) flux* $c D^c V_\alpha / Ds$, the *covariant convected (time) flux* $c D_c V_\alpha / Ds$ and the *co-rotational (time) flux* $\dfrac{c D_r V_\alpha}{Ds}$ defined by[12)]

$$\frac{D^c V_\alpha}{Ds} = \overset{\perp}{g}_\alpha{}^\beta \frac{D V_\beta}{Ds} - u_{\alpha/\beta} V^\beta, \quad \frac{D_c V_\alpha}{Ds} = \overset{\perp}{g}_\alpha{}^\beta \frac{D V_\beta}{Ds} + V^\beta u_{\beta/\overset{\perp}{\alpha}} \quad (V^\beta u_\beta \equiv 0) \tag{22.1}$$

and

$$\frac{D_r V_\alpha}{Ds} = \overset{\perp}{g}_\alpha{}^\beta \frac{DV_\beta}{Ds} - u_{[\alpha/\overset{\perp}{\beta}]} V^\beta = \overset{\perp}{g}_\alpha{}^\beta \frac{DV_\beta}{Ds} + V^\beta u_{[\beta/\overset{\perp}{\alpha}]} \quad (V^\beta u_\beta \equiv 0). \tag{22.2}$$

The 4-velocity of matter at the event point $x^\alpha + dx^\alpha$ is $u^\alpha + u^\alpha{}_{/\beta} dx^\beta$. We assume that dx^α is parallel with V_α at the event point x^α, so that dx^α is spatial. Then $cu^\alpha{}_{/\beta} dx^\beta$ is the rate of change of that infinitesimal vector which is joined to $\mathscr{C}$ (at P^*) and passes through the space time linear element $x^\alpha, x^\alpha + dx^\alpha$. Hence, if we consider the motion $\mathscr{M}$ of $\mathscr{C}$ as motion of transport, we may regard $cu^\alpha{}_{/\beta} V^\beta$ as the *rate of change of transport* for the spatial vector V_α. As a consequence $D^c V^\alpha / D\tau$—cf. (18.6)—may be considered as the *rate of change, relative to* $\mathscr{M}$, *of the spatial vector* V^α.

Definitions (22.1) and (22.2) imply

$$\frac{D^c V_\alpha}{Ds} = \frac{D_c V_\alpha}{Ds} - 2u_{(\alpha/\overset{\perp}{\beta})} V^\beta = \frac{D_r V_\alpha}{Ds} - u_{(\alpha/\overset{\perp}{\beta})} V^\beta. \tag{22.3}$$

Therefore in case $\mathscr{M}$ is locally rigid (in the Born sense, i.e. $u_{(\alpha/\overset{\perp}{\beta})} = 0$) $D^c = D_c = D$; in particular, $D_r V_\alpha / D\tau$ is also a relative rate of change (with respect to $\mathscr{M}$).

In the general case let $\mathfrak{F}_r$ be a locally rigid *co-rotational* ideal fluid, in the sense that for the 4-velocity field u'_α of $\mathfrak{F}_r$ the relations

$$u'_\alpha = u_\alpha, \qquad u'_{(\alpha/\overset{\perp}{\beta})} = 0, \qquad u'_{[\alpha/\overset{\perp}{\beta}]} = u_{[\alpha/\overset{\perp}{\beta}]} \tag{22.4}$$

hold at $\mathscr{E}$. Then $u_{[\alpha/\overset{\perp}{\beta}]} V^\beta$ is the rate of change of transport of V_α for an observer joined to $\mathfrak{F}_r$, and $\dfrac{D_r V_\alpha}{D\tau}$ —cf. (22.2), (18.6)—is the rate of change of V_α relative to $\mathfrak{F}_r$.[13] Perhaps the main reason for considering both D^c and D_c is that (22.1) implies the first of the following identities in the spatial fields V_α and W_α:

$$V_\alpha \frac{D^c W^\alpha}{Ds} + \frac{D_c V_\alpha}{Ds} W^\alpha = \frac{DV_\alpha W^\alpha}{Ds} = V_\alpha \frac{D_r W^\alpha}{Ds} + \frac{D_r V_\alpha}{Ds} W^\alpha. \tag{22.5}$$

The second identity obviously follows from (22.2).

By (22.1), (22.2), and (21.6), *for every spatial field* V_α we have

$$k \frac{D^c k^{-1} V_\alpha}{Ds} = \frac{D^c V_\alpha}{Ds} + u^\beta{}_{/\beta} V_\alpha, \qquad k \frac{D_c k^{-1} V_\alpha}{Ds} = \frac{D_c V_\alpha}{Ds} + u^\beta{}_{/\beta} V_\alpha,$$

$$k \frac{D_r k^{-1} V_\alpha}{Ds} = \frac{D_r V_\alpha}{Ds} + u^\beta{}_{/\beta} V_\alpha \quad (V_\alpha u^\alpha \equiv 0). \tag{22.6}$$

Using convention (17.14), by (17.3) and $(17.4)_1$ we have

$$T_{\alpha\beta}{}^{/\beta} = u_\alpha \left(\frac{DT}{Ds} + T u^\beta{}_{/\beta} + T''^\beta{}_{/\beta} \right) + \frac{DT'_\alpha}{Ds} + T'_\alpha u^\beta{}_{/\beta} + \overset{\perp}{T}_{\alpha\beta}{}^{/\beta} + T A_\alpha + u_{\alpha/\beta} T''^\beta; \tag{22.7}$$

hence by (21.6), $(17.14)_{2,3}$, and (17.17) for $T^{\cdots\alpha} = T''^\alpha$, we respectively have

$$\overset{\perp}{g}_{\alpha\gamma} T^{\gamma\beta}{}_{/\beta} = T A_\alpha + \overset{\perp}{g}_{\alpha\gamma} \overset{\perp}{T}{}^{\gamma\beta}{}_{/\beta} + k \overset{\perp}{g}_{\alpha\gamma} \frac{D k^{-1} T'^\gamma}{Ds} + u_{\alpha/\beta} T''^\beta, \tag{22.8}$$

$$\begin{aligned} -u_\alpha T^{\alpha\beta}{}_{/\beta} &= k \frac{D}{Ds} \frac{T}{k} + \overset{\perp}{T}{}^{\alpha\beta} u_{\alpha/\beta} + T'^\beta A_\beta + T''^\beta{}_{/\beta} = \\ &= k \frac{D}{Ds} \frac{T}{k} + \overset{\perp}{T}{}^{\alpha\beta} u_{\alpha/\beta} + (T'^\beta + T''^\beta) A_\beta + T''^\beta{}_{/\overset{\perp}{\beta}}. \end{aligned} \tag{22.9}$$

Now we consider the case where the mixed part $T'''_{\alpha\beta}$ of $T_{\alpha\beta}$ is symmetric ($T'_\alpha = T''_\alpha$) and we remark that in this case the contribution given by $T'''_{\alpha\beta}$ to the right-hand side of (22.8)—i.e. the sum of the last two terms in it—represents k times the relative rate of change of T'_α/k with respect to an ideal fluid $\mathfrak{F}''$ whose velocity field u''_α locally satisfies the conditions $u''_\alpha = u_\alpha$ and $u''_{\alpha/\overset{\perp}{\beta}} = -u'_{\alpha/\overset{\perp}{\beta}}$.

On the other hand, if $T'''_{\alpha\beta}$ is skewsymmetric ($T'_\alpha = -T''_\alpha$), then by $(22.1)_1$ relation (22.8) becomes

$$\overset{\perp}{g}_{\alpha\gamma} T^{\gamma\beta}{}_{/\beta} = \overset{\perp}{g}_{\alpha\gamma} \overset{\perp}{T}{}^{\gamma\beta}{}_{/\beta} + T A_\alpha + k \frac{D^c k^{-1} T'_\alpha}{Ds} \quad (\text{for } T'_\alpha = T''_\alpha). \tag{22.10}$$

Footnotes to Chapter 2

1) The considerations exposed in this section differ from the analogues made in Synge [1960, Chap. III] in that, unlike Synge, we take into account the possibility for the intrinsic motion of a clock to depend on external non-gravitational actions.

Of course, besides this, there are some other differences of a formal or mathematical kind, mainly related to the uniqueness theorem at the end of this section.

2) Of course the ratio r of the periods of the two hinted clocks, say C_1 and C_2, is not affected by the presence of a third clock C, if any, on the same world line ℓ. So, calling r_i the ratio of the periods of the clocks C_i and C $(i=1,2)$, we have $r = r_1/r_2$.

3) We remember that a scalar system $T_{\alpha\beta\ldots}{}^{\gamma\ldots}$, defined for every frame (x) in S_4, is said to be a *tensor* at the event point $\mathscr{E}$ with the covariant indices $\alpha, \beta, \ldots$ and the contravariant indices $\gamma\ldots$ if, denoting the value of the afore-mentioned scalar system for the frame $(\bar{x})$ by $\bar{T}_{\alpha\beta\ldots}{}^{\gamma\ldots}$, we have

$$\bar{T}_{\alpha\beta\ldots}{}^{\gamma\ldots} = T_{\rho\sigma\ldots}{}^{\tau\ldots} \frac{\partial x^\rho}{\partial \bar{x}^\alpha} \frac{\partial x^\sigma}{\partial \bar{x}^\beta} \cdots \frac{\partial \bar{x}^\gamma}{\partial x^\tau} \cdots$$

where the derivatives $\partial x^\rho/\partial \bar{x}^\alpha$ and $\partial \bar{x}^\alpha/\partial x^\rho$ are evaluated at $\mathscr{E}$.

Tensors are special cases of double tensors whose definition and basic properties are dealt with in the Appendix *A*.

4) We set $dx^2=dx^3=0$ and fix $dx^0>0$. Then for every dx^1 with $|dx^1|<dx^0$ we have $ds^2>0$, so that $(15.4)_3$ holds. Moreover, $(15.4)_3$ takes the form

$$(dx^0)^2-(dx^1)^2=-\gamma_{00}(dx^0)^2-2\gamma_{01}dx^0dx^1-\gamma_{11}(dx^1)^2\,. \tag{*}$$

Since (*) holds for infinitely many values of dx^1, we have $\gamma_{00}=-1$, $\gamma_{01}=0$, and $\gamma_{11}=1$. In a similar way we can prove

$$\gamma_{00}=-1\,,\quad \gamma_{rr}=1\,,\quad \gamma_{0r}=0\,. \tag{**}$$

Now we keep dx^0 (with $dx^0>0$) fixed and we let dx^r vary arbitrarily except that ds^2 must be kept positive. Then $(15.4)_3$ holds and by $(**)_{1,3}$ from $(15.4)_3$ we deduce the equality $(\gamma_{rs}-\delta_{rs})dx^r dx^s=0$. Its first member is a homogeneous function of dx^r, hence the same equality must hold identically, so that $\gamma_{rs}=\delta_{rs}$. Then, by (**) we have $g_{\alpha\beta}=\gamma_{\alpha\beta}$.

5) A domain is a closed set for which every accumulation point is an accumulation point of inner points.

6) The tensor $A_{\alpha\beta}$, which plays an essential role in the Einstein gravitational equations, is also called the Levi Civita tensor because, in its tensorial and final form, it was introduced by Levi Civita [1917]. The quantity previously playing the essential role mentioned above was not a tensor and had led Einstein to obtain rather surprising results which are avoided by the use of $A_{\alpha\beta}$. In order to explain them, Einstein had suggested use of quantum theory.

7) There exists a frame $(\bar{x})$ locally natural at $\mathscr{E}$, so that with obvious notations $\bar{g}_{\alpha\beta}=\delta'_{\alpha\beta}$ and $\overline{\{\alpha,\beta,\gamma\}}=0$, but generally $\bar{u}^r\neq 0$. By a rigid displacement in the Euclidean space $(\bar{x}^1,\bar{x}^2,\bar{x}^3)$, it can be shown in this case that the additional relations $\bar{x}^\alpha=0$, $\bar{u}^1>0$, and $\bar{u}^2=\bar{u}^3=0$ can hold at $\mathscr{E}$. Now let v be the value of $D\bar{x}^1/D\bar{x}^0$ at $\mathscr{E}$ and let us consider the special Lorentz transformation

$$x^1=\frac{\bar{x}^1-v\bar{x}^0}{\sqrt{1-v^2}},\quad x^2=\bar{x}^2,\quad x^3=\bar{x}^3,\quad x^0=\frac{\bar{x}^0-v\bar{x}^1}{\sqrt{1-v^2}}\quad \left(v=\frac{D\bar{x}^1}{D\bar{x}^0}\right). \tag{*}$$

Then $(*)_{1,5}$ obviously imply

$$\frac{Dx^1}{Ds}=\frac{Dx^1}{D\bar{x}^0}\frac{D\bar{x}^0}{Ds}=\left(\frac{D\bar{x}^1}{D\bar{x}^0}-v\right)\frac{1}{\sqrt{1-v^2}}\frac{D\bar{x}^0}{Ds}=0$$

and $(*)_{2,3}$ imply

$$Dx^r/Ds=\bar{u}^r=0 \quad \text{for} \quad r=2,3\,.$$

Therefore $(17.7)_3$ holds for the frame (x). Also so do the equalities $(17.7)_{1,2}$, because they hold for the frame $(\bar{x})$ and because (*) is a Lorentz transformation.

8) The ratio $d\overset{\perp}{s}/dx^0$ was substantially introduced by C. Cattaneo—who called it *standard relative velocity* with respect to the ideal fluid $\mathfrak{F}$—for an extensive study of relative motions in general relativity, made both from the kinematic and dynamic points of view—cf. Cattaneo [1959] and [1960, p. 155].

9) The frame (x) is called *static* if in it $g_{r0}=0=g_{\alpha\beta,0}$.

10) In ordinary (thermodynamic) processes the internal energy varies by ordinary amounts. By the equivalence principle (4.4) the corresponding amounts by which proper mass varies are extremely small.

11) Such an attempt at denying (21.3) is briefly considered by Pratelli [1961, p. 2, lines 1—6; § 2] who asserts what we consider the first principle in the adiabatic case, together with a constitutive equation of the form $T=T(p,\beta)$ and an equation involving entropy which turns out to be incompatible with the former equations.

12) These fluxes are considered by Bressan [1966d, e] in accord with the analogous fluxes used by Truesdell and Toupin in [1960, p. 450] within a 3-dimensional Euclidean space—cf. Appendix *D*.

13) The vectors $-V^\beta u_{(\beta/\overset{\perp}{\alpha})}$ and $D_c V^\alpha/D\tau$ are the rate of change of transport and the relative rate of change of V_α in case the motion of transport is the one of an ideal fluid, $\mathfrak{F}$, whose velocity field u'_α fulfills the conditions $u'_\alpha=u_\alpha$ and $u'_{\alpha/\overset{\perp}{\beta}}=-u_{\beta/\overset{\perp}{\alpha}}$—i.e. $u'_{(\alpha/\overset{\perp}{\beta})}=-u_{(\alpha/\overset{\perp}{\beta})}$, $u'_{[\alpha/\overset{\perp}{\beta}]}=u_{[\alpha/\beta]}$—at $\mathscr{E}$.

Chapter 3

Gravitation and Conservation Equations. Fluids and Elastic Waves

§ 23. The Einstein Gravitation Equations and Basic Consequences

Let h be Cavendish's constant. Then in view of $(16.10)_1$ and (21.1) the Einstein non-cosmological gravitation equations—on which general relativity is based—can be written in the form

$$A_{\alpha\beta} + \frac{8\pi h}{c^4} \mathscr{U}_{\alpha\beta} = 0 \quad \text{where} \quad \mathscr{U}_{\alpha\beta} = \rho u_\alpha u_\beta \ldots \tag{23.1}$$

The tensor $\mathscr{U}_{\alpha\beta}$, which is called *total energy tensor*, reduces to $\rho u_\alpha u_\beta$ only in the case of non-interacting matter. In this case the temporal part of $(23.1)_1$ which by (17.14) is

$$A + 8\pi h c^{-4} \rho \equiv A + 8\pi h c^{-4} \mathscr{U} = 0, \tag{23.2}$$

appears to be very similar to Poisson's equation, provided it is evaluated in a frame very close to a Lorentz frame and provided in addition $c^{-2}\rho$ be thought of as the gravitational mass density [§ 21] as we shall do.

As to orders of magnitude, ρ is larger by the factor c^2 than ordinary magnitudes—cf. (21.1), (21.4). Hence $(23.2)_2$ can be considered as an approximate version of Poisson's equation in every case where the dots in (23.1) are replaced by terms of the same order as c^{-m} with $m \geqslant 2$. This condition will always be fulfilled in the remainder of this tract even when couple stresses are taken into account [§ 87].

By $(16.10)_{3,2}$ from (23.1) we deduce

$$\mathscr{U}_{\alpha\beta}{}^{/\beta} = 0, \quad \mathscr{U}_{[\alpha\beta]} = 0. \tag{23.3}$$

Equation $(23.3)_1$ is called *(energy-momentum) conservation equation.* In special relativity $(23.3)_1$ is postulated instead of (23.1). In special or general relativity it is useful to split $(23.3)_1$ into its temporal and spatial parts. Since $ds = c\,d\tau$—cf. (18.6)—by (22.8) and (22.9) for $T^{\alpha\beta} = \mathscr{U}^{\alpha\beta}$, and by (23.1) and (21.4) these parts can be written as follows:

$$\rho A_\alpha + \cdots \equiv \overset{\perp}{g}_{\alpha\gamma} \mathscr{U}^{\gamma\beta}{}_{/\beta} = 0, \quad k \frac{Dw}{D\tau} + \cdots \equiv -c u_\alpha \mathscr{U}^{\alpha\beta}{}_{/\beta} = 0. \tag{23.4}$$

In general relativity, where $(23.3)_2$ must hold, equations $(23.3)_1$ and (23.4) become

$$\mathscr{U}_{(\alpha\beta)}{}^{/\beta}=0\,,\qquad \overset{\perp}{g}_{\alpha\gamma}\mathscr{U}^{(\gamma\beta)}{}_{/\beta}=0\,,\qquad -cu_\alpha\mathscr{U}^{(\alpha\beta)}{}_{/\beta}=0\,. \tag{23.5}$$

For $\mathscr{U}_{\alpha\beta}=\rho u_\alpha u_\beta$, i.e. for non-interacting matter, equation $(23.4)_{1,2}$ implies that the world lines of material points are geodesics, and equation $(23.4)_{3,4}$ that $Dw/D\tau=0$ holds, so that c^2+w is constant along world lines. By the definition of conventional mass the same holds for kdC—cf. $(21.2)_1$. Hence by (21.4) the gravitational mass $c^{-2}\rho dC$ of any matter element $d\mathscr{C}$ is constant.

The explicit part of the expression $(23.1)_2$ for $\mathscr{U}_{\alpha\beta}$ is substantially due to the requirement that the tensor equations (23.1) in the unknown metric tensor $g_{\alpha\beta}$ (and other quantities) should be in accord with Poisson's equation.[1)]

We may now add that by the mass-energy equivalence principle w is the specific internal energy [§ 21] and $(kdC)Dw$ is the increment of the internal energy of $d\mathscr{C}$ in the proper time $D\tau$; therefore in any case of interacting matter the dots in $(23.1)_2$ must be replaced with terms such that equation $(23.4)_{3,4}$ becomes an acceptable relativistic version of the first principle of thermodynamics or, more generally, of the equation of energy balance.

Similarly, since by $(21.6)_1$ ρA_α is the intrinsic rate of change of momentum (ρu_α)—i.e. the one with respect to a locally natural and proper observer (17.7)—in ordinary units, the terms replacing the dots in $(23.1)_2$ must also render equation $(23.4)_{1,2}$ an acceptable relativistic version [§ 8] of the first Cauchy equation for continuous media in classical physics.

The preceding requirements will be met in all of the remaining sections, where more and more general determinations of $\mathscr{U}_{\alpha\beta}$ will be considered in order to include stresses, thermodynamics with heat conduction, electromagnetism and couple stresses. As to couple stresses—following Bressan [1966a, b] where general kinds of materials are considered—the energy tensor $\mathscr{U}_{\alpha\beta}$ will be so chosen that $(23.3)_2$ constitutes an acceptable relativistic version [§ 8] of the second classical equation of Cauchy for continuous media.

§ 24. The Case of Interacting Matter Capable of Heat Conduction

Let dx^α and δx^α be two infinitesimal spatial linear elements starting at the typical event point $\mathscr{E}$. They determine an oriented parallelogram, $d\tilde{\sigma}$, of area $d\sigma$ and unit normal n_α; $d\tilde{\sigma}$ is represented by the spatial vector $d\sigma_\alpha$ with

$$d\sigma_\alpha=n_\alpha d\sigma=\pm\overset{\perp}{\varepsilon}_{\alpha\beta\gamma}dx^\beta\delta x^\gamma \tag{24.1}$$

where the upper or lower sign holds according to whether the frame (x) is right-handed or not.

On the basis of the classical theory of continuum media and in particular Cauchy's tetrad theorem for stresses, we may assert the existence of a spatial tensor, $X_{\alpha\beta}$, such that for every choice of dx^α and δx^α, the resultant dR_α of the forces exerted by the matter elements adjacent to the negative face of $d\tilde{\sigma}$ on those adjacent to the positive one is expressed by

$$dR_\alpha = X_\alpha^{\ \beta} d\sigma_\beta \quad \text{or} \quad dR_\alpha/d\sigma = X_\alpha^{\ \beta} n_\beta \quad (X^{\alpha\beta} u_\beta = 0 = u_\alpha X^{\alpha\beta}). \tag{24.2}$$

On the basis of the classical theory of heat conduction we may assert the existence at $\mathscr{E}$ of a spatial vector cq^α, to be called the *heat flux vector*, such that for every choice of the surface $d\tilde{\sigma}$ (through $\mathscr{E}$) the heat flux through $d\tilde{\sigma}$, in the sense of n_α, equals $cq^\alpha d\sigma_\alpha$.

We now show that for interacting matter capable of heat conduction the total energy tensor $\mathscr{U}_{\alpha\beta}$ can be given the form

$$\mathscr{U}_{\alpha\beta} = \rho u_\alpha u_\beta + X_{\alpha\beta} + Q_{\alpha\beta} \quad \text{where} \quad Q_{\alpha\beta} = u_\alpha q_\beta + q_\alpha u_\beta \quad (q_\alpha u^\alpha = 0). \tag{24.3}$$

The tensor $Q_{\alpha\beta}$, which takes heat conduction into account and is therefore called *thermodynamic tensor*, was introduced by Eckart in [1940b] in connection with viscous fluids within special relativity. However it works in every case to be considered in this tract.

By (24.3) and convention (17.13) for $\mathscr{U}_{\alpha\beta} = T_{\alpha\beta}$ we have $T'_\alpha = T''_\alpha = q_\alpha$, $T = \rho$ and $\overset{+}{T}_{\alpha\beta} = X_{\alpha\beta}$. Then, on the one hand, by (22.8) for $T_{\alpha\beta} = \mathscr{U}_{\alpha\beta}$ the basic Cauchy equation $(23.4)_{1,2}$ becomes

$$\rho A_\alpha = -\overset{\perp}{g}_{\alpha\gamma} X^{\gamma\beta}{}_{/\beta} - k\left(\overset{\perp}{g}_{\alpha\gamma} \frac{D}{Ds}\frac{q^\gamma}{k} + u_{\alpha/\gamma}\frac{q^\gamma}{k}\right). \tag{24.4}$$

On the other hand, remarking that (21.4) implies $D(\rho/k) = Dw$ and setting

$$c^{-1} k q_{\text{ass}} = -q^\alpha{}_{/\alpha} - q^\alpha A_\alpha = -q^\alpha{}_{/\overset{\perp}{\alpha}} - 2q^\alpha A_\alpha \tag{24.5}$$

—cf. $(24.3)_3$, (17.17)—, by use of $(22.9)_1$ for $T_{\alpha\beta} = \mathscr{U}_{\alpha\beta}$ we see that the basic equation $(23.4)_4$ becomes

$$k\frac{Dw}{Ds} + X^{\alpha\beta} u_{\alpha/\beta} = c^{-1} k q_{\text{ass}} \quad (kq_{\text{ass}} = c u_\alpha Q^{\alpha\beta}{}_{/\beta}). \tag{24.6}$$

Now let us remark that under assumption (24.3) the consequence $(23.3)_2$ of the gravitational equations (23.1) is equivalent to

$$X_{[\alpha\beta]} = 0. \tag{24.7}$$

In the case being considered the classical analogue of (24.7) follows from the equation of balance of moment of momentum for every matter portion, and may be considered as the second Cauchy equation of continuous media. Hence in the same case $(23.3)_2$ is assumed to hold also in special relativity.

Especially for our treatment of acceleration waves [§ 31] in general elastic materials it is useful to remark that, by $(24.2)_{3,4}$ and (17.17) for $T^{\cdot\cdot\cdot\alpha} = X^{\gamma\alpha}$, we can put (24.4) into the form

$$(\rho \overset{\perp}{g}_\alpha{}^\gamma + X_\alpha{}^\gamma) A_\gamma = -\overset{\perp}{g}_{\alpha\gamma} X^{\gamma\beta}{}_{/\overset{\perp}{\beta}} - k\left(\overset{\perp}{g}_{\alpha\gamma} \frac{D}{Ds}\frac{q^\gamma}{k} + u_{\alpha/\gamma}\frac{q^\gamma}{k}\right). \tag{24.8}$$

All terms in (24.8) are spatial and, by (21.4) and (18.6), in locally proper and natural co-ordinates—cf. (17.7)—(24.8) can be written in the form

$$k a_r + \frac{1}{c^2}(w\delta_r{}^s + X_r{}^s) a_s = -X_r{}^s{}_{/s} - \frac{k}{c^2}\left(\frac{D}{D\tau}\frac{cq_r}{k} + v_{r/s}\frac{cq^s}{k}\right) \tag{24.9}$$

where a_r and $v_{r/s}$ are magnitudes of ordinary size—as well as k, w, X_{rs} and cq_r—defined by

$$a_r = c^2 A_r, \qquad v_{r/s} = c u_{r/s}. \tag{24.10}$$

Their physical meanings are obvious. Hence in the absence of electromagnetic phenomena (24.9) differs only by terms of the order of c^{-2} from the equation $k a_r = -X_r{}^s{}_{/s}$ which constitutes the first Cauchy equation of continuous media referred to a classical frame locally freely falling and non-rotating. Hence (24.9) is an acceptable relativistic version of the same equation.

By (24.10) and (18.6), in locally proper and natural co-ordinates (24.6) and (24.5) become

$$k\frac{Dw}{D\tau} + X^{rs} v_{r/s} = k q_{\text{ass}} = -cq^r{}_{/r} - \frac{2}{c^2} cq^r a_r. \tag{24.11}$$

Equation $(24.11)_2$ differs only in the last term, which is of the order of c^{-2}, from the classical expression $k q_{\text{ass}} = -cq^r{}_{/r}$ (in any Euclidean frame) for the heat absorbed per unit time and (proper) volume by heat conduction. In addition equation $(24.11)_1$ is identical with the classical first principle of thermodynamics in any Euclidean frame. Hence the meaning of absorbed heat per unit time and proper mass is acceptable in relativity theory for the quantity q_{ass} defined by (24.5). Furthermore (24.6) is an acceptable version of the first principle.

Thus the relativistic conservation equations (23.1) with the expression (24.3) of $\mathscr{U}_{\alpha\beta}$ have been shown to be acceptable in the case being considered.

§ 25. On the Second Principle of Thermodynamics, the Clausius-Duhem Inequality and Fourier's Law. A Relativistic Proof of the Symmetry of the Heat Conduction Coefficient

We denote by T the absolute temperature at the typical event point $\mathscr{E}$ in $W_{\mathscr{C}}$, and we understand that T is to be measured—according to Kelvin's procedure based on the Carnot cycle—by an observer joined with matter.

The second principle of thermodynamics can be stated as follows: *For every material point P^* of $\mathscr{C}$ there is a function, η, of the local (intrinsic) physical state of $\mathscr{C}$ at P^*, such that along every process physically possible for P^* we have*—cf. $(24.5)_1$

$$T\frac{D\eta}{D\tau} \geqslant \frac{dQ}{D\tau} \quad \textit{where} \quad \frac{dQ}{d\tau} = q_{\text{ass}} + r = r - \frac{c}{k}(q^{\alpha}{}_{/\alpha} + q^{\alpha} A_{\alpha}). \tag{25.1}$$

Thus $dQ/D\tau$ is the total heat absorbed per unit time and conventional mass; it consists of the heat q_{ass} absorbed by heat conduction plus the absorbed heat r due to electromagnetic radiation.

We shall be mainly interested in the case where the physical state of $\mathscr{C}$ at P^* and at the proper time τ may be regarded as determined by the motion and temperature distribution inside a small material neighbourhood $\mathscr{N}(P^*)$ of P^*—cf. the principle of determinism for the stress and the one of local action in Truesdell and Noll [1965, p. 56].

Several relativistic versions of Fourier's law have been proposed and are still used. For reasons considered in the second part of this section and related to a result of Tolman and Ehrenfest [§ 45] most sections on materials presented in this tract may be based on the assumption

$$q^{\alpha} = -\kappa^{\alpha\beta}\theta_{\beta} \quad \textit{with} \quad \theta_{\beta} = T_{/\beta} + TA_{\beta}, \qquad \kappa^{[\alpha\beta]} = 0 = u_{\alpha}\kappa^{\alpha\beta}, \tag{25.2}$$

where the spatial vector $\kappa^{(\alpha\beta)}$ is positive definite, i.e. $\kappa_{\alpha\beta}dx^{\alpha}dx^{\beta} \geqslant 0$ (so that the present *generalized Fourier law* can also be applied to non-conducting materials) *and where $\kappa^{\alpha\beta}$ depends on A_{β} at most through θ_{β}*.

However the same sections on materials can also be based on other relativistic postulates of heat conduction. A set of these postulates is proposed in connection with fluids by Müller [1969]. We shall consider this interesting work in the last part of § 26. Some among the main concequences of its basic assumptions have been investigated. For instance Müller [1970] remarks that the non-relativistic limit ($c \to +\infty$) of the constitutive equations derived from these basic assumptions does not (automatically) comply with the principle of material indifference [§ 81], as one might have hoped.

When constitutive equations are dealt with, we shall be much interested in materials capable of only reversible processes, i.e. processes along which the second principle (25.1) always holds as an equality.

The remainder of this section will not be used in the sequel; it is devoted to discussions and motivations, and in particular to the way in which the different relativistic versions of Fourier's equation mentioned above are related to different relativistic versions of the Clausius-Duhem inequality.

In order to discuss heat conduction in relativity we now assume (25.1) and not (25.2).

In classical physics the second principle can be stated with substantially the same words that we used in relativity theory. However it can also be given an integral form (the Clausius-Duhem integral inequality) which through a combination with the first principle leads us to the following classical Clausius-Duhem (local) inequality, where $\bar{q}^r$ is the heat flux:

$$k\dot{\eta} \geqslant \frac{k\,r}{T} - \left(\frac{1}{T}\bar{q}^r\right)_{/r} = \frac{1}{T}(kr - \bar{q}^r{}_{/r}) + \frac{1}{T^2}\bar{q}^r T_{/r} \quad (\bar{q}^r = c\,q^r). \tag{25.3}$$

Incidentally the classical equation of energy balance reads

$$k\dot{w} + \frac{d\,l^{(i)}}{D\tau} = k(r + \bar{q}_{\text{ass}}) \quad \text{where} \quad \frac{d\,l^{(i)}}{D\tau} = X^{rs} v_{r/s}, \quad k\bar{q}_{\text{ass}} = -\bar{q}^r{}_{/r}. \tag{25.4}$$

Hence inequality (25.3) becomes[2)]

$$k\,T\dot{\eta} \geqslant k(r + \bar{q}_{\text{ass}}) + \frac{1}{T}\bar{q}^r T_{/r}, \qquad k\,T\dot{\eta} \geqslant k\dot{w} + \frac{d\,l^{(i)}}{d\tau} + \frac{1}{T}\bar{q}^r T_{/r}. \tag{25.5}$$

Now let us remark that, in classical physics, by $(25.4)_3$ [the principle $(25.4)_{1,3}$] inequalities (25.3) and $(25.5)_1$ [$(25.5)_2$]—considered in any Euclidean frame—are mutually equivalent, and reversible processes can be defined as those along which any one of the alternatives

$$k\,T\dot{\eta} = kr - \bar{q}^s{}_{/s}, \qquad k\,T\dot{\eta} = k(r + \bar{q}_{\text{ass}}) = k\dot{w} + \frac{d\,l^{(i)}}{D\tau} \tag{25.6}$$

holds. For materials capable of only reversible processes (25.3) or (25.5) implies

$$\bar{q}^r T_{/r} \leqslant 0; \tag{25.7}$$

and the linearity assumption

$$\bar{q}^r = -\bar{\kappa}^{rs} T_{/s}, \quad \text{where} \quad \bar{\kappa}^{[rs]} = 0 \quad \text{and} \quad \partial\bar{\kappa}^{rs}/\partial T_{/s} = 0, \tag{25.8}$$

easily yields by (25.7) that $\bar{\kappa}^{rs}$ must be positive definite:

$$\bar{\kappa}^{rs}\xi_r\xi_s \geqslant 0 \quad \text{for every} \quad \xi^r. \tag{25.9}$$

Hence (25.8) is the (generalized) Fourier law.

Conversely, assume only (25.4) and neither (25.3) nor (25.5). The Fourier law—cf. (25.8) and (25.9)—yields (25.7). Then inequalities (25.3) and (25.5) hold for materials capable of only reversible processes in that by (25.7) we deduce (25.3) from $(25.6)_1$, and (25.5) from $(25.6)_{2,3}$.[3)]

We assume $r=0$ for the sake of simplicity. Then by $(25.3)_3$ we can relativize the mutually equivalent classical versions $(25.3)_1$ and $(25.5)_1$ of the Clausius-Duhem inequality into

$$k\dot{\eta} \geqslant -\left(\frac{1}{T}q^\alpha\right)_{/\alpha} \equiv -\left(\frac{1}{T}q^\alpha\right)_{/\overset{+}{\alpha}} - \frac{1}{T}q^\alpha A_\alpha \tag{25.3b}$$

—cf. (17.17) and (20.2)—and

$$kT\dot{\eta} \geqslant c^{-1} k q_{\mathrm{ass}} + \frac{1}{T}q^\rho T_{/\rho} \tag{25.5b}$$

respectively.[4)] Since the relativistic analogue of (25.4) for $r=0$ is (24.11), inequalities (25.3b) and (25.5b) are not equivalent, unlike (25.3) and (25.5). The theory developed in the remaining chapters is in accordance with (25.3b).

Now we consider the direct relativization (25.3b) of the Clausius-Duhem inequality and we show that it is in accordance with the relativization (25.2) of Fourier's law which, at least at first sight, does not appear very natural. By $(25.3\mathrm{b})_1$ and $(24.5)_1$ and by $(25.2)_2$ and the consequence $q^\alpha u_\alpha = 0$ of $(25.2)_{1,4}$ we have, respectively,

$$Tk\dot{\eta} \geqslant \frac{k}{c} q_{\mathrm{ass}} + q^\alpha A_\alpha + \frac{1}{T} q^\alpha T_{/\alpha} \equiv \frac{k}{c} q_{\mathrm{ass}} + \frac{1}{T} q^\alpha \theta_\alpha \,. \tag{25.10}$$

Obviously *inequalities* (25.3b) *and* (25.10) *are equivalent.*

For materials capable of only reversible processes, characterized by

$$cT\dot{\eta} = q_{\mathrm{ass}}\,, \tag{25.6b}$$

(25.10) is equivalent to the thermal conduction inequality

$$q^\alpha \theta_\alpha \leqslant 0 \qquad (\theta_\alpha = T_{/\overset{+}{\alpha}} + T A_\alpha)\,. \tag{25.7b}$$

Hence *for materials obeying* (25.6b), (25.3b) *implies* (25.7b).

Now it is obvious that (even if $\kappa^{\alpha\beta}$ is allowed to depend on θ_α) *for the aforementioned materials the relativistic version* (25.2) *of Fourier's law implies inequality* (25.3b). More, *for* $r=0$ *that law and the principle* (25.1) *imply* (25.3b) through (25.10).

Theorem 25.1. *Let q^α be a function of the position gradient $\alpha^\rho{}_L$, T, $T_{/\overset{+}{\alpha}}$, and A_α, that is linear in $T_{/\overset{+}{\alpha}}$ and A_α. Then inequality* (25.7b) *implies, in relativity, the version* $(25.2)_1$ *of Fourier's law, with $\kappa^{\alpha\beta}$ spatial and depending at most on $\alpha^\rho{}_L$ and T. Furthermore $\kappa^{\alpha\beta}$ is symmetric if so are the coefficients of $T_{/\overset{+}{\alpha}}$ and A_α in the aforementioned expression of q^α.*

Proof. By the assumption that q^α is linear in A_α and $T_{/\overset{+}{\alpha}}$, and by (25.7b)

$$q^\alpha(TA_\alpha + B_\alpha) \leqslant 0 \quad \text{where} \quad q^\alpha = a^{\alpha\beta} A_\beta + b^{\alpha\beta} B_\beta + c^\alpha \tag{25.11}$$

holds for all spatial vectors A_α and $B_\alpha(=T_{/\overset{\perp}{\alpha}})$ where the tensors $a^{\alpha\beta}$, $b^{\alpha\beta}$, and c^α do not depend on A_α or B_α. We also assume that these tensors are spatial, which is not restrictive. Then, by making A_α and B_α very small, it is clear that c^α must be zero. Hence, setting $X_\alpha = TA_\alpha + B_\alpha$, we can rewrite inequality (25.11) as follows:

$$[(a^{\alpha\beta} - Tb^{\alpha\beta})A_\beta + b^{\alpha\beta}X_\beta]X_\alpha \leqslant 0. \tag{25.11 b}$$

Since A_α and X_α are independent variables, should the tensor equality $a^{\alpha\beta} = Tb^{\alpha\beta}$ not hold, then for some values of A_α and X_α inequality (25.11 b) would obviously be violated. Hence the same equality holds. In other words—cf. $(25.11)_2$—q^α has the form $(25.2)_1$. Furthermore the coefficient $\kappa^{\alpha\beta}$ in (25.2) can surely be chosen spatial for q^α and $T_{/\overset{\perp}{\alpha}}$ are spatial.

Lastly from (25.7 b) and $(25.2)_1$ we deduce that the spatial tensor $\kappa^{(\alpha\beta)}$ is positive definite.

We immediately deduce $(25.2)_3$ if we add the assumptions $a^{[\alpha\beta]} = 0 = b^{[\alpha\beta]}$. q.e.d.

Now we consider the version (25.5 b) of the Clausius Duhem inequality and we show that it is in agreement with the following direct (natural) relativization of the Fourier law (25.8):[5)]

$$q^\alpha = -\kappa^{\alpha\beta} T_{/\beta} \qquad (\kappa^{[\alpha\beta]} = 0 = u_\alpha \kappa^{\alpha\beta}) \quad \text{where } \kappa^{\alpha\beta} \text{ is positive definite}. \tag{25.8 b}$$

To this end we first consider materials capable of only reversible processes —cf. (25.6 b). Then for these materials (25.5 b) is equivalent to the (thermal conduction) inequality

$$q^\alpha T_{/\overset{\perp}{\alpha}} \leqslant 0. \tag{25.7 c}$$

Now it is obvious that *for the afore-mentioned materials the law* (25.8 b) *implies inequality* (25.5 b) even if $\kappa^{\alpha\beta}$ is allowed to depend on $T_{/\overset{\perp}{\alpha}}$.

Theorem 25.2. *Let q^α be a function of the position gradient, T, A_α, and $T_{/\overset{\perp}{\alpha}}$, that is linear in A_α and $T_{\overset{\perp}{\alpha}}$. Then for materials obeying* (25.6b) *inequality* (25.5b) *implies the version* (25.8 b) *of Fourier's law, with $\kappa^{\alpha\beta}$ depending at most on T.*

The proof of this theorem is quite similar with the one of Theorem 25.1 (and simpler).

Historically the relativistic heat conduction law (25.2) was proposed by Eckart [1940 b] in the case of linearly viscous fluids, in special relativity. Eckart motivated this proposal by the following assertion which may be related to an extension of Theorem 25.1 to viscous fluids: Assumption (25.2) for $\kappa^{\alpha\beta} = \kappa \overset{\perp}{g}{}^{\alpha\beta}$ with $\kappa > 0$, is the simpliest one which enables us to deduce $(25.3)_1$ within Eckart's theory.[6)] However Eckart himself wrote in [1940 b] that the explicit form of the heat flux vector q^α still is a matter of discussion.

As a matter of fact some authors prefer the heat conduction assumption (25.2) in accordance with the direct relativization $(25.3\text{b})_1$ of the Clausius-Duhem inequality; others prefer the assumption (25.8 b) which we showed to be in accordance with the indirect relativization (25.5 b) of the same inequality. Among the latter there is Pham Mau Quan—cf. e.g. [1955], [1956]—who proposed the first general theories on fluids capable of heat conduction in general relativity; Bressan in his first works [1963 b] to [1964 b] in general relativity; and Kraniŝ [1966]. Pham Mau Quan apparently preferred (25.8 b) to (25.2) only for reasons of simplicity. This is to be related to the fact (substantially admitted by Eckart himself) that Eckart simply showed that (25.2) is a sufficient condition for the particular relativistic version (25.3 b) of the Clausius-Duhem inequality to hold. The converse of this result—i.e. the assertion that (25.3 b) implies (25.2)—was substantially proved by Stückelberg and Wander [1953]. These authors assume (25.3 b) while, as we showed, another relativistic version of the Clausius-Duhem inequality, (25.5 b), can be accepted, and in connection with it the version (25.8 b) of Fourier's law is outstanding. I think that this fact is responsible for a rather wide acceptance of (25.8 b), or at least may strongly support this acceptance.

There are rather strong motives for using assumption (25.2) instead of (25.8 b) [§ 45]. They are based on Stefan's and Boltzmann's laws on the theormodynamic equilibrium of electromagnetic radiation. Because of these motives and in particular because these laws are in finite terms, which implies their having unique relativizations, Bressan preferred (25.2) to (25.8 b) in his later works [1967 a] and [1966 c, d].[7)] The same point of view is complied with independently by Grot and Eringen [1966].

Unlike the corresponding classical expression, the second of the relativistic expressions (24.5) for q_{ass} includes the term $q^{(a)}_{\text{ass}} = -2ck^{-1}q^\alpha A_\alpha$ in the acceleration. Following Bressan [1964a] let us first show that also when the explicit expression $(25.8\text{b})_1$ of q^α is taken into account, then another term, $q^{(a)}_{\text{ass}}$, without any classical analogue and precisely a term in the deformation velocity arises. We do this in order to present the proof of the symmetry of $\kappa^{\alpha\beta}$ mentioned in the title. Of course we now disregard assumption $(25.2)_3$ but we assume that $\kappa^{\alpha\beta}$ is a spatial tensor.

Relation $(17.19)_1$ holds for $T^{\cdots\alpha\sigma} = \kappa^{\sigma\alpha}$; hence $\kappa^{\alpha\beta}{}_{/\overset{\perp}{\alpha}} = \overset{\perp}{g}{}^\beta{}_\rho \kappa^{\alpha\rho}{}_{/\overset{\perp}{\alpha}} + u^\beta u_{\rho/\alpha}\kappa^{\alpha\rho}$ which by $(17.9)_1$ and (17.3) implies

$$(\kappa^{\alpha\beta} T_{/\beta})_{/\overset{\perp}{\alpha}} = \kappa^{\alpha\beta} T_{/\beta\alpha} + \kappa^{\alpha\beta}{}_{/\overset{\perp}{\alpha}} T_{/\overset{\perp}{\beta}} + u_{\rho/\alpha}\kappa^{\alpha\rho} DT/Ds\,. \tag{25.12}$$

By (25.12), $(25.8\text{b})_1$, and $(20.2)_2$ relation (24.5) becomes

$$\begin{gathered} q_{\text{ass}} = q^{(c)}_{\text{ass}} + q^{(d)}_{\text{ass}} + q^{(a)}_{\text{ass}} \quad \text{where} \quad k q^{(a)}_{\text{ass}} = -2cq^\alpha A_\alpha\,, \\ c^{-1} k q^{(c)}_{\text{ass}} = \kappa^{\alpha\beta} T_{/\beta\alpha} + \kappa^{\alpha\beta}{}_{/\overset{\perp}{\alpha}} T_{/\overset{\perp}{\beta}}\,, \qquad k q^{(d)}_{\text{ass}} = \kappa^{\rho\sigma} u_{\sigma/\overset{\perp}{\rho}} \dot{T}\,. \end{gathered} \tag{25.13}$$

By $(25.13)_3$, in locally proper and natural co-ordinates $q^{(c)}_{\text{ass}}$ looks like the classical analogue of q_{ass} (so it may be called the classical part of q_{ass}).

On the other hand the term $q^{(d)}_{\text{ass}}$ in the velocity gradient and $\dot{T}$ has no classical analogue (as well as $q^{(a)}_{\text{ass}}$).

In case the proposal (25.2) is used instead of (25.8b), by (24.5) the relations $(25.13)_{1,2}$ must be turned into

$$q_{\mathrm{ass}} = q_{\mathrm{ass}}^{(c)} + q_{\mathrm{ass}}^{(d)} + q_{\mathrm{ass}}^{(a)} + \frac{cT}{k} \kappa^{\alpha\beta} A_{\beta/\alpha}, \tag{25.14}$$

$$k q_{\mathrm{ass}}^{(a)} = c[(2\kappa^{\alpha\beta} + \kappa^{\beta\alpha}) T_{/\beta} + T \kappa^{\beta\alpha}{}_{\overset{\perp}{/\beta}} + 2T\kappa^{\alpha\beta} A_\beta] A_\alpha \tag{25.15}$$

while $(25.13)_{3,4}$ are still valid.

Thus the difference $kq_{\mathrm{ass}} - kq_{\mathrm{ass}}^{(c)}$ between the relativistic heat absorbed (per unit time and proper volume) and the corresponding classical heat is given by $(25.13)_{1,2,4}$ or by (25.14), (25.15) and $(25.13)_4$, according to whether (25.8b) or (25.2) is accepted as the relativistic heat conduction law.

Theorem 25.3. *Let q^α be a function of some physical real parameters $p_{(r)}$ that are independent of the local angular velocity $u_{[\alpha/\overset{\perp}{\beta}]}$; and let q^α have the linear axpression $(25.2)_1$ or $(25.8\,\mathrm{b})_1$ with $\kappa_{\alpha\beta}$ spatial. Then $\kappa_{\alpha\beta} = \kappa_{\beta\alpha}$ if and only if q_{ass} is a function of A_α and the $p_{(r)}$ (but not of $u_{[\alpha/\overset{\perp}{\beta}]}$).*

To prove the theorem it suffices to remark that either of the assumptions $(25.2)_1$ and (25.8 b) with $\kappa_{\alpha\beta}$ spatial, implies the dependence of q_{ass} on $q_{\mathrm{ass}}^{(d)}$ through $(25.13)_1$ or $(25.14)_1$; hence q^α depends on $u_{[\alpha/\overset{\perp}{\beta}]}$ if and only if $\kappa^{[\alpha\beta]} \neq 0$.

Let us add that by the objectivity principle (also called principle of material indifference) or simply by its rotational part [§§ 81, 82], if a priori q^α is a function of the above real parameters $p_{(r)}$ and $u_{[\alpha/\overset{\perp}{\beta}]}$, then it is independent of $u_{[\alpha/\overset{\perp}{\beta}]}$. Thus q^α is independent of the history of $u_{[\alpha/\overset{\perp}{\beta}]}$ [§ 80], by which q^α can be said to be *rotationally objective*. The thesis of Theorem 25.3 says that q_{ass} *is rotationally objective if and only if* $\kappa_{\alpha\beta}$ *is symmetric.*

The reader can realize that on the basis of sections 80—82 on the objectivity principle, the extension of Theorem 25.3 to materials with memory is straightforward.

§ 26. On the Paradox of an Infinite Velocity of Heat Propagation from the Classical Point of View and the Relativistic One

Within the classical theory of elasticity let us consider a body $\mathscr{C}$ stationary with respect to the inertial frame (x), in the absence of electromagnetic phenomena. We assume that $\mathscr{C}$ is isotropic and homogeneous in its actual configuration. Then the first principle reads

$$\gamma k \dot{T} = -\bar{q}^r{}_{/r} \quad \text{where} \quad \gamma = \frac{\partial w}{\partial T}. \tag{26.1}$$

Moreover the Fourier law (25.8) holds for $\bar{\kappa}^{rs}=\bar{\kappa}\delta^{rs}$, where $\bar{\kappa}$ is a positive constant, so that by (26.1) we obtain the heat-propagation equation

$$\frac{\gamma k}{\bar{\kappa}}\dot{T}=T_{/r}{}^{r}. \tag{26.2}$$

This equation is parabolic, hence it implies an infinite velocity of heat propagation.

In classical physics gravitational forces are considered as actions at a distance, i.e. they are transmitted instantaneously. Thus they constitute signals travelling with an infinite velocity. It may be added that the propagation velocity of elastic waves in incompressible fluids is also infinite. However these fluids, and more generally all mechanically constrained systems, are considered by various authors as first approximation systems; from a more rigorous point of view the same authors prefer avoiding an infinite propagation velocity in every phenomenon, such as heat propagation, which occurs completely inside matter.

Up to now the problem of avoiding the afore-mentioned paradoxes in classical physics has not yet received any completely satisfactory general solution. A first serious attempt toward such a solution can be found in C. Cattaneo [1948].

More in particular, Cattaneo studies the preceding problem of heat conduction on the basis of statistical mechanics and in the case of a nearly stationary heat propagation within a dilute gas. He first reaches, as a second order approximation of the unknown true heat conduction law, the result

$$\bar{q}_r=-\bar{\kappa}\,T_{/r}+\sigma\,\ddot{T}_{/r} \quad \text{with } \sigma \text{ constant and very small} \tag{26.3}$$

which goes back to Fourier's law, in first order approximation, in the stationary case. By some approximation device based on the smallness of σ and the assumed nearly stationary character of heat propagation, Cattaneo arrives at the relation

$$\bar{q}_r+\bar{\kappa}'\dot{\bar{q}}_r=-\bar{\kappa}_r{}^{s}\,T_{/s} \quad \text{for} \quad \bar{\kappa}^{rs}=\bar{\kappa}\delta^{rs}, \tag{26.4}$$

which he proposes to test (disregarding (26.3)).

In case $\bar{\kappa}$ and $\bar{\kappa}'$ are constant, from the "divergence" of $(26.4)_1$ and $(26.1)_1$ we obtain, instead of (26.2), the heat propagation equation

$$T^{/r}{}_{r}-\frac{1}{V^2}\ddot{T}-\frac{\gamma k}{\bar{\kappa}}\dot{T}=0 \quad \text{where} \quad V=\sqrt{\frac{\bar{\kappa}}{\bar{\kappa}'\gamma k}}. \tag{26.5}$$

It has the form of a "telegraph equation", so that V is the largest possible velocity for heat propagation in the particular problem being considered.

Incidentally P. Vernotte [1958] rediscovered Cattaneo's proposal (26.4) by reasoning directly on the form of the equations being considered. Furthermore an equation similar to (26.4) had been considered and immediately discarded by Maxwell in 1867—cf. Lianis *[1974], p. 300;* hence it was not discussed at all in connection with the above paradox.

From this purely formal point of view let us observe that for $\bar{\kappa}'$ and $\bar{\kappa}^{rs}$ not necessarily constant, $(26.1)_1$ and the "divergence" of $(26.4)_1$ imply

$$\gamma k \dot{T} + \bar{\kappa}' \gamma k \ddot{T} + \dot{\overline{\bar{\kappa}' \gamma k}} \dot{T} - \bar{\kappa}'_{/r} \dot{\bar{q}}^r = \bar{\kappa}^{rs} T_{/sr} + \bar{\kappa}^{rs}{}_{/r} T_{/s}, \tag{26.6}$$

whereas as far as the second derivatives are concerned no such implication would be possible if the term $\bar{\kappa}' \dot{\bar{q}}_r$ in (26.4) had been replaced by $\bar{\kappa}'^{rs} \dot{\bar{q}}_s$.

Let T and $\partial T/\partial x^\alpha$ $(x^0 = t)$ be functions of x^α which are continuous across the surface Σ locally normal to the unit vector n^r, and let $\partial^2 T/\partial x^\alpha \partial x^\beta$ have a discontinuity of the first kind across Σ. In addition, since $\dot{\bar{q}}_r$ appears in (26.4) we assume the continuity of $\bar{q}_r$. Then by (26.4) $\dot{\bar{q}}_r$ is also continuous. Since γ is usually assumed to be continuous, the propagation speed of Σ is given by

$$V^2 = \frac{1}{\gamma k \bar{\kappa}'} \bar{\kappa}^{rs} n_r n_s. \tag{26.7}$$

In the preceding considerations $\mathscr{C}$ has been assumed to be stationary with respect to an inertial space. To treat the case in which $\mathscr{C}$ is continuously deforming, it suffices to use the consequence $-\bar{q}^r{}_{/r} = k\dot{w} + dl^{(i)}/D\tau$ of $(25.4)_{1,3}$ for $r \equiv 0$, instead of $(26.1)_1$. In case $\mathscr{C}$ is elastic (hence capable of only reversible processes), by our assumption that only $\partial^2 T/\partial x^\alpha \partial x^\beta$ can be discontinuous across Σ, $\bar{q}^r{}_{/r}$ is continuous and the discontinuity of $-\dot{\bar{q}}^r{}_{/r}$ again equals that of $-k(\partial w/\partial T)\ddot{T}$. Thus the local propagation speed V of the discontinuity surface Σ is given by (26.7) again.

Let us add that (i) the initial conditions connected with equations (26.5) or (26.6) concern the initial values of both T and $\dot{T}$, and moreover (ii) Newton's law of cooling for surfaces separating different media must be transformed, like $(25.6)_1$ which was transformed into (26.4).

The problem of heat conduction in a body $\mathscr{C}$ stationary with respect to an inertial frame (x) remains unaltered when instead of classical physics we consider special relativity. Indeed in the latter case (x) can be identified with a Lorentz frame and, since $A_\alpha \equiv 0$, every one of the relativistic versions (25.8b) and (25.2) of Fourier's law is equivalent to (25.8) under the condition $(25.3)_3$. It is easy to realize that heat propagation is again governed by equation (26.2), so that it must occur with an infinite speed.

An infinite speed of heat propagation is much more troublesome in relativity than in classical physics, because in relativity it is incompatible with the causality principle, as well as the existence of a signal travelling with a speed greater than c. In accord with the causality principle, in relativity gravitational actions propagate with the speed c and e.g. incompressible fluids are defined as the fluids in which the speed of sound equals c—cf. Lichnerowicz [1955].

By the above reasons in Bressan [1964a] where a relativistic elasticity theory on a thermodynamic basis is constructed, used a modified version of the Fourier law (25.8b) based on a result by Cattaneo—precisely on (26.3). In this version the heat flux vector q_α is still explicitly expressed. The results of Cattaneo [1948] and Vernotte [1958] combined with the versions (25.8b) and (25.2) of the Fourier heat conduction law give rise to various relativistic versions of the same law, even if no explicit expression of q^α is understood to appear in these versions; some of them were explicitly stated by M. Kraniŝ [1966].

The following modified versions of (25.8b) and (25.2) respectively are differential equations in q_α constituting natural relativizations of Cattaneo's proposal $(26.4)_1$

$$q^\alpha + \kappa' \overset{\perp}{g}{}^\alpha{}_\beta \dot{q}^\beta = -\kappa^{\alpha\beta} T_{/\beta}, \qquad q^\alpha + \kappa' \overset{\perp}{g}{}^\alpha{}_\beta \dot{q}^\beta = -\kappa^{\alpha\beta} \theta_\beta \qquad (\kappa' > 0) \tag{26.8}$$

where (20.2) and $(25.2)_{3,4}$ hold, and where the spatial tensor $\kappa^{\alpha\beta}$ is positive definite.

Indeed, in the particular problem of heat conduction considered at the outset of this section, which has meaning also in special relativity, both $(26.8)_1$ and $(26.8)_2$ become (26.4) for $c\bar{\kappa}' = \kappa'$ and $\bar{\kappa}^{\alpha\beta} = c\kappa^{\alpha\beta}$. Then (26.6) and (26.7) also hold, where V is the maximum speed of heat propagation in the particular problem being considered. In addition (26.7) gives the propagation speed V of discontinuity surfaces for $\partial^2 T/\partial x^\alpha \partial x^\beta$ in case $\mathscr{C}$ is continuously deforming. Then, by the causality principle ($V \leqslant c$) and (26.7) the inequality

$$\kappa^{\alpha\beta} n_\alpha n_\beta \leqslant \gamma k \kappa' \qquad \left(\gamma = \frac{\partial w}{\partial T}\right) \tag{26.9}$$

must hold for every spatial unit vector n_α. It is natural to assert that condition (26.9) must hold as a strict inequality because the strict inequality $V < c$, and not $V \leqslant c$, is usually considered to be the relativistic analogue of the condition $V < +\infty$ in classical physics.

The above brief outline of a relativistic theory of heat conduction based on either of proposals $(26.8)_{1,2}$, is substantially in agreement with M. Kraniŝ [1966]. However let us remark that this author explicitly considered the isotropic case and proposes the equations obtained from (26.8) by replacing q^α with a not necessarily spatial vector $\hat{q}^\alpha$ and $\kappa^{\alpha\beta}$ with $\kappa \overset{\perp}{g}{}^{\alpha\beta}$. Then he identifies $\hat{q}_{\overset{\perp}{\beta}}$ with the heat flux vector.

Thus in comparison with either of the proposals $(26.8)_{1,2}$ he has four differential scalar equations instead of three such equations plus an equation in finite terms.

M. Kraniŝ takes into account from the relativistic point of view also Newton's law of cooling at a surface separating two media and Fick's law of diffusion.

In Lianis [1974] the constitutive equation (26.4) is remarked to be equivalent to a linear memory functional in which the kernel $e^{-t/\bar{\kappa}'}$ plays the role of a relaxation modulus; and from this point of view an interesting attempt is made to extend the Cattaneo-Vernotte theory to relativistic visco-elastic dielectrics.

In connection with the paradox of infinite velocity of heat-propagation let us now consider Müller [1969]. Following this author in part, we first consider the most general form of $\mathscr{U}_{\alpha\beta}$ which is a Lorentz invariant, in case $\mathscr{U}_{\alpha\beta}$ (is symmetric and) is a function of k, u_α, $u_{\alpha/\beta}$, T, and $T_{/\alpha}$, linear in $T_{/\alpha}$ and $u_{\alpha/\beta}$: [8)]

$$\mathscr{U}_{\alpha\beta}=[p-(\nu-\tfrac{2}{3}\bar{\mu})c u^{\rho}{}_{/\rho}]g_{\alpha\beta}+(\rho+p+c^3 t_{12} u^{\rho}{}_{/\rho})u_\alpha u_\beta-2\bar{\mu}c u_{(\alpha/\overset{+}{\beta})}+\bar{Q}_{\alpha\beta}\,, \quad (26.10)$$

where

$$\begin{aligned}\bar{Q}_{\alpha\beta}=&-\kappa(u_\alpha T_{/\overset{+}{\beta}}+u_\beta T_{/\overset{+}{\alpha}})+(c\bar{\mu}+c^3 t_{40})(A_\alpha u_\beta+u_\alpha A_\beta)\\ &+t_{01}c^2\dot{T}g_{\alpha\beta}+\left(\frac{2\kappa}{c}+c^2 t_{11}\right)\dot{T}u_\alpha u_\beta\end{aligned} \quad (26.11)$$

and where the coefficients p, ν, $\bar{\mu}$, $\rho(=kc^2+kw)$, κ, t_{11}, t_{01}, t_{04}, and t_{12} are functions of k and T. [9)]

By $(24.3)_{2,3}$ the first term on the right hand side of (26.11) is $Q_{\alpha\beta}$ for $q_\alpha=-\kappa T_{/\overset{+}{\alpha}}$; furthermore in case

$$c\bar{\mu}+c^3 t_{40}=-\kappa T \qquad (t_{40}=-c^{-3}\kappa T-c^{-2}\bar{\mu})\,, \quad (26.12)$$

we have—cf. $(25.2)_2$

$$\bar{Q}_{\alpha\beta}=Q_{\alpha\beta}+\left(\frac{2\kappa}{c}+c^2 t_{11}\right)\dot{T}u_\alpha u_\beta+t_{01}\dot{T}g_{\alpha\beta} \quad \text{for} \quad q^\alpha=-\kappa\theta^\alpha\,. \quad (26.13)$$

Hence at least in this case $\bar{Q}_{\alpha\beta}$ can be called the thermodynamic tensor (for t_{11} and t_{01} small).

Let us assume that the body $\mathscr{C}$ being referred to is at rest with respect to a stationary frame—$g_{\alpha\beta,0}=0$ whence $DA_\alpha/Ds=0$—and that $\theta_\alpha\equiv T_{/\overset{+}{\alpha}}+TA_\alpha\equiv 0$. Then $\dot{T}\equiv 0$ and by (26.13) the thermodynamic tensor $\bar{Q}_{\alpha\beta}$ vanishes identically in accord with a result by Tolman and Ehrenfest [§ 45], which shows that in relativistic equilibrium $\theta_\alpha=0$ (and not $T_{/\overset{+}{\alpha}}=0$) holds. [10)]

Now we consider the rest of $\mathscr{C}$ with respect to a Minkowskian frame in special relativity, whence $u_{\alpha/\beta}\equiv 0\equiv k_{/\alpha}$. Then the first principle $(23.4)_4$ under definitions (26.10) and (26.11) becomes—cf. (22.9), (18.6)

$$k\frac{\partial w(T,k)}{\partial T}\dot{T}+\frac{d}{d\tau}\left[\frac{2\kappa}{c}\dot{T}-(t_{01}-c^2 t_{11})\dot{T}\right]=\left(\frac{\kappa}{c}T_{,r}\right)_{,s}\delta^{rs}\,. \quad (26.14)$$

If the squared speed V^2 of propagation of disturbances in the temperature is positive, i.e.

$$\frac{1}{V^2}=\frac{2}{c^2}-c\frac{t_{01}-c^2 t_{11}}{\kappa}>0\,, \quad (26.15)$$

then these disturbances, or more precisely the discontinuity surfaces for $\partial^2 T/\partial x^\alpha \partial x^\beta$, propagate with the finite (real) speed given by (26.15).

Relation $(26.15)_{1,2}$ could be motivated just by the requirement due to the causality principle, that the afore-mentioned speed of propagation must be finite. Furthermore on the basis of experiment we can assert the relations $\mu>0$ and $\bar{\mu}\geqslant 0$.

In Müller [1969] the *empirical temperature* ϑ is used as a primitive, within special relativity. Its physical meaning is not discussed from the operational point of view. However the equilibrium of the fluid being considered is defined there as a state of uniform and time-independent temperature. Hence, incidentally, ϑ cannot coincide with T [§ 45]. Now, following this author completely, we consider an entropy vector, η_α such that $-u^\alpha \eta_\alpha$ is the entropy density (possibly $\neq k\eta$) and $\eta_{\alpha}^{\perp}$ is the entropy flux. Furthermore we assume that η_α is—as well as $\mathscr{U}_{\alpha\beta}$—a function of k, ϑ, $\vartheta_{/\alpha}$, u_α, and $u_{\alpha/\beta}$[11] linear in $\vartheta_{/\alpha}$ and u_α. By the requirement of Lorentz invariance η_α has the form

$$\eta_\alpha=(k\eta+s_{01}\dot{\vartheta}+s_{02}u^{\rho}{}_{/\rho})u_\alpha+s_{10}\vartheta_{/\alpha}+s_{20}A_\alpha \tag{26.16}$$

and $\mathscr{U}_{\alpha\beta}$ has the one, say $\mathscr{U}_{\alpha\beta}^{(M)}$, obtained from (26.10) and (26.11) by replacing T with ϑ.

Müller's relativistic version of the second principle is

$$\eta^{\alpha}{}_{/\alpha}\geqslant 0 \tag{26.17}$$

—cf. Stückelberg and Wander's assumption $(a)_3$ in footnote 1 in Chapter 4. This author added (26.17) with this condition: Let (ϑ, k) be the values of the equilibrium variables of a body in a particular equilibrium; in a neighbourhood of any such equilibrium there exist other equilibria characterized by $(\vartheta+d\vartheta,\ k+dk)$ and these can be reached from the equilibrium (ϑ, k) by thermodynamic processes even when the body is closed, i.e. when the body is supply free $(r\equiv 0)$ and the entropy flux vanishes at the boundary. The increase of entropy of the body—i.e. the integral of $-u^\alpha\eta_\alpha$ over the volume of the body—which is associated with these processes is finite. Müller also assumed the possibility of certain equilibria.

From the validity of (26.17) for all fields $k(x)$, $\vartheta(x)$, and $u_\alpha(x)$ which are solutions of the system

$$\mathscr{U}_{\alpha\beta}^{(M)/\beta}=0\,,\qquad \dot{k}+ku^{\alpha}{}_{/\alpha}=0\,,\qquad u^\alpha u_\alpha=-1 \tag{26.18}$$

and from the above additions to (26.17) one derives restrictions on constitutive equations. Among them—cf. [1969, pp. 277, 279] and footnote 8 in Chapter 3

$$s_{10}+s_{01}=\frac{c^3 t_{11}-c t_{01}}{T}\,,\qquad s_{20}+\frac{c^2}{T}(\bar{\mu}-t_{40}c^2)\geqslant 0\,,\qquad s_{20}-\frac{c^2}{T}(\bar{\mu}+c^2 t_{40})\leqslant 0$$

$$\bar{\mu}+t_{40}c^2\leqslant 0\,,\qquad s_{20}\leqslant 0\,,\qquad -s_{10}=\frac{c\kappa}{T}\geqslant 0\,,\qquad \bar{\mu}\geqslant 0\,. \tag{26.19}$$

Furthermore one proves that relation (26.15) is compatible with all of the foregoing assumptions (also in connection with $\mathscr{U}^{(M)}_{\alpha\beta}$).

Let us add that in Müller's theory one has, in the rest frame—cf. [1969, p. 279]

$$\text{energy flux} \;=\; -c\kappa\,\vartheta_{/r} + c^2(\bar{\mu} + t_{40}c^2)A_r\,, \tag{26.20}$$

$$\text{entropy flux} = -\frac{c\kappa}{T}\vartheta_{/r} + s_{20}A_r\,. \tag{26.21}$$

In general the ratio of these fluxes is not the classical one, i.e. $1/T$. However this ratio is $1/T$, in case $A_\alpha = 0$.

Müller's solution of the paradox is given in an exact relativistic theory whose basic assumptions are not all usual, but they look reasonable and of a rather general kind. As we already said in § 25 these assumptions are being tested through their consequences, and their relations to the principle of material indifference were recently found not to be as one might have hoped—see Müller [1970]. However actual experiment seems to allow us scant departure from the principle of material indifference—cf. § 81.

It is remarkable that by using Chernikov's relativistic kinetic theory Alts and Müller [1972] proved that the entropy flux does not have the form ordinarily assumed in the Clausius-Duhem inequality.

§ 27. On the Local Spatial Physical Isotropy of S_4

We shall consider any relativistic constitutive equation of the (completely) Eulerian type in locally natural and proper co-ordinates; we mean an equation of the form

$$T_{(0)}{}^{ab\cdots} = \Phi^{ab\cdots}(T_{(1)}{}^{rs\cdots}, \ldots, T_{(n)}{}^{rs\cdots}) \tag{27.1}$$

where $T_{(0)}{}^{\alpha\beta\cdots}, \ldots, T_{(n)}{}^{\alpha\beta}$ are spatial tensors, hence they are determined by those of their components which are present in (27.1).

It is easy to see that *the above tensor valued function is invariant under local spatial rotations—or transformations between locally natural and proper frames—if and only if, for every orthogonal matrix* $\|\rho^a{}_r\|$, (27.1) *implies*

$$T'_{(0)}{}^{ab\cdots} = \Phi^{ab\cdots}(T'_{(1)}{}^{rs\cdots}, \ldots, T'_{(n)}{}^{rs\cdots}) \tag{27.1 b}$$

where

$$T'_{(\alpha)}{}^{ab\cdots} = \rho^a{}_r\,\rho^b{}_s \ldots T_{(\alpha)}{}^{rs\cdots} \qquad (\alpha = 0, \ldots, n)\,. \tag{27.2}$$

In classical physics the constitutive equations of the form (27.1) are usually referred to Galilean frames and incidentally they also hold in classical locally natural frames [§ 19]. Furthermore they are invariant under (spatial) rotations because of the physical isotropy of inertial spaces. The same holds in special relativity. The above invariance property of (27.1) is also asserted in general relativity, where it mirrors a local spatial physical isotropy of S_4.

Incidentally we now want to show that in general relativity the afore-mentioned invariance property can be reached in a different way. Among other things we do so because this enables us to show an aspect of Fock's conjecture [§ 10] which is interesting from the conceptual point of view in connection with constitutive equations.

To the above end let us remark that in classical physics an orientation can be chosen in a way independent of phenomena, so that it is physically meaningful to assert the validity of (27.1) for a particular material, referred e.g. to any Galilean frame having the chosen orientation and connected with given units, even if $\Phi^{ab\cdots}$ is not invariant under spatial rotations. Analogously in special relativity we can associate with every inertial space $S_{(i)}$ an orientation $\Omega(S_{(i)})$ independently of phenomena, so that, even in the absence of the afore-mentioned invariance property for $\Phi^{ab\cdots}$, it is physically meaningful to assert the validity of a constitutive equation of the form (27.1) in any locally natural and proper frame which, locally, is connected with given units and has the orientation $\Omega(S_{(i)})$ where $S_{(i)}$ is the locally co-moving inertial space.

Now we consider general relativity and for the moment accept Fock's conjecture [§ 10] by which also in general relativity we have ∞^{10} privileged harmonic frames connected with given units and distributed in ∞^3 classes of mutually stationary frames. These classes can be thought of as characterizing the inertial spaces. Then in general relativity we can choose one of these privileged harmonic frames (ξ) in a way independent of phenomena just as we can in special relativity.

Let ξ^α be the co-ordinates in (ξ) of an event point, $\mathscr{E}$, at which the frame (x) —which equation (27.1) is referred to—is locally proper and natural; furthermore let $D\xi^r/D\xi^0$ be the velocity of matter at $\mathscr{E}$ in the frame (ξ).

The unit vectors of the spatial axes of the frame (x) at $\mathscr{E}$ generally are not mutually orthogonal for the observer (ξ) in spite of their being so with respect to our chronometry (15.2). These unit vectors are characterized in the frame (ξ) by 9 real parameters. If these parameters, ξ^α, and $D\xi^r/D\xi^0$ are regarded as implicit arguments of $\Phi^{ab\cdots}$—cf. (27.1)—, it is certainly physically meaningful to consider a constitutive equation of the form (27.1), even if $\Phi^{ab\cdots}$ is not invariant for spatial rotations with respect to its explicit arguments.

We have shown that Fock's conjecture makes constitutive equations of the above kind physically meaningful. However we want to keep the bulk of this tract independent of this conjecture. Then no satisfactory general criterion to recognize and represent absolute positions, 4-velocities, and orientations is known in general relativity. Therefore in connection with some tensor-valued physical magnitudes $\mathscr{M}_0, \ldots, \mathscr{M}_n$ the requirement that constitutive equations should be stated in a physically meaningful way practically forces us to make assertions like the following:

For every particle P^* of the continuous body $\mathscr{C}$ there is a set of functions $\Phi^{ab\cdots}$ of the form (27.1) *which are the constitutive equations of P^*, relative to the magnitudes $\mathscr{M}_0, \ldots, \mathscr{M}_n$, in every locally proper and natural frame.*

The above assertion obviously implies the invariance of $T_{(0)}{}^{ab\cdots}$—cf. (27.1)—for the spatial rotations, and its independence of place and 4-velocity—i.e. of ξ^α and $D\xi^r/D\xi^0$.

We conclude that when Fock's conjecture or something like that is not accepted, then in general relativity the requirement that constitutive equations should be physically meaningful (and in particular independent of phenomena) practically has the effect due in classical physics to the principles holding for both local and global physical equations—i.e. the principles of physical homogeneity, physical isotropy, and physical indistinguishability of inertial space, and the principle of physical homogeneity of time.

§ 28. Free Energy and Relativistic Thermodynamics for Possibly Viscous Fluids

We introduce the *specific free energy* $\mathscr{F}=w-T\eta$. Then the second principle $(25.1)_{1,2}$ for $r\equiv 0$ is equivalent to $D\mathscr{F}\leqslant Dw-\eta DT-dQ$ and, by the first principle (24.6), also to[12)]

$$kD\mathscr{F}\leqslant -k\eta DT-dl^{(i)} \qquad (\mathscr{F}=w-T\eta,\ dl^{(i)}/Ds=X^{\alpha\beta}u_{\alpha/\beta})\,. \tag{28.1}$$

Let us remark that (28.1) holds as an equality if and only if, $(25.1)_1$ does.

We say that the body $\mathscr{C}$ is a possibly viscous fluid at its material point P^* if the following conditions (a) and (b) hold:

(a) *In any locally natural frame* (x) *the values at* P^* *of the magnitudes* w, η, *and* X^{rs} *are functions of* T, k, $u_{r/s}$, *and the N parameters* $q_1,\ldots,q_N$ *which are assumed to be the components in* (x) *of a set of spatial tensors possibly including scalars. These function are independent of* (x).

(b) *The values at* P^* *of* T, k, $u_{r/s}$, $q_1,\ldots,q_N$, *and the (Römer) time derivatives* $\grave{T}$, $Du_{\rho/\sigma}/Ds$, $\grave{q}_1,\ldots,\grave{q}_N$ *are not subject to any mechanical or physical constraint or any kinematic or mathematical condition (*e.g. *consisting of a definition).* (Of course $\grave{k}=-ku_{r/s}$.)

$\mathscr{C}$ is said to be a *non-viscous [(ordinary) viscous] fluid at* P^*, *if conditions* (a) *and* (b) *are fulfilled and* X^{rs} *is independent of* $u_{r/s}$ *[(*X^{rs} *is independent of* q_1 *to* q_r *but) depends on* $u_{r/s}$ *effectively].*

The independence of certain functions on (x) asserted in condition (a) implies that the functions that express (the values at P^* of) w, η, $\mathscr{F}=\tilde{\mathscr{F}}(T,k,u_{r/s},q_1,\ldots,q_N)=w-T\eta$, and $X^{\alpha\beta}$ are invariant for spatial rotations [§ 27]. In the present case, where electromagnetic phenomena and couple stresses are disregarded, the spatial tensor $X^{\alpha\beta}$ is symmetric by (24.7). Under assumptions (a) and (b) we can decompose $X^{\alpha\beta}$ into the *reversible stress* $X^{\alpha\beta}_{\text{rev}}$ and the *irreversible* stress $X^{\alpha\beta}_{\text{irr}}$ by means of the function $\mathscr{F}=\tilde{\mathscr{F}}(T,k,u_{\alpha/\beta}^{\perp},q)$ as follows:

$$X^{\alpha\beta}_{\text{rev}} = k^2 \frac{\partial \tilde{\mathscr{F}}}{\partial k} \overset{\perp}{g}{}^{\alpha\beta} = X^{\beta\alpha}_{\text{rev}} \quad (\text{hence } u_\alpha X^{\alpha\beta}_{\text{rev}} = 0), \tag{28.2}$$

$$X^{\alpha\beta} = X^{\alpha\beta}_{\text{rev}} + X^{\alpha\beta}_{\text{irr}} \quad (X^{[\alpha\beta]}_{\text{irr}} = 0 = u_\alpha X^{\alpha\beta}_{\text{irr}}). \tag{28.3}$$

Remark that by $(17.20)_2$ and $(21.3)_1$ the reversible work is

$$\frac{dl^{(i)}_{\text{rev}}}{Ds} = X^{\alpha\beta}_{\text{rev}} u_{\alpha/\beta} = k^2 \frac{\partial \tilde{\mathscr{F}}}{\partial k} u^\alpha{}_{/\alpha} = -k \frac{\partial \tilde{\mathscr{F}}}{\partial k} \dot{k}. \tag{28.4}$$

Hence by (28.2) and (28.3), inequality (28.1)—equivalent to $(25.1)_1$—becomes

$$k \left[\frac{\partial \tilde{\mathscr{F}}}{\partial T} \dot{T} + \sum_{i=1}^{N} \frac{\partial \tilde{\mathscr{F}}}{\partial q_i} \dot{q}_i + \frac{\partial \tilde{\mathscr{F}}}{\partial u_{r/s}} \frac{D}{Ds} u_{r/s} + \eta \dot{T} \right] + X^{\alpha\beta}_{\text{irr}} u_{(\alpha/\beta)} \leqslant 0. \tag{28.5}$$

Let $\mathscr{C}$ fulfill conditions (a) and (b) and let us fix the values of $q_1, \ldots, q_N$, T, k, and $u_{\alpha/\overset{\perp}{\beta}}$. Then by condition (b) the values of $\dot{q}_1, \ldots, \dot{q}_N$, $\dot{T}$, and $Du_{r/s}/Ds$ are arbitrary, so that (28.5) yields

$$\eta = -\frac{\partial \tilde{\mathscr{F}}}{\partial T}, \quad \frac{\partial \tilde{\mathscr{F}}}{\partial u_{\alpha/\overset{\perp}{\beta}}} = 0, \quad \frac{\partial \tilde{\mathscr{F}}}{\partial q_i} = 0 \quad (i = 1, \ldots, N). \tag{28.6}$$

Hence (28.5) simplifies into

$$X^{\alpha\beta}_{\text{irr}} u_{(\alpha/\beta)} \leqslant 0; \tag{28.7}$$

furthermore (28.6) and $(28.1)_2$ imply

$$\mathscr{F} = \tilde{\mathscr{F}}(T,k), \quad \eta = \tilde{\eta}(T,k), \quad w = \tilde{w}(T,k). \tag{28.8}$$

We may conclude that *for a possibly viscous fluid the magnitudes* $\mathscr{F}$, η, *and* w *are functions of* T *and* k, *the temperature-entropy relation* $(28.6)_1$ *holds, and in addition the second principle* $(25.1)_1$*—equivalent to any of the inequalities* $(28.1)_1$ *and* (28.5)*—takes the form* (28.7) *and holds as an equality whenever* $(28.1)_1$ *does.*

The preceding result is compatible with viscosity and in particular with ordinary linear viscosity, i.e. with the condition that $X^{\alpha\beta}_{\text{irr}}$ should be a linear function $\eta^{\alpha\beta\gamma\delta} u_{\gamma/\delta}$ of the deformation velocity $u_{(\alpha/\overset{\perp}{\beta})}$, where $\eta^{\alpha\beta\gamma\delta} (= \eta^{\beta\alpha\gamma\delta} = \eta^{\alpha\beta\delta\gamma})$ is a spatial tensor depending only on T and k. By (28.7) the quadratic form $-\eta^{\alpha\beta\gamma\delta} \xi_{\alpha\beta} \xi_{\gamma\delta}$ is positive definite.

Now let $\mathscr{C}$ be a non-viscous fluid at P^*, so that, besides conditions (a) and (b) we have that

(c) *At* P^* $X^{\alpha\beta}$ *is at most a function of* $q_1, \ldots, q_N$, T, *and* k *(and of* $g_{\alpha\beta}$ *and* u^α*).* Then by (28.7) and condition (b)

$$X^{\alpha\beta}_{\text{irr}} \equiv 0. \tag{28.9}$$

As a consequence the relation (28.7) always holds as an equality. Then by a conclusion made below (28.8) the second principle of thermodynamics $(25.1)_1$ must also hold at P^* as an equality along every process physically possible for $\mathscr{C}$. Furthermore by $(28.6)_1$ and by $(28.2)_1$, $(28.3)_1$, and (28.9) we respectively have

$$\eta = -\frac{\partial \tilde{\mathscr{F}}}{\partial T}, \qquad X^{\alpha\beta} = k^2 \frac{\partial \tilde{\mathscr{F}}}{\partial k} \overset{\perp}{g}{}^{\alpha\beta}. \tag{28.10}$$

We conclude that *any non-viscous fluid is capable of only reversible processes and has the constitutive equations* (28.8) *and* (28.10).

Of course the theory of viscous or non-viscous fluids can be simplified by assuming (28.8) and the validity of conditions (a) and (b) for $N=0$ at the outset. However an application of our more general approach will be given in § 29.

By the second principle $(25.1)_1$ we have $T D\eta \geqslant dQ$. Furthermore Helmoltz's postulate asserts that, under a constant configuration, the heat dQ absorbed during an increase DT of T is positive. Then $D\eta > 0$, so that by $(28.6)_1$ $\partial^2 \tilde{\mathscr{F}}/\partial T^2 < 0$. Consequently equation $(28.6)_1$ defines T as a function of η and k. Then we can use e.g. the following functions of η and k instead of the functions (28.8) of T and k;

$$T = \hat{T}(\eta, k), \qquad w = \hat{w}(\eta, k) = \tilde{w}[\hat{T}(\eta, k), k]. \tag{28.11}$$

Furthermore we can disregard for the moment definition $(28.2)_1$ and replace it by

$$X^{\alpha\beta}_{\text{rev}} = k^2 \frac{\partial \hat{w}}{\partial k} \overset{\perp}{g}{}^{\alpha\beta} = X^{\beta\alpha}_{\text{rev}}. \tag{28.12}$$

By (24.6) the second principle $(25.1)_{1,2}$ for $r=0$ becomes

$$k D w \leqslant k T D\eta - X^{\alpha\beta} u_{\alpha/\beta} Ds. \tag{28.13}$$

Then, under conditions (a) and (b) or (a) and (c), we can apply to (28.11) the reasoning which led us from (28.1) to (28.7) and (28.8) and to (28.10), respectively. Thus we see that for viscous [non-viscous fluids] the first [both] of the following relations hold:

$$T = \frac{\partial \hat{w}}{\partial \eta}, \qquad X^{\alpha\beta} = k^2 \frac{\partial \hat{w}}{\partial k} \overset{\perp}{g}{}^{\alpha\beta} \tag{28.14}$$

and that $(28.2)_1$ and (28.12) are equivalent.

§ 29. On Non-Viscous Fluids in the Presence of Heat Conduction and on Perfect Gases

Let $\mathscr{C}$ be a *non-viscous fluid* at P^* so that (28.10) holds. By introducing the pressure $p=3^{-1}X^{\alpha}{}_{\alpha}$, from $(28.10)_2$ and the equality $\overset{\perp}{g}{}^{\alpha}{}_{\alpha}=3$ we have

$$p=k^2\frac{\partial\tilde{\mathscr{F}}}{\partial k}\qquad (p=\tfrac{1}{3}X^{\alpha}{}_{\alpha}). \tag{29.1}$$

Using (28.14) instead of (28.10), the constitutive equations of our non-viscous fluid can be written[13)]

$$T=\frac{\partial\hat{w}}{\partial\eta},\qquad p=k^2\frac{\partial\hat{w}}{\partial k},\qquad w=\hat{w}(\eta,k). \tag{29.2}$$

Let us now remark that by $(28.10)_2$ and $(29.1)_1$ $X_{\alpha\beta}=p\overset{\perp}{g}_{\alpha\beta}$ holds. Furthermore by (17.9′) and $(20.6\,\mathrm{b})_3$

$$\overset{\perp}{g}_{\alpha\gamma}\overset{\perp}{g}{}^{\gamma\beta}{}_{/\overset{\perp}{\beta}}=0. \tag{29.3}$$

Hence for non-viscous fluids the dynamic equations (24.8) become

$$(\rho+p)A_{\alpha}=-p_{/\overset{\perp}{\alpha}}-k\left(\overset{\perp}{g}_{\alpha\gamma}\frac{D}{Ds}\frac{q^{\gamma}}{k}+u_{\alpha/\gamma}\frac{q^{\gamma}}{k}\right)\qquad (X_{\alpha\beta}=p\overset{\perp}{g}_{\alpha\beta}). \tag{29.4}$$

Incidentally the conclusion stated below formula (28.10) can be applied, for example, to show that *if for a material without internal constraints $X^{\alpha\beta}$, η, and $\mathscr{F}$ are functions of T, k, and $k_{/\alpha}$, then these functions are constant with respect to $k_{/\alpha}$.*[14)]

Indeed from $(21.3)_{2,3}$ we have

$$0=(ku^{\alpha})_{/\alpha\beta}=k_{/\alpha\beta}u^{\alpha}+k_{/\alpha}u^{\alpha}{}_{/\beta}+(ku^{\alpha}{}_{/\alpha})_{/\beta}; \tag{29.5}$$

hence, since $k_{/\alpha\beta}=k_{/\beta\alpha}$,

$$\frac{D}{Ds}k_{/\beta}=k_{/\beta\alpha}u^{\alpha}=-k_{/\alpha}u^{\alpha}{}_{/\beta}-k_{/\beta}u^{\alpha}{}_{/\alpha}-ku^{\alpha}{}_{/\alpha\beta}. \tag{29.6}$$

The value of $(u^{\alpha}{}_{/\alpha})_{/\beta}$ at a given event point, $\mathscr{E}$ is independent of the values of k, $k_{/\alpha}$, and $u^{\alpha}{}_{/\beta}$ at $\mathscr{E}$—cf. (29.5); hence by (29.6) the same independence holds for the value of $Dk_{/\beta}/Ds$ at $\mathscr{E}$. Then the conditions that at P^* no internal constraints are present and that η, $\mathscr{F}$, and $X^{\alpha\beta}$ are functions of T, k, $k_{/\alpha}$ (and $g^{\alpha\beta}$, u^{α}) imply the validity of the conditions (a) to (c) considered in § 28 for $N=4$ and $q_{(j+1)}=k_{/j}$ $(j=0,\ldots,3)$. By the conclusion written after (28.10) the material of $\mathscr{C}$ at P^* has the constitutive equations (28.8) and (28.10), i.e. it is a non-viscous fluid. In particular (29.1) holds, and the magnitudes $\mathscr{F}$, η, and p are independent of $k_{/\alpha}$; hence they cannot depend on $k_{/\overset{\perp}{\alpha}}$.

In classical physics a perfect gas is often defined as a fluid which is free of constraints and has the constitutive equations

$$p = \mathfrak{r} k T \qquad (\mathfrak{r} = R/M = \text{const}), \tag{29.7}$$

where R is the universal constant of gases and M the molecular weight. Furthermore one proves that *a non-viscous fluid is a perfect gas if and only if, the following conditions* (a) *and* (b) *hold:*

(a) *The ratio p/k depends only on the absolute temperature T.*

(b) *w depends only on T.*

The two relativizations of (29.7) performed by turning k into the conventional mass density $k^* dC^*/dC$ [§ 21] and the gravitational mass density $c^{-2}\rho$ are not equivalent. Instead, the analogous two relativizations of conditions (a) and (b) taken together, are equivalent. More in particular, in relativity we have $\rho/k = w + c^2$, so that $p/k = (p/\rho)(w + c^2)$. Then conditions (a) and (b), where k is understood to be the conventional mass density, are obviously equivalent to (b) and the following condition:

(a′) *The ratio p/ρ depends only on the absolute temperature T.*

Conditions (a) and (b), and in relativity especially conditions (a′) and (b), follow directly—unlike (29.7)—from the molecular theory of perfect gases (where these gases are considered to consist of particles with a vanishing mutual energy). Hence in relativity it is natural to call *perfect gas* an elastic fluid fulfilling conditions (a′) and (b)—or (a) and (b)—cf. Bressan [1963c]. Then we can prove that in relativity (29.7) *holds for perfect gases where k is the conventional mass density*—and not a law of the form $p = \mathfrak{r}\rho T$ with $\mathfrak{r} = \text{const}$.

Indeed assume conditions (a′) and (b), whence (a) follows. By condition (a) and $(29.1)_1$ we have $k\partial\tilde{\mathscr{F}}/\partial k = g(T)$; hence $\tilde{\mathscr{F}}$ has the form $\mathscr{F} = \tilde{\mathscr{F}}(T, k) = g(T)\log k + g_1(T)$. Then, taking $(28.1)_2$ and $(28.10)_1$ into account, we have

$$w = \mathscr{F} - T\frac{\partial\tilde{\mathscr{F}}}{\partial T} = [g(T) - Tg'(T)]\log k + g_1(T) - Tg_1'(T). \tag{29.8}$$

By condition (b) w is a function of T. Then by (29.8) $g(T) \equiv T g'(T)$; hence $g(T)$, i.e. $k\partial\mathscr{F}/\partial k$, is a linear function, $\mathfrak{r}\,T$, of T. Thus by $(29.1)_1$ we have $(29.7)_1$.

q.e.d.

§ 30. Acceleration Waves in Non-Viscous Fluids in the Absence of Electromagnetic Phenomena

Acceleration waves are dealt with in well known relativistic treatises—e.g. Lichnerowicz [1955, p. 40]—which are not very recent. Consequently considering these waves might appear to be outside of the scope of this tract. However our relativistic theory of these waves constitutes a preliminary for more recent and general theories of acceleration waves in fluids, as the one including polarizations (electromagneto-striction) to be dealt with in §§ 46, 47. Therefore we are going to present it. We strive to do so in a way very similar to the corresponding classical treatment; furthermore we try to put in evidence some aspects of the relativistic results concerning acceleration waves in fluids, which are of special interest in connection with the relativistic acceleration waves in general elastic materials [§§ 66, 67, 68].

We assume that $u_{\alpha/\beta}$ has a discontinuity $\Delta u_{\alpha/\beta}$ of the first kind across the time-like hypersurface Σ having the equation $f(x^s)=0$, while $u^\rho, g_{\alpha\beta}, \partial g_{\alpha\beta}/\partial x^\gamma$, and $\partial^2 g_{\alpha\beta}/\partial x^\gamma \partial x^\delta$ are continuous everywhere. We fix an event point $\mathscr{E}$ on Σ and assume that our frame is locally natural and proper at $\mathscr{E}$—cf. (17.7).

Then the velocity of propagation V of Σ (at $\mathscr{E}$) has the same expression as in classical physics for matter locally at rest, so that by (18.6) and (16.1)

$$V^2 = c^2 \frac{(f_{,0})^2}{\delta^{hi} f_{,h} f_{,i}} = c^2 \frac{(f_{,\rho} u^\rho)^2}{\overset{\perp}{g}{}^{\alpha\beta} f_{/\alpha} f_{/\beta}}. \tag{30.1}$$

By $(17.4)_3$ $u^\alpha \Delta u_{\alpha/\beta}=0$, so that by $(17.7)_{3,2}$ $0=\Delta u^0{}_{/\beta}=\Delta \partial u^0/\partial x^\beta$. Hence $u^0{}_{/\beta}$ and $\partial u^0/\partial x^\beta$ are continuous at $\mathscr{E}$ (as well as the Christoffel symbols). Then on the basis of the Hugoniot-Hadamard theorem, at $\mathscr{E}$ there exists a vector λ^ρ with $\lambda^0=0$ for which—cf. (16.1)

$$\Delta u_{r/s} = \lambda_r N_s, \quad c\Delta u_{r/0} = -\lambda_r V, \quad \text{where } N_r = (\delta^{hi} f_{,h} f_{,i})^{-1/2} f_{,r}. \tag{30.2}$$

From the physical point of view it is reasonable to consider the propagation of the discontinuity surface Σ in a non-viscous fluid as an adiabatic phenomenon. Thus (29.2) holds, and the equality $\eta=\text{const}$ can be assumed in space time; furthermore the heat flux vector q^α vanishes identically: Then at $\mathscr{E}$ the Cauchy equation (29.4) of non-viscous fluids becomes, by (17.7),

$$(\rho+p) u_{r,0} = -p_{,r} \equiv -\phi'(k) k_{,r} \quad \text{where } p=\phi(k)=k^2 \frac{\partial \hat{w}}{\partial k}. \tag{30.3}$$

At $\mathscr{E}$ the continuity equation $(21.3)_3$ takes the form

$$k_{,0} + k u^r{}_{,r} = 0. \tag{30.4}$$

The analogues of $(30.2)_{1,2}$, for k are

$$c\Delta k_{,0} = -\lambda_{(4)} V, \quad \Delta k_{,r} = \lambda_{(4)} N_r. \tag{30.5}$$

We can write the equations (30.3) and (30.4) in the limit as either side of Σ at $\mathscr{E}$ is approached, and then subtract the resulting equalities. Thus we obtain

$$(\rho+p)\Delta u_{r,0} = -\phi'(k)\Delta k_{,r}, \quad \Delta k_{,0} + k\Delta u^r{}_{,r} = 0. \tag{30.6}$$

Then, by (30.2) and (30.5) we have

$$\frac{1}{c}(\rho+p)\lambda_r V = \phi'(k)\lambda_{(4)} N_r, \quad \frac{1}{c}\lambda_{(4)} V = k\lambda^r N_r. \tag{30.7}$$

For $\lambda^r N_r = 0$, by $(30.7)_2$ we have either $\lambda_{(4)} = 0$ or $V = 0$. By $(30.7)_1$ the latter alternative implies $\lambda_{(4)} = 0$, which is the former. For $\lambda_{(4)} = 0$, by (30.5) $\Delta k_{,\alpha} = 0$, so that by $(30.6)_1$ $\Delta u_{r,0} = 0$, i.e. the equality $\lambda^r N_r = 0$ is incompatible with Σ being a discontinuity surface for the intrinsic acceleration. Therefore, we must have $\lambda^r N_r \neq 0$. In addition, since $N_r N^r = 1$, by multiplying $(30.7)_1$ by $c V N_r$, from $(30.7)_{1,2}$ we deduce

$$(\rho + p) V^2 \lambda^r N_r = c\phi'(k) \lambda_{(4)} V = c^2 \phi'(k) k \lambda^r N_r . \tag{30.8}$$

Then by $(30.3)_{3,4}$, (21.4) and condition $\lambda^r N_r \neq 0$ we have

$$V^2 = \frac{c^2 k}{\rho + p} \frac{dp}{dk} = \left(1 + \frac{w}{c^2} + \frac{p}{c^2 k}\right)^{-1} \frac{dp}{dk} \quad \text{with } p = k^2 \frac{\partial \hat{w}}{\partial k}(k, \eta). \tag{30.9}$$

In the adiabatic case ($\eta = \text{const}$) from (21.4) and $(30.9)_3$ we obtain

$$\frac{d\rho}{dk} = w + c^2 + k \frac{\partial \hat{w}}{\partial k} = \frac{\rho + p}{k}. \tag{30.10}$$

By (30.9) and (30.10) the relativistic velocity of propagation V has the expressions

$$\left(1 + \frac{w}{c^2} + \frac{p}{c^2 k}\right)^{-1/2} \bar{V} = V = c \sqrt{\left(\frac{dp}{d\rho}\right)_{\eta = \text{const}}} \quad \text{where } \bar{V} = \sqrt{\left(\frac{dp}{dk}\right)_{\eta = \text{const}}} . \tag{30.11}$$

Equality $(30.11)_3$ coincides with the classical expression of the classical velocity of propagation $\bar{V}$ of elastic waves. Formally $(30.11)_2$ is a direct relativistic analogue of $(30.11)_3$. However in relativity k satisfies the differential equation satisfied by the mass density in classical physics, while ρ does not. Considering k as a kinematic quantity, i.e. considering the constitutive equation $(29.2)_2$ as a relation among pressure, entropy, and strain, we see that equality $(30.11)_1$ affords the relativistic correction to the classical propagation velocity $\bar{V}$.

Bressan [1963c] emphasizes this point of view, it being substantially the only one from which elastic waves in general elastic materials are usually dealt with in classical physics—cf. e.g. Bressan (1963d); a relativistic formula quite similar to $(30.11)_1$ holds for the principal waves along the acoustic axes of a general elastic materials [§ 68].

It must be added that $\bar{V}$ has a special physical interest when the reference physical state is identified with the present one, so that $w = 0$ and $c^2 k = \rho$. Calling $V^{(c)}$ this determination of $\bar{V}$, from $(30.11)_1$ we have

$$V = V^{(c)} \left(1 + \frac{p}{\rho}\right)^{-1/2} \simeq V^{(c)} - \frac{p}{2\rho} V^{(c)} \quad \left(V^{(c)} = \sqrt{\frac{dp}{dk}} \text{ for } w = 0\right). \tag{30.12}$$

As to the order of magnitude of this correction, at the center of the sun we may reasonably expect to have $V^{(c)} - V \simeq 0.4\,\text{m/sec} = 1.5\,\text{km/h}.$[15)]

Since no signal can travel faster than light, we must have $V<c$, which by $(30.11)_2$ yields the following restriction on the constitutive equation $(30.3)_{3,4}$:

$$\left(\frac{dp}{d\rho}\right)_{\eta=\text{const}}<1,\quad \text{i.e.}\quad \frac{d\rho_0}{dp}>0 \quad \text{for} \quad \rho=\rho_0+p\,. \tag{30.13}$$

Cattaneo [1969] attempts to deduce $(30.13)_1$ from assumptions that are not *ad hoc.* He says in effect that on the basis of Bernoulli's classical theorem $\left(\mu\frac{v^2}{2}+p+V=\text{const}\right.$ where V is the energy density of the forces at a distance$\left.\right)$ p may have the role of an energy density, and that in the case of non-viscous fluids p can be interpreted as the "energy density due to the stress". Then by $(30.13)_3$ only ρ_0 ($+$const) is the part of ρ due to internal energy. Hence $(30.13)_2$ appears to be the natural relativization of a well known classical relation.

§ 31. On Elastic Waves in Perfect Gases

Let c_p [c_v] be the specific heat at constant pressure [at constant proper volume] per unit conventional mass. We now consider a perfect gas, and as usual we assume c_v constant. Then, by essentially the same reasoning as in classical physics, we can deduce the relation $c_p=c_v+R$ (where R is evaluated in connection with a particular reference state p^*, T^*) whence c_p is also constant. By the same reasoning as in classical physics the equation of the adiabatic processes of perfect gases turns out to be

$$p=p^*\left(\frac{k}{k^*}\right)^{\gamma} \quad \text{where} \quad \gamma=\frac{c_p}{c_v}=1+\frac{R}{c_v} \qquad (R=R_{p^*,T^*})\,. \tag{31.1}$$

Incidentally since conventional mass depends on the reference state p^*, T^*, the same holds for c_v, c_p, and $R(=c_p-c_v)$. Since c_v and c_p vary proportionally, γ is constant by $(31.1)_2$.

In the case of perfect gases, by $(31.1)_1$ equalities $(30.11)_3$ and $(30.12)_{3,1}$, can be written

$$\bar{V}=\sqrt{\frac{\gamma p}{k}}, \qquad V^{(c)}=c\sqrt{\frac{\gamma p}{\rho}}, \qquad V=c\sqrt{\frac{\gamma p}{\rho+p}}\,. \tag{31.2}$$

As was hinted in connection with (24.8), Bressan [1963d] emphasizes the role of inertial mass hold by the spatial tensor $c^{-2}(\rho\overset{\perp}{g}{}^{\alpha\beta}+X^{\alpha\beta})$ in connection with acceleration waves in general elastic materials. For ideal fluids this spatial tensor becomes $c^{-2}(\rho+p)\overset{\perp}{g}_{\alpha\beta}$. Then, on the basis of $(31.2)_{2,3}$ it can be said that for perfect gases the relativistic velocity of propagation V of elastic waves differs from its classical analogue $V^{(c)}$ in that the gravitational mass density (which in classical physics equals the inertial mass density) is replaced by the relativistic inertial mass density [for fluids] $c^{-2}(\rho+p)$—cf. (29.4).

Let us mention some other results concerning perfect gases and waves propagation. Along every adiabatic process of any ideal fluid w is a function of p and k. Furthermore for a perfect gas a law of the form

$$w=\left(\frac{1}{K}-1\right)\frac{p}{k} \qquad (K=\text{const}) \tag{31.3}$$

is used. E. Lamla [1912] remarked that in special relativity the *propagation speed V of shock waves is greater than c, if* (31.3) *holds for* $K=2$.

By means of statistical considerations A.H. Taub [1948] proved that *if in connection with an ideal fluid, w is a function of p and k, then*

$$w\geqslant\frac{3}{2}\frac{p}{k}+c^2\left[1+\frac{9}{4}\left(\frac{p}{c^2k}\right)^2\right]^{1/2}-c^2\,. \tag{31.4}$$

By this fundamental inequality *the law* (31.3) *implies* $K\geqslant 5/3$ which strengthens the afore-mentioned result by Lamla. Taub also stated the Rankine-Hugoniot equations for shock waves in special relativity; furthermore, on the basis of the inequality $K\geqslant 5/3$ he proved that *the law* (31.3), *where K is either a constant or a slowly varying function of p/k, effectively implies* $V<c$.

§ 32. On the Importance of the Thermodynamic Tensor

In some papers the Einstein equation (23.1) is regarded as valid in circumstances when heat is conducted, for $\mathscr{U}_{\alpha\beta}=\rho u_\alpha u_\beta+X_{\alpha\beta}$, i.e. in connection with $(24.3)_1$ for $Q_{\alpha\beta}\equiv 0$. Although these papers are very few it seems to me useful to try to show—cf. Bressan [1967a, § 4]—that the inclusion of a thermodynamic tensor in the expression of $\mathscr{U}_{\alpha\beta}$ is essential in order to apply Einstein's equations to heat conduction.

The Einstein equation (23.1) for $\mathscr{U}_{\alpha\beta}=\rho u_\alpha u_\beta+X_{\alpha\beta}$ implies (24.6) for $Q_{\alpha\beta}\equiv 0$, i.e.

$$k\,Dw+dl^{(i)}=0 \quad\text{where}\quad dl^{(i)}=X^{\alpha\beta}u_{(\alpha/\beta)}Ds\,. \tag{32.1}$$

We assume that the element $d\mathscr{C}$ of $\mathscr{C}$ is a non-viscous fluid, more, a perfect gas. For the moment we do not take the mass-energy principle into account, so that we do not assert that the magnitude w introduced by means of definition (21.4) is the specific internal energy.

From experiment we know that $d\mathscr{C}$ can undergo an infinitesimal Carnot cycle $d\mathfrak{C}$ at the ends of which the pressure takes the same value; and the same holds for the proper volume $d\mathfrak{C}$ of $d\mathscr{C}$, hence for k. This is also compatible with the constitutive equations $(28.8)_1$ and $(29.1)_1$ of non-viscous fluids in that they afford a relation of the form

$$p=\tilde{p}(T,k), \tag{32.2}$$

whence T can (and even must) take the same value at the ends of $d\mathfrak{C}$. According to classical thermodynamics the work $\Delta l^{(i)}(\Delta l^{(i)}=\int dl^{(i)})$ done by internal forces in $d\mathfrak{C}$ does not vanish. Furthermore the difference between the classical and relativistic values of $\Delta l^{(i)}$ has the same order of magnitude as $c^{-2}\Delta l^{(i)}$—cf. $(32.1)_2$, (18.6), $(24.10)_2$, and $(25.4)_2$. Hence in relativity we may assert that $\Delta l^{(i)}\neq 0$ holds along $d\mathfrak{C}$. Then, since $(32.1)_1$ holds along $d\mathfrak{C}$, we have $\Delta w\neq 0$ in $d\mathfrak{C}$, so that—since $\Delta k=0$—by (21.4) $\Delta\rho\neq 0$ holds along $d\mathfrak{C}$, i.e. the gravitational mass $c^{-2}\rho\, dC$ of $d\mathscr{C}$ changes in the cycle $d\mathfrak{C}$. The sign of $\Delta\rho$ along $d\mathfrak{C}$ is the one of Δw; hence by $(32.1)_1$ $\Delta\rho \gtrless 0$ holds according as $\Delta l^{(i)} \lessgtr 0$, that is according as in $d\mathfrak{C}$ the matter element $d\mathscr{C}$ yields or absorbs heat.

Since a Carnot cycle $\mathfrak{C}$ of ordinary size, along which heat is absorbed can be thought of as the union of very many infinitesimal Carnot cycles of the same kind and forming a lattice, the inequality $\Delta\rho\neq 0$ must also hold along $\mathfrak{C}$ according to the theory being considered.

By the preceding conclusion it is natural to observe, first, that by repeating the cycle $\mathfrak{C}$ very many times we should be able to render the total mass change directly measurable. However nobody seems actually willing even to consider seriously such an experiment. Second, according to the theory being considered the afore-mentioned change of mass is not accompanied by any energy production or energy loss. This contrasts with every experiment including a change of mass, and with the mass-energy equivalence principle, which is confirmed by the same experiments. We conclude that, at least when a law of the form (32.2) is accepted, the inclusion of a thermodynamic tensor into the expression (24.3) of $\mathscr{U}_{\alpha\beta}$ is essential for dealing with fluids capable of heat conduction.

Now we presuppose, instead of (32.2), a law of the form

$$p=\bar{p}(T,\rho) \tag{32.3}$$

with

$$\frac{\partial\bar{p}}{\partial T}\neq 0\neq\frac{\partial\bar{p}}{\partial\rho}.$$

An infinitesimal Carnot cycle can be described, within classical physics, by means of a graph in the plane p, T. In relativity we denote by $d\mathfrak{C}'$ the infinitesimal process described by the same graph. At the ends of $d\mathfrak{C}'$ each of the magnitudes T, p, and ρ takes the same value—cf. (32.3).

If along $d\mathfrak{C}'$ we had $\Delta k=0$, besides $\Delta\rho=0$, then by (21.4) $\Delta w=0$ would hold. By (32.1) this implies $\Delta l^{(i)}=0$ along $d\mathfrak{C}'$, up to an infinitesimal of an order greater than 1 (here as the principal infinitesimal we can assume e.g. the supremum of the difference in pressure between two states belonging to $d\mathfrak{C}'$). However, by a comparison with classical physics similar to the comparison made in the case (32.2), we cannot have $\Delta l^{(i)}=0$ along $d\mathfrak{C}'$ up to infinitesimals of the first order. Hence the inequality $\Delta k \neq 0$ must hold along $d\mathfrak{C}'$.

The preceding conclusion can be extended to an ordinary size Carnot cycle $\mathfrak{C}'$ just as was done (in connection with the cycle $\mathfrak{C}$) in the case (32.2). Then by having $d\mathscr{C}$ describe $\mathfrak{C}'$ very many times in a suitable sense, the initial mass of $d\mathscr{C}$ could be concentrated in an arbitrarily preassigned extremely small volume. In general relativity this proved not to be possible—cf. Cattaneo [1960, p. 210].

Thus, for the case (32.2) the same conclusion asserted for the case (32.1) holds: *The use of a thermodynamic tensor is essential to deal with heat conduction on the basis of the Einstein gravitational equation or the energy-momentum conservation equations, at least in the case when a constitutive equation of the form* (32.1) *or* (32.2) *is accepted.*

As far as I know no constitutive equation for p essentially different from (32.1) and (32.2) has been considered in connection with relativistic fluids.

Furthermore Eckart's tensor $Q_{\alpha\beta}$—cf. $(24.3)_2$—is sufficient for the relativistic treatment of heat conduction, and no better thermodynamic tensor is known, at least in general relativity.

§ 33. Some Historical Remarks on Relativistic Theories of Fluids and Hints at Some Further Results

Non-viscous fluids capable only of adiabatic transformations, and with negligible polarizations, are dealt with rather extensively in special and general relativity in not so recent treatises—cf. Tolman [1949, §§ 85, 126 to 129], Fock [1964, § 32] and Lichnerowicz [1955, Chapter IX, p. 99]. In these treatments a reduced constitutive equation, e.g. $p=\hat{p}(\rho)$, and the specification $\rho u_\alpha u_\beta + p \overset{\perp}{g}_{\alpha\beta}$ of $\mathscr{U}_{\alpha\beta}$ are used.

Particular relativistic fluids capable of heat conduction, and called perfectly perfect fluids, were considered by Van Dantzig [1939] and [1940].

As was hinted e.g. in § 24, the decisive step toward a general relativistic theory of materials capable of heat conduction was performed by Eckart [1940b] with the introduction of the thermodynamic tensor $Q_{\alpha\beta}$. This work refers to viscous fluids in special relativity, but it can be easily extended to more general materials and to general relativity.

The first works where general fluids capable of heat conduction were explicitly dealt with in general relativity (in the absence or presence of electromagnetic phenomena)—cf. Pham [1955], [1956]—included both the Einstein gravitation equations with the thermodynamic tensor, and the so called continuity principle of heat. Unfortunately, from the physical point of view this principle turns out to be unacceptable in the presence of viscosity or Joule heat. Furthermore, for viscous fluid it is acceptable if taken alone, but unacceptable if also the afore-mentioned Einstein gravitation equations are postulated—cf. Bressan [1964a, pp. 11—15].

The theory presented in the preceding sections is in accord with Eckart [1940b] and with works which are also more or less based on it, such as Hughes [1961] (within special relativity).

We mention the theory of relativistic fluids of Landau and Lifschitz [1959] last of all only because it is not used in this tract.

Historical hints concerning the afore-mentioned fluids in the presence of electromagnetic phenomena will be given in § 44. Now let us mention that Stückelberg and Wanders [1953] consider the equations of motion for a viscous fluid in general or special relativity. This fluid may consist of more then one component, and in it diffusion phenomena and chemical reactions can take place.

In case the fluid consists of one component the basic equations considered by Stückelberg and Wanders [1953] are

$$\mathscr{U}_{\alpha\beta}{}^{/\beta}=0\,, \quad (k u^{\alpha})_{/\alpha}=0\,, \quad \eta^{\alpha}{}_{/\alpha}-i=0 \quad \left(\eta^{\alpha}=k\eta u^{\alpha}+\frac{1}{T}q^{\alpha}\right) \tag{33.1}$$

where i is the *density of entropy production* and is assumed to be positive—compare (33.1) with Müller's assumption (26.17).

The authors start with some thermodynamic and linearity hypotheses on the relations $(33.1)_{1,2,3}$. Then through some considerations based, among other things, on a certain "canonical" transformation of variables, they reach certain expressions for $\mathscr{U}_{\alpha\beta}$ (assumed to be equal $\mathscr{U}_{\beta\alpha}$) and for η^{α} and i. The authors also conclude that (i) the set of the six scalar equations $(33.1)_{1,2,3}$ is possible only in a manifold $\{x^{\alpha}\}$ where only one co-ordinate is time like, and that (ii) *the bulk and shear viscosities have the sign of absolute temperature, and the heat conductivity κ is positive.*

Stückelberg [1962] adds to the assumptions made by Stückelberg and Wanders [1953] an equilibrium condition that enables him, among other things, to assert that the enthalpy density and the elastic modulus have the same sign as T, that the specific heat is non-negative, and that the velocity of propagation V of elastic waves is smaller than or equal to c. The last result is deduced independently of the causality principle.

Footnotes to Chapter 3

[1] Bressan [1972e] deals with bodies capable of heat conduction (but not of couple stress) and electromagnetic phenomena, so that $\mathscr{U}_{\alpha\beta}$ is expressed by (43.5) below. In Part 2 he postulates conservation equations, the Maxwell equations, and a natural relativistic analogue, Postulate 2, of Poisson's equation; furthermore he deduces gravitation equations from the postulates above and an additional postulate which is surely true and asserts the existence of a possibility, so that it would not be explicitly writtén according to common practice.

In Part 1 Bressan [1972e] had considered a relativistic condition in any natural (possibly nonproper) frame that differs by very small terms from a condition that in classical physics is equivalent to Poisson's equation. On the one hand (the choice of) those small terms cannot be justified directly. On the other hand that relativistic condition alone implies gravitation equations, so that it includes conservation equations too. Postulate 2 in Part 2 is weak enough to be independent of the other postulates considered in Part 2 and to allow us to justify it satisfactorily as a relativistic version of Poisson's equation.

2) The implications of inequality $(25.5)_2$ have been deeply studied, especially by B. Coleman.

3) This reasoning can be extended e.g. to the linearly viscous fluids for which the contribution of $v_{r/s}$ to the stress has the form $-x^{rslm} v_{l/m}$ with $x^{rslm} \xi_{rs} \xi_{lm}$ positive definite in the symmetric variables ξ_{rs}.

4) In the relativistic version (25.3b) of the Clausius-Duhem inequality $(25.3)_1$ for $r \equiv 0$, the classical divergence $(T^{-1}\bar{q}^r)_{/r}$ is relativized into a space-time divergence, which is in accordance with the point of view emphasized by Cattaneo [1963] and motivated, as far as kinematics is concerned, in Appendix C. Other authors relativize classical divergences into spatial divergences.

In the case $r \equiv 0$ the relativistic version (24.11) of the first principle was derived from conservation equations $(23.3)_1$ and—cf. (25.4)—in this version the classical divergence $\bar{q}^r{}_{/r}$ appears to have been relativized into $\bar{q}^r{}_{/r} + 2c^{-2}\bar{q}^r a_r$; hence neither into a spatial divergence nor into a space-time divergence $(c^{-1}q^\alpha{}_{/\alpha})$, i.e. into none of the most natural relativizations of classical divergences.

5) This is the first time that the present discussion on (25.5b) and its connection with (25.8b) is published. The same holds for the last part of this section (on a relativistic equivalent of the symmetry of $\kappa^{\alpha\beta}$).

6) An admittedly non-essential difference between Eckart's theory of fluids and most others, including ours, consists in Eckart's use of Kelvin's hypothesis—cf. Eckart [1940a, p. 268].

7) Eringen [1970] asserts that the relativistic law of heat conduction (25.8b) (for $\kappa^{\alpha\beta} = \kappa \overset{\perp}{g}{}^{\alpha\beta}$) used by Pham and Bressan, is incompatible with the second principle. This assertion is not decisive for making a choice between (25.2) and (25.8b); indeed (i) it is true in case one identifies the second principle with the particular relativistic version (25.3b) of the Clausius-Duhem inequality, as Eringen does, but (ii) the assertion is wrong if we replace (25.3b) with (25.5), i.e. (25.8b) is compatible with the relativistic version (25.5b) of the Clausius-Duhem inequality. Let us add that, a priori, to identify the second principle with (25.5b) is not less natural than accepting the version (25.2) of Fourier's law, for the natural (direct) relativization (25.8b) of Fourier's law implies just (25.5b) in the case of materials capable of only reversible processes.

Let us add that Alts and Müller [1972] introduce the heat potential as a variable whose gradient is proportional to the heat flux (in the case of equilibrium). In the typical case this variable is distinct from T. See also Liu and Müller [1972].

8) Müller starts with an "empirical temperature" ϑ. Then he defines the absolute temperature $T = T(\vartheta, k)$ in the case of equilibrium.

9) As far as Lorentz invariance is taken into account, the coefficients p to t_{12} can be assigned no special physical meanings. However by comparing (26.10) with (24.3) we see that ρ is the total internal energy per unit conventional mass, p the pressure, and $\bar{\kappa} = c\kappa$ the heat conductivity. One can also see that $\bar{\mu}$ is the shear viscosity and ν is the bulk viscosity.

10) Müller [1969] does not consider this topic or in particular (26.12).

11) The constitutive equations for $\mathscr{U}^{(M)}_{\alpha\beta}$ and η_α are in accordance with the principle of equipresence —see Truesdell and Toupin [1960]—according to which a quantity present as an independent variable in one constitutive equation should be so present in all.

12) Our use of inequality of (28.1) in this section and e.g. in [§ 46] and [§ 47] has been strongly influenced by some papers of B.D. Coleman on classical thermodynamics.

13) In some textbooks of classical physics the constitutive equations such as (28.10) or (29.2) below are deduced under the hypothesis that at P^* (i) $\mathscr{C}$ is a non-viscous fluid and (ii) $\mathscr{C}$ is capable of only reversible processes. In § 28 hypothesis (ii) was proved to follow from (i).

14) This is remarked in Eglit [1965, § 2] in classical physics and in Bressan [1966d] in general relativity, and in connection with a more general expression for the work of internal contact forces, which is compatible with couple stresses.

15) With an aim of making a rough estimate of an order of magnitude, in Bressan [1963d, p. 25] it is assumed that at the center of the sun we have $p = 1.2 \times 10^{17}$ dynes cm^{-2} and $c^{-2}\rho = 76$ gr cm^{-3}, whence $p\rho^{-1} = 1.2 \times 10^{17}(76 \times 9 \times 10^{20})^{-1} \simeq 0{,}2 \times 10^{-5}$.

At the center of the sun the equation of adiabatic processes for perfect gases is usually considered to hold: $p\rho^{-\gamma} = p_0 \rho_0{}^{-\gamma}$ with $\gamma = 1.4$. In the case being considered the Laplace formula $V^{(c)} = c\sqrt{\gamma p/\rho}$ can be accepted. Then $V^{(c)} \simeq (1.4 \times 1.2 \times 10^{17}/76)^{1/2} \simeq (0.02 \times 10^{17})^{1/2} \simeq 4 \times 10^7$ cm/sec, whence $V^{(c)} p(2\rho)^{-1} \simeq 40$ cm/sec. Hence $V^{(c)} - V \simeq 0.4$ m/sec.

Chapter 4

Electromagnetism from the Eulerian Point of View. Polarizable Fluids

§ 34. Introductory Considerations. The Ohm Law and the Relations Between the Electric and Magnetic Fields and the Respective Inductions

In the present theory of electromagnetism the interactions between field and matter are described following the Einstein-Minkowski-Abraham pattern, i.e. we include a suitable term, ${}^sE_{\alpha\beta}$, in the total energy tensor $\mathscr{U}_{\alpha\beta}$.

The main use of the corresponding conservation equations made in this tract is the derivation of constitutive equations for elastic polarizable materials in the cases of fluids [§ 46] and solids [§ 72]. For this deduction only the temporal part $(23.4)_{3,4}$ of conservation equations—i.e. the equation of energy balance—is relevant.

The interactions between field and matter gave and are still giving rise to discussions. They mainly concern whether Minkowski's tensor ${}^1E_{\alpha\beta}$ or Abraham's ${}^2E_{\alpha\beta}$ is to be used. From a certain point of view these tensors are mutually incompatible. However both of them are in agreement with physical phenomena as far as effectively performed experiments are concerned—cf. Brevik [1970a, b].

Let us add that Dixon and Eringen [1965] studied (in classical physics) an interesting molecular model for a polarizable body. Their results induced Grot and Eringen [1966] to criticize strongly both the use of ${}^1E_{\alpha\beta}$ and the one of ${}^2E_{\alpha\beta}$, and to be very pessimistic about future theories of polarizable media in general relativity.[1)]

However Pitteri [1973] shows that Grot and Eringen's criticism of the use of ${}^1E_{\alpha\beta}$ or ${}^2E_{\alpha\beta}$ and their pessimistic assertions—cf. footnote 1 in Chapter 4—are not based on logically rigorous arguments and that in fact the use of ${}^1E_{\alpha\beta}$ or ${}^2E_{\alpha\beta}$ is in accord with Dixon and Eringen's model.

Grot and Eringen's objection to the use of Minkowski's and Abraham's tensors—cf. the preceding footnote 1 in Chapter 4—is only based on the difference (by non negligible terms) between either of the (spatial) forces ${}^1\overset{\perp}{K}_r$ and ${}^2\overset{\perp}{K}_r$ corresponding to those tensors—cf. $(36.2)_{1,3}$ below—and the ponderomotive force $\bar{f}_r$ calculated by means of Dixon and Eringen's model. The insufficiency of this basis appears from a certain possibility of exchanging terms within the expression of the total energy tensor $\mathscr{U}_{\alpha\beta}$—cf. (36.9) below. Unlike Grot and Eringen [1966], Pitteri [1973] takes this possibility into account and introduces a suitable tensor ${}^8E_{\alpha\beta}$ of electromagnetic energy, that, unlike Eringen's tensor ${}^EE_{\alpha\beta}$, can be accepted in general relativity. The corresponding spatial force ${}^8\overset{\perp}{\bar{K}}_r$ equals $\bar{f}_r$ up to terms that are negligible as well as the analogues for ${}^EE_{\alpha\beta}$ (i.e. those forming ${}^EK_r-\bar{f}_r$) and are of the same kind as the latter. Hence ${}^8E_{\alpha\beta}$ is acceptable and to the same extent as ${}^EE_{\alpha\beta}$. Then by the afore-mentioned possibility of exchanging terms, ${}^2E_{\alpha\beta}$ and ${}^1E_{\alpha\beta}$ also appear to be in agreement with the afore-mentioned model. Pitteri's result is interesting especially in that it dissipates the pessimism about general relativity mentioned in the preceding footnote.

Therefore the theory of electromagnetism in special or general relativity presented in this book is as reliable as are other common non-quantistic relativistic theory of continuous media (possibly restricted to special relativity).

Let E_α be the proper electric field, H_α the proper magnetic field, D_α and B_α the respective inductions, P_α and M_α the respective polarizations per unit proper volume, and in addition π_α and μ_α the electric and magnetic polarizations respectively per unit conventional mass—i.e. proper reference mass per unit actual proper volume [§ 21]. Then, given a suitable choice of units,

$$D_\alpha=E_\alpha+P_\alpha,\quad B_\alpha=H_\alpha+M_\alpha,\quad P_\alpha=k\pi_\alpha,\quad M_\alpha=k\mu_\alpha, \tag{34.1}$$

and

$$E_\alpha u^\alpha=0=H_\alpha u^\alpha,\quad P_\alpha u^\alpha=0=M_\alpha u^\alpha. \tag{34.2}$$

We denote the proper density of true electric charge by $\bar{\rho}$ and the one of ordinary true electric current by j_α. Then the proper density of true electric 4-current, J_α, may be defined by means of the conditions

$$\bar{\rho}=-u^\alpha J_\alpha,\quad j_\alpha=c\overset{\perp}{g}_\alpha{}^\beta J_\beta. \tag{34.3}$$

Ohm's law can be written in the form

$$\overset{\perp}{g}{}^{\alpha\beta}J_\beta=\bar{\sigma}^{\alpha\beta}E_\beta \quad\text{where}\quad \bar{\sigma}_{[\alpha\beta]}=0=\bar{\sigma}_{\alpha\beta}u^\beta. \tag{34.4}$$

The so called *conductivity tensor* $\bar{\sigma}^{\alpha\beta}$ is a function of the local physical state of matter. It is usually regarded as independent of E_β. However this restriction is not necessary.

A spatial tensor $T_{\alpha\beta}$ is (called) *non-singular* if

$$\overset{\perp}{\varepsilon}_{\alpha\beta\gamma}\overset{\perp}{\varepsilon}_{\rho\sigma\tau}T^{\alpha\rho}T^{\beta\sigma}T^{\gamma\tau}\neq 0. \tag{34.5}$$

In locally proper and pseudo-Euclidean co-ordinates—cf. (17.7)—this condition is equivalent to $\det\|T_{rs}\|\neq 0$. Then there exist the inverse $\|T_{rs}^{-1}\|$ of the matrix $\|T_{rs}\|$. This inverse naturally defines a spatial tensor $T_{\alpha\beta}^{-1}$. We call this tensor the *inverse* of the spatial tensor $T_{\alpha\beta}$.

For the material element $d\mathscr{C}$ being considered one often assumes the existence of two spatial symmetric and non-singular tensors $\eta_{\alpha\beta}$ and $\mu_{\alpha\beta}$, which are functions of the (local) physical state of $d\mathscr{C}$ and fulfill in every physically possible process the conditions

$$D_\alpha=\eta_{\alpha\beta}E^\beta\,,\qquad B_\alpha=\mu_{\alpha\beta}H^\beta\qquad(\eta_{[\alpha\beta]}=0=\mu_{[\alpha\beta]},\ \eta_{\alpha\beta}u^\beta=0=\mu_{\alpha\beta}u^\beta)\,. \tag{34.6}$$

Constitutive equations such as (34.4) and (34.6) will be considered from a more complete point of view in §§ 46, 72, 80, 81, 82.

Now we introduce the *first electromagnetic tensor* $F_{\alpha\beta}$

$$F_{\alpha\beta}=(E_\alpha,B_\alpha)_{\alpha\beta}=2u_{[\alpha}E_{\beta]}+\overset{\perp}{\varepsilon}_{\alpha\beta}{}^{\gamma}B_\gamma\qquad(F_{\beta\alpha}=-F_{\alpha\beta})\,. \tag{34.7}$$

It is understood that E_α, H_α, D_α, and B_α are measured with respect to an observer whose 4-velocity γ^α equals u^α in $W_{\mathscr{C}}$, so that $F_{\alpha\beta}$ is completely determined in $W_{\mathscr{C}}$ by (34.7). That $F_{\alpha\beta}$ is determined, i.e. is independent of γ^α, outside matter is also stated by experiment. Hence if one prefers to assume fields and inductions as primitive notions, then, besides the existence of E_α to B_α everywhere, one has to postulate e.g. the existence of a tensor, $F_{\alpha\beta}$, that satisfies (34.7) outside matter for every choice of γ^α. Furthermore (34.7) can be used to define E_α and B_α in $W_{\mathscr{C}}$ for $\gamma^\alpha\neq u^\alpha$. (The analogous definitions of D_α and H_α are possible by means of the tensor $f_{\alpha\beta}$ below.)

We understand $(34.7)_2$ as a convention holding for all (spatial) vectors E_α and B_α, i.e. disregarding the physical meaning assigned to them above. Thus it is meaningful to introduce the *polarization tensor* $P_{\alpha\beta}$ and the *second electromagnetic tensor* $f_{\alpha\beta}$ as follows:

$$P_{\alpha\beta}=(P_\alpha,-M_\alpha)_{\alpha\beta}\,,\qquad f_{\alpha\beta}=(D_\alpha,H_\alpha)_{\alpha\beta}=F_{\alpha\beta}+P_{\alpha\beta}\,. \tag{34.8}$$

Since the spatial tensors $\eta_{\alpha\beta}$ and $\mu_{\alpha\beta}$ are non-singular, their inverses $\eta^{-1}_{\alpha\beta}$ and $\mu^{-1}_{\alpha\beta}$ exist. On the basis of $(17.4)_2$ and $(20.5)_2$ we respectively derive from (34.7)

$$E_\alpha=F_{\alpha\beta}u^\beta\,,\qquad B_\alpha=\tfrac{1}{2}\overset{\perp}{\varepsilon}_\alpha{}^{\rho\sigma}F_{\rho\sigma}\,. \tag{34.9}$$

By $(34.8)_2$ and convention $(34.7)_2$, by $(34.6)_1$ and the inverse $H_\gamma=\mu^{-1}_{\gamma\delta}B^\delta$ of $(34.6)_2$, and by (34.9) we respectively have

$$\begin{aligned}f_{\alpha\beta}&=2u_{[\alpha}D_{\beta]}+\overset{\perp}{\varepsilon}_{\alpha\beta}{}^{\gamma}H_\gamma=2u_{[\alpha}\eta_{\beta]\rho}E^\rho+\overset{\perp}{\varepsilon}_{\alpha\beta}{}^{\gamma}\mu^{-1}_{\gamma\delta}B^\delta\\&=2u_{[\alpha}\eta_{\beta]\rho}F^{\rho\sigma}u_\sigma+\tfrac{1}{2}\overset{\perp}{\varepsilon}_{\alpha\beta}{}^{\gamma}\mu^{-1}_{\gamma\delta}\overset{\perp}{\varepsilon}{}^{\delta\rho\sigma}F_{\rho\sigma}\,.\end{aligned} \tag{34.10}$$

Hence the relations $(34.6)_{1,2}$ can be written concisely in the form

$$f_{\alpha\beta}=\eta_{\alpha\beta}{}^{\rho\sigma}F_{\rho\sigma}\qquad(\eta_{[\alpha\beta]}{}^{\rho\sigma}=0=\eta_{\alpha\beta}{}^{[\rho\sigma]}) \tag{34.11}$$

where, by taking the skewsymmetry of $F_{\rho\sigma}$ into account, we understand

$$\eta_{\alpha\beta}{}^{\rho\sigma}=u_{[\alpha}(\eta_{\beta]}{}^{\rho}u^\sigma-\eta_{\beta]}{}^{\sigma}u^\rho)+\tfrac{1}{2}\overset{\perp}{\varepsilon}_{\alpha\beta}{}^{\gamma}\overset{\perp}{\varepsilon}{}^{\rho\sigma\delta}\mu^{-1}_{\gamma\delta}\,. \tag{34.12}$$

In the case of a non-viscous fluid the laws (34.4) and (34.6) become

$$\overset{\perp}{g}_{\alpha}{}^{\beta} J_{\beta} = \bar{\sigma} E_{\alpha}, \quad D_{\alpha} = \eta E_{\alpha}, \quad B_{\alpha} = \mu H_{\alpha} \quad (\eta_{\alpha\beta} = \eta \overset{\perp}{g}_{\alpha\beta}, \ldots). \tag{34.13}$$

Then by $(20.5)_1$ relations (34.12) and (34.11) respectively simplify to

$$\eta_{\alpha\beta}{}^{\rho\sigma} = \eta u_{[\alpha}(\overset{\perp}{g}_{\beta]}{}^{\rho} u^{\sigma} - \overset{\perp}{g}_{\beta]}{}^{\sigma} u^{\rho}) + \frac{1}{2\mu}(\overset{\perp}{g}_{\alpha}{}^{\rho} \overset{\perp}{g}_{\beta}{}^{\sigma} - \overset{\perp}{g}_{\alpha}{}^{\sigma} \overset{\perp}{g}_{\beta}{}^{\rho}), \tag{34.14}$$

$$f_{\alpha\beta} = 2\eta u_{[\alpha} F_{\beta]\sigma} u^{\sigma} + \frac{1}{\mu} F_{\alpha\beta}. \tag{34.15}$$

Of course (34.15) can be derived directly from $(34.10)_1$, $(34.13)_{2,3}$, (34.9), and $(20.5)_1$.

§ 35. On the Maxwell Equations in Space Time

Our theory will be based mainly on the following tensor form of Maxwell's equations:

$$\varepsilon^{\alpha\beta\gamma\delta} F_{\beta\gamma/\delta} = 0, \quad f_{\alpha\beta}{}^{/\beta} = J_{\alpha} \text{—cf. (34.7), (34.8)}. \tag{35.1}$$

However, with a view to some marginal considerations and according to common practice, we introduce the proper density $J'_{\alpha} = -P_{\alpha\beta}{}^{/\beta}$ of polarization 4-current; and on the basis of $(34.8)_{2,3}$ we turn $(35.1)_2$ into

$$F_{\alpha\beta}{}^{/\beta} = J''_{\alpha} \quad \text{where} \quad J''_{\alpha} = J_{\alpha} + J'_{\alpha}, \quad J'_{\alpha} = -P_{\alpha\beta}{}^{/\beta}. \tag{35.2}$$

We call J''_{α} the proper density of total 4-current. The proper densities $\bar{\rho}'$, j'_{α} of polarization charge and ordinary polarization current and the proper densities $\bar{\rho}''$ and j''_{α} of total charge and ordinary total current are defined in terms of J'_{α} and J''_{α} respectively as $\bar{\rho}$ and j_{α} are defined in terms of J_{α} by means of (34.3).

Let us deduce from (35.1) the four customary Maxwell equations extended to the case of deforming, rotating, and accelerated media. To this end we first introduce the dual $*F_{\alpha\beta}$ of $F_{\alpha\beta}$:

$$*F_{\alpha\beta} = \tfrac{1}{2} \varepsilon_{\alpha\beta}{}^{\gamma\delta} F_{\gamma\delta}. \tag{35.3}$$

Since $2F^{*}_{\alpha\beta}{}^{/\beta} \equiv \varepsilon_{\alpha\beta}{}^{\gamma\delta} F_{\gamma\delta}{}^{/\beta} \equiv \varepsilon_{\alpha}{}^{\beta\gamma\delta} F_{\beta\gamma/\delta}$, equation $(35.1)_1$ becomes

$$*F_{\alpha\beta}{}^{/\beta} = 0 \quad (2*F_{\alpha\beta}{}^{/\beta} \equiv \varepsilon_{\alpha}{}^{\beta\gamma\delta} F_{\beta\gamma/\delta}). \tag{35.4}$$

In locally proper and pseudo-Euclidean co-ordinates ($g_{\alpha\beta}=\delta'_{\alpha\beta}$, $u^\alpha=\delta^\alpha{}_0=-u_\alpha$) $(34.7)_1$ is equivalent to $(34.7)_3$ and

$$F_{r0}=E_r\,,\qquad F_{12}=B_3\,,\qquad F_{23}=B_1\,,\qquad F_{31}=B_2\,, \tag{35.5}$$

and $(34.7)_3$ and (35.3) imply ($F^*_{10}=\varepsilon_{10}{}^{23}F_{23}=-\varepsilon_{01}{}^{23}F_{23}=-F_{23}$, ${}^*F_{12}=\varepsilon_{12}{}^{30}F_{30}=-\varepsilon_{1230}F_{30}=F_{30},\ldots$, i.e.)

$$\begin{aligned}&{}^*F_{10}=-F_{23}\,,\qquad {}^*F_{20}=-F_{31}\,,\qquad {}^*F_{30}=-F_{12}\,;\\ &{}^*F_{12}=F_{30}\,,\qquad {}^*F_{23}=F_{10}\,,\qquad {}^*F_{31}=F_{20}\,.\end{aligned} \tag{35.6}$$

Since by convention $(34.7)_2$ condition $(34.7)_1$ is equivalent to $(34.7)_3$ and (35.5) for all spatial vectors E_α and B_α, by (35.5) and (35.6) we have ${}^*F_{r0}=-B_r$, ${}^*F_{12}=E_3,\ldots$. Therefore

$${}^*F_{\alpha\beta}=(-B_\alpha,E_\alpha)_{\alpha\beta}\quad \textit{for all spatial vectors } E_\alpha \textit{ and } B_\alpha\,. \tag{35.7}$$

Then ${}^{**}F_{\alpha\beta}=(-E_\alpha,-B_\alpha)_{\alpha\beta}=-F_{\alpha\beta}$. Hence

$$F_{\alpha\beta}=-\tfrac{1}{2}\varepsilon_{\alpha\beta}{}^{\gamma\delta}\,{}^*F_{\gamma\delta}=-{}^{**}F_{\alpha\beta}\,. \tag{35.8}$$

Relations (35.7), (35.8), and $(35.4)_2$ are useful for reducing several calculations on the left hand side of $(35.1)_1$ where (34.7) holds, to calculations on the left hand side of $(35.2)_1$, and viceversa.

Now we wish to write the preceding Maxwell equations, in the general case of deforming media, in a form which becomes the familiar one for $u_{\alpha/\beta}=0$. To this end let us remark that by (34.7) and (17.13), for $T_{\alpha\beta}=F_{\alpha\beta}$ we have $T=0$, $T'_\alpha=-T''_\alpha=-E_\alpha$, $\overset{\perp}{T}_{\alpha\beta}=\overset{\perp}{\varepsilon}_{\alpha\beta}{}^{\gamma}B_\gamma$, so that both identities (22.9) and equality (22.10) hold, i.e. we have

$$-u_\alpha F^{\alpha\beta}{}_{/\beta}=E^\alpha{}_{/\overset{\perp}{\alpha}}+2\Omega^\gamma B_\gamma\quad\text{where}\quad 2\Omega^\gamma=\overset{\perp}{\varepsilon}{}^{\alpha\beta\gamma}u_{\alpha/\beta}\,, \tag{35.9}$$

$$\overset{\perp}{g}_{\alpha\rho}F^{\rho\beta}{}_{/\beta}=\overset{\perp}{g}_{\alpha\rho}(\overset{\perp}{\varepsilon}{}^{\rho\beta\gamma}B_\gamma)_{/\beta}-k\,\frac{D^c}{Ds}\,\frac{E_\alpha}{k}\,. \tag{35.10}$$

Of course the spatial vector Ω^γ—cf. $(35.9)_2$—is the local [Römer] angular velocity.

By $(17.20)_1$ and (20.4), we easily deduce

$$\overset{\perp}{\varepsilon}{}^{\alpha\beta\gamma}{}_{/\beta}=u_{\sigma/\overset{\perp}{\beta}}\,\varepsilon^{\sigma\alpha\beta\gamma}+A_\beta\,\overset{\perp}{\varepsilon}{}^{\alpha\beta\gamma}\,, \tag{35.11}$$

so that

$$\overset{\perp}{g}_{\alpha\rho}\,\overset{\perp}{\varepsilon}{}^{\rho\beta\gamma}{}_{/\beta}B_\gamma=\overset{\perp}{g}_{\alpha\rho}u_{\sigma/\overset{\perp}{\beta}}B_\gamma\varepsilon^{\sigma\rho\beta\gamma}+\overset{\perp}{\varepsilon}_\alpha{}^{\beta\gamma}A_\beta B_\gamma\,. \tag{35.12}$$

Since $\overset{\perp}{g}_{\alpha\rho} u_{\sigma/\overset{\perp}{\beta}} B_\gamma$ is a spatial tensor, the first term in the right-hand side of (35.12) vanishes. Thus (35.10) becomes

$$\overset{\perp}{g}_{\alpha\rho} F^{\rho\beta}{}_{/\beta} = \overset{\perp}{\varepsilon}_\alpha{}^{\beta\gamma}(B_{\gamma/\beta}+B_\gamma A_\beta) - k\frac{D^c}{Ds}\frac{E_\alpha}{k}. \tag{35.13}$$

Since (34.7) implies (35.9) and (35.13) for all spatial fields E_α and B_α, (35.7) implies

$$\begin{aligned} -u_\alpha {}^*F^{\alpha\beta}{}_{/\beta} &= 2\Omega^\alpha E_\alpha - B^\alpha{}_{/\overset{\perp}{\alpha}} \\ \overset{\perp}{g}_{\alpha\rho} {}^*F^{\rho\beta}{}_{/\beta} &= k\frac{D^c}{Ds}\frac{B_\alpha}{k} + \overset{\perp}{\varepsilon}_\alpha{}^{\beta\gamma}(E_{\gamma/\beta}+E_\gamma A_\beta). \end{aligned} \tag{35.14}$$

By (35.14) the Maxwell equation $(35.1)_1$, turned into $(35.4)_1$, takes the form

$$B^\alpha{}_{/\overset{\perp}{\alpha}} = 2\Omega^\alpha E_\alpha, \qquad k\frac{D^c}{Ds}\frac{B_\alpha}{k} + \overset{\perp}{\varepsilon}_\alpha{}^{\beta\gamma}(E_{\gamma/\beta}+E_\gamma A_\beta) = 0. \tag{35.15}$$

Furthermore, by (34.3) and the analogues of (35.9) and (35.13) for $f_{\alpha\beta}$, the Maxwell equation $(35.1)_2$ becomes

$$D^\alpha{}_{/\overset{\perp}{\alpha}} + 2\Omega^\alpha H_\alpha = \bar{\rho}, \qquad \overset{\perp}{\varepsilon}_\alpha{}^{\beta\gamma}(H_{\gamma/\beta}+H_\gamma A_\beta) = k\frac{D^c D_\alpha}{Dk} + \frac{1}{c} j_\alpha. \tag{35.16}$$

Likewise (35.9), (35.13) and the analogue for J''_α of the relations (34.3) on J_α imply the equivalence of the Maxwell equation $(35.2)_1$ to

$$E^\alpha{}_{/\overset{\perp}{\alpha}} + 2\Omega^\alpha B_\alpha = \bar{\rho}'', \qquad \overset{\perp}{\varepsilon}_\alpha{}^{\beta\gamma}(B_{\gamma/\beta}+B_\gamma A_\beta) = k\frac{D^c E_\alpha}{Dk} + \frac{1}{c} j''_\alpha. \tag{35.17}$$

Of course in considering electromagnetic formulas such as (35.15) to (35.17) outside matter, we understand that u^α is any unit vector field—to be interpreted as the one describing the motion of the ideal fluid $\mathfrak{F}$ joined to the observer—and that k is any solution of the continuity equation $(21.3)_3$.

In the case $u_{\alpha/\overset{\perp}{\beta}}=0$, by $(21.3)_3$ and $(17.20)_2$ we have $\dot{k}=0$ and, by $(22.1)_1$, $D^c T\ldots=(DT\ldots)^\perp$. Hence *if the motion of matter, or of* $\mathfrak{F}$, *is locally rigid* $(u_{(\alpha/\overset{\perp}{\beta})}=0)$, *rotationless* $(u_{[\alpha/\overset{\perp}{\beta}]}=0)$, *and with a vanishing intrinsic acceleration* $(u_{\alpha/\beta}u^\beta=0)$, *then in locally natural and proper co-ordinates*—cf. (17.7)—*the Maxwell equations* (35.15)—(35.16) *take the most familiar form (in the rationalized Gaussian system)*

$$\begin{aligned} &B^r{}_{,r}=0, \qquad \frac{1}{c}\frac{\partial}{\partial\tau}B_i + \overset{\perp}{\varepsilon}_i{}^{rs}E_{s,r}=0 \\ &D^r{}_{,r}=\bar{\rho}, \qquad \overset{\perp}{\varepsilon}_i{}^{rs}H_{s,r} = \frac{1}{c}\frac{\partial}{\partial\tau}D_i + \frac{1}{c}j_i. \end{aligned} \tag{35.18}$$

These equations—with $\overset{\perp}{\varepsilon}_i{}^{rs} = \mathscr{E}_{0i}{}^{rs}$ according to $(20.4)_2$—were asserted to hold in a classical Euclidean frame for matter stationary with respect to the absolute space $S_{(ab)}$ [§ 11].

Now let a classical physicist judge a matter element to be stationary with respect to the (Euclidean) space S', and let ω^α be the angular velocity of S' (with respect to inertial spaces) and $\mathfrak{g}^r$ its local gravity acceleration, i.e. Newton's universal attraction force plus dragging force per unit mass. Then a relativist, who uses a locally proper and natural frame is willing to accept equations (35.15) to (35.16) written in this frame for $\Omega^r = c\omega^r$ and $c^2 A_r = -\mathfrak{g}_r$:

$$
\begin{aligned}
&B^r{}_{,r} = \frac{2}{c}\omega^r E_r, && \frac{1}{c}\frac{\partial}{\partial\tau}B_i + \overset{\perp}{\varepsilon}_i{}^{rs}\left(E_{s,r} - \frac{1}{c^2}E_s\mathfrak{g}_r\right),\\
&E^r{}_{,r} + \frac{2}{c}\omega^r B_r = \bar{\rho}, && \varepsilon_i{}^{rs}\left(B_{s,r} - \frac{1}{c^2}B_s\mathfrak{g}_r\right) = \frac{1}{c}\frac{\partial E_i}{\partial\tau} + \frac{1}{c}j_i.
\end{aligned}
\tag{35.19}
$$

In case S' is joined to the earth, ω^r is the angular velocity of the earth. In case S' is the privileged inertial space $S_{(ab)}$, ω^r vanishes and $\mathfrak{g}_r$ reduces to Newton's universal attraction force [§ 11].

Incidentally, in § 11 we briefly hinted at a modified classical theory including electromagnetism and gravitation, which is to general relativity what the ordinary classical electromagnetic theory—based on equations (35.18) for matter stationary with respect to $S_{(ab)}$—is to special relativity. In this modified theory the electromagnetic equations for matter stationary with respect to $S_{(ab)}$ are (35.19) for $\omega^r \equiv 0$.

We incidentally note that $(35.1)_2$ implies the conservation equation

$$
J^\alpha{}_{/\alpha} = 0 \qquad \left(J^\alpha = \bar{\rho}u^\alpha + \frac{1}{c}j^\alpha,\ u_\alpha j^\alpha = 0\right)
\tag{35.20}
$$

for electric charge. Indeed from $f_{(\alpha\beta)} = 0$—cf. (34.10)—we deduce $f_{\alpha\beta}{}^{/\alpha\beta} = -f_{\beta\alpha}{}^{/\alpha\beta}$. Then, in accordance with (16.9) we have

$$
2f_{\alpha\beta}{}^{/\alpha\beta} = f_{\alpha\beta}{}^{/\alpha\beta} - f_{\alpha\beta}{}^{/\beta\alpha} = f_{\rho\beta}R^\rho{}_\alpha{}^{\alpha\beta} + f_{\alpha\rho}R^\rho{}_\beta{}^{\beta\alpha} = f_{\rho\beta}(R^\rho{}_\alpha{}^{\alpha\beta} - R^\rho{}_\alpha{}^{\alpha\beta}) = 0,
$$

which by $(35.1)_2$ yields $(35.20)_1$.

By $(35.20)_2$ and (21.6) for $T_{\cdots}^{\cdots} = \bar{\rho}$ we can write (35.20) in the forms

$$
k\frac{D}{Ds}\frac{\bar{\rho}}{k} + \frac{1}{c}j^\alpha{}_{/\alpha} = 0 = \frac{D\bar{\rho}}{Ds} + \bar{\rho}u^\alpha{}_{/\alpha} + \frac{1}{c}j^\alpha{}_{/\alpha} \qquad (j^\alpha{}_{/\alpha} = j^\alpha{}_{/\overset{\perp}{\alpha}} + j^\alpha A_\alpha).
\tag{35.21}
$$

Equality $(35.21)_3$ follows from $(35.20)_3$ and (17.17).

§ 36. On the Electromagnetic Energy Tensor $E_{\alpha\beta}$. Some Requirements for it in the Absence of Polarization. Its Indeterminancy in the Presence of Polarization

In order to take electromagnetic phenomena into account within the schemes set up in § 23 for special and general relativity, a suitable term, $E_{\alpha\beta}$, to be called *electromagnetic energy tensor*, is included among the terms filling the dots in the expression $(23.1)_2$ of the total energy tensor $\mathscr{U}_{\alpha\beta}$. More in particular, we have to generalize definition (24.3) into

$$\mathscr{U}_{\alpha\beta}=\rho u_\alpha u_\beta+X_{\alpha\beta}+Q_{\alpha\beta}+E_{\alpha\beta}, \tag{36.1}$$

where the tensor $E_{\alpha\beta}$ is so chosen that conservation equation $(23.3)_1$ and the symmetry conditions $(23.3)_2$ should be acceptable physical equations in case electromagnetic phenomena take place together with mechanic and thermic phenomena. This requirement for the choice of $E_{\alpha\beta}$ implies that, after setting

$$K_\alpha=-E_{\alpha\beta}{}^{/\beta} \quad \text{and} \quad \Pi^{(e)}=cu^\alpha E_{\alpha\beta}{}^{/\beta}, \quad \text{whence} \quad K_\alpha=\overset{\perp}{K}_\alpha+c^{-1}\Pi^{(e)}u_\alpha, \tag{36.2}$$

$\overset{\perp}{K}_\alpha$ and $\Pi^{(e)}$ should be acceptable as the proper densities of the *(ordinary) ponderomotive force* and the power spent by the field on matter respectively. We may call K_α *density of ponderomotive 4-force.*

The above power density, $\Pi^{(e)}$, is the sum of the Joule heat $E^\alpha j_\alpha$ and the (microscopic) work $k\,d\lambda/D\tau$ of the field for polarization changes, per unit time and proper volume.

$$\Pi^{(e)}=E^\alpha j_\alpha+k\frac{d\lambda}{D\tau} \quad \text{or} \quad \frac{1}{c}\Pi^{(e)}=E^\alpha J_\alpha+k\frac{d\lambda}{Ds}. \tag{36.3}$$

Unlike the work of (internal) stresses, the work $k\,d\lambda\,dC$ is not characterized by the macroscopic motion of $\mathscr{C}$ and some force elements (such as E_α, H_α, D_α and B_α). However it is a work in the sense of classical mechanics if a molecular model is considered for $\mathscr{C}$ [§ 42].

By the consequence $(23.3)_2$ of (23.1), and by (36.1), (17.13), and the symmetry of $Q_{\alpha\beta}$, *in general relativity we have*

$$X_{[\alpha\beta]}+\overset{\perp}{E}_{[\alpha\beta]}=0=E'''_{[\alpha\beta]}. \tag{36.4}$$

Obviously $(36.4)_1$ must be acceptable as the second Cauchy equation of continuous media in that $(36.1)_1$ is certainly such for $E_{\alpha\beta}=0$. Hence $\overset{\perp}{E}_{[\alpha\beta]}$ must be acceptable as the intrinsic moment of the electromagnetic actions on matter.

In special relativity $(36.4)_1$ is also to be assumed, but so far we have no compelling reason for assuming $(36.4)_2$. This condition will be discussed in § 40 after certain interesting determinations of $\Pi^{(e)}$ are presented.

Here we set forth two requirements on $E_{\alpha\beta}$ which from the physical point of view are necessary in the case $P_\alpha \equiv M_\alpha \equiv 0$. In this case on the one hand the expression of the proper force $\overset{\perp}{K}_\alpha dC$ acting on the element $d\mathscr{C}$ when it has the charge $e = \bar{\rho}\, dC$ and the current $j_\alpha dC$ must be in accord with the experiments on a test particle, $\bar{P}$, having the charge e and the 4-velocity dx^α/ds such that $e\, dx^\alpha/ds = J^\alpha dC$. The particle $\bar{P}$ is acted on by the Lorentz force—see footnote 8 in Chapter 1 [§ 4]. By this, (18.6), and (34.3) *we must have*

$$\overset{\perp}{K}_\alpha dC = e\left(E_\alpha + \overset{\perp}{\varepsilon}_{\alpha\beta}{}^{\gamma}\frac{dx^\beta}{ds}H_\beta\right) = (-E_\alpha u^\beta J_\beta + \overset{\perp}{\varepsilon}_\alpha{}^{\beta\gamma} J_\beta H_\gamma)\, dC\,. \tag{36.5a}$$

On the other hand, in the absence of polarizations the afore-mentioned microscopic work $k\, dC\, d\lambda$ vanishes, so that by $(36.2)_3$ and $(36.3)_2$ *we must have*

$$-cu^\alpha K_\alpha = \Pi^{(e)} = cE^\alpha J_\alpha \qquad (P_{\alpha\beta} \equiv 0)\,. \tag{36.5b}$$

Let us add that, for $P_\alpha \equiv M_\alpha \equiv 0$, i.e. for $P_{\alpha\beta} \equiv 0$—cf. $(34.8)_1$—, by $(35.2)_3$ we have $J'_\alpha = 0$, hence by $(35.2)_2$ $J''_\alpha = J_\alpha$; furthermore by $(34.1)_2$ $H_\alpha = B_\alpha$ holds for $M_\alpha = 0$. We conclude that by $(36.2)_3$ and (36.5a) the preceding two requirements (36.5a) and (36.5b) are summed up by the condition

$$K_\alpha = \mathscr{L}_\alpha \quad \text{for} \quad P_{\alpha\beta} \equiv 0\,, \quad \text{where} \quad \mathscr{L}_\alpha = u_\alpha E^\beta J_\beta - E_\alpha u^\beta J_\beta + \overset{\perp}{\varepsilon}_\alpha{}^{\beta\gamma} J_\beta B_\gamma\,. \tag{36.6}$$

The vector $\mathscr{L}_\alpha$, to be called the proper density of Lorentz 4-force, can be given the following expression

$$\mathscr{L}_\alpha = F_{\alpha\beta} J^\beta \qquad \left(\overset{\perp}{\mathscr{L}}_\alpha = \bar{\rho}\, E_\alpha + \frac{1}{c}\overset{\perp}{\varepsilon}_\alpha{}^{\beta\gamma} j_\beta B_\gamma, \quad -cu^\alpha \mathscr{L}_\alpha = E^\alpha j_\alpha\right). \tag{36.7}$$

Indeed $(36.6)_3$ is equivalent to $(36.7)_1$ by (34.7).

Now let us assume that at a given material point, P^*, of $\mathscr{C}$ both $\mathscr{U}_{\alpha\beta}$ and $\mathscr{U}_{\alpha\beta} + \Delta\mathscr{U}_{\alpha\beta}$ are acceptable total energy tensors, where, as is natural, we understand that

$$\Delta\mathscr{U}_{\alpha\beta} = \Delta\rho\, u_\alpha u_\beta + \Delta X_{\alpha\beta} + \Delta Q_{\alpha\beta} + \Delta E_{\alpha\beta} \qquad (\Delta\rho = k\,\Delta w, \quad \Delta Q_{\alpha\beta} = u_\alpha \Delta q_\beta + \Delta q_\alpha u_\beta) \tag{36.8}$$

holds and that e.g. $X_{\alpha\beta}$ and $X_{\alpha\beta} + \Delta X_{\alpha\beta}$ are given by (generally) different constitutive equations. Then, since the stress tensor is spatial and the thermodynamic tensor is mixed and symmetric, the Einstein gravitation equations $(23.1)_1$ and convention (17.13) imply, in general relativity, $\Delta\mathscr{U}_{\alpha\beta} = 0$, i.e. by (36.1)

$$\Delta\rho = -\Delta E, \quad \Delta X_{\alpha\beta} = -\Delta\overset{\perp}{E}_{\alpha\beta}, \quad \Delta E'''_{[\alpha\beta]} = 0, \quad \Delta q_\alpha = -\Delta E'_\alpha (= -\Delta E''_\alpha)\,. \tag{36.9}$$

Conversely in general or special relativity let $\mathscr{U}_{\alpha\beta}$—cf. (36.1)—be an acceptable total energy tensor for P^*, and let (36.9) hold. Then $\mathscr{U}_{\alpha\beta}+\Delta\mathscr{U}_{\alpha\beta}$ is also an acceptable total energy tensor for P^*.

The possibility of changing $E_{\alpha\beta}$ under the condition of performing suitable changes in certain thermo-mechanical counterparts was emphasized by Gyorgyi in [1954]. Also $\Delta E'''_{\alpha\beta}$ was left arbitrary. Following Bressan [1966c] to [1967b], where Eckart's thermodynamic tensor is used, we imposed the limitation $(36.4)_2$ upon the choice of $E_{\alpha\beta}$ (also in the presence of polarization). Other such limitations will be considered [§§ 37, 40] as consequences of some physically interesting requirements, which in some cases may contribute to give a physical content to some debated question such as the one concerning Minkowski's and Abraham's determinations of $E_{\alpha\beta}$.

§ 37. On Some Widely Used Instances of the Electromagnetic Energy Tensor $E_{\alpha\beta}$ and Some Instances of $E^{\alpha\beta}{}_{/\beta}$

We now consider some particular instances of $E_{\alpha\beta}$, to be denoted by ${}^rE_{\alpha\beta}$ where $r=1,2,\ldots$. Then it is natural to denote by ${}^rK_\alpha$ and ${}^r\Pi^{(e)}$ the instances of K_α and $\Pi^{(e)}$ respectively—cf. $(36.2)_{1,2}$—for $E_{\alpha\beta}={}^rE_{\alpha\beta}$. Among the instances of $E_{\alpha\beta}$ hinted at above there is Minkowski's ${}^1E_{\alpha\beta}$, given by

$$ {}^1E_{\alpha\beta}=[F_{\alpha\beta},f_{\alpha\beta}]_{\alpha\beta}=-F_{\alpha\gamma}f^{\gamma}{}_{\beta}-\tfrac{1}{4}F_{\gamma\delta}f^{\gamma\delta}g_{\alpha\beta}\quad (\text{hence } E_\alpha{}^\alpha=0)\,. \tag{37.1} $$

We shall understand $(37.1)_2$ as a convention holding for all skewsymmetric tensors $F_{\alpha\beta}$ and $f_{\alpha\beta}$.

By convention (17.13) we generally have ${}^1E'_\alpha\neq{}^1E''_\alpha$—see $(37.3)_{3,4,5}$ below. The analogue does not occur with Abraham's tensor ${}^2E_{\alpha\beta}$, given by

$$ {}^2E_{\alpha\beta}={}^1E_{\alpha\beta}+({}^1E''_\alpha-{}^1E'_\alpha)u_\beta={}^1Eu_\alpha u_\beta+2u_{[\alpha}{}^1E''_{\beta]}+{}^1\overset{\perp}{E}_{\alpha\beta}\quad (\text{hence } {}^2E_\alpha{}^\alpha=0). \tag{37.2} $$

Thus, unlike ${}^1E_{\alpha\beta}$, the tensor ${}^2E_{\alpha\beta}$ meets requirement $(36.4)_2$; hence it can be used in our theory of general relativity where Eckart's thermodynamic tensor is included.

From convention (17.13), (34.7), (34.10), and (37.1) we deduce—see (a) to (e) below—

$$ \begin{aligned} &{}^1E=\tfrac{1}{2}(E_\gamma D^\gamma+H_\gamma B^\gamma)\,,\qquad {}^1\overset{\perp}{E}_{\alpha\beta}={}^1E\overset{\perp}{g}_{\alpha\beta}-E_\alpha D_\beta-H_\alpha B_\beta\,,\\ &{}^1E'_\alpha=\overset{\perp}{\varepsilon}_{\alpha\rho\sigma}D^\rho B^\sigma\,,\qquad {}^1E''_\alpha=\mathscr{P}_\alpha=\overset{\perp}{\varepsilon}_{\alpha\rho\sigma}E^\rho H^\sigma\,,\qquad {}^1E'_\alpha-{}^1E''_\alpha=\overset{\perp}{\varepsilon}_\alpha{}^{\rho\sigma}(E_\rho M_\sigma+P_\rho B_\sigma)\,, \end{aligned} \tag{37.3} $$

so that we can make both ${}^1E_{\alpha\beta}$ and ${}^2E_{\alpha\beta}$—cf. (37.2)—explicit and we see that by $(37.3)_5$ $c^2\mathscr{P}_\alpha$ is Poynting's vector, the physical meaning of which will appear from (41.1) or (41.2) below.

The expression (35.5) of $F_{\alpha\beta}=(E_\alpha, B_\alpha)_{\alpha\beta}$ and its analogue for $f_{\alpha\beta}=(D_\alpha, H_\alpha)_{\alpha\beta}$ imply, for $g_{\alpha\beta}=\delta'_{\alpha\beta}$ and $u^\alpha=\delta_0{}^\alpha$,

$$F_{\gamma\delta}f^{\gamma\delta}=F_{rs}f^{rs}-2F_{r0}f^r{}_0=2(B_rH^r-E_rD^r), \tag{a}$$

$$F_{1\gamma}f^\gamma{}_1=F_{10}f_{10}-F_{1r}f_1{}^r=E_1D_1-B_2H_2-B_3H_3, \tag{b}$$

$$F_{1\gamma}f^\gamma{}_2=F_{10}f_{20}-f_2{}^rF_{1r}=E_1D_2+H_1B_2, \tag{c}$$

$$F_{0\gamma}f^\gamma{}_0=-F_{r0}f^r{}_0=-E_rD^r. \tag{d}$$

From (37.1) we deduce $(37.3)_1$ by (a) and (d) ($g_{00}=-1$), and $(37.3)_2$ by (b), (c), and $(37.3)_1$.

From (34.7) and definition $(37.3)_5$, and from (34.10) respectively we deduce

$$F_{0\gamma}f^\gamma{}_1=F_{r0}f_1{}^r=E_r\overset{\perp}{\varepsilon}_1{}^{rs}H_s=-u_0\mathscr{P}_1, \tag{d}$$

$$F_{1\gamma}f^\gamma{}_0=F_{1r}f^r{}_0=\overset{\perp}{\varepsilon}_{1r}{}^sB_sD_r, \tag{e}$$

which on the basis of (37.1) yield $(37.3)_{3,4}$.

As is well known (37.1) implies

$$-{}^1E_{\alpha\beta}{}^{/\beta}=F_{\alpha\gamma}J^\gamma+\tfrac{1}{4}(F_{\gamma\delta}f^{\gamma\delta}{}_{/\alpha}-F_{\gamma\delta/\alpha}f^{\gamma\delta}) \tag{37.4}$$

which by $(36.7)_1$ and $(34.8)_{2,3}$ becomes—cf. $(36.6)_3$ and $(36.2)_1$

$${}^1K_\alpha=-{}^1E_{\alpha\beta}{}^{/\beta}=\mathscr{L}_\alpha+\tfrac{1}{4}(F_{\beta\gamma}P^{\beta\gamma}{}_{/\alpha}-P^{\beta\gamma}F_{\beta\gamma/\alpha}). \tag{37.5}$$

For the ease of the reader we prove (37.4). From (37.1) and $(35.1)_2$ we easily deduce

$$-{}^1E_{\alpha\beta}{}^{/\beta}=F_{\alpha\gamma}J^\gamma+F_{\alpha\gamma/\beta}f^{\gamma\beta}+\tfrac{1}{4}(F_{\gamma\delta}f^{\gamma\delta}{}_{/\alpha}+F_{\gamma\delta/\alpha}f^{\gamma\delta}). \tag{f}$$

Since $F_{\alpha\beta}$ and $f_{\alpha\beta}$ are skewsymmetric, we have $F_{\alpha\gamma/\beta}f^{\gamma\beta}=f^{\beta\gamma}F_{\gamma\alpha/\beta}$. By this and $(35.1)_1$

$$2F_{\alpha\gamma/\beta}f^{\gamma\beta}=f^{\beta\gamma}(F_{\alpha\beta/\gamma}+F_{\gamma\alpha/\beta})=-f^{\beta\gamma}F_{\beta\gamma/\alpha}=-f^{\gamma\delta}F_{\gamma\delta/\alpha}. \tag{g}$$

Lastly (f) and (g) imply (37.4).

Now one could directly read, in § 40, the deduction of the expression (40.9) for ${}^1\Pi^{(e)}$ and the subsequent reasonings in which the electromagnetic tensors ${}^5E_{\alpha\beta}$ to ${}^4E_{\alpha\beta}$ are introduced and are shown to have the basic properties (40.21). However we now consider the instance ${}^3E_{\alpha\beta}$ of $E_{\alpha\beta}$ which is very simple and much used, in order to show the interesting expression (37.8) below for the ponderomotive 4-force connected with it, and for some other marginal remarks [§ 4]. Furthermore we shall consider Dällenbach's tensor ${}^4E_{\alpha\beta}$—cf. (37.12) below—to present an instance of $E_{\alpha\beta}$ which is not simple according to Definition 39.1 below.

By $(34.7)_1$ and $(34.8)_2$ $f_{\alpha\beta}$ becomes $F_{\alpha\beta}$ by replacing—in its definition $(34.8)_2$—D_α with E_α and H_α with B_α. By $(37.1)_1$ the same replacements turn ${}^1E_{\alpha\beta}$ into ${}^3E_{\alpha\beta}=[F_{\alpha\beta},F_{\alpha\beta}]_{\alpha\beta}$. Thus from (37.3)—or by taking (37.1) into account and using locally proper and natural co-ordinates—we deduce

$$ {}^3E_{\alpha\beta}=[F_{\alpha\beta},F_{\alpha\beta}]_{\alpha\beta}={}^3E(u_\alpha u_\beta+\overset{\perp}{g}_{\alpha\beta})+2\,{}^3E'_{(\alpha}u_{\beta)}-E_\alpha E_\beta-B_\alpha B_\beta \tag{37.6} $$

where

$$ {}^3E=\tfrac{1}{2}(E_\gamma E^\gamma+B_\gamma B^\gamma)\,,\qquad {}^3E'_\alpha=\overset{\perp}{\varepsilon}_\alpha{}^{\beta\gamma}E_\beta B_\gamma={}^3E''_\alpha\,. \tag{37.7} $$

The replacements $f_{\alpha\beta}\rightarrowtail F_{\alpha\beta}$ and $J_\alpha\rightarrowtail J''_\alpha$ turn the Maxwell equation $(35.1)_2$ into $(35.2)_1$, the tensor ${}^1E_{\alpha\beta}$ into ${}^3E_{\alpha\beta}$—cf. $(37.6)_1$, $(37.1)_1$—and consequently equality (37.4) into $-{}^3E_{\alpha\beta}{}^{/\beta}=F_\alpha{}^\gamma J''_\gamma$.

Then the last equality is implied by equations $(35.1)_1$ and $(35.2)_1$ as (37.4) is implied by $(35.1)_{1,2}$. From the same equality and (34.7) we deduce

$$ {}^3K_\alpha=-{}^3E_{\alpha\beta}{}^{/\beta}=F_\alpha{}^\beta J''_\beta=u_\alpha E^\beta J''_\beta+(-E_\alpha u^\beta J''_\beta+\overset{\perp}{\varepsilon}_\alpha{}^{\beta\gamma}J''_\beta B_\gamma)\,. \tag{37.8} $$

Thus the ponderomotive 4-force connected with ${}^3E_{\alpha\beta}$ is the analogue for the (proper density of) total 4-current J''_α, of the Lorentz force $\mathscr{L}_\alpha$ due to the true 4-current J_α. In particular from $(36.2)_{1,2}$ and the analogue of (34.3) for J''_α we deduce the analogues for ${}^3K_\alpha$ of $(36.7)_{2,3}$:

$$ {}^3\overset{\perp}{K}_\alpha=\bar{\rho}''E_\alpha+\frac{1}{c}\overset{\perp}{\varepsilon}_\alpha{}^{\beta\gamma}j_\beta B_\gamma\,,\qquad {}^3\Pi^{(e)}=E^\alpha j''_\alpha\,. \tag{37.9} $$

Let us further remark that by (37.1), (37.2), (37.3), (37.6), and (37.7) we have

$$ {}^1E_{\alpha\beta}={}^2E_{\alpha\beta}={}^3E_{\alpha\beta}\quad\text{for}\quad P_{\alpha\beta}=0\,,\quad\text{i.e.}\quad P_\alpha=M_\alpha=0\,. \tag{37.10} $$

Furthermore the comparison of (37.5) with (36.6) shows that ${}^1E_{\alpha\beta}$ meets the requirements on $E_{\alpha\beta}$ considered in § 36 in the case of identically vanishing polarizations and summed up just be (36.6). Thus the same holds for ${}^2E_{\alpha\beta}$ and ${}^3E_{\alpha\beta}$, and the afore-mentioned requirements are equivalent to the condition

$$ E_{\alpha\beta}{}^{/\beta}={}^3E_{\alpha\beta}{}^{/\beta}\,,\quad\text{i.e.}\quad K_\alpha={}^3K_\alpha,\quad\text{for}\quad P_{\alpha\beta}\equiv 0\,, \tag{37.11} $$

where we prefer to use ${}^3E_{\alpha\beta}$ rather than ${}^1E_{\alpha\beta}$ or ${}^2E_{\alpha\beta}$ because ${}^3E_{\alpha\beta}$ can be defined using only one of the tensors $F_{\alpha\beta}$ and $f_{\alpha\beta}$.

For a forthcoming marginal remark we remember that Dällenbach [1919] proposed the following instance of $E_{\alpha\beta}$:

$$ {}^4E_{\alpha\beta}=-F_{\alpha\gamma}f^\gamma{}_\beta-\tfrac{1}{2}\left(\int_{F_{\gamma\delta}{}^{(0)}}^{F_{\gamma\delta}} f'_{\gamma\delta}DF'^{\gamma\delta}+\tfrac{1}{2}f_{\gamma\delta}{}^{(0)}F^{(0)\gamma\delta}\right)g_{\alpha\beta} \tag{37.12} $$

where the integral is extended to an arc of W_{P*}, of ends $\mathscr{E}_0$ and $\mathscr{E}$, where $F_{\alpha\beta}{}^{(0)}$ and $f_{\alpha\beta}{}^{(0)}$ are the values of $F_{\alpha\beta}$ and $f_{\alpha\beta}$ respectively at $\mathscr{E}_0$, and where $F'_{\alpha\beta}$ and $f'_{\alpha\beta}$ are the values of $F_{\alpha\beta}$ and $f_{\alpha\beta}$ respectively at the typical point of the arc.

The case when $f_{\alpha\beta}$ is a linear homogeneous function of $F_{\rho\sigma}$ is important. This occurs e.g. in case (34.6) holds with $\eta_{\alpha\beta}$ and $\mu_{\alpha\beta}$ constant, so that (34.11) holds with $\eta_{\alpha\beta}{}^{\rho\sigma}$ constant. Hence the afore-mentioned linear case obviously occurs for $P_\alpha \equiv M_\alpha \equiv 0$ in that then $\eta_{\alpha\beta} \equiv \mu_{\alpha\beta} \equiv \overset{\perp}{g}_{\alpha\beta}$ holds.

On the basis of (37.9) and (37.1) it is easy to realize that in the linear (and homogeneous) case being considered we have ${}^4E_{\alpha\beta} = {}^1E_{\alpha\beta}$. This and (37.8) imply that, in particular, for $P_\alpha \equiv M_\alpha \equiv 0$ the tensor ${}^4E_{\alpha\beta}$ also meets the requirement on $E_{\alpha\beta}$ considered in § 36 in the absence of polarizations and summed up by condition (36.6) or (37.11).

Incidentally, to explain the physical meaning of ${}^4E$ in the general case, let us remark that by convention $(17.13)_1$ ${}^4E = {}^4E_{\alpha\beta} u^\alpha u^\beta$. Then by (37.12) we have—cf. (a) and (d) in the beginning of § 37

$$ {}^4E = E_\gamma D^\gamma - \int_{E_\gamma{}^{(0)}}^{E_\gamma} D'_\gamma D E'^\gamma + \int_{H_\gamma{}^{(0)}}^{H_\gamma} H'_\gamma D B'^\gamma - \tfrac{1}{2}(D_\gamma{}^{(0)} E^{(0)\gamma} - B_\gamma{}^{(0)} H^{(0)\gamma}) ; $$

hence $D\,{}^4E/Ds = E_\gamma D D^\gamma/Ds + H_\gamma D B^\gamma/Ds$.

§ 38. On Isotropic Functions and Tensors

This section serves as a preliminary for the next section and others where polarizable materials are dealt with.

Every tensor $\rho_{\alpha\beta}$ of rank 2 determines a linear vector transformation of equation $w_\alpha = \rho_{\alpha\beta} v^\beta$. Of course this vector transformation—as well as the tensor $\rho_{\alpha\beta}$—is called a *spatial rotation* if we have

$$ \rho_{\alpha\gamma} \rho^\gamma{}_\beta = g_{\alpha\beta} , \quad u^\alpha = \rho^\alpha{}_\beta u^\beta , \quad \det \|\rho_{\alpha\beta}\| < 0 \tag{38.1} $$

which in locally proper and pseudo-Euclidean co-ordinates—cf. $(17.7)_{1,3}$—is equivalent to

$$ \rho_{rh} \rho^h{}_s = \delta_{rs} , \quad \rho_{00} = -1 , \quad \rho_{0r} = \rho_{r0} = 0 , \quad \det \|\rho_{rs}\| = 1 . \tag{38.2} $$

Obviously the tensor $g_{\alpha\beta}$ is a spatial rotation; furthermore, if $\rho_{\alpha\beta}$ and $\sigma_{\alpha\beta}$ are such, then the same holds for $\rho_\alpha{}^\gamma \sigma_{\gamma\beta}$ and $\rho'_{\alpha\beta}$ with $\rho'_{\alpha\beta} = \rho_{\beta\alpha}$. By (38.1) $\rho_{\alpha\beta}$ and $\rho'_{\alpha\beta}$ determine mutually inverse transformations.

By $(38.1)_2$ the spatial rotation $w_\alpha = \rho_\alpha{}^\beta v_\beta$ leaves the temporal part $-u_\alpha u_\beta v^\beta$ of every vector v_β unchanged and rotates its spatial part $\overset{\perp}{g}_{\alpha\beta} v^\beta$. Similar properties hold for the *transform* $\rho^{\alpha_1}{}_{\beta_1} \dots \rho^{\alpha_n}{}_{\beta_n} T^{\beta_1 \dots \beta_n}$ of every tensor $T^{\beta_1 \dots \beta_n}$ of rank n by the same linear vector transformation.

If the tensor $T^{\beta_1 \dots \beta_n}$ is invariant for spatial rotations (relative to the 4-velocity u^α) we call it *spatially isotropic.*

Incidentally, if $T^{\beta_1 \dots \beta_n}$ is so for every choice of the time-like unit vector u^α (any such vector can be assumed as 4-velocity outside matter), it is called *isotropic.*

Obviously *spatially isotropic tensors [isotropic tensors] are the tensors whose components are invariant for transformations between locally proper and pseudo-Euclidean [locally pseudo-Euclidean] frames.*

Let us remark that $\overset{\perp}{g}_{\alpha\beta}$ and $\overset{\perp}{\varepsilon}_{\alpha\beta\gamma}$ are spatially isotropic and $g_{\alpha\beta}$ and $\varepsilon_{\alpha\beta\gamma\delta}$ are isotropic tensors; the product of two isotropic tensors is isotropic and the result of contracting indices in an isotropic tensor is an isotropic tensor; the analogue holds for spatially isotropic tensors.

In a 3-dimensional cartesian frame of a Euclidean space the most general isotropic tensor T_{rs} of rank 2 (invariant for the rotations $w^r = \rho^r{}_s v^s$) has the form $H\,\delta_{rs}$. Thence it is easy to see that *in relativity the most general spatially isotropic tensor $T_{\alpha\beta}$ and isotropic tensor $T'_{\alpha\beta}$ of rank 2 have the respective forms*

$$T_{\alpha\beta} = \mathscr{H}\,\overset{\perp}{g}_{\alpha\beta} + \mathscr{K}\,u_\alpha u_\beta\,, \qquad T'_{\alpha\beta} = \mathscr{H}\,g_{\alpha\beta} \quad (\mathscr{H}, \mathscr{K} \text{ constants}). \tag{38.3}$$

Definition 38.1. *Let us consider a function $T_{\dots} = f_{\dots}(\mathscr{U}_{\dots}, \mathscr{V}_{\dots}, \dots)$ whose arguments and values are tensors of any non-negative rank attached to the event point $\mathscr{E}$ (in S_4). We say that this function is spatially isotropic if (understood as an operation) it commutes with the operation of transforming tensors by means of an arbitrary spatial rotation $w_\alpha = \rho_\alpha{}^\beta v_\beta$*—cf. (38.1).

Incidentally, if the function $T_{\dots} = f_{\dots}(\mathscr{U}_{\dots}, \mathscr{V}_{\dots}, \dots)$ is spatially isotropic in case an arbitrary time-like unit vector is assumed as local 4-velocity, this function is called *isotropic.*

For instance $T_\alpha = f_\alpha(U_{\gamma_1 \dots \gamma_n})$ is a spatially isotropic function of $\mathscr{U}_{\gamma_1 \dots \gamma_n}$ if for every $\rho_\alpha{}^\beta$ fulfilling (38.1) we have

$$f_\alpha(\mathscr{U}'_{\gamma_1 \dots \gamma_n}) = \rho_\alpha{}^\beta f_\beta(\mathscr{U}_{\gamma_1 \dots \gamma_n}) \quad \text{where} \quad \mathscr{U}'_{\gamma_1 \dots \gamma_n} = \rho_{\gamma_1}{}^{\delta_1} \dots \rho_{\gamma_n}{}^{\delta_n} \mathscr{U}_{\delta_1 \dots \delta_n}\,. \tag{38.4}$$

By Definition 38.1 *the domain of a spatially isotropic function is closed under spatial rotations.*

E.g. if $\mathscr{U}_{\gamma_1 \dots \gamma_n}$ belongs to the domain of the isotropic function $f_\alpha(\mathscr{U}_{\gamma_1 \dots \gamma_n})$, then for every $\rho_{\alpha\beta}$ fulfilling (38.1), $\mathscr{U}'_{\gamma_1 \dots \gamma_n}$ also belongs to this domain, under condition $(38.4)_2$.

Theorem 38.1. *A function whose arguments and value are tensors of non-negative rank attached to an event point $\mathscr{E}$, is form invariant (not only invariant in the sense of tensor calculus) for transformations between locally natural and proper [between locally natural] co-ordinate systems if, and only if it is spatially isotropic* [*is isotropic—see below Definition* 38.1].

This obvious theorem was substantially asserted in § 27 in the special but important case in which the above arguments and value are spatial tensors.

Theorem 38.2. *By convention* $(34.7)_2$ $(E_\alpha, B_\alpha)_{\alpha\beta}$ *is a spatially isotropic function (of the not necessarily spatial) vectors* E_α *and* B_α, *i.e.*

$$\rho_\alpha{}^{\alpha'}\rho_\beta{}^{\beta'}(E_{\alpha'}, B_{\alpha'})_{\alpha'\beta'} = (\rho_\alpha{}^{\alpha'}E_{\alpha'}, \rho_\alpha{}^{\alpha'}B_{\alpha'})_{\alpha\beta} \tag{38.5}$$

for every solution $\rho_{\alpha\beta}$ *of* (38.1).

Proof. We locally assume $g_{\alpha\beta} = \delta'_{\alpha\beta}$ and $u^\alpha = \delta^\alpha{}_0$, so that $(38.2)_{1,2,5}$ hold. Then it is easy to realize the equality

$$\rho_\alpha{}^{\alpha'}\rho_\beta{}^{\beta'}\overset{\perp}{\varepsilon}_{\alpha'\beta'}{}^{\gamma} = \overset{\perp}{\varepsilon}_{\alpha\beta}{}^{\gamma'}\rho_{\gamma'}{}^{\gamma} \tag{38.6}$$

So by $(34.7)_2$ and $(38.1)_2$ we have

$$\rho_\alpha{}^{\alpha'}\rho_\beta{}^{\beta'}(E_{\alpha'}, B_{\alpha'})_{\alpha'\beta'} = u_\alpha \rho_\beta{}^{\beta'}E_{\beta'} - \rho_\alpha{}^{\alpha'}E_{\alpha'}u_\beta + \overset{\perp}{\varepsilon}_{\alpha\beta}{}^{\gamma'}\rho_{\gamma'}{}^{\gamma}B_\gamma, \tag{38.7}$$

hence just (38.5). q.e.d.

The following theorem immediately follows from Definition 38.1.

Theorem 38.3. *If the tensor* $T_{\dots}^{\dots}$ *is a spatially isotropic function of other tensors which in turn are spatially isotropic functions of the tensors* $\mathscr{U}^{(1)\dots}_{\dots}, \dots, \mathscr{U}^{(n)\dots}_{\dots}$, *then* $T_{\dots}^{\dots}$ *is a spatially isotropic function of* $\mathscr{U}^{(1)}, \dots, \mathscr{U}^{(n)}$.

Incidentally, the analogue holds for isotropic tensors.

The next theorem is a straightforward generalization of a useful theorem—due to Noll [1955, § 16]—in the classical theory of materials.

Theorem 38.4. *We assume that the tensor* $T_{\dots}$ *of rank* r_0 *is a function* $f_{\dots}(\dots)$ *defined on a set S of n-tuples of tensors of given ranks* $r_1, \dots, r_n$, *and that* $f_{\dots}(\dots)$ *is continuous in S together with its first partial derivatives—which are tensors of ranks* $r_0 + r_1, \dots, r_0 + r_n$—*cf.* Appendix A, § A 3. *Furthermore let S be closed under spatial rotations. Then the function* $f_{\dots}(\dots)$ *is spatially isotropic [is isotropic] if and only if its first partial derivatives are spatially isotropic [are isotropic].*

For the ease of the reader we prove the part of Theorem 38.4 which concerns spatial isotropy—which part obviously implies the other—in the case $T_{\dots}^{\dots} = T_\alpha = f_\alpha(\mathscr{U}_{\delta_1\dots\delta_n})$; the extension of this proof to the general case is obvious.

We assume that $f_\alpha(\dots)$ is spatially isotropic, so that (38.4) holds for every $\mathscr{U}_{\delta_1\dots\delta_n}$ in the domain S of f_α and for every $\rho_\alpha{}^\beta$ fulfilling (38.1). By $(38.4)_2$ from $(38.4)_1$ we deduce

$$\rho_\alpha{}^\beta \frac{\partial f_\beta(\mathscr{U}_{\delta_1\dots\delta_n})}{\partial \mathscr{U}_{\delta_1\dots\delta_n}} = \frac{\partial}{\partial \mathscr{U}_{\delta_1\dots\delta_n}} f_\alpha(\rho_{\gamma_1}{}^{\delta_1}\dots\rho_{\gamma_n}{}^{\delta_n}\mathscr{U}_{\delta_1\dots\delta_n}) = \rho_{\gamma_1}{}^{\delta_1}\dots\rho_{\gamma_n}{}^{\delta_n}\frac{\partial f_\alpha(\mathscr{U}'_{\gamma_1\dots\gamma_n})}{\partial \mathscr{U}'_{\gamma_1\dots\gamma_n}} \tag{a}$$

where $\partial f_\alpha / \partial \mathscr{U}'_{\gamma_1\dots\gamma_n}$ means $\partial f_\alpha / \partial \mathscr{U}_{\delta_1\dots\delta_n}$ evaluated at $\mathscr{U}'_{\gamma_1\dots\gamma_n}$. By $(38.1)_1$ (a) is equivalent to

$$\frac{\partial f_\alpha(\mathscr{U}'_{\gamma_1\dots\gamma_n})}{\partial \mathscr{U}'_{\gamma_1\dots\gamma_n}} = \rho_\alpha{}^\beta \rho_{\delta_1}{}^{\gamma_1}\dots\rho_{\delta_n}{}^{\gamma_n}\frac{\partial f_\beta(\mathscr{U}_{\delta_1\dots\delta_n})}{\partial \mathscr{U}_{\delta_1\dots\delta_n}}. \tag{b}$$

Hence the tensor $\partial f_\beta / \partial \mathscr{U}_{\delta_1 \ldots \delta_n}$ is a spatially isotropic function of $\mathscr{U}_{\delta_1 \ldots \delta_n}$.

Now we conversely assume this spatial isotropy, i.e. that (b) holds (in S) for every $\rho_\alpha{}^\beta$ fulfilling (38.1). Then by $(38.1)_1$, $(a)_1$ holds, so that the difference

$$\Delta_\alpha = \rho_\alpha{}^\beta f_\beta(\mathscr{U}_{\delta_1 \ldots \delta_n}) - f_\alpha(\rho_{\gamma_1}{}^{\delta_1} \ldots \rho_{\gamma_n}{}^{\delta_n} \mathscr{U}_{\delta_1 \ldots \delta_n}) \tag{c}$$

is independent of $\mathscr{U}_{\delta_1 \ldots \delta_n}$. Then, setting

$$\mathscr{U}^{(0)}_{\delta_1 \ldots \delta_n} = \mathscr{U}_{\delta_1 \ldots \delta_n}, \qquad \mathscr{U}^{(p)}_{\delta_1 \ldots \delta_n} = \rho_{\delta_1}{}^{\gamma_1} \ldots \rho_{\delta_n}{}^{\gamma_n} \mathscr{U}^{(p-1)}_{\gamma_1 \ldots \gamma_n} \; (p = 1, 2, \ldots) \tag{d}$$

and

$$\rho_{\alpha\beta}{}^{(0)} = \rho_{\alpha\beta}, \qquad \rho_{\alpha\beta}{}^{(p)} = \rho_\alpha{}^\gamma \rho_{\gamma\beta}{}^{(p-1)} \; (p = 1, 2, \ldots), \tag{e}$$

we have

$$\rho_\alpha{}^{(p)\beta} f_\beta(\mathscr{U}^{(p)}_{\delta_1 \ldots \delta_n}) = f_\alpha(\mathscr{U}_{\delta_1 \ldots \delta_n}) + p\Delta_\alpha. \tag{f}$$

By (d) $\mathscr{U}^{(p)}_{\delta_1 \ldots \delta_n}$ $(p = 1, 2, \ldots)$ belongs to the set σ' described by the tensor $\mathscr{U}'_{\gamma_1 \ldots \gamma_n}$—cf. $(38.4)_2$—when $\rho_\gamma{}^\delta$ describes the set of spatially isotropic tensors—cf. (38.1). Then σ' is closed and bounded. Furthermore, by the assumed continuity of the function $f_\alpha(\mathscr{U}'_{\delta_1 \ldots \delta_n})$ this function is bounded in σ'. Then the set of the tensors $\rho_\alpha{}^{(q)\beta} f_\beta(\mathscr{U}^{(p)}_{\beta_1 \ldots \beta_n})$ $(p, q = 1, 2, \ldots)$ and in particular the set described by the left hand side of (f) for $p = 1, 2, \ldots$ are also bounded. Then by (f) we must have $\Delta_\alpha = 0$. This occurs for every $\rho_{\alpha\beta}$ fulfilling (38.1), so that by (c) $f_\beta(\mathscr{U}_{\delta_1 \ldots \delta_n})$ is a spatially isotropic function. q.e.d.

§ 39. Some Uniqueness Properties of the Electromagnetic Energy Tensor $E_{\alpha\beta}$. On its Arbitrariness in Connection with Heat Conduction

The purpose of this section—which is not essential for understanding the subsequent sections—is to discuss to what extent $E_{\alpha\beta}$ is arbitrary and in particular to show that from the physical point of view the mixed part $E'''_{\alpha\beta}$ of $E_{\alpha\beta}$ is determined (in a suitable sense).

For $r = 1$ to 3 and $E_{\alpha\beta} = {}^r E_{\alpha\beta}$—cf. (37.1) to (37.2) and (37.6)—it is easy to see that $E_{\alpha\beta}$ fulfills the following reasonable conditions (a) to (d):

(a) *The electromagnetic energy tensor $E_{\alpha\beta}$ is a universal invariant function*

$$E_{\alpha\beta} = \phi_{\alpha\beta}(F_{\rho\sigma}, f_{\rho\sigma}, g_{\rho\sigma}, u_\rho) = \bar{\phi}_{\alpha\beta}(E_\rho, H_\rho, D_\rho, B_\rho, g_{\rho\sigma}, u_\rho) \tag{39.1}$$

of the arguments indicated, a function having continuous first and second partial derivatives;[2)]

(b) *We have* $E_{\alpha\beta}=0$ *for* $F_{\alpha\beta}=f_{\alpha\beta}=0$.

(c) *At least in the empty part* $\mathbf{R}_4^{(0)}$ *of space time—where* $E_{\alpha\beta}$ *cannot depend on* u_α*—we have* $E_{[\alpha\beta]}=0$.

(d) *For every non-polarizable matter element* $d\mathscr{C}$ *and every process physically possible for* $d\mathscr{C}$ *(locally rigid and with vanishing rotation and intrinsic acceleration, i.e. with* $u_{\alpha/\beta}=0$*) we have* $E_{\alpha\beta}{}^{/\beta}={}^3E_{\alpha\beta}{}^{/\beta}$—cf. (37.6).

Condition (d) is implied by the requirements on $E_{\alpha\beta}$ considered in § 36 and summed up by condition (36.6) or (37.11).

Condition (a) implies that $\phi_{\alpha\beta}$ and $\bar{\phi}_{\alpha\beta}$ are form invariant for transformations between locally natural and proper frames—cf. footnote 2 in Chapter 4. Thence we deduce by Theorem 38.1 the following condition:

(a′) ϕ *and* $\bar{\phi}$—cf. (39.1)—*are spatially isotropic functions of* $F_{\rho\sigma}$ *and* $f_{\rho\sigma}$, *and of* E_ρ, H_ρ, D_ρ *and* B_ρ *respectively—cf. Definition 38.1.*

Let us show that *the conditions that* ϕ *and* $\bar{\phi}$ *should be spatially isotropic are equivalent.* Indeed let $\phi_{\alpha\beta}$ be spatially isotropic. By Theorem 38.2 $F_{\rho\sigma}$ and $f_{\rho\sigma}$—cf. (34.7), $(34.8)_2$—are spatially isotropic functions of E_ρ, H_ρ, D_ρ, and B_ρ. Hence by (39.1) and Theorem 38.3. $E_{\alpha\beta}$ also is such a function. In other words $\bar{\phi}$—cf. (39.1)—is spatially isotropic.

Now let $\phi_{\alpha\beta}$ be a spatially isotropic function. Since $F_{\alpha\beta}$ is such a function of E_α and B_α and moreover $F_{\alpha\beta}$ determines E_α and B_α—cf. (34.9)—E_α and B_α are spatially isotropic functions of $F_{\alpha\beta}$, as is easy to see. Hence such is $\phi_{\alpha\beta}$. q.e.d.

Dällenbach's tensor ${}^4E_{\alpha\beta}$—cf. (37.12)—meets conditions (b) to (d) but not condition (a), e.g. in the case of magnetic hysteresis. This tensor has a more complex expression than ${}^rE_{\alpha\beta}$ $(r=1,2,3)$. On the other hand every material—even those with memory—can be dealt with by means of ${}^rE_{\alpha\beta}$ $(r=1,2,3)$. Then a criterion of simplity induces us to pay attention especially to the instances of $E_{\alpha\beta}$ for which conditions (a) to (d)—fulfilled by ${}^rE_{\alpha\beta}$ $(r=1,2,3)$—hold. This is useful, among other things, in that (i) the same conditions give rise to certain uniqueness theorems to be considered shortly and (ii) for proving these theorems e.g. condition (a)—not met by ${}^4E_{\alpha\beta}$—is essential.

Conditions (a) to (d) imply—cf. Appendix *B*—that

(e) *We have* $E_{\alpha\beta}={}^3E_{\alpha\beta}$—cf. (37.6)—*for* $P_\alpha=M_\alpha=0$.[3)]

By (37.6) from (a) and (e) we immediately derive (b) to (d). Therefore the following definition can be stated.

Definition 39.1. *We call any tensor* $E_{\alpha\beta}$ *expressed in terms of* $F_{\alpha\beta}, f_{\alpha\beta}, g_{\alpha\beta}$, *and* u^α *and fulfilling the conditions* (a) *to* (d)*—which are equivalent to* (a) *and* (e)*— admissible and simple electromagnetic tensor.*

Incidentally Bressan [1967b] proved the following theorem—see Appendix B, Theorem B 2.

Theorem 39.1. *Let* $\Pi^{(e)}$ *be assigned as a function of* $F_{\alpha\beta}, f_{\alpha\beta}, F_{\alpha\beta/\gamma}$ *and* $f_{\alpha\beta/\gamma}$ *—so that by* $(35.1)_2$ *and* (35.2) $\Pi^{(e)}$ *may depend on* J_α *and* J'_α. *Then there is at most one admissible and simple electromagnetic energy tensor* $E_{\alpha\beta}$ *for which* $(36.2)_2$, *i.e.* $cu_\alpha E^{\alpha\beta}{}_{/\beta}=\Pi^{(e)}$, *holds for every matter element* $d\mathscr{C}$ *and every process* $\mathscr{P}$ *physically possible for* $d\mathscr{C}$.

The tensor $E_{\alpha\beta}$ mentioned in Theorem 39.1 is considered to be simple. This condition is essential for the validity of Theorem 39.1—cf. Bressan [1965, § 8]. Indeed, let us consider special relativity for the sake of simplicity. Then the general solution of the equation $T^{\alpha\beta}{}_{/\beta}=0$ is

$$T^{\alpha\beta}=\varepsilon^{\alpha\lambda\mu\gamma}\varepsilon^{\beta\rho\sigma\delta}\chi_{\lambda\mu\rho\sigma/\gamma\delta} \tag{*}$$

where χ is an arbitrary function of x^ρ. Since $E_{\xi\eta}$ and $f_{\xi\eta}$ are functions of x^ρ, we can set

$$\chi_{\lambda\mu\rho\sigma}=h_{\lambda\mu\rho\sigma}(F_{\xi\eta},f_{\xi\eta}), \qquad h_{\lambda\mu\rho\sigma}(F_{\xi\mu},F_{\xi\eta})\equiv 0, \tag{**}$$

where $h_{\lambda\mu\rho\sigma}$ is an arbitrary spatially isotropic function that vanishes for $F_{\xi\eta}=f_{\xi\eta}$. Then, generally, $T^{\alpha\beta}\not\equiv 0$; furthermore, for every admissible and simple electromagnetic energy tensor $E_{\alpha\beta}, E_{\alpha\beta}+T_{\alpha\beta}$ is not simple, but it is admissible in that it fulfills conditions (b)—(d).

Now we want to show that within the theory of general relativity based on the Einstein equations (23.1), on the expression (36.1) of the total energy tensor $\mathcal{U}_{\alpha\beta}$, and on either of the relativistic versions (25.8b) and (25.2) of Fourier's (heat conduction) law, the mixed part $E'''_{\alpha\beta}$ of every admissible and simple electromagnetic energy tensor $E_{\alpha\beta}$ is completely determined under condition (a).

To the above end we assume that along the world line W_{P*} of P^* we have

$$T_{/\alpha}=0, \qquad A_\alpha=0 \quad (\text{on } W_{P*}) \tag{39.2}$$

which is physically possible. Then according to either of the heat conduction hypotheses (25.8b) and (25.2) we have $q^\alpha=0$ on W_{P*}, hence also $Q_{\alpha\beta}=0$—cf. $(24.3)_2$.

In our theory, where the version (25.2) of the Fourier law is preferred, (and in many other relativistic theories including heat conduction and electromagnetism) the vanishing of q^α (hence of $Q_{\alpha\beta}$) is determined by the values of $T_{/\alpha}$ and A_α. Furthermore, from the physical point of view the variables $E_\alpha, P_\alpha, H_\alpha$, and M_α are completely independent of $T_{/\alpha}$ and A_α. Hence every value possible for $E'''_{\alpha\beta}$—cf. (39.1)—is compatible with (39.2) so that $E'''_{\alpha\beta}$ can take this value at an event point, $\mathscr{E}$, on W_{P*}.

Now we consider, besides $\mathcal{U}_{\alpha\beta}$—cf. (36.1)—any acceptable total energy tensor $\mathcal{U}_{\alpha\beta}+\Delta\mathcal{U}_{\alpha\beta}$—cf. (36.8). Then, as we know, (36.9) holds in general relativity. Since (39.2) implies $q^\alpha=0$, we have $\Delta q^\alpha=0$ on W_{P*}. This, $(36.9)_{4,5}$, and $(17.13)_4$ imply that $(36.9)_3$ can be strengthened into

$$\Delta E'''_{\alpha\beta}=0\,. \tag{39.3}$$

Untill otherwise noted we shall understand that $E_{\alpha\beta}$ is admissible and simple [Definition 39.1]—which is practically assumed, e.g. in discussions concerning Minkowski's and Abraham's tensors. Then the result above means that *if for a given choice of $E'''_{\alpha\beta}$ our relativistic theory is true—so that $E'''_{[\alpha\beta]}=0$ must hold according to $(36.4)_2$—, then no other choice of $E'''_{\alpha\beta}$ is acceptable.* Authors seem to agree that if a symmetric choice of $E'''_{\alpha\beta}$ is possible, then it must be Abraham's

${}^2E'''_{\alpha\beta}$, i.e.—cf. (37.2), $(37.3)_5$

$$E'''_{\alpha\beta}=\mathscr{P}_\alpha u_\beta+u_\alpha\mathscr{P}_\beta,\ \text{i.e.}\ E'_\alpha=\mathscr{P}_\alpha=E''_\alpha\quad(\mathscr{P}_\alpha=\overset{\perp}{\varepsilon}_\alpha{}^{\beta\gamma}E_\beta H_\gamma) \tag{39.4}$$

which is based on Poynting's vector $c\mathscr{P}_\alpha$ and implies $(36.4)_2$. I believe that this choice is the best among all possible non-symmetric determinations of $E'''_{\alpha\beta}$, especially because some instances of $E_{\alpha\beta}$ with $E'''_{\alpha\beta}={}^2E'''_{\alpha\beta}$, i.e. ${}^6E_{\alpha\beta}$ and ${}^7E_{\alpha\beta}$, have the physically meaningful expressions (40.18) and give rise to the satisfactory versions (43.8)—(43.9) of the equation of energy balance in the case of deforming, rotating, and accelerated media. This pushes us to prefer the instance ${}^2E_{\alpha\beta}$ of $E_{\alpha\beta}$ even in special relativity where $(36.4)_2$ may not hold, at least in case Eckart's thermodynamic tensor $Q_{\alpha\beta}$ is accepted.

Let us now consider any theory where the mixed part of the tensor $\mathscr{U}_{\alpha\beta}-E_{\alpha\beta}$ which takes the properties of matter into account, is connected with heat conduction. In classical physics thermal heat conduction is zero for $T_{/r}\equiv 0$; the same holds in relativity for $T_{/\overset{\perp}{\alpha}}+TA_\alpha\equiv 0$ (by a result of Tolman) [§ 45]. Then even in case an exchange of terms between $E'''_{\alpha\beta}$ and $\mathscr{U}'''_{\alpha\beta}-E'''_{\alpha\beta}$ is allowed—which involves the possibility of considering terms of non-thermal heat conduction—it is meaningful and natural to consider (as privileged) the expression of $E'''_{\alpha\beta}$ corresponding to that determination of $\mathscr{U}'''_{\alpha\beta}-E'''_{\alpha\beta}$ which vanishes for $T_{/\alpha}+TA_\alpha=0$. Then the reasoning above leading us to (39.3) and somehow to (39.4) can be repeated.

Let us remark that condition (39.4) rules out the coupling of $Q_{\alpha\beta}$ with the much used tensor ${}^3E_{\alpha\beta}$—cf. (37.6) (this does not prevent ${}^3E_{\alpha\beta}$ from having a remarkable auxiliary role even in case (39.4) is presupposed). In the remaining sections $(36.4)_2$ will almost always be assumed, but most of these sections will be compatible with every choice of $E'''_{(\alpha\beta)}$.

§ 40. Some Historical Hints. Basic General Energetic Properties of Minkowski's Tensor and the Instances ${}^5E_{\alpha\beta}$ to ${}^7E_{\alpha\beta}$ of $E_{\alpha\beta}$

Besides ${}^1E_{\alpha\beta}$ to ${}^4E_{\alpha\beta}$ other instances of $E_{\alpha\beta}$ have been used in order to obtain certain instances of the quantities $E_{[\alpha\beta]}$, $\overset{\perp}{K}_\alpha$, and $\Pi^{(\varepsilon)}$—cf. (36.2).[4)]

Various instances of $E_{\alpha\beta}$ have been introduced in order to apply relativistic thermodynamics to reversible or non-reversible polarizable media in more or less general cases. For instance Kneissler [1949] introduced a symmetric instance of $E_{\alpha\beta}$ useful in a magnetostatic case. This instance was improved by Sommerfeld and Bopp [1950], who considered the stationary magnetic case $(E_\alpha\equiv 0)$.[5)]

An extensive contribution to the theory of polarizable media is due to G. A. Kluitenberg, S. R. de Groot, and P. Mazur [1953, I, II] and [1955, IV]. In the last work the authors consider an instance of $E_{\alpha\beta}$ which allows them to apply consistently in classical physics the first two principles of thermodynamics to continuous media which are isotropically and linearly polarizable and magnetizable.[6)]

G.M. Rancoita [1959] proposed an instance of $E_{\alpha\beta}$ fit for dealing with the same media in special relativity, and he dealt explicitly with fluids.[7)]

Let us mention that Toupin [1956] constructs a classical static theory for general elastic dielectrics capable of finite deformations. Truesdell and Toupin [1960] present the same results within a thermodynamic framework including an instance of $E_{\alpha\beta}$ useful in connection with the convected time flux.[8)]

More recent and general works on this subject are Sedov [1965b] and Bressan [1966c,d,e] and [1967b].[9)]

Our presentation is mainly based on Bressan [1966c] to [1967b] especially in connection with the point of view emphasized in § 39, with instances of $E_{\alpha\beta}$ explicitly used—cf. (40.11) and (40.18) below—and with results (in particular the result (40.21) presented in this section is taken from Bressan [1967b]).

We wish to develop (37.5). To this end we first remark that by (34.7) and its analogue for $P_{\alpha\beta}$—cf. (34.8)—we have

$$F_{\gamma\delta}=2u_{[\gamma}E_{\delta]}+\overset{\perp}{\varepsilon}_{\gamma\delta}{}^{\rho}B_{\rho}, \qquad P_{\gamma\delta}=2u_{[\gamma}P_{\delta]}-\overset{\perp}{\varepsilon}_{\gamma\delta}{}^{\sigma}M_{\sigma}. \tag{40.1}$$

Since $u_{\gamma}u^{\gamma}{}_{/\alpha}=E_{\delta}u^{\delta}=u_{\gamma}P^{\gamma}=0$,

$$2u_{[\gamma}E_{\delta]}(u^{[\gamma}P^{\delta]})_{/\alpha}=u_{\gamma}u^{\gamma}E_{\delta}P^{\delta}{}_{/\alpha}=-E_{\rho}P^{\rho}{}_{/\alpha} \tag{40.2a}$$

holds. Furthermore, since the tensor $\overset{\perp}{\varepsilon}_{\gamma\delta}{}^{\rho}M_{\sigma}$ is spatial, $(20.6a)_1$ and $(20.5)_2$ imply

$$\overset{\perp}{\varepsilon}_{\gamma\delta}{}^{\rho}B_{\rho}(-\overset{\perp}{\varepsilon}{}^{\gamma\delta\sigma}M_{\sigma})_{/\alpha}=-2B_{\rho}M^{\rho}{}_{/\alpha}. \tag{40.3}$$

From the identity $u_{[\gamma}E_{\delta]}\overset{\perp}{\varepsilon}{}^{\gamma\delta\sigma}M_{\sigma}=0$ we deduce

$$u_{[\gamma}E_{\delta]}(-\overset{\perp}{\varepsilon}{}^{\gamma\delta\sigma}M_{\sigma})_{/\alpha}=(u_{[\gamma}E_{\delta]})_{/\alpha}\overset{\perp}{\varepsilon}{}^{\gamma\delta\sigma}M_{\sigma}=u_{\gamma/\alpha}\overset{\perp}{\varepsilon}{}^{\gamma\rho\sigma}E_{\rho}M_{\sigma}. \tag{40.4a}$$

The formulas (40.2b) to (40.4b) obtained in the order from (40.2a) to (40.4a) by exchanging the couples (E_{ρ},B_{ρ}) and (P_{ρ},M_{ρ}) hold. In particular we have

$$-u_{[\gamma}P_{\delta]}(\overset{\perp}{\varepsilon}{}^{\gamma\delta\rho}B_{\rho})_{/\alpha}=(u_{[\gamma}P_{\delta]})_{/\alpha}\overset{\perp}{\varepsilon}{}^{\gamma\delta\rho}B_{\rho}=u_{\gamma/\alpha}\overset{\perp}{\varepsilon}{}^{\gamma\rho\sigma}P_{\rho}B_{\sigma}. \tag{40.4b}$$

From (40.1) to (40.4a) and (40.2b) to (40.4b) we deduce

$$\tfrac{1}{2}(F_{\gamma\delta}P^{\gamma\delta}{}_{/\alpha}-P^{\gamma\delta}F_{\gamma\delta/\alpha})=P^{\rho}E_{\rho/\alpha}-E^{\rho}P_{\rho/\alpha}+M^{\rho}B_{\rho/\alpha}-B^{\rho}M_{\rho/\alpha}+u_{\gamma/\alpha}\overset{\perp}{\varepsilon}{}^{\gamma\rho\sigma}(E_{\rho}M_{\sigma}+P_{\rho}B_{\sigma}). \tag{40.5}$$

By (40.5), $(34.1)_2$, and (37.5) *the proper density* ${}^{1}K_{\alpha}$ *of the ponderomotive 4-force connected with Minkowski's tensor, has the expression*

$$\begin{aligned}{}^{1}K_{\alpha}=-{}^{1}E_{\alpha\beta}{}^{/\beta}=\mathscr{L}_{\alpha}&+\tfrac{1}{2}(P^{\rho}E_{\rho/\alpha}-E^{\rho}P_{\rho/\alpha}+M^{\rho}H_{\rho/\alpha}-H^{\rho}M_{\rho/\alpha})\\&+u_{\gamma/\alpha}\overset{\perp}{\varepsilon}{}^{\gamma\rho\sigma}(E_{\rho}M_{\sigma}+P_{\rho}B_{\sigma})\text{—cf. (36.7)}\end{aligned} \tag{40.6}$$

in the general case of a deforming, rotating, and accelerated continuous medium. If polarizations vanish, then ${}^1K_\alpha$ equals the proper density of Lorentz 4-force $\mathscr{L}_\alpha$.

Incidentally, covariant derivatives can be replaced with ordinary partial derivatives in the second term on the right hand side of $(40.6)_2$, even using general co-ordinates.

By $(36.2)_2$ and by $(40.6)_2$, $(36.7)_3$ and $(20.20)_1$ *we have, in the general case mentioned above,*

$$\begin{aligned}\frac{1}{c}\,{}^1\Pi^{(e)} &= u^\alpha\,{}^1E_{\alpha\beta}{}^{/\beta} \\ &= E^\alpha J_\alpha + \tfrac{1}{2}(E^\rho \dot{P}_\rho - P^\rho \dot{E}_\rho + H^\rho \dot{M}_\rho - M^\rho \dot{H}_\rho) - \overset{\perp}{\varepsilon}{}^{\gamma\rho\sigma} A_\gamma (E_\rho M_\sigma + P_\rho B_\sigma)\,.\end{aligned} \tag{40.7}$$

From $(34.1)_3$ and (21.6) we deduce

$$\frac{E^\rho \dot{P}_\rho - P^\rho \dot{E}_\rho}{2} = E^\rho \dot{P}_\rho - \frac{D}{Ds}\frac{E^\rho P_\rho}{2} = k\left(E^\rho \dot{\pi}_\rho - \frac{D}{Ds}\frac{E^\rho P_\rho}{2k}\right) - \frac{E^\rho P_\rho}{2} u^\alpha{}_{/\alpha}\,. \tag{40.8}$$

By (40.8) and its analogue for H^ρ, M_ρ, and μ_ρ—cf. $(34.1)_4$—(40.7) becomes

$$\frac{1}{c}\,{}^1\Pi^{(e)} = u^\alpha\,{}^1E_{\alpha\beta}{}^{/\beta} = E^\alpha J_\alpha + k\frac{d_1\lambda}{Ds} - k\frac{D}{Ds}\frac{\mathscr{R}}{k} - \mathscr{R} u^\alpha{}_{/\alpha} - \overset{\perp}{\varepsilon}{}^{\gamma\rho\sigma} A_\gamma (E_\rho M_\sigma + P_\rho B_\sigma) \tag{40.9}$$

where

$$\frac{d_1\lambda}{Ds} = E^\rho \frac{D\pi_\rho}{Ds} + H^\rho \frac{D\mu_\rho}{Ds}\,, \qquad \mathscr{R} = \tfrac{1}{2}(E^\rho P_\rho + H^\rho M_\rho)\,. \tag{40.10}$$

Let us remark that by $(40.10)_1$ $k\,d_1\lambda\,dC$ is the (microscopic) work made by the electromagnetic field on the dipoles in $d\mathscr{C}$ (in the proper time Ds) with respect to a locally natural and proper frame. Incidentally we call the specific work $d_1\lambda$ microscopic because it is not related to any macroscopic velocity of matter or charge.

The terms $E^\alpha J_\alpha$ and $k\,d_1\lambda/Ds$ have interesting physical meanings, unlike the remaining terms on the right hand side of $(40.9)_2$. Then it is natural to try to make the latter terms disappear by changing ${}^1E_{\alpha\beta}$—to which ${}^1\Pi^{(e)}$ is related—into a suitable admissible and simple [Definition 39.1] electromagnetic energy tensor. To achieve this we set

$${}^5E_{\alpha\beta} = W(u_\alpha u_\beta + \overset{\perp}{g}_{\alpha\beta}) + 2u_{(\alpha}\mathscr{P}_{\beta)} - E_\alpha D_\beta - H_\alpha B_\beta\,, \tag{40.11}$$

where

$$W = \tfrac{1}{2}(E_\gamma E^\gamma + H_\gamma H^\gamma) \quad (\text{and } \mathscr{P}_\alpha = \overset{\perp}{\varepsilon}_\alpha{}^{\beta\gamma} E_\beta H_\gamma)\,. \tag{40.12}$$

We shall denote ${}^1E_{\alpha\beta} - {}^5E_{\alpha\beta}$ by $\mathscr{R}_{\alpha\beta}$, which is compatible with $(40.10)_2$ and convention (17.13). Indeed by comparing (40.11) with (37.3) and by taking $(40.10)_2$

and (40.12) into account, we deduce the second of the equalities

$$\mathscr{R}_{\alpha\beta} = {}^1E_{\alpha\beta} - {}^5E_{\alpha\beta} = \mathscr{R}(u_\alpha u_\beta + \overset{\perp}{g}_{\alpha\beta}) + \overset{\perp}{\varepsilon}_\alpha{}^{\rho\sigma}(E_\rho M_\sigma + P_\rho B_\sigma) u_\beta . \tag{40.13}$$

Then by $(22.9)_1$ for $T_{\alpha\beta} = \mathscr{R}_{\alpha\beta}$ and (17.13) we have

$$-u^\alpha \mathscr{R}_{\alpha\beta}{}^{/\beta} = k \frac{D}{Ds} \frac{\mathscr{R}}{k} + \mathscr{R} \overset{\perp}{g}{}^{\alpha\beta} u_{\alpha/\beta} + A^\alpha \overset{\perp}{\varepsilon}_\alpha{}^{\rho\sigma}(E_\rho M_\sigma + P_\rho B_\sigma) . \tag{40.14}$$

From (40.9), (40.13), (40.14), and (20.12) we deduce—cf. $(40.10)_1$—

$$\frac{1}{c} {}^5\Pi^{(e)} = u^\alpha\, {}^5E_{\alpha\beta}{}^{/\beta} = E^\alpha J_\alpha + k \frac{d_1\lambda}{Ds} . \tag{40.15}$$

The instance ${}^5E_{\alpha\beta}$ of $E_{\alpha\beta}$ is of the desired kind.

However in order to deal with fluids and solids two other instances of $E_{\alpha\beta}$ (${}^6E_{\alpha\beta}$ and ${}^7E_{\alpha\beta}$) are useful—see (40.18) below. They are connected with the instances $d_2\lambda$ and $d_3\lambda$ of $d\lambda$ that constitute the specific works of the electric and magnetic fields on the electric and magnetic dipoles, evaluated with respect to a locally co-rotational frame [§ 22] and a locally convected frame (i.e. with respect to the body) respectively:

$$d_2\lambda = E_\alpha D_\tau \pi^\alpha + H_\alpha D_\tau \mu^\alpha , \qquad d_3\lambda = E_\alpha D^c \pi^\alpha + H_\alpha D^c \mu^\alpha \tag{40.16}$$

—cf. (34.1), (22.2), and (22.1). The comparison of (40.16) with $(40.10)_1$ on the basis of $(34.1)_{3,4}$, $(22.1)_1$, and (22.3) yields

$$k \frac{d_1\lambda}{Ds} = k \frac{d_3\lambda}{Ds} + (E^\alpha P^\beta + H^\alpha M^\beta) u_{\alpha/\beta} , \qquad k \frac{d_2\lambda}{Ds} = k \frac{d_3\lambda}{Ds} + (E^\alpha P^\beta + H^\alpha M^\beta) u_{(\alpha/\beta)} . \tag{40.17}$$

At this point it is natural to set

$$\begin{aligned} {}^6E_{\alpha\beta} &= W(u_\alpha u_\beta + \overset{\perp}{g}_{\alpha\beta}) + 2u_{(\alpha}\mathscr{P}_{\beta)} - E_{(\alpha} D_{\beta)} - H_{(\alpha} B_{\beta)} \\ {}^7E_{\alpha\beta} &= W(u_\alpha u_\beta + \overset{\perp}{g}_{\alpha\beta}) + 2u_{(\alpha}\mathscr{P}_{\beta)} - E_\alpha E_\beta - H_\alpha H_\beta , \end{aligned} \tag{40.18}$$

in that (40.18) and (40.11) yield

$${}^5E_{\alpha\beta} = {}^7E_{\alpha\beta} - E_\alpha P_\beta - H_\alpha M_\beta , \qquad {}^6E_{\alpha\beta} = {}^7E_{\alpha\beta} - E_{(\alpha} P_{\beta)} - H_{(\alpha} M_{\beta)} , \tag{40.19}$$

which by (34.2), $(22.9)_1$, and (40.17) implies

$$\begin{aligned} u_\alpha({}^5E^{\alpha\beta} - {}^7E^{\alpha\beta})_{/\beta} &= (E^\alpha P^\beta + H^\alpha M^\beta) u_{\alpha/\beta} = k(d_1\lambda - d_3\lambda)/Ds , \\ u_\alpha({}^6E^{\alpha\beta} - {}^7E^{\alpha\beta})_{/\beta} &= (E^\alpha P^\beta + H^\alpha M^\beta) u_{(\alpha/\beta)} = k(d_2\lambda - d_3\lambda)/Ds . \end{aligned} \tag{40.20a}$$

Thence by $(36.2)_2$ we have

$$ck\frac{d_1\lambda}{Ds} - {}^5\Pi^{(e)} = ck\frac{d_2\lambda}{Ds} - {}^6\Pi^{(e)} = ck\frac{d_3\lambda}{Ds} - {}^7\Pi^{(e)}. \tag{40.20b}$$

Lastly by (40.20b) and (40.15) *the instance* ${}^{4+r}E_{\alpha\beta}$ *of* $E_{\alpha\beta}$ *is connected with the specific work* $d_r\lambda$ *and the proper density of Joule heat* $E_\alpha J^\alpha$ *by the basic energetic relation*

$$\frac{1}{c}\,{}^{4+r}\Pi^{(e)} = u_\alpha\,{}^{4+r}E^{\alpha\beta}{}_{/\beta} = E_\alpha J^\alpha + k\frac{d_r\lambda}{Ds} \quad (r=1,2,3). \tag{40.21}$$

This result and Theorem 39.1 imply that *for* $r=1,2,3$ ${}^{4+r}E_{\alpha\beta}$ *is the only admissible and simple electromagnetic energy tensor which satisfies* $(40.21)_2$.

This uniqueness property increases the interest for the instances ${}^5E_{\alpha\beta}$ to ${}^4E_{\alpha\beta}$ which satisfy the simple and physically meaningful relations (40.21). Since *these instances have the same mixed part* (${}^2E'''_{\alpha\beta}$) *of Abraham's tensor* ${}^2E_{\alpha\beta}$, the results just presented lead us to prefer Abraham's tensor to Minkowski's even in special relativity, because the acceptance of any among the tensors ${}^5E_{\alpha\beta}$ to ${}^7E_{\alpha\beta}$ is physically incompatible (in the sense explained in § 39) with the use of any instance of $E_{\alpha\beta}$ whose mixed part is not ${}^2E'''_{\alpha\beta}$.[10)]

§ 41. Some Versions of Poynting's Theorem for Moving Media

From (40.11), (40.18), (40.21), and the validity of $(22.9)_1$ for $T_{\alpha\beta} = {}^{4+r}E_{\alpha\beta}$, we have

$$k\frac{D}{Ds}\frac{W}{k} + \mathscr{P}^\alpha{}_{/\alpha} + \mathscr{P}^\alpha A_\alpha + {}^{4+r}\overset{\perp}{E}{}^{\alpha\beta} u_{\alpha/\beta} + E^\alpha J_\alpha + \frac{d_r\lambda}{Ds} = 0 \quad (r=1,2,3). \tag{41.1}$$

Then, for $u_{\alpha/\beta}=0$—whence $A_\alpha=0$, $\mathscr{P}^\alpha{}_{/\alpha} = \mathscr{P}^\alpha{}_{/\overset{\perp}{\alpha}}$, and $d_1\lambda = d_2\lambda = d_3\lambda$—the quantities W and $\mathscr{P}^\alpha$ can surely be interpreted as the proper density (in Römer units) of the electromagnetic energy superposed to the matter element $d\mathscr{C}$ and the proper density of the corresponding momentum respectively. Indeed under this interpretation, in the time lapse $(s, s+Ds)$ the increment $DW = kD(Wk^{-1})$ of the proper electromagnetic energy WdC superposed to the matter element $d\mathscr{C}$ equals the electromagnetic energy $-\mathscr{P}^\alpha{}_{/\alpha}dC\,Ds$ entering the region dC minus the Joule heat $E^\alpha J_\alpha dC\,Ds$ produced in $d\mathscr{C}$ and the microscopic work $kd_r\lambda\,dC\,Ds$ done in $d\mathscr{C}$. (For the sake of simplicity impressed electric currents are assumed to be absent.)

We now consider the general case, so that $u_{\alpha/\beta}$ may not vanish. Then, by the mass-energy equivalence principle, $-\mathscr{P}^\alpha A_\alpha dC\,Ds$ is the work done in the time lapse $(s, s+Ds)$ and in $d\mathscr{C}$ by the dragging forces acting on the radiating electromagnetic energy, with respect to a locally natural and proper frame—see Appendix C.

The term $-\mathscr{P}^\alpha{}_{/\alpha}dC\,Ds$ has the same meaning as in the case $u_{\alpha/\beta}=0$ provided relativistic space time divergences are considered as natural relativizations of classical divergences according to a point of view emphasized by Cattaneo [1963].

We follow this point of view by the reasons pointed out in Appendix C. Of course this does not prevent us from believing that for comparing a relativistic theory involving constitutive equations and (41.1), with the corresponding classical theory, it may be useful to replace the sum $\mathscr{P}^{\alpha}{}_{/\alpha}+\mathscr{P}^{\alpha}A_{\alpha}$ in equality (41.1) with $\mathscr{P}^{\alpha}{}_{/\overset{\perp}{\alpha}}+2\mathscr{P}^{\alpha}A_{\alpha}$. The same holds for the forthcoming analogue (41.2) of (41.1) for ${}^{3}E_{\alpha\beta}$.

By analogy with the classical theory of continuous media we consider the quantity $-{}^{4+r}\overset{\perp}{E}{}^{\alpha\beta}u_{\alpha/\beta}dC\,Ds$ as the work (evaluated with respect to an observer locally stationary with respect to matter) done in $(s, s+Ds)$ on (the electromagnetic energy in) $d\mathscr{C}$ by the electromagnetic pressures ${}^{4+r}\overset{\perp}{E}_{\alpha\beta}$ exerted from outside $d\mathscr{C}$. By (41.1) this work, as well as the work $-\mathscr{P}^{\alpha}A_{\alpha}dC\,Ds$ of the dragging force, increases the electromagnetic energy superposed to the matter element $d\mathscr{C}$.

We conclude that a physical meaning has been considered for all terms in the relation (41.1) which constitutes *the Poynting theorem for deforming media in connection with the electromagnetic energy tensor* ${}^{4+r}E_{\alpha\beta}$.

Incidentally, since by $(37.7)_{2,3}$ ${}^{3}E'_{\alpha}$ equals ${}^{3}E''_{\alpha}$, from $(22.9)_1$ for $T_{\alpha\beta}={}^{3}E_{\alpha\beta}$ and $(37.8)_2$ we deduce the following version of Poynting's theorem for the instance ${}^{3}E_{\alpha\beta}$ of $E_{\alpha\beta}$:

$$k\frac{D}{Ds}\frac{{}^{3}E}{k}+{}^{3}E'^{/\alpha}_{\alpha}+{}^{3}E'_{\alpha}A^{\alpha}+{}^{3}\overset{\perp}{E}{}^{\alpha\beta}u_{\alpha/\beta}+E^{\alpha}J''_{\alpha}=0\text{—cf. }(35.2)_{2,3},\,(37.7)_1\,. \tag{41.2}$$

In this version the electromagnetic energy ist ${}^{3}E$, the heat flux vector is ${}^{3}E'_{\alpha}$; furthermore, on the one hand, no instance of the microscopic power $k\,d\lambda/Ds$ appears; on the other hand, so to speak, the Joule heat $E^{\alpha}J''_{\alpha}$ due to the total electric current and not only to the true current is present in (41.2).

§ 42. W as the Proper Density of Non-Material Electromagnetic Energy

We consider an inertial spatial frame x^r in special relativity or classical physics. Furthermore we assume that matter is stationary with respect to this frame and consists of some conductors and dielectrics whose boundaries are the connected surfaces σ_i $(i=1,\dots,n)$.

We take the electrostatic case $(H_r=B_r=0)$ into account, and besides the true charge per unit proper volume $\bar{\rho}$, we consider the true surface charge per unit proper area $\bar{\rho}_{\sigma}$ as a function defined on the surface σ_i.

By $P_r dC$ we can understand a suitable time average of the resultant of the microscopic electric dipoles belonging to $d\mathscr{C}$. Usually P_r is assumed to be a function of E_r and the absolute temperature T (and the local position gradient in the case of deforming media).

Now we assume that an ideal experimenter has some charge dispersed at infinity and that he carries into finitely distant parts of space the quantity de of this charge, which gives the fields $\bar{\rho}$, $\bar{\rho}_{\sigma}$, D_r, E_r, and P_r the respective increments

$d\bar{p}$, $d\bar{p}_\sigma$, dD_r, and dP_r (in the whole space S_3). To do so the experimenter must perform against the (non-molecular) electric forces described by the field E_r the macroscopic work

$$d\mathscr{L}_E = \int_{S_3} E_r dD^r dS \text{—cf. Stratton [1941, p. 110 (29)]}, \tag{42.1}$$

using suitable forces external to the system being considered. Hence the aforementioned electric forces do the (macroscopic) work $-d\mathscr{L}_E$ on the charge de carried to the finite.

During the above transport of charges, E_r increases by dE_r. This causes P_r to increase by dP_r. We have $P^r dC = (x_r^+ - x_r^-) de^+$ where de^+ is the total positive charge belonging to the microscopic dipoles in $d\mathscr{C}$; x_r^+ is (the time average of the position of) the center of the same charges; and x_r^- is the analogue for the negative charges. Thus the work done by the field E_r on the dipoles in $d\mathscr{C}$ is

$$k\,d\lambda_E = de^+ E^r (dx_r^+ - dx_r^-) = E^r dP_r dC\,. \tag{42.2}$$

Hence by (42.1), (42.2), and $(34.1)_1$ the total (macroscopic and microscopic) work done by the electric field when the charge de is carried from the infinite to the finite, is

$$\int_{S_3} k\,d\lambda_\sigma dC - d\mathscr{L}_E = -\int_{S_3} dW_1 dC \quad \text{where} \quad 2W_1 = E_r E^r. \tag{42.3}$$

Hence W_1 can be considered, in the first place, as the *total (potential) energy* of the electric field per unit (proper) volume.

Now let us remark that it is customary to consider as the proper density of electrostatic energy a quantity W_1' for which

$$dW_1' = E_r dD^r \quad \text{whence} \quad dW_1' = dW_1 + k\,d\lambda_E \quad \text{and} \quad d\mathscr{L}_E = \int_{S_3} dW_1' dS. \tag{42.4}$$

The integral of dW_1' over S_3 is usually called *energy stored by the electrostatic field*—cf. Stratton [1941, pp. 110—111]. Its part consisting of the integral of $-k\,d\lambda_E$ over S_3 contributes to the internal energy of matter. Therefore in the second place it is natural to call the remaining part, i. e. the integral of dW_1 over S_3, the *non-material electric energy.*

Let us add that W_1 has a universal and simple expression—cf. $(42.3)_2$—whereas the expression of W_1' depends on the material being considered, and only in special linear cases has an expression in finite terms: $W_1' = 2^{-1} E_r D^r$.

Through similar considerations for the case of stationary currents (on bodies which are stationary with respect to the frame x^r) the quantity $W_2 = 2^{-1} H_r H^r$ can be shown to have the meaning of the proper density of *total energy*, or *non-material energy*, of the magnetic field.[11)]

We conclude that W—cf. $(40.12)_1$—can be considered as the proper density of the *total energy*, or *non-material energy*, of the electromagnetic field.

§ 43. On the Equations of Gravitation and Energy Balance in the Presence of Electromagnetic Phenomena

As was said in § 36, the electromagnetic energy tensor $E_{\alpha\beta}$, the stress tensor $X_{\alpha\beta}$, and the proper density ρ of gravitational mass have arbitrary parts which are linked by conditions (36.8) and (39.3).

We decide that, in case we use the instance ${}^sE_{\alpha\beta}$ of $E_{\alpha\beta}$, the corresponding instances of $X_{\alpha\beta}$, $\rho[\rho=k(w+c^2)]$, and w are to be denoted by ${}^sX_{\alpha\beta}$, ${}^s\rho$, and ${}^sw$ respectively. Then besides asserting on the basis of (40.19)

$$ {}^6X_{\alpha\beta}={}^5X_{\alpha\beta}-E_{[\alpha}P_{\beta]}-H_{[\alpha}M_{\beta]}\,, \quad {}^7X_{\alpha\beta}={}^5X_{\alpha\beta}+E_\alpha P_\beta+H_\alpha M_\beta\,, \tag{43.1} $$

by $(36.8)_2$, $(36.9)_1$, (40.19), and (17.13) we can write

$$ {}^s\rho={}^5\rho\,, \quad {}^sw={}^5w \quad (s=6,7)\,, \quad {}^5E'''_{\alpha\beta}={}^6E'''_{\alpha\beta}={}^7E'''_{\alpha\beta}\,. \tag{43.2} $$

Let us incidentally add that by $(37.2)_1$, $(37.3)_6$, $(40.10)_2$, and $(40.13)_2$ Abraham's tensor ${}^2E_{\alpha\beta}$ is related to ${}^5E_{\alpha\beta}$ by

$$ {}^2E={}^5E+\tfrac{1}{2}(E_\alpha P^\alpha+H_\alpha M^\alpha),\ {}^2\overset{\perp}{E}_{\alpha\beta}={}^5\overset{\perp}{E}_{\alpha\beta}+(E_\rho P^\rho+H_\rho M^\rho)\overset{\perp}{g}_{\alpha\beta},\ {}^2E'''_{\alpha\beta}={}^5E'''_{\alpha\beta}, \tag{43.3} $$

so that by $(36.9)_{1,2}$ we have

$$ {}^2\rho={}^5\rho-\tfrac{1}{2}(E_\alpha P^\alpha+H_\alpha M^\alpha)\,, \quad {}^2X_{\alpha\beta}={}^5X_{\alpha\beta}-(E_\rho P^\rho+H_\rho M^\rho)\overset{\perp}{g}_{\alpha\beta}\,. \tag{43.4} $$

We postulate in general [special] relativity that *gravitation equations* $(23.1)_1$ [*conservation equations* $(23.3)_3$] *hold for*

$$ \mathscr{U}_{\alpha\beta}={}^s\mathscr{U}_{\alpha\beta}={}^s\rho u_\alpha u_\beta+q_\alpha u_\beta+u_\alpha q_\beta+{}^sX_{\alpha\beta}+{}^sE_{\alpha\beta} \tag{43.5} $$

with $s=5$. Then the same holds for $s=2,6,7$. Indeed $(43.2)_{3,4}$ and $(\Delta q^\alpha=0)$ $(43.3)_3$ imply $(36.9)_{3,4,5}$ for $\Delta E_{\alpha\beta}={}^sE_{\alpha\beta}-{}^5E_{\alpha\beta}$ $(s=2,6,7)$; furthermore ${}^s\rho$ and ${}^sX_{\alpha\beta}$ have been determined by the condition that $(36.9)_{1,2}$ should hold $(s=2,6,7)$. Then $\Delta\mathscr{U}_{\alpha\beta}={}^s\mathscr{U}_{\alpha\beta}-{}^5\mathscr{U}_{\alpha\beta}=0$.

By (40.11) and (40.18), for $\mathscr{U}_{\alpha\beta}={}^s\mathscr{U}_{\alpha\beta}$ $(s=5,6,7)$ the second Cauchy equation $(23.3)_2$ and $(43.4)_2$ imply

$$ {}^2X_{[\alpha\beta]}={}^5X_{[\alpha\beta]}=E_{[\alpha}D_{\beta]}+H_{[\alpha}B_{\beta]}\,, \quad {}^6X_{[\alpha\beta]}=0={}^7X_{[\alpha\beta]}\,. \tag{43.6} $$

We shall explicitly consider conservation equations $(23.3)_1$ for $\mathscr{U}_{\alpha\beta}={}^s\mathscr{U}_{\alpha\beta}$ $(s=6,7)$ which are useful for dealing with fluids (from the Eulerian point of view) and solids (from the Lagrangian point of view) respectively.

We now consider equation $(23.4)_4$ of energy balance for $\mathscr{U}_{\alpha\beta}={}^s\mathscr{U}_{\alpha\beta}$. Then, by $(22.9)_1$ for $T_{\alpha\beta}={}^s\mathscr{U}_{\alpha\beta}$—cf. (43.5)—(21.4), and $(24.3)_3$ we have

$$ k{}^s\dot{w}+q^\alpha{}_{/\alpha}+q^\alpha A_\alpha+{}^sX^{\alpha\beta}u_{\alpha/\beta}=u_\alpha{}^sE^{\alpha\beta}{}_{/\beta}\,. \tag{43.7} $$

Furthermore (40.21) holds for $s=n+4=5,6,7$. Hence—cf. $(43.2)_2$—*the balance of energy is described by the local equation*

$$k\dot{w}+\frac{d_n l^{(i)}}{Ds}=k\left(\frac{dQ}{Ds}+\frac{d_n\lambda}{Ds}\right)\qquad (w={}^5w;\ n=2,3)\,, \tag{43.8}$$

where—cf. (24.5)

$$\frac{d_n l^{(i)}}{Ds}={}^{n+4}X^{\alpha\beta}u_{\alpha/\beta}\ (n=1,2,3)\,,\quad k\frac{dQ}{Ds}=J_\alpha E^\alpha-q^\alpha{}_{/\alpha}-q^\alpha A_\alpha=J_\alpha E^\alpha+\frac{k}{c}q_{\mathrm{ass}}\,. \tag{43.9}$$

We consider equation $(23.4)_4$ of energy balance, for the expression (36.1) of $\mathscr{U}_{\alpha\beta}$. Then by $(22.9)_1$ and (21.4) this equation implies

$$k\frac{D}{Ds}\left(w+\frac{E}{k}\right)+(q^\alpha+E''^\alpha)_{/\alpha}+(q^\alpha+E'^\alpha)A_\alpha+(X^{\alpha\beta}+\overset{\perp}{E}{}^{\alpha\beta})u_{\alpha/\beta}=0\,. \tag{43.10}$$

Now we assume the equality $E'_\alpha=E''_\alpha$ which was deduced in general relativity—cf. $(36.4)_2$—but not in special relativity.[12] Then (43.10) says: *Per unit proper time the increase of the proper total energy (i.e. internal energy plus electromagnetic energy) $(kw+E)dC$ in $d\mathscr{C}$ equals the total heat flux $-(q^\alpha+E'^\alpha)_{/\alpha}dC$ entering $d\mathscr{C}$ and due both to heat conduction and electromagnetic radiation, plus the work $-(q^\alpha+E'^\alpha)A_\alpha dC$ done by the dragging force on the energy transferred by the total heat flux and evaluated with respect to a locally proper and natural frame*—see Appendix C—*plus the work $-(X^{\alpha\beta}+\overset{\perp}{E}{}^{\alpha\beta})u_{\alpha/\beta}$ done by the mechanical stresses (pressures) $X^{\alpha\beta}$ and the electromagnetic stresses $E^{\alpha\beta}$ exerted on matter and electromagnetic energy in dC.*

In our theory the following instances of (43.10) hold—cf. $(40.12)_1$:

$$k\frac{D}{Ds}\left({}^s w+\frac{W}{k}\right)+(q^\alpha+\mathscr{P}^\alpha)_{/\alpha}+(q^\alpha+\mathscr{P}^\alpha)A_\alpha+({}^sX^{\alpha\beta}+{}^s\overset{\perp}{E}{}^{\alpha\beta})u_{\alpha/\beta}=0\qquad (s=2,5,6,7). \tag{43.11}$$

Footnotes to Chapter 4

[1] Grot and Eringen [1966] claim (on p. 613) that "the seemingly fond wish to use the Minkowski stress tensor or its symmetric part is erroneous, unless the material has no polarization or magnetization" and they propose, among other things, to add an interaction term and a body couple. Unfortunately the resulting relativistic theory is based on Eringen's nonsymmetric tensor of electromagnetic energy ${}^E E_{\alpha\beta}$, so that it is restricted to special relativity.

Grot and Eringen [1966, p. 664] say that "it must be mentioned that for accelerating frames and curved spaces where the special theory fails, new unified theories are needed employing the fundamental ideas of the general theory of relativity. Such a grandiose plan presently is out of our reach."

[2] The function $\phi_{\alpha\beta}$—or $\bar{\phi}_{\alpha\beta}$—is universal in the sense that it does not depend on the material being considered. It is invariant in the sense that it does not depend on the frame being considered. Then by condition (a), $E_{\alpha\beta}$ is a function of $F_{\rho\sigma}$ and $f_{\rho\sigma}$, invariant under the co-ordinate transformations which leave $g_{\rho\sigma}$ and u_ρ invariant.

3) A similar assertion concerning only empty space ($F_{\alpha\beta} \equiv f_{\alpha\beta}$) was proved first by Fock in special relativity and then by Jankiewicz [1962] in general relativity. Our theorem—which is Theorem B.1—is proved in Appendix B using the procedures employed by Bressan [1965, § 8] for a similar theorem concerning the case $F_{\alpha\beta} \not\equiv f_{\alpha\beta}$ and matter stationary with respect to an inertial space in special relativity.

4) Let us remember that $\overset{\perp}{K}_\alpha$ may be regarded as the proper density of the ponderomotive force (on the basis of the first Cauchy equation $(23.4)_{1,2}$ and (36.2), $\bar{E}_{[\alpha\beta]}$ as the proper density of intrinsic moment of momentum (on the basis of the second Cauchy equation $(36.4)_1$), and $\Pi^{(e)}$ as the proper density of electromagnetic energy on the basis of Poynting's theorem [§ 41].

5) Sommerfeld and Bopp [1950] assume $E_\alpha \equiv 0$ and $E_{\alpha\beta} = H_{(\alpha} B_{\beta)} - \frac{1}{2} H_\gamma B^\gamma \overset{\perp}{g}_{\alpha\beta}$.

6) Kluitenberg, Groot and Mazur in effect set $E_{\alpha\beta} = F_{\alpha\gamma} f_\beta{}^\gamma - \frac{1}{4} F_{\gamma\delta} f^{\gamma\delta} g_{\alpha\beta} + 2u^\gamma F_{[\gamma\delta} f^\delta{}_{\alpha]} u_\beta$.

7) Rancoita [1959] in effect assumes $E_{\alpha\beta} = {}^3E u_\alpha u_\beta + 2u_{(\alpha} \mathscr{P}_{\beta)} + ({}^3E - H_\rho B^\rho) \overset{\perp}{g}_{\alpha\beta} - E_\alpha D_\beta - H_\alpha B_\beta$—cf. $(37.7)_1$, $(37.3)_5$.

8) Truesdell and Toupin [1960], assume $H_\alpha \equiv B_\alpha \equiv 0$ and $E_{\alpha\beta} = {}^3E_{\alpha\beta} - P_\alpha E_\beta$—cf. (37.7).

9) The main differences between the theories developed by Sedov and Bressan are the following: (i) The first is framed within special relativity and is completely based on Minkowski's tensor—cf. footnote 10 in Chapter 4—whereas the second, presented in general relativity, is substantially based on Abraham's tensor which according to the views written in § 39 cannot be replaced with Minkowski's as far as the mixed part of $E_{\alpha\beta}$ is concerned, (ii) in Sedov's theory, for instance, some terms which can be connected to couple stresses appear in the equation of energy balance as terms of heat production; and they have no dynamic counterpart on $X_{[\alpha\beta]}$ (which in this theory vanishes identically in the absence of the electromagnetic field) whereas in Bressan's theory the same terms appear in the energetic equation and the dynamic equation, and are equal to the corresponding terms of the classical theory according to G. Grioli [1960] or Toupin [1962], up to very small terms.

10) By $(37.2)_1$ and (36.2) and by $(37.3)_6$

$$\text{(a)} \quad {}^2\Pi^{(e)} - {}^1\Pi^{(e)} = -c u^\alpha [({}^1E'_\alpha - {}^1E''_\alpha) u_\beta]^{/\beta} = c A^\alpha \overset{\perp}{\varepsilon}_\alpha{}^{\rho\sigma} (E_\rho M_\sigma + P_\rho B_\sigma).$$

Sedov [1965b] substantially proves in special relativity that ${}^1\Pi^{(e)}$ has at $\mathscr{E}$ our expression for ${}^2\Pi^{(e)}$—which by (a) is c times the right hand side of $(40.7)_1$ without the term in A_γ—provided e.g. P_ρ and M_ρ are meant everywhere as polarizations relative to the inertial space orthogonal to u^α at $\mathscr{E}$.

Constitutive equations for polarizations are simpler if these are relative to the co-moving frame. Hence formulas on these polarizations appear more useful. Among them, for $s = 2, 4, 5, 6$ the expression for ${}^s\Pi^{(e)}$, which is related to Abraham's tensor, is simpler than the one for $-c u^\alpha [{}^sE_{\alpha\beta} + ({}^1E'_\alpha - {}^1E''_\alpha) u_\beta]^{/\beta}$, which is the analogue for Minkowski's tensor, in that only the latter expression includes the term in A_γ that appears in (40.7). This is a motive to prefer Abraham's to Minkowski's tensor, that is based (only) on special relativity.

11) Cf. Bressan [1966c, § 4]. In this paper the point of view presented in the present section is emphasized. Also deforming bodies and couple stresses are taken into account.

12) We have ${}^sE'_\alpha = {}^sE''_\alpha$ for $s = 2, 3, 5, 6, 7$, but the inequality ${}^sE'_\alpha \neq {}^sE''_\alpha$ may hold for $s = 1, 4$—cf. $(37.3)_6$, (37.12).

Chapter 5

On Media Capable of Electromagnetic Phenomena from the Eulerian Point of View. Magneto-Elastic Waves in Ideal Conductors

§ 44. Introduction

The first medium we consider is a black body. We use it to show why we prefer $\theta_\alpha = T_{/\dot{\alpha}} + TA_\alpha$ to $T_{/\dot{\alpha}}$ in relativistic versions of the Fourier law. In § 46, [§ 47] we lay down the foundations of a theory for electrically and magnetically polarizable fluids in the absence [presence] of viscosity. Lastly we take into account ideal conductors and after a few considerations (including dynamic equations) which hold for materials of any kind, we restrict ourselves to (possibly polarizable) non-viscous ideal fluids, and we deal with magneto-elastic waves (in particular Alfvén waves). The same waves in general magneto-elastic materials, and certain constancy properties for ideal conductors of any kind, will be presented in Chapter 8.

Let us now extend the historical hints on fluids given in § 33 and concerning the case $E_\alpha \equiv H_\alpha \equiv 0$, to the case where electromagnetic phenomena are present. Kluitenberg, De Groot, and Mazur [1953] extended Eckart's theory [1940b] of viscous fluids capable of heat conduction within special relativity, to a mixture of an arbitrary number of chemical components; besides heat conduction they took viscous flows, diffusion, and chemical reactions into account. Kluitenberg, De Groot, and Mazur [1955] dealt with (electrically and magnetically) polarizable systems within special relativity, under the assumption that these systems are isotropic as far as polarizations are concerned. The same isotropy assumption is made by Rancoita [1959], who deals with fluids within special relativity using a new electromagnetic energy tensor—cf. footnote 7 in Chapter 4. This assumption is avoided in Bressan [1966, d, e] where the foundations of a general theory of polarizable materials are stated in general relativity and, among other things, the constitutive equations for elastic fluids and finitely deformable solids are deduced. These works will be followed in [§§ 46, 47], and in Chapter 8. Hughes [1961] is concerned with heat conduction and Joule heat in polarizable fluids within special relativity, in accord with Eckart [1940b]. Alfvén's classical theory of magneto-elastic waves in fluids with an infinite electric conductivity has been successfully tested by experimenters. Ideal conductors and magneto-elastic waves in fluids were studied in classical physics, among others, by Lüst [1959] and Carstiou [1963]; in relativity they were studied, among others, by Mrs. Y. Fourer Bruhat [1959], [1960], and [1966], Saini [1961a), b)], Taub [1963], Pratelli [1965] and Schöpf [1965b].[1)]

Some generalizations—to be spoken of in Chapter 8—of these researches are the foundations of piezo-elasticity (for solids) in general relativity of Bressan [1966e], the linear elasticity for non-polarizable ideal conductors due to Pichon [1965][2)], and the theory of magneto-elastic waves in polarizable elastic ideal conductors presented by Bressan [1972a]—see Chapter 8.

Of the Russian school we mention only Sedov [1965b] on polarizable materials, as a basis for works of the afore-mentioned kinds.

§ 45. Black Body and Absolute Temperature in Thermodynamic Equilibrium

We consider a disordered black body radiation R' in thermodynamic equilibrium at the absolute temperature T. Let $\mathscr{F}$ be an ideal fluid with respect to which the average total flux of the electromagnetic energy of R' is zero. This condition determines the 4-velocity u^α of $\mathscr{F}$ in the space-time region R'_0 effectively occupied by R'.

In R'_0 the total energy tensor $\mathscr{U}_{\alpha\beta}$ can be identified with $E_{\alpha\beta}={}^1E_{\alpha\beta}=\cdots={}^7E_{\alpha\beta}$, and this tensor has the form

$$E_{\alpha\beta}=\rho u_\alpha u_\beta+p\overset{\perp}{g}_{\alpha\beta} \qquad (3p=\overset{\perp}{E}_\rho{}^\rho,\ \rho=E)\,, \tag{45.1}$$

Then by (24.3) and $(29.4)_2$ we see that the dynamic equations $(23.4)_{1,2}$ for radiation coincide with those of a non-viscous fluid with $\kappa\equiv 0$ (hence $q_\alpha\equiv 0\equiv Q_{\alpha\beta}$). Then the dynamic equations $(29.4)_1$ hold for R' in the simplified form

$$(\rho+p)A_\alpha=-p_{/\overset{\perp}{\alpha}}\,. \tag{45.2}$$

According to Stefan and Boltzmann the density ρ and the pressure p of our electromagnetic radiation are related to T by the classical relations

$$\rho=3p=aT^4 \qquad (a=\text{const})\,. \tag{45.3}$$

Since $P_\alpha=M_\alpha=0$ in R_0, $E_{\alpha\beta}$—cf. (45.1)—must coincide with ${}^1E_{\alpha\beta}$. Then (45.1) and $(37.1)_3$ yield $(45.3)_1$ in (special or general) relativity. Furthermore $(45.3)_2$ is in finite terms; hence it has to be accepted in relativity without any changes. From (45.2) and (45.3), we deduce – cf. (25.2)

$$TA_\alpha=-T_{/\overset{\perp}{\alpha}}, \quad \text{i.e.} \quad \theta_\alpha=0\,. \tag{45.4}$$

Now let us assume that R' is the rigidly moving radiation enclosed in a hollow of $\mathscr{C}$ (black body). Incidentally the motion of $\mathscr{F}$ in R'_0 is stationary, that is an $\mathscr{F}$-fixed frame (x) can be found, for which $g_{\alpha\beta,0} \equiv 0$ holds in R_0'. Then *the "pocket temperature"* $T_{(p)} = T\sqrt{-g_{00}}$ *is constant in* R_0':

$$T_{(p)} = T\sqrt{-g_{00}} = \text{const} \quad \textit{for} \quad g_{\alpha\beta,0} \equiv 0 \text{—cf. (16.1).} \tag{45.5}$$

This result was first proved—on the basis of Tolman [1930]—by Tolman and Ehrenfest [1930] in the static case $(g_{\alpha\beta,0} = 0 = g_{0r})$—cf. Tolman [1949, p. 318]. It was generalized to the stationary case by Stückelberg and Wander [1953].

Now we deduce (45.5) from (45.4). Since in R_0' the stationary frame (x) is $\mathscr{F}$-fixed, we have $u^\beta = \delta_0{}^\beta / \sqrt{-g_{00}}$, hence $u_\alpha = g_{\alpha 0}/\sqrt{-g_{00}}$. Furthermore $(45.5)_3$ implies $u_{\alpha,0} = 0$. Now it is easy to see that

$$A_\alpha = u_{\alpha/\beta} u^\beta = u_{\alpha/0} u^0 = (u_{\alpha,0} - \{\alpha 0, 0\} u^0) u^0 = g_{00,\alpha}/2 g_{00} = (\sqrt{-g_{00}})_{,\alpha}/\sqrt{-g_{00}} \tag{45.6}$$

holds in R_0'. Let us add that the assumed constancy of T implies $T_{/\alpha} u^\alpha = DT/Ds = 0$, so that $T_{/\overset{\perp}{\alpha}} = T_{/\alpha} = T_{,\alpha}$. Then (45.5) and $(45.4)_2$ imply

$$0 = \theta_\alpha = T A_\alpha + T_{,\alpha} = [T(\sqrt{-g_{00}})_{,\alpha} + \sqrt{-g_{00}}\, T_{,\alpha}]/\sqrt{-g_{00}}\,. \tag{45.7}$$

Thence $(45.5)_2$ follows. q.e.d.

The system $\mathscr{C} + R'$ (in the case considered above) constitutes a thermometer—cf. (45.3). In case $\mathscr{C} + R'$ is very small, it is called a pocket thermometer.

Now let us consider any body $\mathscr{C}$. At least from the theoretical point of view it is reasonable to assume that two arbitrary points P_1 and P_2 of $\mathscr{C}$, extremely near to each other, can be put into mutual thermal contact through a pocket thermometer.

We assume that (at least around P_1 and P_2) the motion of $\mathscr{C}$ is stationary and that $\mathscr{C}$ is in thermodynamic equilibrium—so that a pocket temperature, $T_{(p)}$, can be defined. Relation (45.4) holds inside the thermometer—hence $(45.5)_2$ also does. Thus, if the temperatures of $\mathscr{C}$ at P_1 and P_2 were not in agreement with $(45.4)_1$, then a heat flux would arise, for which in some place heat would move towards warmer regions without the concomitance of any other facts, such as the performance of some work. This is against the second principle of thermodynamics.

We conclude that—in contrast with a result of classical physics—in relativity absolute temperature is not constant within bodies in (local) thermodynamic equilibrium, as was first remarked by Tolman [1930], but it fulfills condition (45.4), so that the pocket temperature is constant—cf. (45.5).

The (generalized) results by Tolman and Ehrenfest mentioned above are of interest especially in general relativity for cosmological problems—cf. Tolman [1949, p. 330]. However they are relevant also in special relativity for accelerated bodies in (local) thermodynamic equilibrium.

On the basis of the results presented in this section I prefer the relativistic version (25.2) of Fourier's law to the direct relativization (25.8b) of the same law.

Let us remark that *if for some continuously differentiable f, at all equilibria of $\mathscr{C} + R'$ the equality $\rho = f(T)$ holds in R_0' and* (45.4) *holds in $\mathscr{C}$—and hence in R'—, then, besides* $(45.3)_1$, *we have* $(45.3)_{2,3}$. Indeed by (45.2) and $(45.3)_1$ and by $(45.4)_1$ we respectively deduce $4 f(T) A_\alpha = -f'(T)\, T_{/\overset{\perp}{\alpha}} = f'(T)\, T A_\alpha$.

§ 46. Polarizable Non-Viscous Fluids

By the basic assumptions made in § 43 equation (43.8) of energy balance hold for $n=1,2,3$ under the definitions (43.9) and $(40.10)_1$.

For arbitrary materials in the presence of electromagnetic phenomena the second principle is expressed by (25.1) for $r=cJ^\alpha E_\alpha$:

$$T\frac{D\eta}{Ds}\geqslant\frac{dQ}{Ds} \quad \text{where} \quad k\frac{dQ}{Ds}=J^\alpha E_\alpha-q^\alpha{}_{/\alpha}-q^\alpha A_\alpha\,. \tag{46.1}$$

By replacing q_{ass} with $dQ/D\tau$ and $dl^{(i)}(=X^{\alpha\beta}u_{\alpha/\beta}Ds)$ with $d_n l^{(i)}-d_n\lambda$ in our deduction of inequality $(28.1)_1$ from the definition $\mathscr{F}=w-T\eta$ and the first two principles $kDw+dl^{(i)}=c^{-1}\cdot kq_{\text{ass}}Ds$ and $TD\eta\geqslant c^{-1}q_{\text{ass}}Ds$—cf. $(28.1)_2$, $(24.6)_2$, and (25.1)—we obtain a deduction of the inequality

$$kD\mathscr{F}\leqslant -k\eta DT-d_n l^{(i)}+kd_n\lambda \qquad (n=1,2,3) \tag{46.2}$$

—cf. $(40.10)_1$, (40.16)—from the same definition and the generalizations (43.8) and (46.1) of those principles. Of course (46.2) holds as an equality if and only if (46.1) does.

By $(40.16)_1$ and (43.9) inequality (46.2) for $n=2$ becomes

$$k\dot{\mathscr{F}}\leqslant -k\eta\dot{T}-X^{\alpha\beta}u_{(\alpha/\beta)}+k\left(E_\alpha\frac{D_\tau\pi^\alpha}{Ds}+H_\alpha\frac{D_\tau\mu^\alpha}{Ds}\right) \quad \text{for} \quad X^{\alpha\beta}={}^6X^{\alpha\beta}\,. \tag{46.3}$$

In addition we set

$$W'=E_\alpha\pi^\alpha+H_\alpha\mu^\alpha\,, \qquad \bar{\mathscr{F}}=\mathscr{F}-W'\,, \qquad \bar{w}=w-W'=\bar{\mathscr{F}}+T\eta \tag{46.4}$$

—cf. $(28.1)_2$. From $(46.3)_1$, $(46.4)_{1,2}$, $(22.5)_2$, and $(34.1)_{3,4}$ we have

$$k\dot{\bar{\mathscr{F}}}+k\eta\dot{T}+X^{\alpha\beta}u_{(\alpha/\beta)}+P^\alpha\frac{D_\tau E_\alpha}{Ds}+M^\alpha\frac{D_\tau H_\alpha}{Ds}\leqslant 0\,. \tag{46.5}$$

We say that the body $\mathscr{C}$ is a *polarizable non-viscous fluid* at its material point P^* if at P^* (i) the magnitudes $\bar{\mathscr{F}}$, $X^{\alpha\beta}$, π^α, and μ^α are functions of $g_{\alpha\beta}$, u^α, T, k, E_α, and H_α, and (ii) constraints are absent in the sense that $\dot{T}$, $u_{\alpha/\overset{\perp}{\beta}}$, $\dot{\pi}^\alpha$, and $\dot{\mu}^\alpha$ are physically independent of one another and of T, k, π^α, and μ^α ($\dot{k}=-ku^\alpha{}_{/\alpha}$) —i.e. arbitrary values of the quantities $T,\ldots,\mu^\alpha$, $\dot{T},\ldots,\dot{\mu}^\alpha$ (fulfilling obvious conditions such as $E_\alpha u^\alpha=0$) can hold for $\mathscr{C}$ at P^* in a suitable physically possible process.

We assume that $\mathscr{C}$ is a polarizable non-viscous fluid at P^*. Then, by $(22.2)_1$ and the version $\dot{k}=-k\overset{\perp}{g}{}^{\alpha\beta}u_{\alpha/\beta}$ of $(21.3)_1$—cf. (20.12)—inequality (46.5) becomes

$$k\left(\frac{\partial\bar{\mathscr{F}}}{\partial T}+\eta\right)\dot{T}+\left(X^{\alpha\beta}-k^2\frac{\partial\bar{\mathscr{F}}}{\partial k}\overset{\perp}{g}{}^{\alpha\beta}\right)u_{(\alpha/\beta)}$$

$$+k\left(\frac{\partial\bar{\mathscr{F}}}{\partial E_\alpha}+\pi^\alpha\right)\dot{E}_\alpha+k\left(\frac{\partial\bar{\mathscr{F}}}{\partial H_\alpha}+\mu^\alpha\right)\dot{H}_\alpha-(P^\alpha E^\beta+M^\alpha H^\beta)u_{[\alpha/\overset{\perp}{\beta}]}\leqslant 0\,. \tag{46.6}$$

Since the quantities $\dot{T}$, $u_{\alpha/\overset{\perp}{\beta}}$, $\dot{E}_\alpha$, $\dot{H}_\alpha$ can assume arbitrary values, (46.6) implies

$$\eta=-\frac{\partial\bar{\mathscr{F}}}{\partial T}\,,\qquad X^{\alpha\beta}=k^2\frac{\partial\bar{\mathscr{F}}}{\partial k}\overset{\perp}{g}{}^{\alpha\beta}\qquad\text{where}\qquad \bar{\mathscr{F}}=\bar{\mathscr{F}}(T,k,E_\alpha,H_\alpha,g_{\alpha\beta},u^\alpha)\,, \tag{46.7}$$

$$\pi^\alpha=-\frac{\partial\bar{\mathscr{F}}}{\partial E_\alpha}\,,\qquad \mu^\alpha=-\frac{\partial\bar{\mathscr{F}}}{\partial H_\alpha}\,, \tag{46.8}$$

and

$$P^{[\alpha}E^{\beta]}+M^{[\alpha}H^{\beta]}=0\,,\qquad\text{hence}\qquad \frac{\partial\bar{\mathscr{F}}}{\partial E_{[\alpha}}E^{\beta]}+\frac{\partial\bar{\mathscr{F}}}{\partial H_{[\alpha}}H^{\beta]}=0\,. \tag{46.9}$$

Furthermore by (46.7) to (46.9) inequality (46.6)—equivalent to (46.5), hence to (46.2) for $n=2$—must hold as an equality. We conclude that a *polarizable elastic fluid has constitutive equations of the form* (46.7) *and* (46.8), *where condition* $(46.9)_2$ *holds identically; in addition such a fluid is capable of only reversible processes.*

Theorem 46.1. a) *Condition* $(46.9)_2$ *holds identically if and only if* $\bar{\mathscr{F}}(T,k,E_\alpha,H_\alpha,g_{\alpha\beta},u^\alpha)$ *is a spatially isotropic function of* E_α *and* H_α [Definition 38.1].

b) *Condition* $(46.9)_2$ *is equivalent to the possibility of writing* $(46.7)_3$ *in the form*

$$\bar{\mathscr{F}}=\bar{\mathscr{F}}(T,k,\mathscr{M},\mathscr{N},\mathscr{P})\ \text{where}\ 2\mathscr{M}=E^\alpha E_\alpha\,,\ \mathscr{N}=E^\alpha H_\alpha\,,\ 2\mathscr{P}=H^\alpha H_\alpha\,. \tag{46.10}$$

When locally proper and natural co-ordinates are considered, this theorem appears equivalent to known theorems on isotropic functions of vectors belonging to a 3-dimensional Euclidean space.

For the ease of the reader we prove Theorem 46.1. Let $\bar{E}_\alpha$ and $\bar{M}_\alpha$ be arbitrary spatial vectors and let $\rho_\alpha{}^\lambda=\rho_\alpha{}^\lambda(t)$ be an arbitrary spatial rotation [§ 38] which is defined for $0\leqslant t\leqslant 1$, which is attached to a point of S_4 [§ A1] independent of t, and which is a continuously differentiable function of t in $[0,1]$ such that condition $\rho_\alpha{}^\lambda(0)=\delta_\alpha{}^\lambda$ holds. Now we set

(a) $\quad\bar{\mathscr{F}}=\bar{\mathscr{F}}(T,k,E_\alpha,H_\alpha,g_{\alpha\beta},u^\alpha)=\bar{\mathscr{F}}(t)\quad$ where $\quad E_\alpha=\rho_\alpha{}^\lambda\bar{E}_\lambda\,,\quad H_\alpha=\rho_\alpha{}^\lambda\bar{H}_\lambda\,.$

Since T, k, $g_{\alpha\beta}$, u^α, $\bar{E}_\alpha$, and $\bar{H}_\alpha$ are constants, we have

(b) $$\dot{\bar{\mathscr{F}}}=\left(\frac{\partial\bar{\mathscr{F}}}{\partial E_\alpha}\bar{E}_\lambda+\frac{\partial\bar{\mathscr{F}}}{\partial H_\alpha}\bar{H}_\lambda\right)\dot{\rho}_\alpha{}^\lambda\qquad\left(\dot{f}=\frac{df}{dt}\right).$$

From $(38.1)_1$ we deduce $\dot\rho^\alpha{}_\gamma \rho^{\gamma\beta} + \rho^\alpha{}_\gamma \dot\rho^{\gamma\beta} = 0$; hence, by $(38.1)_1$ again, $\dot\rho^\alpha{}_\gamma g^{\gamma\lambda} + \rho^\alpha{}_\gamma \dot\rho^{\gamma\beta} \rho_\beta{}^\lambda = 0$ also holds; these two equalities can be written in the form

$$\text{(c)} \qquad \rho_{(\alpha\lambda}\dot\rho^{\lambda\beta)} = 0\,, \qquad \dot\rho_\alpha{}^\lambda = -\rho_{\alpha\gamma}\dot\rho^{\gamma\beta}\rho_\beta{}^\lambda\,.$$

By $(c)_2$ and $(a)_{3,4}$ we have $\dot\rho_\alpha{}^\lambda \bar E_\lambda = -\rho_{\alpha\gamma}\dot\rho^{\gamma\beta}\rho_\beta{}^\lambda \bar E_\lambda = -\rho_{\alpha\gamma}\dot\rho^{\gamma\beta} E_\beta$ and the analogue for $\bar H_\lambda$. Then $(b)_1$ becomes

$$\text{(d)} \qquad \dot{\bar{\mathscr F}} = -\left(\frac{\partial \bar{\mathscr F}}{\partial E_\alpha} E_\beta + \frac{\partial \bar{\mathscr F}}{\partial H_\alpha} H_\beta\right)\rho_{\alpha\gamma}\dot\rho^{\gamma\beta}\,.$$

Let us first assume the identical validity of $(46.9)_2$. Then by (d) we have $\dot{\bar{\mathscr F}} \equiv 0$. Hence the quantity $\bar{\mathscr F}$ expressed by (a) is independent of the spatial rotation $\rho_\alpha{}^\lambda$. We conclude that $\bar{\mathscr F}$ is a spatially isotropic function of E_α and H_α [Definition 38.1].

Now we conversely assume this spatial isotropy of $\bar{\mathscr F}$; thence $\dot{\bar{\mathscr F}} \equiv 0$ follows. For $t=0$, $\rho_\alpha{}^\lambda = \delta_\alpha{}^\lambda$. Then by $(c)_1$ and the consequence $0 = \dot\rho^\alpha{}_\beta u^\beta$ of $(38.1)_2$, $\dot\rho_{\alpha\beta}$ is a spatial skewsymmetric tensor for $t=0$. We can choose such a tensor arbitrarily (for $t=0$), as it becomes apparent by considering locally pseudo-Euclidean co-ordinates and remembering well known results of the theory of rigid motions in 3-dimensional space. Then the identity $\dot{\bar{\mathscr F}} \equiv 0$ and (d) imply the identical validity of $(46.9)_2$. Thus part a) is completely proved.

In order to prove part b) we first remark that (46.10) obviously implies the spatial isotropy of $\bar{\mathscr F}$ with respect to E_α and H_α. Now we conversely assume this isotropy and we consider four arbitrary spatial vectors E_α, H_α, $\bar E_\alpha$, and $\bar H_\alpha$ for which $E^\alpha E_\alpha = \bar E^\alpha \bar E_\alpha$, $H^\alpha H_\alpha = \bar H^\alpha \bar H_\alpha$, and $E^\alpha H_\alpha = \bar E^\alpha H_\alpha$ hold—cf. $(46.10)_{2,3,4}$.

Then there is a spatial rotation $\rho_\alpha{}^\beta$ for which $(a)_{3,4}$ hold. As a consequence $\bar{\mathscr F}(T,k,E_\alpha,H_\alpha,g_{\alpha\beta},u^\alpha) = \bar{\mathscr F}(T,k,\bar E_\alpha,\bar H_\alpha,g_{\alpha\beta},u^\alpha)$. Then $\bar{\mathscr F}$ has the form (46.10). q.e.d.

For non-viscous fluids $\bar{\mathscr F}$ is a function of $g_{\alpha\beta}$, u_α, T, k, E_α, and H_α. Of course this is understood to hold in all co-ordinate systems. Then by Theorem 38.1 $\bar{\mathscr F}$ is a spatially isotropic function. By Theorem 46.1 a) this implies $(46.9)_2$. Thus the above thermodynamic deduction of $(46.9)_2$ from (46.6) is superabundant.

However let us consider special relativity, or a theory of general relativity where Fock's conjecture [§ 10] is accepted. Then we can assert the possibility of determining a privileged class of frames, which is absolute in the sense that it is independent of phenomena. Let (x) belong to this class. We can now give the constitutive function for $\bar{\mathscr F}$ in a way independent of phenomena by determining $\bar{\mathscr F}$ (for $\mathscr C$ at P^*) in (x) as a function of $g_{\alpha\beta}$, u_α, T, k, E_α, H_α (and x^α). This alone does by no means imply the spatial isotropy of this function, which by Theorem 46.1 a) is equivalent to $(46.9)_2$. Hence within the theory being considered the above thermodynamic deduction of $(46.9)_2$ is fully interesting.

The above definition of non-viscous fluid was set up without choosing any particular frame (x) because we wish to keep the main part of our theory independent of Fock's conjecture.

By (46.10) the constitutive equations (46.8) take the form

$$\pi^\alpha = -\frac{\partial\bar{\mathscr{F}}}{\partial\mathscr{M}}E^\alpha - \frac{\partial\bar{\mathscr{F}}}{\partial\mathscr{N}}H^\alpha,\ \mu^\alpha = -\frac{\partial\bar{\mathscr{F}}}{\partial\mathscr{N}}E^\alpha - \frac{\partial\bar{\mathscr{F}}}{\partial\mathscr{P}}H^\alpha \text{ where } \bar{\mathscr{F}} = \bar{\mathscr{F}}(T,k,\mathscr{M},\mathscr{N},\mathscr{P}). \tag{46.11}$$

Now we remark that by $(46.4)_1$ and (46.10) we have

$$W' = -2\left(\mathscr{M}\frac{\partial\bar{\mathscr{F}}}{\partial\mathscr{M}} + \mathscr{N}\frac{\partial\bar{\mathscr{F}}}{\partial\mathscr{N}} + \mathscr{P}\frac{\partial\bar{\mathscr{F}}}{\partial\mathscr{P}}\right), \tag{46.11b}$$

so that by $(46.7)_{1,2}$ and $(46.4)_{2,3,4}$ η, $p(=k^2\partial\bar{\mathscr{F}}/\partial k)$, $\mathscr{F}$, w, and $\bar{w}$ also are functions of T, k, $\mathscr{M}$, $\mathscr{N}$, and $\mathscr{P}$.

If the fluid is polarizable completely, i.e. both electrically and magnetically, then it is reasonable to assume that equations $(46.11)_{1,2}$ can be solved with respect to E_α and H_α. Then (i) the magnitudes $\mathscr{F}$, $X^{\alpha\beta}$, E_α, and H_α are functions of $g_{\alpha\beta}$, u^α, T, k, π^α, and μ^α, and (ii) the magnitudes $\grave{T}$, $u_{\alpha/\overset{+}{\beta}}$, $\grave{E}_\alpha$, and $\grave{H}_\alpha$ are physically independent of one another and of T, k, E_α, and H_α. Now, setting

$$2\mathscr{M}' = \pi^\alpha\pi_\alpha, \quad \mathscr{N}' = \pi^\alpha\mu_\alpha, \quad 2\mathscr{P}' = \mu^\alpha\mu_\alpha, \tag{46.12}$$

we can deduce from inequality (46.3) the constitutive equations

$$E_\alpha = k\left(\frac{\partial\tilde{\mathscr{F}}}{\partial\mathscr{M}'}\pi_\alpha + \frac{\partial\tilde{\mathscr{F}}}{\partial\mathscr{N}'}\mu_\alpha\right), \quad H_\alpha = k\left(\frac{\partial\tilde{\mathscr{F}}}{\partial\mathscr{N}'}\pi_\alpha + \frac{\partial\tilde{\mathscr{F}}}{\partial\mathscr{P}'}\mu_\alpha\right) \tag{46.13}$$

where $\quad \mathscr{F} = \tilde{\mathscr{F}}(T,k,\mathscr{M}',\mathscr{N}',\mathscr{P}')$

and the analogues of $(46.7)_{1,2}$ for $\mathscr{F}$—cf. $(46.13)_3$—by the analogue of the foregoing deduction of (46.11) from (46.5).

Equations (46.11)—or (46.8)—hold also in case the fluid is only electrically [magnetically] polarizable, i.e. $\mu_\alpha \equiv 0$ and $\bar{\mathscr{F}} = \bar{\mathscr{F}}(T,k,\mathscr{M})$ [$\pi_\alpha \equiv 0$ and $\bar{\mathscr{F}} = \bar{\mathscr{F}}(T,k,\mathscr{P})$] hold. In this case, it is quite reasonable to assume that $(46.11)_1$ [$(46.11)_2$] can be solved with respect to E_α [H_α] which appears to be a function of T, k, and π^α [μ^α]. Then, since $W' = \pi^\alpha E_\alpha$ [$W' = \mu^\alpha H_\alpha$], the same holds for W', and by $(46.7)_{1,2}$ and $(46.4)_2$ also for η, $X^{\alpha\beta}$, and $\mathscr{F}$.

At this point it is not difficult to deduce from (46.3) the constitutive equations $(46.13)_1$, [$(46.13)_2$], $\mathscr{F} = \tilde{\mathscr{F}}(T,k,\mathscr{M}')$ [$\mathscr{F} = \tilde{\mathscr{F}}(T,k,\mathscr{P}')$], and the analogues of $(46.7)_{1,2}$ for $\mathscr{F}$—as (46.11) was deduced from (46.5).

We conclude that the constitutive equations of a fluid which is polarizable either completely or only electrically [magnetically] are the analogues of $(46.7)_{1,2}$ for $\mathscr{F}$ ($\eta = -\partial\tilde{\mathscr{F}}/\partial T$, $X^{\alpha\beta} = k^2(\partial\tilde{\mathscr{F}}/\partial k)\overset{+}{g}{}^{\alpha\beta}$) and either (46.13) or

$$E_\alpha = k\frac{\partial\tilde{\mathscr{F}}(T,k,\mathscr{M}')}{\partial\mathscr{M}'}\pi_\alpha \quad \left[H_\alpha = k\frac{\partial\tilde{\mathscr{F}}(T,k,\mathscr{P}')}{\partial\mathscr{P}'}\mu_\alpha\right]. \tag{46.13b}$$

The use of $\bar{\mathscr{F}}$ and (46.5) is usually preferred to that of $\mathscr{F}$ and (46.3) for, as it is now apparent, it gives rise to a more unified theory.

By $(46.4)_{2,3,4}$ $w=\mathscr{F}+T\eta$ and $\bar{w}=\bar{\mathscr{F}}+T\eta$; hence inequalities (46.2) and (46.5) can be written

$$k D w - k T D\eta + d_n l^{(i)} - k d_n \lambda \leqslant 0 \quad (n=1,2,3)\,, \tag{46.14}$$

$$k\dot{\bar{w}} - k T\dot{\eta} + X^{\alpha\beta} u_{(\alpha/\beta)} + P^{\alpha}\frac{D_{\mathrm{r}} E_{\alpha}}{Ds} + M^{\alpha}\frac{D_{\mathrm{r}} H_{\alpha}}{Ds} \leqslant 0 \qquad (X^{\alpha\beta} = {}^{6}X^{\alpha\beta}) \tag{46.15}$$

The Helmholtz postulate—see the last part of § 28—says that *the specific heat dQ/DT under constant configuration ($u_{\alpha/\dot{\beta}}=0$) and constant fields ($\dot{E}_{\alpha}=\dot{H}_{\alpha}=0$) is positive.*

As a consequence, by $(46.1)_1$ and $(46.7)_1$ $\partial^2\bar{\mathscr{F}}/\partial T^2<0$, so that by $(46.7)_1$ we can express T, $\bar{w}$, $\bar{\mathscr{F}}$, and p $[T, w, \mathscr{F}$, and $p]$ as functions of η, k, E_{α}, and H_{α} $[\eta, k, \mu_{\alpha}$, and $\pi_{\alpha}]$.

The comparison of (46.15) with (46.5) and the one of $(46.4)_3$ with $(46.4)_2$ shows that we can deduce the formulas obtained from (46.6) to (46.12) by replacing T, η, $\bar{\mathscr{F}}$, and $\mathscr{F}$ with η, $-T$, $\bar{w}$, and w respectively. Thus e. g. in correspondence to $(46.7)_{1,2}$, $(46.11)_3$, and $(46.11)_{1,2}$ we respectively have

$$T=\frac{\partial\bar{w}}{\partial\eta}\,, \qquad X^{\alpha\beta}=k^2\frac{\partial\bar{w}}{\partial k}\overset{\perp}{g}{}^{\alpha\beta}\,, \qquad \bar{w}=\bar{w}(\eta,k,\mathscr{M},\mathscr{N},\mathscr{P})\,, \tag{46.16}$$

$$\pi_{\alpha}=-\frac{\partial\bar{w}}{\partial\mathscr{M}}E_{\alpha}-\frac{\partial\bar{w}}{\partial\mathscr{N}}H_{\alpha}\,, \qquad \mu_{\alpha}=-\frac{\partial\bar{w}}{\partial\mathscr{N}}E_{\alpha}-\frac{\partial\bar{w}}{\partial\mathscr{P}}H_{\alpha}\,. \tag{46.17}$$

§ 47. Polarizable Viscous Fluid

We say that $\mathscr{C}$ is a *polarizable viscous fluid* at P^* if at P^* (i) the magnitudes $\bar{\mathscr{F}}$, $X^{\alpha\beta}$, π^{α}, and μ^{α} are functions (of class $C^{(1)}$) of $(g_{\alpha\beta}, u_{\alpha},)$ T, k, $u_{(\alpha/\dot{\beta}}$, E_{α}, H_{α}, $D_{\mathrm{r}}E_{\alpha}/Ds$, and $D_{\mathrm{r}}H_{\alpha}/Ds$, and (ii) constraints are absent, i. e. the time derivatives of these arguments are physically independent of one another and of the arguments themselves with an exception for $\dot{k}$ ($\dot{k}=-ku^{\alpha}{}_{/\alpha}$).[3]

We call such a fluid *of the first kind* or *of the second kind* according as some of $\bar{\mathscr{F}}$, $X^{\alpha\beta}$, π^{α}, and μ^{α} depend on $D_{\mathrm{r}}E^{\alpha}/Ds$ or $D_{\mathrm{r}}H^{\alpha}/Ds$ effectively or not.

We assume that $\mathscr{C}$ is a polarizable viscous fluid at P^* and we introduce—cf. $(46.7)_2$, (46.8)—the reversible stress $X^{\alpha\beta}_{\mathrm{rev}}$, the reversible electric and magnetic polarizations $\pi^{\alpha}_{\mathrm{rev}}$, $\mu^{\alpha}_{\mathrm{rev}}$ and the corresponding irreversible quantities:

$$X_{\text{rev}}^{\alpha\beta}=k^2\frac{\partial\bar{\mathscr{F}}}{\partial k}\overset{\perp}{g}{}^{\alpha\beta}, \qquad \pi_{\text{rev}}^{\alpha}=-\frac{\partial\bar{\mathscr{F}}}{\partial E_\alpha}, \qquad \mu_{\text{rev}}^{\alpha}=-\frac{\partial\bar{\mathscr{F}}}{\partial H_\alpha}, \tag{47.1}$$

$$X^{\alpha\beta}=X_{\text{rev}}^{\alpha\beta}+X_{\text{irr}}^{\alpha\beta}, \qquad \pi^{\alpha}=\pi_{\text{rev}}^{\alpha}+\pi_{\text{irr}}^{\alpha}, \qquad \mu^{\alpha}=\mu_{\text{rev}}^{\alpha}+\mu_{\text{irr}}^{\alpha}. \tag{47.2}$$

Let us remark that $\bar{\mathscr{F}}$ is a function of $(g_{\alpha\beta}, u_\alpha)$ T, k, $u_{\alpha/\overset{\perp}{\beta}}$, E_α, H_α, $\grave{E}_\alpha$, and $\grave{H}_\alpha$—cf. (20.2)—and (using a procedure introduced by B. Coleman) that $\bar{\bar{\mathscr{F}}}$ is a linear form in the time derivatives $\grave{T}$, $\grave{k}$, $Du_{\alpha/\overset{\perp}{\beta}}/Ds, \ldots, \grave{\grave{H}}_\alpha$ of the preceding arguments T to $\grave{H}_\alpha$, whose coefficients, as well as η, $X^{\alpha\beta}$, π^α, and μ^α, depend at most on the same arguments. We fix T to $\grave{H}_\alpha$ and consider arbitrary values for $\grave{T}$, $Du_{\alpha/\beta}/Ds$, $\grave{\grave{E}}_\alpha$, and $\grave{\grave{H}}_\alpha$. Thus we see on the basis of (46.5) that (i) *$\bar{\bar{\mathscr{F}}}$ is independent of $\grave{E}_\alpha$, $\grave{H}_\alpha$, and $u_{\alpha/\beta}$*, so that (46.6) holds, and (ii) *the temperature entropy relation* $(46.7)_1$ *holds.*

By the consequence $\grave{k}=-k\overset{\perp}{g}{}^{\alpha\beta}u_{\alpha/\beta}$ of $(21.3)_1$ and by $(47.1)_1$ we have $k\grave{k}\,\partial\bar{\mathscr{F}}/\partial k=-X_{\text{rev}}^{\alpha\beta}u_{(\alpha/\beta)}$. Thus by the assertions (i) and (ii) above, and by $(47.1)_{2,3}$, (47.2), and $(22.2)_1$ we can put (46.5) into the form

$$X_{\text{irr}}^{\alpha\beta}u_{(\alpha/\beta)}+P_{\text{irr}}^{\alpha}\frac{D_\tau E_\alpha}{Ds}+M_{\text{irr}}^{\alpha}\frac{D_\tau H_\alpha}{Ds}-(P_{\text{rev}}^{\alpha}E^\beta+M_{\text{rev}}^{\alpha}H^\beta)u_{[\alpha/\beta]}\leqslant 0, \tag{47.3}$$

where e. g. $P_{\text{rev}}^{\alpha}=k\pi_{\text{rev}}^{\alpha}$. The fourth term in the right-hand side of (47.3) is a linear form in $u_{[\alpha/\beta]}$ whose coefficients are independent of $u_{\alpha/\beta}$, $D_\tau E_\alpha/Ds$, and $D_\tau H_\alpha/Ds$ —cf. $(47.1)_{2,3}$. Then (47.3) implies

$$P_{\text{rev}}^{[\alpha}E^{\beta]}+M_{\text{rev}}^{[\alpha}H^{\beta]}=0 \tag{47.4}$$

which by $(47.1)_{2,3}$ and $(34.1)_{3,4}$ is equivalent to $(46.9)_2$. Then, by Theorem 46.1 a) $\bar{\mathscr{F}}$ is a spatially isotropic function of E_α and H_α, and by Theorem 46.1 b) $\bar{\mathscr{F}}$ has the expression (46.10). By (47.4) inequality (47.3) is equivalent to

$$X_{\text{irr}}^{\alpha\beta}u_{(\alpha/\beta)}+P_{\text{irr}}^{\alpha}\frac{D_\tau E_\alpha}{Ds}+M_{\text{irr}}^{\alpha}\frac{D_\tau H_\alpha}{Ds}\leqslant 0. \tag{47.5}$$

By (47.5) *for a polarizable viscous fluid we have*

$$X_{\text{irr}}^{\alpha\beta}=P_{\text{irr}}^{\alpha}=M_{\text{irr}}^{\alpha}=0 \quad \textit{for} \quad u_{(\alpha/\overset{\perp}{\beta})}=\frac{D_\tau E_\alpha}{Ds}=\frac{D_\tau H_\alpha}{Ds}=0. \tag{47.6}$$

Let us prove, e. g., the equality $X_{\text{irr}}^{11}=0$ under conditions $(47.6)_{4,5,6}$. To this end we make all of the quantities present in those conditions vanish except $u_{1/1}$. Then (47.5) becomes $X_{\text{irr}}^{11}u_{1/1}\leqslant 0$. Thus for $u_{1/1}<0$ $[u_{1/1}>0]$ we have $X_{\text{irr}}^{11}\geqslant 0$ $[X_{\text{irr}}^{11}\leqslant 0]$. Then by the obvious continuity of X_{irr}^{11} we have $X_{\text{irr}}^{11}=0$ for $u_{1/1}=0$.

Now let the fluid be of the first kind. Then (47.5) implies

$$P_{\text{irr}}^{\alpha}=0, \qquad M_{\text{irr}}^{\alpha}=0, \qquad X_{\text{irr}}^{\alpha\beta}u_{(\alpha/\beta)}\leqslant 0. \tag{47.7}$$

From $(47.7)_{1,2}$, $(34.1)_{3,4}$, $(47.1)_{2,3}$, and $(47.2)_{2,3}$ we deduce (46.8). Summing up *for a polarizable fluid of the first kind we have* $(47.1)_1$, $(47.7)_3$, *and*

$$\begin{aligned} &\eta = -\frac{\partial \bar{\mathscr{F}}}{\partial T}, \quad \pi_\alpha = -\frac{\partial \bar{\mathscr{F}}}{\partial \mathscr{M}} E_\alpha - \frac{\partial \bar{\mathscr{F}}}{\partial \mathscr{N}} H_\alpha, \\ &\mu_\alpha = -\frac{\partial \bar{\mathscr{F}}}{\partial \mathscr{N}} E_\alpha - \frac{\partial \bar{\mathscr{F}}}{\partial \mathscr{P}} H_\alpha, \quad \bar{\mathscr{F}} = \bar{\mathscr{F}}(T,k,\mathscr{M},\mathscr{N},\mathscr{P}). \end{aligned} \tag{47.8}$$

Since $(47.8)_1$ holds also for polarizable viscous fluids (of either kind), by the Helmholtz postulate [§ 46] and by (47.1) and $(46.4)_{3,4}$ *for such a fluid* (i) *T, $\bar{\mathscr{F}}$, $\bar{w}$, and $p_{\text{rev}} = 3^{-1} X^\alpha_{\text{rev}\alpha}$ [π^α_{rev} and μ^α_{rev}] are functions of η, k, $\mathscr{M}$, $\mathscr{N}$, $\mathscr{P}$* [of η, k, E_α, H_α] (ii) *the inverse* $(46.16)_1$ *of* $(47.8)_1$ *holds, and* (iii) *also* (46.17) *hold in case the fluid is of the first kind*—cf. $(47.7)_{1,2}$, $(47.8)_{2,3}$.

The assumption that for a completely polarizable viscous fluid E_α and H_α are functions of ($g_{\alpha\beta}, u_\alpha$) T, k, $u_{(\alpha/\overset{\perp}{\beta})}$, π^α, μ^α, $D_\tau E_\alpha/Ds$, and $D_\tau H_\alpha/Ds$ is reasonable at least in case the fluid is of the first kind. Under this assumption we can apply to $\mathscr{F}$ (and $w = T\eta + \mathscr{F}$) on the basis of inequality (46.3) an analogue of the argument above concerning polarizable viscous fluids, based on inequality (46.5).

We call a polarizable viscous fluid *linear* if for it $X^{\alpha\beta}$, π^α, and μ^α are linear in $u_{(\alpha/\overset{\perp}{\beta})}$, $D_\tau E_\alpha/Ds$, and $D_\tau H_\alpha/Ds$. Then by (47.1) and (47.2) the same holds for the quantities $X^{\alpha\beta}_{\text{irr}}$, π^α_{irr}, and μ^α_{irr}. Furthermore by (47.6) the afore-mentioned linear expressions of these quantities are homogeneous. After they are substituted in the left hand side of (47.5), that side becomes a quadratic form, $\mathscr{P}_2$, in $u_{(\alpha/\overset{\perp}{\beta})}$, $D_\tau E_\alpha/Ds$, and $D_\tau H_\alpha/Ds$, which by (47.5) itself is negative semidefinite.

In case the fluid is only electrically polarizable we may have, in particular,

$$P^\alpha_{\text{irr}} = a \frac{D_\tau E^\alpha}{Ds} \quad \text{with} \quad a = a(T,k,\mathscr{M}) > 0. \tag{47.9}$$

In order to show a motive for considering fluids of the second kind we now consider a cylindrical vessel $\mathscr{V}$ which is uniformly rotating around its axis z supposed to be vertical. Let $\mathscr{V}$ be full of a viscous fluid polarizable only electrically, and let there be a constant uniform electric field orthogonal to z.

If the fluid is of the first kind, then P_α is parallel to E_α and molecularly invariable. Then P_α rotates with respect to the fluid. One may be unwilling to exclude that P_α could be dragged by the fluid, which therefore cannot be of the first kind. Such a dragging follows from (47.9). Indeed in the vessel $D_\tau E_\alpha/Ds = -u_{[\alpha/\beta]} E^\beta$—cf. $(22.2)_1$. Furthermore the definition $(35.9)_2$ of the angular velocity Ω_γ implies $u_{[\alpha/\overset{\perp}{\beta}]} = \overset{\perp}{\varepsilon}_{\alpha\beta}{}^\gamma \Omega_\gamma$. Then (47.9) becomes

$$P^\alpha_{\text{irr}} = a \overset{\perp}{\varepsilon}{}^{\alpha\beta\gamma} \Omega_\beta E_\gamma \quad (a > 0). \tag{47.10}$$

§ 48. The Cauchy Equations in the Presence of Heat Conduction and an Electromagnetic Field; Preliminaries for Ideal Conductors

In order to state the general version (48.1) of the dynamic equations, we remark that by (43.5), in the case when ${}^sE_{\alpha\beta}\equiv 0$ the equality $\mathcal{U}_{\alpha\beta}={}^s\mathcal{U}_{\alpha\beta}$ becomes equivalent to $(24.3)_1$ for $X_{\alpha\beta}={}^sX_{\alpha\beta}$ and in this case the dynamic equations $(23.4)_{1,2}$ were put into the form (24.8). Furthermore by (40.11) and (40.18), ${}^sE_{\alpha\beta}u^\beta={}^sE_{\beta\alpha}u^\beta=\mathcal{P}_\alpha$ for $s=5,6,7$, where $c\mathcal{P}_\alpha$ is Poynting's vector—cf. $(40.12)_2$.

Then by (40.11), (40.12), and (40.18) the dynamic equations $(23.4)_{1,2}$ in the general case can be obtained from (24.8) by the substitutions

$$\rho\to\rho+W,\qquad X_{\alpha\beta}\to{}^sX_{\alpha\beta}+{}^s\overset{\perp}{E}_{\alpha\beta},\qquad\text{and}\qquad q_\alpha\to q_\alpha+\mathcal{P}_\alpha .$$

The result is

$$\begin{aligned}[](\rho+W)\overset{\perp}{g}_\alpha{}^\gamma+{}^sX_\alpha{}^\beta+{}^s\overset{\perp}{E}_\alpha{}^\beta]A_\beta=&-\overset{\perp}{g}_{\alpha\gamma}({}^sX^{\gamma\beta}+{}^s\overset{\perp}{E}{}^{\gamma\beta})_{/\overset{\perp}{\beta}}\\&-k\left(\overset{\perp}{g}_{\alpha\gamma}\frac{D}{Ds}\frac{q^\gamma+\mathcal{P}^\gamma}{k}+u_{\alpha/\gamma}\frac{q^\gamma+\mathcal{P}^\gamma}{k}\right).\end{aligned}\tag{48.1}$$

We now assume that the body $\mathscr{C}$ with which we are dealing is an *ideal conductor*, i.e. its electric conductivity is infinite $(\sigma_{\alpha\beta}=\infty)$. A reasonable consequence of this assumption is—cf. (34.4) and (40.12)

$$E_\alpha\equiv 0,\qquad\text{hence}\qquad \mathcal{P}_\alpha\equiv 0,\qquad 2W=H_\gamma H^\gamma .\tag{48.2}$$

As a consequence the Maxwell equations (35.15) and (35.17) become

$$\begin{aligned}&B^\alpha{}_{/\overset{\perp}{\alpha}}=0,\qquad \frac{D^c}{Ds}\frac{B_\alpha}{k}=0,\\&2\Omega^\alpha B_\alpha=\bar\rho'',\qquad \overset{\perp}{\varepsilon}_\alpha{}^{\beta\gamma}(B_{\gamma/\beta}+B_\gamma A_\beta)=\frac{1}{c}j''_\alpha,\end{aligned}\tag{48.3}$$

where $c\Omega^\alpha$ is the spatial angular velocity—cf. $(35.9)_2$—and $\bar\rho''$ $[j''_\alpha]$ the proper density of ordinary total charge [current].

Incidentally, since $B_\alpha=H_\alpha+M_\alpha$, by comparing $(48.3)_4$ with $(35.16)_2$ for $E_\rho\equiv D_\rho\equiv 0$ we see that j''_α bears to the proper density j_α of ordinary true current the relation

$$j''_\alpha=j_\alpha+c\varepsilon_\alpha{}^{\beta\gamma}(M_{\gamma/\beta}+M_\gamma A_\beta)\qquad(E_\rho\equiv D_\rho\equiv 0).\tag{48.3b}$$

By $(22.1)_1$ and (22.6) we can turn $(48.3)_2$ into

$$\overset{\perp}{g}_\alpha{}^\beta\frac{DB_\beta}{Ds}-u_{\alpha/\beta}B^\beta+u^\beta{}_{/\beta}B_\alpha=0\qquad(B_\alpha u^\alpha=0).\tag{48.4}$$

We also assume that $\mathscr{C}$ has a zero thermic conductivity, so that $q_\alpha \equiv 0$. Then by $(40.18)_2$ and (48.2), for $s=7$ (48.1) becomes

$$[(\rho+2W)\overset{\perp}{g}_\alpha{}^\beta + {}^7X_\alpha{}^\beta - H_\alpha H^\beta]A_\beta = -W_{/\overset{\perp}{\alpha}} + \overset{\perp}{g}_{\alpha\gamma}(H^\gamma H^\beta - {}^7X^{\gamma\beta})_{/\overset{\perp}{\beta}} \quad (q^\alpha \equiv 0). \tag{48.5}$$

This equation is useful for dealing with solid piezoelastic ideal conductors [§§ 76—78].

Let $\mathscr{C}$ be a magnetizable non-viscous fluid [§ 46], whose constitutive equations are (46.16). Then, since $(46.10)_{2,3,4}$ and $(48.2)_{1,3}$ yield $\mathscr{M}=\mathscr{N}=0$, $\mathscr{P}=W$, we have (understanding $X_{\alpha\beta}={}^6X_{\alpha\beta}$)

$$X^{\alpha\beta} = p\overset{\perp}{g}^{\alpha\beta}, \quad p=k^2\frac{\partial\bar{w}}{\partial k}, \quad \bar{w}=\bar{w}(\eta,k,W), \quad \mathscr{P}=W \tag{48.6}$$

and—cf. $(34.1)_{2,4}$ and $(46.17)_2$

$$B_\alpha = \mu H_\alpha \quad \text{where} \quad \mu = 1-k\frac{\partial\bar{w}}{\partial W}, \quad \mu=\mu(\eta,k,W). \tag{48.7}$$

By $(40.18)_1$, (48.2), $(48.5)_2$, (48.6), and (48.7) equation (48.1) for $s=6$ becomes

$$[(\rho+2W+p)\overset{\perp}{g}_\alpha{}^\beta - H_\alpha B^\beta]A_\beta = -(p+W)_{/\overset{\perp}{\alpha}} + \overset{\perp}{g}_\alpha{}^\gamma\left(\frac{1}{\mu}B_\gamma B^\beta\right)_{/\overset{\perp}{\beta}}. \tag{48.8}$$

§ 49. Dynamic Discontinuity Equations for Magneto-Elastic Acceleration Waves in Magnetizable Fluids

We consider the magnetizable non-viscous fluid $\mathscr{C}$ introduced in § 48, so that (48.2) to (48.8) hold. Let σ_3 be an acceleration wave, let the event point $\mathscr{E}$ belong to σ_3, and let the frame (x) be locally natural at $\mathscr{E}$. We denote discontinuity across σ_3 at $\mathscr{E}$ by Δ, the discontinuity parameter of u_α $[B_\alpha]$ by λ_α $[\beta_\alpha]$, the one of the conventional mass density k by $\lambda_{(4)}$, and the propagation speed of σ_3 by V. Formulas (30.1), (30.2), (30.4), (30.5), and (30.7) still hold. Furthermore

$$c\Delta\frac{DB_\alpha}{Ds} = -V\beta_\alpha, \quad \Delta B_{\alpha,b} = \beta_\alpha N_b. \tag{49.1}$$

By $(49.1)_1$ and (30.2) the discontinuity equation corresponding to $(48.4)_1$ is

$$-\frac{V}{c}\beta_{\overset{\perp}{\alpha}} - \lambda_\alpha N_\beta B^\beta + \lambda^\beta N_\beta B_\alpha = 0, \quad \text{hence} \quad \beta^r = \frac{2c}{V}B^{[r}\lambda^{s]}N_s. \tag{49.2}$$

From $(49.1)_2$ and $(49.2)_2$ we deduce the first of the equalities

$$V\Delta B^r{}_{,b}=2cB^{[r}\lambda^{s]}N_sN_b, \qquad V\Delta(B^rB^b)_{,b}=2cB^bN_bN_sB^{[r}\lambda^{s]}. \tag{49.3}$$

We have $(49.3)_2$, for $(49.3)_1$ yields

$$\frac{V}{2c}\Delta(B^rB^b)_{,b}=B^{[r}\lambda^{s]}N_sN_bB^b+B^rB^{[b}\lambda^{s]}N_sN_b.$$

Discontinuity waves constitute an adiabatic phenomenon, so that our assumption characterized by $(48.5)_2$ is reasonable. Then we can consider the entropy η as a constant (in $W_{\mathscr{C}}$), so that by $(48.6)_{2,3}$ and $(48.7)_2$

$$p_{,r}=p'_k k_{,r}+p'_w W_{,r}, \qquad \mu_{,r}=\mu'_k k_{,r}+\mu'_w W_{,r} \qquad \left(p'_k=\frac{\partial p}{\partial k},\ldots\right). \tag{49.4}$$

By $(48.2)_3$, $(48.7)_1$, and $(49.4)_2$

$$W_{,\beta}=\left(\frac{B^\gamma B_\gamma}{2\mu^2}\right)_{,\beta}=\frac{1}{\mu^2}B_cB^c{}_{,\beta}-\frac{2W}{\mu}(\mu'_k k_{,\beta}+\mu'_w W_{,\beta});$$

hence

$$\left(1+\frac{2W}{\mu}\mu'_w\right)W_{,\beta}=\frac{1}{\mu^2}B_cB^c{}_{,\beta}-\frac{2W}{\mu}\mu'_k k_{,\beta}.$$

Furthermore $(30.5)_2$ and $(30.7)_2$ yield

$$V\Delta k_{,r}=ck\lambda^sN_sN_r. \tag{49.5}$$

Then by $(49.3)_1$ and $(48.7)_1$

$$\begin{aligned}V\Delta W_{,b}&=\frac{2\mu c}{\mu+2W\mu'_w}H_cH^{[c}\lambda^{s]}N_sN_b-\frac{2cW}{\mu+2W\mu'_w}\mu'_k kN_s\lambda^sN_b\\&=\frac{2c}{\mu+2W\mu'_w}[B_{[c}H^cN_{s]}-Wk\mu'_kN_s]\lambda^sN_b.\end{aligned} \tag{49.6}$$

By (49.4) and $(17.4)_1$ equation (48.8) becomes, in locally natural and proper co-ordinates,

$$\begin{aligned}[(\rho+2W+p)\delta_{rs}-H_rB_s]u^s{}_{,0}&=\frac{1}{\mu}(B_rB^b)_{,b}-(\mu^{-2}\mu'_kB_rB^b+p'_k\delta_r{}^b)k_{,b}\\&-[\mu^{-2}\mu'_wB_rB^b+(p'_w+1)\delta_r{}^b]W_{,b}.\end{aligned} \tag{49.7}$$

Using $(30.2)_2$, $(48.7)_1$, $(49.3)_2$, and (49.5) we develop partially the discontinuity equation corresponding to (49.7):

$$[(\rho+2W+p)\delta_{rs}-H_r B_s]\frac{V^2}{c^2}\lambda^s=2H^b N_b \lambda_{[r} B^{s]} N_s$$

$$+(\mu_k' H_r H^b+p_k'\delta_r{}^b)k N_b N_s \lambda^s+[\mu_w' H_r H^b+(p_w'+1)\delta_r{}^b]\frac{V}{c}\Delta W_{,b}\,. \tag{49.8}$$

Incidentally by $(17.4)_2$ $u^\alpha \Delta u_{\alpha/\beta}=0$, which yields the first of the equalities

$$\lambda_0=0\,,\qquad \beta_0=-B^r\lambda_r\,,\qquad \beta^r N_r=0\,. \tag{49.9}$$

By $(48.4)_2$ $u^\alpha \Delta B_{\alpha/\beta}+B^\alpha \Delta u_{\alpha/\beta}=0$; hence by $(49.1)_2$ and $(30.2)_1$ $u^\alpha\beta_\alpha+B^\alpha\lambda_\alpha=0$, which yields $(49.9)_2$. We deduce $(49.9)_3$ from $(48.3)_1$ and $(49.1)_2$. It is interesting that $(49.9)_3$ also follows from $(49.2)_2$, i.e. from $(48.3)_2$. We conclude that the discontinuity equation corresponding to $(48.3)_1$ follows from the one corresponding to $(48.3)_2$.

§ 50. Magneto-Elastic Acceleration Waves in Magnetizable Non-Viscous Fluids

We now assume that σ_3 is an *Alfvén wave*, i.e. a discontinuity wave for acceleration for which the discontinuity vector λ^α (of u^α) is perpendicular to both the field H_α and the propagation direction N_α. Then by (49.6)

$$\Delta W_{,b}=0\qquad (N_s\lambda^s=0,\ H_s\lambda^s=0)\,, \tag{50.1}$$

so that by $(48.7)_1$ equation (49.8) in λ^r reduces to

$$(\rho+2W+p)\frac{V^2}{c^2}\lambda^r=\mu a^2\lambda^r\qquad\text{where}\qquad a=H^b N_b\,,\qquad 2W=H_l H^l\,. \tag{50.2}$$

Then $(50.2)_1$ has a proper solution if and only if $V=V_A$ where—cf. (21.4)

$$V_A=c\sqrt{\frac{\mu}{\rho+H_l H^l+p}}\,|H^b N_b|\simeq\sqrt{\frac{\mu}{k}}\,|H^l N_l|\,. \tag{50.3}$$

The right hand side of $(50.3)_2$ constitutes the expression according to classical physics for the speed V_A of Alfvén waves in the direction N_l.

To calculate the speeds of the other magneto-elastic waves we assume

$$H_2=H_3=0\,,\qquad H_1\geqslant 0\,,\qquad N_3=0=\lambda^3\,, \tag{50.4}$$

where $(50.4)_{1,2,3,4}$ can be thought of as restrictions on the frame (x) alone. Furthermore $(50.4)_5$ is a consequence of $(50.4)_{1,2,3,4}$ and equation (49.8) for $r=3$, in the typical case. Indeed by (49.6) this equation can be given the form

$$\left(q_{rs}\frac{V^2}{c^2} - p_{rs}\right)\lambda^s = 0 \tag{50.5}$$

where—cf. $(48.7)_1$ and $(50.2)_2$

$$\begin{aligned} q_{rs} &= (\rho + 2W + p)\delta_{rs} - H_r B_s, \\ p_{rs} &= \mu a^2 \delta_{rs} - a B_r N_s + k(\mu'_k a H_r + p'_k N_r) N_s \\ &\quad - [\mu'_w H_r H^b + (p'_w + 1)\delta_r{}^b] \frac{2N_b}{\mu + 2W\mu'_w} [H^l N_{[l} B_{s]} + W k \mu'_k N_s]. \end{aligned} \tag{50.6}$$

Hence $q_{3l} = p_{3l} = 0 = q_{l3} = p_{l3}$ $(l=1,2)$. Then, in the typical case, $\lambda^3 \neq 0$ implies $\lambda^r = \delta^r{}_3$, i.e. that σ_3 is again an Alfvén wave. Hence $(50.4)_5$ holds.

Under conditions (50.4) equation $(50.2)_1$ in λ^r has a proper solution if and only if

$$\begin{vmatrix} p_{11} - q_{11}\dfrac{V^2}{c^2} & p_{12} \\ p_{21} & p_{22} - q_{22}\dfrac{V^2}{c^2} \end{vmatrix} = 0, \quad \text{i.e.} \quad \frac{V^4}{c^4} - \beta\frac{V^2}{c^2} + \gamma = 0 \tag{50.7}$$

where

$$\beta = \frac{p_{11}}{q_{11}} + \frac{p_{22}}{q_{22}}, \qquad \gamma = \frac{p_{11}p_{22} - p_{12}p_{21}}{q_{11}q_{22}}. \tag{50.8}$$

In the particular case $\mu \equiv 1$, from (50.4) and $(50.6)_2$ we deduce[4)]

$$p_{rs} = a^2 \delta_{rs} - a H_r N_s + k p'_k N_r N_s - 2 N_r H^1 N_{[1} H_{s]} \tag{50.9}$$

and one obtains—see Schöpf [1965 b, p. 121] or below

$$\beta c^2 = V_H{}^2 + f V_S{}^2, \qquad \gamma c^4 = V_A{}^2 V_S{}^2, \tag{50.10}$$

where

$$V_H{}^2 = c^2 \frac{H_l H^l}{\rho + p + H_l H^l}, \qquad V_S{}^2 = \frac{c^2 k}{\rho + p}\frac{\partial p}{\partial k}, \qquad f = \frac{\rho + p + (H^l N_l)^2}{\rho + p + H^l H_l}, \tag{50.11}$$

so that by $(50.3)_1$ V_H is V_A for N_α parallel with H_α, V_S has the same expression—cf. (30.9)—as the speed of sound in ordinary non-viscous fluids $[F_{\alpha\beta} \equiv 0]$ (however V_S also depends on H_α), and f is a factor whose classical limit equals 1.

For $\mu\equiv 1$ (50.10) *holds*. Indeed by $(50.2)_{2,3}$ and (50.4)

(a) $$H_1 N_1 = a, \quad 2W N_1{}^2 = a^2, \quad a^2 + H_1{}^2 N_2{}^2 = 2W.$$

Then, by $(50.6)_1$, (50.4), and $(50.11)_3$

(b) $$q_{11} = \rho + p, \quad q_{12} = q_{21} = 0, \quad q_{22} = \rho + p + H_l H^l = (\rho + p + a^2)/f;$$

by (50.9) and (a)

(c) $$\begin{aligned} p_{11} &= a^2 - a^2 + k p'_k N_1{}^2 = k p'_k N_1{}^2, \\ p_{12} &= -a H_1 N_2 + k p'_k N_1 N_2 + N_1 H^1 H_1 N_2 = k p'_k N_1 N_2 = p_{21}, \\ p_{22} &= a^2 + k p'_k N_2{}^2 + N_2 H_1 N_2 H_1 = k p'_k N_2{}^2 + 2W, \end{aligned}$$

and by $(50.8)_1$, (a), (b), and (c)

(d) $$\begin{aligned} \beta q_{11} q_{22} = p_{11} q_{22} + p_{22} q_{11} &= k p'_k N_1{}^2 (\rho + p + 2W) + (k p'_k N_2{}^2 + 2W)(\rho + p) \\ &= k p'_k (\rho + p) + k p'_k N_1{}^2 2W + 2W(\rho + p) \\ &= k p'_k (\rho + p + a^2) + 2W(\rho + p) = k p'_k f q_{22} + 2W q_{11}. \end{aligned}$$

By (d) and $(\mathrm{b})_{1,4}$

(e) $$\beta = k p'_k \frac{f}{q_{11}} + \frac{2W}{q_{22}} = \frac{k p'_k}{\rho + p} f + \frac{H_l H^l}{\rho + p + H_l H^l},$$

which by $(50.11)_{1,2}$ yields $(50.10)_1$. By $(50.8)_2$, (c), and (a)

$$\gamma q_{11} q_{22} = p_{11} p_{22} - p_{12}{}^2 = k p'_k N_1{}^2 (k p'_k N_2{}^2 + H_1{}^2) - (k p'_k)^2 N_1{}^2 N_2{}^2 = k p'_k a^2.$$

Furthermore $(\mathrm{b})_4$, $(50.2)_2$, and $(50.3)_1$ for $\mu = 1$ yield $V_A{}^2 = c^2 a^2 / q_{22}$. Hence by $(\mathrm{b})_1$ and $(50.11)_2$ we have $(50.10)_2$. q.e.d.

By (50.10) and $(50.11)_3$ the first two of the relations

$$\gamma > 0, \quad \beta > 0, \quad V_A{}^2 = V_H{}^2 \xi, \quad f = \frac{\rho + p + H_1{}^2 \xi}{\rho + p + H_1{}^2} \quad (\xi = N_1{}^2) \tag{50.12}$$

hold. The third follows from $(50.11)_1$, $(50.4)_{1,2}$ and $(50.3)_1$. Lastly $(50.11)_3$ and $(50.4)_{1,2}$ yield $(50.12)_4$. By (50.10) and $(50.12)_{3,4}$

$$\Delta(\xi) \equiv (\beta^2 - 4\gamma) c^4 = (V_H{}^2 + f V_S{}^2)^2 - 4\xi V_H{}^2 V_S{}^2, \quad \Delta(1) = (V_H{}^2 - V_S{}^2)^2 \geqslant 0. \tag{50.13}$$

Since $\rho \gg H_1{}^2$, by $(50.13)_{1,2}$ and $(50.12)_4$

$$\frac{d\Delta(\xi)}{d\xi} = \frac{H_1{}^2}{\rho + p + H_1{}^2} 2(V_H{}^2 + f V_S{}^2) V_S{}^2 - 4 V_H{}^2 V_S{}^2 < 0 \quad (0 \leqslant \xi \leqslant 1). \tag{50.14}$$

By $(50.13)_3$ and (50.14) $\Delta(1)\geqslant 0$ and $\Delta(\xi)>0$ for $0\leqslant\xi<1$. Then by $(50.12)_{1,2}$ and $(50.13)_{1,3}$ *equation* $(50.7)_2$ *in* V^2 *has two positive solutions; they coincide if and only if*

$$V_A=V_H=V_S \quad (\text{i. e. } V_H=V_S \text{ and } H_\alpha \| N_\alpha)\,. \tag{50.15}$$

By the preceding discussion we can have $(50.4)_{1,2,3,4}$ but $\lambda^3\neq 0$ only in case V_A solves equation (50.7) in V, which certainly occurs in the case (50.15).

Footnotes to Chapter 5

1) Pratelli [1965] assumes $\mu=\text{const}$ and considers incompressible or compressible fluids that are ideal conductors ($\sigma\equiv 0$). This author makes a distinction between manifolds which are the locus of path lines and the locus of lines of force for the magnetic field, and manifolds which are variously crossed by path lines and by lines of force.

Schöpf [1965b] assumes $\mu=1$, and considers barotropic fluids (with $\sigma\equiv 0$). He deals with them using his relativistic theory for (adiabatic processes of) general elastic bodies—see [1964a].

Our treatment [§§ 49, 50] is partly similar to the one of Schöpf [1965b] and our results for fluids become those of Schöpf for $\mu=1$.

Let us remark that Mrs. Bruhat [1966] assumes no restriction on μ. Furthermore she deals also with waves in non-ideal conductors for which electromagnetostriction phenomena are substantially taken into account by assuming that internal energy depends on the product $\mu\varepsilon$, where ε is the dielectric constant.

2) Pichon [1959] substantially generalizes the relativistic elasticity theory presented by Synge [1958] —where a (purely mechanical) linear isotropic relation between rate of stress and rate of strain is assumed—to the case where the electromagnetic field is present (possibly with $\varepsilon\mu>1$).

3) This definition of polarizable viscous fluids is in accord ith Truesdell's principle of equipresence—cf. Truesdell and Toupin [1960, p. 703].

4) By $(48.7)_1$ and $(34.1)_{2,4}$ $\mu\equiv 1$ implies $\mu_\alpha\equiv 0$. Then $\partial\bar{w}/\partial\mathscr{P}=0$ by $(46.17)_2$, so that by $(48.6)_{2,3,4}$ $\partial p/\partial w=0$.

Part II

Materials from the Lagrangian Point of View

Chapter 6

Kinematics and Stresses from the Lagrangian Point of View

§ 51. Historical Hints at Relativistic Theories of Elastic and More General Materials

The present section serves as an introduction to Chapter 6 and Part II. This part is devoted to the Lagrangian point of view, whose importance in the relativistic theories of materials was felt especially towards 1960, that is in the years when people attached in various ways the problem of constructing more or less general theories of elasticity in general relativity. Therefore the historical hints to be considered in this section—some of which were briefly given in § 13—will mainly concern relativistic theories of elasticity.

The first paper on (linear) elasticity in (special) relativity is Herglotz [1911] —cf. § 6—and the same subject was dealt with in some well known textbook.[1] In spite of this, for a long time practically no correct papers on relativistic elasticity appeared; probably this is related to the lack of suitable theories of kinematics in general relativity from the Lagrangian point of view.[2]

Synge [1959] and Rayner [1963] presented two theories of linear adiabatic elasticity in general relativity and they calculated the speed of propagation for elastic acceleration waves.[3] The former tried to avoid the difficulty of considering a „natural state“ and was followed in this by Bennoun [1964a, b].[4] The ideas of Rayner, who made an essential use of the Lie derivative, were developed by Schöpf [1964a]. Schöpf assumes the elastic strain-stress relation for finite deformations and proves that a certain variational principle holds in general relativity for elastic bodies in the adiabatic case. We shall prove this principle [§§ 69, 70] and also some extensions of its—following Bressan and Pitteri—to materials capable of couple stress or of any order n greater than 1 [§§ 95—99].

Parallel with the afore-mentioned researches and from a different point of view, Bressan [1963b] and [1964a, b] states the foundations of a general theory of materials, and more precisely constructs Lagrangian kinematics in a wide sense[5] using the theory of double tensors due to Bortolotti and Van der Waerden.[6] In Bressan [1964a], among other things, a theory of elasticity based on Einstein's gravitation equations and the first two principles of thermodynamics is presented. The same theory is applied in Bressan [1963d] to study elastic waves. Furthermore in Bressan [1964b] materials with memory are dealt with in general relativity on the basis of Fermi transport, and a relativistic version of the principle of material (frame) indifference is formulated and applied. In Bressan [1966a to e] the afore-mentioned elasticity theory is extended to materials capable of couple-stress and electro-magneto striction. Couple stresses are dealt with from a variational point of view by Bressan [1972b]. This paper is extended to general materials of order 2 or greater than 2 by Pitteri [1975a, b].

The afore-mentioned works by Bressan, on which the main part of the remainder of this tract is based, aim among other things at achieving as much similarity (from the local point of view) with the corresponding classical theories as possible, in order to obtain an expressive (familiar) picture of phenomena and to compare classical and relativistic results carefully. Thus, following substantially the same papers, in Chapter 7 elastic materials will be introduced as materials whose stress is a function of absolute temperature and position gradient; furthermore on the basis of the first two principles of thermodynamics the constitutive equations of these materials will be derived.

Furthermore following substantially Bressan [1963d] we shall strive to deal with elastic waves [§§ 65—67] in a way very similar to classical theories and to show carefully the difference in physical content between classical and relativistic results. For example we generalize the classical use of the polarization quadric by introducing an "inertial-mass quadric", and we prove some theorems on principal waves.

Among the other papers related to relativistic elasticity (and having in part a preparatory character) let us mention Ehlers [1961], Edelen [1963] and [1964], and Schmutzer [1964].[7)]

Among the papers concerned with a relativistic general theory of continuum media possibly with a thermodynamic basis, let us mention Bennoun [1965] on non-isentropic disturbances etc...., Schöpf [1965a] on elastic waves, Schöpf [1965b] on ideal conductors and magneto-elastic waves, Bragg [1965] and Söderholm [1970] on the principle of material indifference—cf. § 79 and § 81, 2nd part; furthermore we have to mention Sedov's school and in particular Berdichewski [1966].[8)] Let us add Maugin [1971] on variational relativistic electromagnetism.

In the first section of any chapter of part II some historical hints concerning the specific subject dealt with in the chapter can be found.

In connection with the comparison of classical and relativistic theories from a non-local point of view let us note Friedrichs [1928] and Toupin [1957/58].

The present chapter—based on Bressan [1963b], [1964b], and [1966d, e]—affords the kinematics for the whole part II, except for section 90 and the subsequent ones in that these sections are based on second-order Lagrangian kinematics.

§ 52. On the Representation of the Motion $\mathcal{M}$ of $\mathcal{C}$

We consider regular motions for the body $\mathcal{C}$; in particular we exclude rupture, tear and slip surfaces. Thus we can think of $\mathcal{C}$ as a set of material particles. We can also consider (space time) S_4 as a twice continuously differentiable manifold.

Let $\mathscr{P}^*$ be any process physically possible for the universe (which includes $\mathscr{C}$). In correspondence to $\mathscr{P}^*$ we consider the metric tensor $g^*_{\alpha\beta}$, the world tube $W^*_{\mathscr{C}}$ of $\mathscr{C}$, a co-moving admissible co-ordinate system in $W^*_{\mathscr{C}}$, say (y), and the intersection S^*_3 of $W^*_{\mathscr{C}}$ with the hypersurface $y_0=0$.

For every material point P^* of $\mathscr{C}$, the co-ordinate y^L of the intersection of S^*_3 with the world line $W^*_{P^*}$ of P^* will be called the L-th *material co-ordinate* of P^* $(L=1,2,3)$.

This system of material co-ordinates—which depends on $\mathscr{P}^*$ and (y)—characterizes a topology on $\mathscr{C}$. We can assume that *under the afore-mentioned regularity conditions for* $\mathscr{M}$ *this topology is independent of* $\mathscr{P}^*$ *and* (y). We mean that it is absolute—i.e. independent of phenomena—unlike e.g. the metric and curvature of S_4.

The material co-ordinates y^L and their increments dy^L allow us to characterize material points and infinitesimal linear material elements. The existence of the above (absolute) topology is sufficient to give these entities a physical meaning. We now call the physical state Σ^* and the configuration of $\mathscr{C}$ in S^*_3 within the process $\mathscr{P}^*$ *reference physical state* and *reference configuration* respectively. We denote the spatial metric on S^*_3 by ds^{*2}:

$$ds^{*2}=a^*_{LM}\,dy^L\,dy^M \quad \text{with} \quad a^*_{rs}=\hat{a}^*_{rs}(y^1,y^2,y^3)=\overset{\downarrow}{g}{}^*_{rs}(0,y^1,y^2,y^3) \tag{52.1}$$

where Latin indices run from 1 to 3—cf. $(15.1)_2$. Of course ds^{*2}, to be called *reference metric* or *material metric*, depends only on $\mathscr{P}^*$ and S^*_3.

Sometimes the body $\mathscr{C}$ associated with metric ds^{*2} will be considered as a Riemannian space and identified with S^*_3. Furthermore capital Latin letters will be used as material indices, i. e. indices of tensors attached to P^* or indices connected with P^* and belonging to double tensors attached to P^* and to any event point—see Appendix A.

We consider an arbitrary (regular) motion, $\mathscr{M}$, physically possible for $\mathscr{C}$ and we represent it in the admissible space time frame (x) by means of equations of the form

$$x^\alpha=\hat{x}^\alpha(t,y^1,y^2,y^3)\,[=\hat{x}^\alpha(t,y^L)=\hat{x}^\alpha(y^\Lambda),\,t=y^0] \quad \text{with} \quad \partial x^0/\partial t>0 \tag{52.2}$$

so that t is a time parameter increasing towards future.[9] We assume that

(i) the transformation $(52.2)_1$ of (t,y^L) into x^α is one-to-one and twice continuously differentiable everywhere with an exception for some time-like hypersurfaces (so that their normals are space-like) where $\partial^2\hat{x}^\alpha/\partial y^\Lambda\partial y^\Sigma$ may have discontinuities of the first kind (which is compatible with acceleration waves),

(ii) for fixed t, the point x^α given by $(52.2)_1$ describes a space-like hypersurface, say $S_3(t)$, when y^L describes S^*_3, while for y^L fixed in S^*_3, x^α describes the world line of the material point y^L, and

(iii) besides the afore-mentioned one-to-one hypothesis on $(52.2)_1$, we have

$$\frac{\partial(x^1,x^2,x^3)}{\partial(y^1,y^2,y^3)}\neq 0\,, \qquad \frac{\partial(x^0,x^1,x^2,x^3)}{\partial(t,y^1,y^2,y^3)}\neq 0\,. \tag{52.3}$$

Since $S_3(t)$ is space-like, $g_{\rho\sigma}x^\rho{}_L x^\sigma{}_L=0$ implies $x^\rho{}_L \equiv \partial x^\rho/\partial y^L=0$, which contrasts $(52.3)_1$. Now we easily see that

$$g_{\rho\sigma}\frac{\partial x^\rho}{\partial t}\frac{\partial x^\sigma}{\partial t}<0\,,\qquad g_{\rho\sigma}x^\rho{}_L x^\sigma{}_L>0 \quad \text{where} \quad x^\rho{}_L=\frac{\partial x^\rho}{\partial y^L}\,. \tag{52.4}$$

Equations $(52.2)_1$ are determined by the motion $\mathscr{M}$ of $\mathscr{C}$ only up to a substitution of the form

$$t=t(\tilde{t},y^1,y^2,y^3) \quad \text{with} \quad \frac{\partial t}{\partial \tilde{t}}>0 \quad [t=\hat{t}(x),\tilde{t}=\hat{\tilde{t}}(x)]\,. \tag{52.5}$$

It is possible to choose $t=\hat{t}(x)$ in such a way that, along every world line, t is the arclength $(Dt=Ds)$. If s and s' are such determinations of t, then

$$s'=s(t,y^1,y^2,y^3)+\sigma(y^1,y^2,y^3)\,. \tag{52.6}$$

Of course along the motion $(52.2)_1$ we have

$$u^\rho=\frac{\partial x^\rho}{\partial t}\frac{Dt}{Ds} \quad \text{where} \quad \frac{Ds}{Dt}=\left(-g_{\rho\sigma}\frac{\partial x^\rho}{\partial t}\frac{\partial x^\sigma}{\partial t}\right)^{1/2}>0\,. \tag{52.7}$$

Generally it is not possible to choose $t=\hat{t}(x)$ in such a way that the relation

$$u_L^\dagger=0, \quad \text{where} \quad u_L^\dagger=u_\rho x^\rho{}_L, \tag{52.8}$$

holds everywhere. However by (52.5) we have

$$\frac{\partial\tilde{x}^\rho}{\partial y^L}=\frac{\partial x^\rho}{\partial y^L}+\frac{\partial x^\rho}{\partial t}\frac{\partial t(\tilde{t},y)}{\partial y^L} \quad \text{where} \quad x^\rho=\tilde{x}^\rho(\tilde{t},y)=x^\rho[t(\tilde{t},y),y]\,. \tag{52.9}$$

Hence

$$\tilde{u}_L^\dagger=0 \quad \left(\tilde{u}_L^\dagger=u_\rho\frac{\partial\tilde{x}^\rho}{\partial y^L}=u_L^\dagger+u_\rho\frac{\partial x^\rho}{\partial t}\frac{\partial t}{\partial y^L}=u_L^\dagger+u_\rho u^\rho\frac{Ds}{Dt}\frac{\partial t}{\partial y^L}\right) \tag{52.10}$$

holds if and only if

$$\frac{\partial t(\tilde{t},y)}{\partial y^L}=u_L^\dagger\frac{Dt}{Ds} \quad (u_L^\dagger=u_\rho x^\rho{}_L)\,. \tag{52.11}$$

Hence, given an event point $\mathscr{E}$, we can certainly choose function $(52.5)_1$ in such a way that $(52.10)_1$ holds at $\mathscr{E}$, i.e. in every physical situation the local validity of (52.8) can be realized by a suitable choice of the time parameter $t=\hat{t}(x)$. More, it is easy to see that the local condition (52.8), which only says that $S_3(t)$ is orthogonal to u^ρ at $\mathscr{E}$, is compatible with the validity of the condition $Dt=Ds$ everywhere. In addition, both conditions are obviously independent of the choice of the frames (y) and (x).

In case (52.8) holds, we say that the time parameter $\hat{t}(x)$ is *time orthogonal.* Incidentally, by $(52.9)_1$ the analogue of $(52.4)_2$ for $\tilde{x}(t,y)$ implies condition

$$\left(\frac{\partial t(\tilde{t},y)}{\partial y^L}\right)^2 \leqslant -g_{\rho\sigma}x^\rho{}_L x^\sigma{}_L \Big/ g_{\rho\sigma}\frac{\partial x^\rho}{\partial t}\frac{\partial x^\sigma}{\partial t} \qquad (>0) \tag{52.12}$$

on substitution $(52.5)_1$.

We also assume that the material frame (y) is right-handed (which has an obvious physical meaning). Then we say that the space-time frame (x) is right-handed if the first of the inequalities

$$\frac{\partial(x^1,x^2,x^3)}{\partial(y^1,y^2,y^3)}>0\,, \qquad \frac{\partial(x^0,x^1,x^2,x^3)}{\partial(t,y^1,y^2,y^3)}>0 \tag{52.13}$$

holds. By $(52.2)_5$, $(52.13)_1$ *implies* $(52.13)_2$.

Indeed there is an admissible right handed frame (x) for which $\partial x^0/\partial y^L=0$ holds (locally). Then by $(52.2)_5$ and $(52.13)_1$ we have

$$\frac{\partial(x^0,\ldots,x^3)}{\partial(t,y^1,y^2,y^3)}=\frac{\partial x^0}{\partial t}\,\frac{\partial(x^1,x^2,x^3)}{\partial(y^1,y^2,y^3)}>0\,. \tag{52.14}$$

Let us now remark that if the first three of the vectors $v_{(1)}{}^\rho,\ldots,v_{(4)}{}^\rho$, tangent to S_4 at $\mathscr{E}$, determine a space like hyperplane and $v_{(4)}{}^\rho$ is time-like and pointing towards future, then the 4-tuple $v_{(1)}{}^\rho,\ldots,v_{(4)}{}^\rho$ is linearly independent, and its orientation is independent of the value of $v_{(4)}{}^\rho$ and is invariant under projection of $v_{(1)}{}^\rho$ to $v_{(3)}{}^\rho$ on a same space like hyperplane. In connection with this it is known that, if (x) and $(\bar{x})$ are admissible frames in S_4, then the two conditions

$$\frac{\partial(\bar{x}_1,\bar{x}_2,\bar{x}_3)}{\partial(x_1,x_2,x_3)}>0\,, \qquad \frac{\partial(\bar{x}^0,\ldots,\bar{x}^3)}{\partial(x^0,\ldots,x^3)}>0 \tag{52.15}$$

are equivalent. Furthermore by (52.14) conditions $(52.13)_{1,2}$ are equivalent for a particular admissible frame $(\partial x^0/\partial y^L=0)$. Then $(52.13)_{1,2}$ are equivalent for every admissible frame (x).

§ 53. Lagrangian Spatial Derivative and Absolute Derivative of a Double Tensor Field with Respect to the Motion $\mathscr{M}$ of $\mathscr{C}$

We consider a double tensor field,

$$T_{\alpha\cdots}{}^{\beta\cdots}{}_{L\cdots}{}^{M\cdots}=T_{\cdots}^{\cdots}=T_{\cdots}^{\cdots}(x,t,y) \tag{53.1}$$

depending on the time parameter t. Given the motion $\mathscr{M}$ of $\mathscr{C}$ by means of $(52.2)_1$ and the event point x'^α, there is a time parameter $\tilde{t}=\tilde{t}(x)$ which is time orthogonal at x'^{α}. Furthermore x'^{α} determines the number $\bar{t}'=\bar{t}(x')$ and the material point y'^{L} such that $x'^{\alpha}=x^{\alpha}(\bar{t}',y'^{L})$—cf. (52.2).

We consider definition $(52.9)_3$, the mapping $(52.9)_2$ of y^L into x^ρ for $\bar{t}=\bar{t}'$, and the total derivative of the field (53.1) at x'^{α} based on this mapping—cf. Appendix A, (A 2.6); furthermore we call this derivative *Lagrangian spatial (covariant) derivative* and denote it by $T_{\cdots}^{\cdots}{}_{|A}$:

$$T_{\alpha\cdots}{}^{\beta\cdots}{}_{L\cdots}{}^{M\cdots}{}_{|A}=[T_{\cdots}^{\cdots}(\ldots)_{t=t(\bar{t},y)}]_{;A} \quad \text{for} \quad \tilde{u}_L^{\dagger}\equiv u_\rho \tilde{x}^\rho{}_L=0\,. \tag{53.2}$$

By (A 2.6) and $(52.9)_{2,3}$ from (53.2) we deduce

$$T_{\cdots}^{\cdots}{}_{|A}=T_{\cdots}^{\cdots}{}_{/\rho}\frac{\partial\tilde{x}^\rho}{\partial y^A}+\tilde{T}_{\cdots}^{\cdots}(\ )_{/A} \quad \text{for} \quad \tilde{T}^{\cdots}{}_{\cdots}(x,\bar{t},y)=[T_{\cdots}^{\cdots}(x,t,y)]_{t=t(\bar{t},y)} \tag{53.3}$$

which by (A 2.4), (A 2.6), and $(52.9)_1$ becomes

$$T_{\cdots}^{\cdots}{}_{|A}=T_{\cdots}^{\cdots}{}_{;A}+\left(T_{\cdots}^{\cdots}{}_{/\rho}\frac{\partial x^\rho}{\partial t}+\frac{\partial T_{\cdots}^{\cdots}}{\partial t}\right)\frac{\partial t(\bar{t},y)}{\partial y^A}\,. \tag{53.4}$$

Since $Dy^L/Ds=0$, the absolute derivative along world lines of the double tensor field (53.1) obviously is

$$\frac{DT_{\alpha\cdots}{}^{\beta\cdots}{}_{L\cdots}{}^{M\cdots}}{Ds}=\left(T_{\cdots}^{\cdots}{}_{/\rho}\frac{\partial x^\rho}{\partial t}+\frac{\partial T_{\cdots}^{\cdots}}{\partial t}\right)\frac{Dt}{Ds}\,. \tag{53.5}$$

We can assume $(53.2)_3$, i.e. $(52.10)_1$, which yields (52.11). By this formula and (53.5), (53.4) becomes the following equality whose left hand side depends—cf. (53.2)—on the particular time parameter $\bar{t}(x)$ and not on $\hat{t}(x)$ which is arbitrary, whereas its right hand side depends on $\hat{t}(x)$ and not on $\bar{t}(x)$:

$$T_{\cdots}^{\cdots}{}_{|A}=T_{\cdots}^{\cdots}{}_{;A}+\frac{DT_{\cdots}^{\cdots}}{Ds}u_A^{\dagger} \quad (u_A^{\dagger}=u_\rho x^\rho{}_A)\,. \tag{53.6}$$

Hence in particular $T_{\cdots}^{\cdots}{}_{|A}=T_{\cdots}^{\cdots}{}_{;A}$ for $u_A^{\dagger}=0$, so that the *Lagrangian spatial derivative of the field* (53.1) *is determined by the motion $\mathcal{M}$ of $\mathcal{C}$* (in that it does not depend on the choice of the locally time-orthogonal parameter $\bar{t}(x)$).

In general $T_{\cdots}^{\cdots}{}_{|A}\neq T_{\cdots}^{\cdots}{}_{;A}$ for $u_A^{\dagger}\neq 0$, and $T_{\cdots}^{\cdots}{}_{;A}$ is independent of the time parameter if and only if $DT_{\cdots}^{\cdots}/Ds=0$.

The *space-time position gradient* $x^\rho{}_L$—which appears to be a double tensor on the basis of (A 2.5)—is spatial[10)] if and only if, the time parameter is time orthogonal—cf. (52.8). Hence $x^\rho{}_L$ depends on this parameter, and because of this it is not very satisfactory from the physical point of view. Therefore we consider the *(first) spatial position gradient* $\alpha^\rho{}_L$:

$$\alpha^\rho{}_L=\overset{\perp}{g}{}^\rho{}_\sigma x^\sigma{}_L=\overset{\perp}{x}{}^\rho{}_L \text{—cf. the last footnote—}. \tag{53.7}$$

Since $x^\rho{}_L = x^\rho{}_{;L}$—cf. (A 2.5)—from (53.7), $(17.6)_1$, and $(53.6)_2$ we deduce

$$\alpha^\rho{}_L = x^\rho{}_{;L} + u^\rho u_\sigma x^\sigma{}_L = x^\rho{}_{;L} + \frac{Dx^\rho}{Ds} u_L^\dagger = x^\rho{}_L + u^\rho u_L^\dagger . \tag{53.8}$$

After having fixed the frame (x) and the index ρ (i.e. $\rho = 0$), we define the scalar field $f(x) = x^\rho$. Then from $(53.8)_{1,2}$ and (53.6) we deduce $\alpha^\rho{}_L = f(x)_{|L}$ which proves that $\alpha^\rho{}_L$ *is determined by* $\mathscr{M}$.

By (53.8) $\alpha^\rho{}_A = x^\rho{}_{;A} + (\partial x^\rho/\partial t)(Dt/Ds) u_A^\dagger$. Furthermore (53.5) and (A 2.6) hold. Then we can put (53.6) into the form

$$T_{\dots}^{\dots}{}_{|A} = T_{\dots}^{\dots}{}_{/\rho} \alpha^\rho{}_A + T_{\dots}^{\dots}{}_{/A} + \frac{\partial T_{\dots}^{\dots}}{\partial t} \frac{Dt}{Ds} u_A^\dagger . \tag{53.9}$$

Let us observe, on the basis of (53.6) and (A 2.3) to (A 2.6), that if (it locally occurs that) the frames (x) and (y) are geodesic and $\hat{t}(x)$ is time orthogonal, then

$$T_{\dots}^{\dots}{}_{|A} = T_{\dots}^{\dots}{}_{,\rho} x^\rho{}_A + T_{\dots}^{\dots}{}_{,A} \qquad (\text{for } g_{\alpha\beta,\gamma} = 0 = a^*_{LM,A},\ u_L^\dagger = 0) . \tag{53.10}$$

By (53.2) or (53.6) all theorems on total derivatives involving only first order derivatives, hold for spatial Lagrangian derivatives also (whereas e.g. the commutation theorems for $T_{\dots}^{\dots}{}_{|AB}$ and $\tilde{T}_{\dots}^{\dots}{}_{|A}$ are fairly different from the analogues for $T_{\dots}^{\dots}{}_{;AB}$—see § 90. In particular from (A 2.7) we deduce

$$g_{\alpha\beta|A} = g_\alpha{}^\beta{}_{|A} = g^{\alpha\beta}{}_{|A} = 0 = a^*_{LM|A} = a^*_L{}^M{}_{|A} = a^{*LM}{}_{|A} , \tag{53.11}$$

so that operation "$_{|A}$" commutes with the one of raising or lowering indices of double tensors. Furthermore the usual differentiation rule of products holds for the operation "$_{|A}$":

$$(T_{\alpha\beta}{}^{LM} \theta^\beta{}_M)_{|A} = T_{\alpha\beta}{}^{LM}{}_{|A} \theta^\beta{}_M + T_{\alpha\beta}{}^{LM} \theta^\beta{}_{M|A} . \tag{53.12}$$

The analogues of properties a) and b) for the operation "$_{;A}$" considered in Appendix A 2, hold for the operation "$_{|A}$". They can be presented as follows:

a) *If* (53.1) *is a space-time tensor field [a material time dependent field], i.e.* $T_{\dots}^{\dots}$ *has no Latin [Greek] indices, and is independent of* y^L *and* t *[of* x^ρ*], then*—cf. (53.9)

$$T_{\alpha\dots}{}^{\beta\dots}(x)_{|A} = T_{\dots}^{\dots}{}_{/\rho} \alpha^\rho{}_A , \qquad T_{L\dots}{}^{M\dots}(t,y)_{|A} = T_{\dots}^{\dots}{}_{/A} + \frac{\partial T_{\dots}^{\dots}}{\partial t} \frac{Dt}{Ds} u_A^\dagger . \tag{53.13}$$

b) *If the double tensor fields* $T_{\dots}^{\dots}(x,t,y)$—cf. (53.1)—*and* $\theta(x,t,y)$ *coincide along* $\mathscr{M}$ *[where e.g.* $\theta_{\dots}^{\dots}(x,t,y)$ *may be independent of either* x*, or* t *and* y*], then*

$$T_{\dots}^{\dots}(x,t,y)_{|A} = \theta_{\dots}^{\dots}(x,t,y)_{|A} \quad (\text{for } T_{\dots}^{\dots}[x(t,y),t,y] = \theta[x(t,y),t,y]) . \tag{53.14}$$

We introduce the *(first right) Cauchy Green tensor* C_{LM}:

$$C_{LM} = \alpha_{\rho L}\alpha^{\rho}{}_{M} = \alpha_{\rho L} x^{\rho}{}_{M} = \overset{\perp}{g}_{\rho\sigma} x^{\rho}{}_{L} x^{\sigma}{}_{M} \tag{53.15}$$

—cf. (53.7) and $(17.6)_4$—and (recursively) the n-th *spatial position gradient* $\alpha^{\rho}{}_{L_1\ldots L_n}$ and n-th *(right) Cauchy Green tensor* $C_{AL_1\ldots L_n}$ *for* $n=2,3,\ldots$:

$$\alpha^{\rho}{}_{L_1\ldots L_n} = \overset{\perp}{g}{}^{\rho}{}_{\sigma}\alpha^{\sigma}{}_{L_1\ldots L_{n-1}|L_n}, \qquad C_{AL_1\ldots L_n} = \alpha_{\rho A}\alpha^{\rho}{}_{L_1\ldots L_n} = \alpha_{\rho A}\alpha^{\rho}{}_{L_1\ldots L_{n-1}|L_n}. \tag{53.16}$$

Obviously C_{LM} is symmetric and $\alpha^{\rho}{}_{L_1\ldots L_n}$ is spatial $(n=1,2,\ldots)$.

The use of Lagrangian spatial derivatives in the definitions (53.16) is essential to ensure independence of the choice of $\hat{t}(x)$, hence acceptability from the physical point of view.

For the relativistic theory of first-order (or simple) materials—cf. footnote 15 in Chapter 1.—which includes ordinary relativistic elasticity and piezo-elasticity and will be presented in §§ 63, 72, only $\alpha^{\rho}{}_{L}$ and C_{LM} are essential.

However tensors (53.16) will also be used in this theory for some theorems of a nonfundamental character. The tensors (53.16) for $n=2$ are essential for the relativistic treatment of polar materials [§§ 91—98]. For dealing with the same materials some further theorems on the operation "$_{|A}$"—cf. e.g. Theorem 90.1—are particularly useful. Of course, in § 99 the treatment of materials of any order n requires the use of all tensors (53.16).

§ 54. Polar Decomposition of the Position Gradient $\alpha^{\rho}{}_{L}$ and Principal Axes of Strain

We assume that, locally, $\hat{t}(x)$ is time orthogonal, the frame (x) is natural and proper and the frame (y) is geodesic and Cartesian, i.e.—cf. $(15.3)_{2,3}$

$$\begin{aligned} &u^{\dagger}_{A}=0, \quad u^{\rho}=\delta^{\rho}_{0}, \quad g_{\alpha\beta}=\delta'_{\alpha\beta}, \\ &g_{\alpha\beta,\gamma}=0=a^{*}_{LM,A}, \quad a^{*}_{LM}=\delta_{LM} \quad (\text{at } \mathscr{E}) \end{aligned} \tag{54.1}$$

(or simply that the local conditions $(54.1)_{1,2,3}$ hold at $\mathscr{E}$). Then the position gradient $\alpha^{\rho}{}_{L}$ and the Cauchy-Green tensor C_{LM}—cf. $(53.7)_1$, (53.15), $(17.8)_1$—have the same expressions as in classical physics:

$$\alpha^{r}{}_{L} = x^{r}{}_{L}, \quad \alpha^{0}{}_{L}=0; \quad C_{LM} = \delta_{rs} x^{r}{}_{L} x^{s}{}_{M}. \tag{54.2}$$

From $(54.2)_3$ we deduce

$$C_{LM}\,\delta y^{L}\,\delta y^{M} = \delta_{rs}\,\delta x^{r}\,\delta x^{s} \geqslant 0 \quad \text{where} \quad \delta x^{r} = x^{r}{}_{L}\,\delta y^{L}. \tag{54.3}$$

Furthermore $(54.2)_3$ and $(52.13)_1$ imply $\det\|C_{LM}\| = (\det\|x^{r}{}_{L}\|)^2 > 0$. Then the quadratic form $C_{LM}\,\delta y^{L}\,\delta y^{M}$ is strictly positive-definite.

At this point the classical theory of first-order local deformation can be repeated in general relativity substantially with the same words. This holds in particular for the polar decompositions of $\alpha^{r}{}_{L}$ and the principal axes of stretching.

However let us explicitly remark that there is one strictly positive-definite symmetric (material) tensor—i.e. one *pure deformation*—$\mathscr{D}_{LM}$ for which

$$\mathscr{D}_{LA}\mathscr{D}^{A}{}_{M}=C_{LM} \qquad (\mathscr{D}_{[LM]}=0=C_{[LM]}). \tag{54.4}$$

Then its inverse $\overset{-1}{\mathscr{D}}_{LM}$ exists, and

$$\alpha^{\rho}{}_{L}=R^{\rho}{}_{A}\mathscr{D}^{A}{}_{L} \quad \text{where} \quad R^{\rho}{}_{A}=\alpha^{\rho}{}_{M}\overset{-1}{\mathscr{D}}{}^{M}{}_{A}, \qquad \overset{-1}{\mathscr{D}}{}^{M}{}_{A}\mathscr{D}^{A}{}_{L}=a^{*M}{}_{L}. \tag{54.5}$$

From $(54.5)_2$, $(53.15)_1$, and $(54.4)_1$ respectively we deduce

$$R^{\rho}{}_{A}R_{\rho B}=\alpha^{\rho}{}_{R}\alpha_{\rho S}\overset{-1}{\mathscr{D}}{}^{R}{}_{A}\overset{-1}{\mathscr{D}}{}^{S}{}_{B}=C_{RS}\overset{-1}{\mathscr{D}}{}^{R}{}_{A}\overset{-1}{\mathscr{D}}{}^{S}{}_{B}=a^{*}_{AB}. \tag{54.6}$$

Then, under conditions $(54.1)_{2,3,6}$, (i) the matrix $\|R^{r}{}_{A}\|$ is orthogonal and (ii) $R^{0}{}_{A}=0$ —cf. $(54.2)_2$, $(54.5)_2$. Thus, summing up,

$$R^{\rho}{}_{L}R_{\rho M}=a^{*}_{LM}, \qquad R^{\rho}{}_{A}R^{\sigma A}=\overset{\perp}{g}{}^{\rho\sigma}, \qquad u_{\rho}R^{\rho}{}_{L}=0 \tag{54.7}$$

holds in every frame. A double tensor $R^{\rho}{}_{L}$, fulfilling (54.7) will be called *rotation (double tensor)*.

It is clear that the pure deformation $\mathscr{D}_{LA}$ and the rotation for which the (right) polar decomposition $(54.5)_1$ holds, are only those given by $(54.4)_1$ and $(54.5)_2$, i.e. the polar decomposition (54.5) is unique.

Incidentally we can define the *left Cauchy-Green (spatial) tensor* $C'_{\rho\sigma}$ and the pure deformation $\mathscr{D}'_{\rho\sigma}$ by means of conditions

$$C'^{\rho\sigma}=\alpha^{\rho}{}_{A}\alpha^{\sigma A}=C'^{\sigma\rho}, \qquad \mathscr{D}'^{\rho}{}_{\lambda}\mathscr{D}'^{\lambda\sigma}=C'^{\rho\sigma}, \qquad \mathscr{D}'_{[\rho\sigma]}=0=u^{\rho}\mathscr{D}'_{\rho\sigma} \tag{54.8}$$

and in addition—using (54.1)—we can prove (as in classical physics) that

$$\alpha^{\rho}{}_{L}=\mathscr{D}'^{\rho}{}_{\sigma}R^{\sigma}{}_{L}, \qquad \mathscr{D}'^{\rho\sigma}=R^{\rho}{}_{L}R^{\sigma}{}_{M}\mathscr{D}^{LM} \tag{54.9}$$

and that the left polar decomposition $(54.9)_1$ is unique.

Now we can deal with the principal triad of strain in the same way as in classical physics. We can introduce, first, the principal stretchings $\mathscr{D}_{(r)}$ by the relations

$$\det\|\mathscr{D}_{LM}-\mathscr{D}_{(r)}a^{*}_{LM}\|=0, \qquad 0<\mathscr{D}_{(1)}\leqslant\mathscr{D}_{(2)}\leqslant\mathscr{D}_{(3)} \tag{54.10}$$

and then an orthonormal triple, $\lambda^{(r)}{}_{L}(=\lambda_{(r)L})$, of proper vectors of $\mathscr{D}_{LM}$:

$$\mathscr{D}_{LH}\lambda_{(r)}{}^{H}=\mathscr{D}_{(r)}\lambda_{(r)L}, \qquad \lambda_{(r)L}\lambda_{(r)}{}^{L}=\delta_{rs}, \qquad \lambda^{(r)}{}_{L}\lambda_{(r)M}=a^{*}_{LM}. \tag{54.11}$$

Incidentally, for $a^*_{LM}=\delta_{LM}$ we have $\lambda_{(r)L}=\lambda_{(r)}{}^L$, so that $(54.11)_2$ is obviously equivalent to $(54.11)_3$. By the tensor form of $(54.11)_{2,3}$ this equivalence holds in every frame.

The triple above, to be called *principal triad of strain*, is unique if and only if $\mathscr{D}_{(1)}<\mathscr{D}_{(2)}<\mathscr{D}_{(3)}$.

An axis whose unit vector λ^L is parallel with $\mathscr{D}^L{}_M\lambda^M$ (is an axis of a principal triad of strain and) is called a *principal axis of strain.*

For every (material) vector v_L (attached to y^L) the conditions

$$v_{(r)}=\lambda_{(r)}{}^A v_A, \qquad v_L=v_{(a)}\lambda^{(a)}{}_L \tag{54.12}$$

are equivalent. Indeed $(54.12)_1$ and $(54.11)_3$ imply $v_{(r)}\lambda^{(r)}{}_L=\lambda^{(r)}{}_L\lambda_{(r)}{}^A v_A=a^{*A}_L v_A$, i.e. $(54.12)_2$. Furthermore $(54.12)_2$ and $(54.11)_2$ imply $\lambda^A{}_{(r)}v_A=\lambda^A{}_{(r)}\lambda^{(a)}{}_A v_{(a)}=\delta_r{}^a v_{(a)}$, i.e. $(54.12)_1$.

Now remark that $(54.11)_{1,3}$ imply $\sum_r \mathscr{D}_{(r)}\lambda^{(r)}{}_L\lambda^{(r)}{}_M=\sum_r \mathscr{D}_{LH}\lambda_{(r)}{}^H\lambda^{(r)}{}_M=\mathscr{D}_{LH}a^{*H}{}_M$, i.e. $(54.13)_1$ below. Furthermore $(54.13)_1$ and $(54.12)_1$ imply $(54.13)_2$ below.

$$\mathscr{D}_{LM}=\sum_r \mathscr{D}_{(r)}\lambda^{(r)}{}_L\lambda^{(r)}{}_M, \qquad \mathscr{D}_{LM}v^M=\sum_r \mathscr{D}_{(r)}v_{(r)}\lambda^{(r)}{}_L. \tag{54.13}$$

By $(54.5)_1$, $(54.11)_1$, $(54.7)_1$ and $(54.11)_2$ *we have, for* $r\neq s$,

$$\alpha^\rho{}_L\alpha_{\rho M}\lambda_{(r)}{}^L\lambda_{(s)}{}^M=\mathscr{D}_{(r)}\mathscr{D}_{(s)}R^\rho{}_A R_{\rho B}\lambda_{(r)}{}^A\lambda_{(s)}{}^B=\mathscr{D}_{(r)}\mathscr{D}_{(s)}\lambda_{(r)}{}^A\lambda_{(s)A}=0; \tag{54.14}$$

hence *the directions* $\alpha^\rho{}_L\lambda^L{}_{(r)}$ *are mutually orthogonal.*

Incidentally let us remark that in every physical situation we can choose the frames (x) and (y) in such a way that the following holds at the arbitrarily preassigned event point $\mathscr{E}$:

(i) $g_{\alpha\beta}=\delta'_{\alpha\beta}$, $u^\rho=\delta^\rho{}_0$, $a^*_{LM}=\delta_{LM}$, i.e. $(54.1)_{2,3,6}$,

(ii) the co-ordinate lines of (y) are tangent to a principal triad of strain, $\lambda^{(r)}{}_L$,

(iii) the image of this triad, at $\mathscr{E}$, in a manifold of S_4 that is time orthogonal at $\mathscr{E}$ (e.g. the totally geodesic one) is the triad of the co-ordinate lines x^1, x^2, x^3, so that $\alpha^s{}_L\lambda^{(r)L}\geqslant 0$ for $r=s$ and $r\neq s$ respectively.

By $(\text{i})_3$ and (ii) $\alpha^s{}_r=\alpha^{sr}=\alpha^s{}_A\lambda^{(r)A}$; hence, by (iii) $\alpha^r{}_r>0$ and $\alpha^r{}_L=0$ $(r\neq L)$. In addition, by (ii) and $(\text{i})_3$, $\lambda^{(r)}{}_L=\delta^r{}_L$, so that, by $(54.13)_1$, $\mathscr{D}_{AM}=\mathscr{D}_{(L)}\delta_{AL}$ (with $\mathscr{D}_{(L)}>0$) which yields $\alpha^\rho{}_L=R^\rho{}_A\mathscr{D}_{(L)}\delta^A{}_L$ by $(54.5)_1$ and $(\text{i})_3$. As a consequence $R^r{}_L>0$ for $L=r$ and $R^r{}_L=0$ $(r\neq L)$. In addition $R^0{}_L=0$ by $(\text{i})_{1,2}$ and hence $R^r{}_A R_{rB}=\delta_{AB}$ by (54.6) and $(\text{i})_3$. We conclude that $R^r{}_A=\delta^r{}_A$ $(R^0{}_A=0)$ holds in every physical situation for a suitable choice of the frames (x) and (y).

The analogue can be realized in classical physics in case both of the Euclidean frames (x^1, x^2, x^3) and (y) are allowed to be chosen arbitrarily. However in classical physics—as well as in special relativity, or in general relativity when Fock's conjecture [§ 10] is accepted—it is physically meaningful to compare (locally) the orientations of the two frames in every case, whereas in our theory of general relativity—which is independent of Fock's conjecture—such a comparison is physically meaningful only in case the reference process $\mathscr{P}^*$ of the universe coincides with the actual one $\mathscr{P}$, at least before a suitable space-like surface, say S_3 (which only need to be placed after S_3^*). How such a comparison can be realized will be said in § 55 with a view to dealing with materials with memory.

Now let us note that, generally, the afore-mentioned (partial) coincidence of $\mathscr{P}^*$ and $\mathscr{P}$ does not hold, and this is important in connection with constitutive equations especially if they are related to materials with memory. Thus the rotation $R^\rho{}_L$ has no physical meaning in our theory. However this causes no trouble in our treatment of constitutive equations because of the principle of material indifference [§ 81] and in particular by the part of it which mirrors the local spatial physical isotropy of space time.

§ 55. Fermi Transport

This section is essential for our theory of materials with memory [§ 80] but will not be applied at all to deal with elasticity or electromagnitostriction.

Let V^ρ and W^ρ be spatial vectors defined along W_{P*}. In accord with Synge [1960, p. 15] we say e.g. that V^ρ undergoes *Fermi transport* if $\grave{V}^\rho (= DV^\rho/Ds)$ has zero spatial components:

$$\overset{\perp}{g}{}^\rho{}_\sigma \grave{V}^\sigma = 0 \quad (\text{and } u_\rho V^\rho = 0, \text{ whence } \grave{V}^\sigma u_\sigma = -V^\sigma A_\sigma). \tag{55.1}$$

By $(55.1)_{1,3}$ and $(17.6)_1$ the explicit form of Fermi transport is[11)]

$$\grave{V}^\rho = V^\sigma A_\sigma u^\rho. \tag{55.2}$$

It is easy to see that *in case the spatial vectors V^ρ and W^ρ undergo Fermi transport, $V^\rho W_\rho$ is constant. Hence in the same case V^ρ and W^ρ have constant moduli and form a constant angle.*

Obviously, given the event point $\mathscr{E}^*$ on W_{P*}, and the spatial vector V_ρ^* attached to $\mathscr{E}^*$, there is exactly one vector V^ρ which is defined along W_{P*}, equals V_ρ^* at $\mathscr{E}$, and fulfills (55.2). Thence we deduce $(55.1)_{1,3}$, hence $D(u^\rho V_\rho)/Ds = 0$ which easily yields $(54.1)_2$. Thus V_ρ is spatial along W_{P*} and undergoes Fermi transport.

We assume that the three vectors $f_{(r)\rho}$ undergo Fermi transport along W_{P*} and constitute a spatial cartesian (i.e. orthonormal) triad at $\mathscr{E}^*$, so that they always constitute such a triad:

$$\grave{f}_{(r)}{}^\rho = f_{(r)}{}^\sigma A_\sigma u^\rho, \qquad u_\rho f_{(r)}{}^\rho = 0, \qquad f_{(r)}{}^\rho f_{(s)\rho} = \delta_{rs}. \tag{55.3}$$

We call such a triad a *Fermi triad*.

It is easy to see that the most general vector undergoing Fermi transport (along W_{P*}) is a vector having constant components in the Fermi triad $f_{(r)\rho}$. In particular two Fermi triads (along W_{P*}) have constant mutual orientations.

In locally proper and natural co-ordinates equations (55.1) in V^ρ become $\grave{V}^r=0$, $V^0=0$. Thus [§ 19] the classical analogue of the vector V^ρ which undergoes Fermi transport is an ordinary vector, V^r, having zero time-derivative with respect to a cartesian frame which is locally freely falling and non-rotating. Then the vector V^r is constant with respect to inertial spaces. Hence a Fermi triad is the analogue for special or general relativity, of a classical frame whose orientation (with respect to inertial spaces) is constant, and whose origin undergoes an arbitrary motion.

The usefulness of Fermi triads mainly consists in their allowing us to compare the oriented directions of spatial vectors attached to points of a same world line, W_{P*}, and the orientations of spatial cartesian frames attached to points of W_{P*}. In particular we now assume that the actual process $\mathscr{P}$ (being considered) coincides with the reference process $\mathscr{P}^*$ at least before—i.e. in the past of—the space-like hypersurface S_3^* [§ 52], and that $\mathscr{E}^*$ belongs to S_3^*. Then we can think of the local reference triad $\lambda^L{}_{(A)}$, considered in § 54, as attached to $\mathscr{E}^*$ and we can compare it with the Cartesian spatial triad $l_{(A)}{}^\rho = R^\rho{}_L \lambda_{(A)}{}^L$ attached to the point $\mathscr{E}$ of W_{P*} as follows: the orientation of the latter with respect to the former is given, as well as in classical physics, by the intrinsic components of $R^\rho{}_L$ with respect to the Fermi frame $f_{(r)}{}^\rho$ which coincides with $\lambda_{(A)}{}^L$, assumed to equal $l_{(A)}{}^\rho$, at $\mathscr{E}^*$:

$$R_{(r)(A)} = f_{(r)\rho}\, l_{(A)}{}^\rho \qquad (l_{(A)}{}^\rho \equiv R^\rho{}_L \lambda_{(A)}{}^L;\ f_{(A)}{}^\rho = R^\rho{}_L \lambda_{(A)}{}^L \text{ at } \mathscr{E}^*)\,. \tag{55.4}$$

Using $(17.8)_{1,2,3}$, from $(55.3)_{2,3}$ we easily deduce

$$f_{(r)\rho}\, f^{(r)}{}_\sigma = \overset{\perp}{g}_{\rho\sigma}\,; \tag{55.5}$$

and by $(55.4)_1$, (55.5), $(55.4)_2$, $(54.7)_1$, and $(54.11)_2$ we have, respectively,

$$R_{(r)(A)} R^{(r)}{}_{(B)} = f_{(r)\rho}\, f^{(r)}{}_\sigma\, l_{(A)}{}^\rho\, l_{(B)}{}^\sigma = l_{(A)}{}^\rho\, l_{(B)\rho} = R^\rho{}_L R_{\rho M} \lambda_{(A)}{}^L \lambda_{(B)}{}^M = \lambda_{(A)}{}^L \lambda_{(B)L} = \delta_{AB}\,. \tag{55.6}$$

Thus, since $R^{(r)}{}_{(A)} = R_{(r)(A)}$, $\|R_{(r)(A)}\|$ is an orthogonal matrix; it allows us to determine the Euler angles of the frame $l_{(A)}{}^\rho$ at $\mathscr{E}$ with respect to the frame $f_{(r)\rho}$—or $\lambda_{(A)}{}^L$—at $\mathscr{E}^*$, the rotation angle ϕ, and (for $\phi \neq 2K\pi$) the rotation axis.

§ 56. On the Dilation Coefficients for Line, Volume, and Surface Elements, and the Ratio dC/dC^*

For the material line element δy^L we set

$$\delta x^\rho = \frac{\partial x^\rho}{\partial y^L}\,\delta y^L\,, \quad \delta y^L = |\delta y|\,\lambda^L\,, \quad \delta \overset{\perp}{x}{}^\rho = |\delta \overset{\perp}{x}|\, l^\rho\,, \quad \lambda^L \lambda_L = 1 = l^\rho l_\rho\,, \tag{56.1}$$

so that the quantity $|\delta y|$ $[|\delta \overset{+}{x}|]$ defined by $(56.1)_{2,4}$ $[(56.1)_{3,5}]$ is the modulus of $\delta y^L [\delta \overset{+}{x}^\rho]$. Notations such as $|\delta y|$ or $|\delta \overset{+}{x}|$ will be used in connection with other material or spatial vectors.

By $(56.1)_1$ and $(53.7)_1$ we have $\delta \overset{+}{x}^\rho = \alpha^\rho{}_L \delta y^L$. Then, on the one hand, the coefficient $\delta_\lambda{}^{(l)}$ of line dilation (at y^L) in the direction λ_L has the expressions

$$\delta_\lambda{}^{(l)} = \frac{|\delta \overset{+}{x}| - |\delta y|}{|\delta y|} = \frac{|\delta \overset{+}{x}|}{|\delta y|} - 1 = (C_{LM} \lambda^L \lambda^M)^{1/2} - 1 = (\mathscr{D}_{AL} \lambda^L \mathscr{D}^A{}_M \lambda^M)^{1/2} - 1\,. \quad (56.2)$$

Indeed by $(56.1)_1$ and $(53.15)_1$ we have $|\delta \overset{+}{x}|^2 = \alpha^\rho{}_L \delta y^L \alpha_{\rho M} \delta y^M = C_{LM} \lambda^L \lambda^M |d y|^2$, whence $(56.2)_3$. Lastly $(56.2)_4$ follows from $(54.4)_{1,2}$.

On the other hand $(53.7)_1$ and $(56.1)_{1,2,3}$ imply $|\delta \overset{+}{x}|\, l^\rho = \alpha^\rho{}_L \lambda^L |d y|$, which by $(56.2)_{1,2}$ yields

$$l^\rho = \frac{1}{1 + \delta_\lambda{}^{(l)}} \alpha^\rho{}_L \lambda^L \qquad (\delta \overset{+}{x}^\rho = \alpha^\rho{}_L \delta y^L)\,. \quad (56.3)$$

We now consider the matter element $d\mathscr{C}$ containing the material point y^L (or P^*), its proper volume dC^* $(dC^* > 0)$ in the reference configuration, and the one dC $(dC > 0)$ when P^* is at the event point $\mathscr{E}$ of W_{P^*}.

In order to prove the equality

$$\mathscr{D} = \frac{dC}{dC^*} \quad \text{where} \quad \mathscr{D} = \sqrt{\frac{-g}{a^*}} \left| \frac{\partial(x^0, \ldots, x^3)}{\partial(t, y^1, y^2, y^3)} \right| \frac{Dt}{Ds} > 0 \qquad (a^* = \det \|a^*_{LM}\|)\,, \quad (56.4)$$

we identify $d\mathscr{C}$ with the parallelepiped $(y^L, dy^L, \delta y^L, \partial y^L)$ for the sake of simplicity. Then, denoting by ε^*_{HLM} the material Ricci tensor, i.e. the one in S^*_3, we have

$$dC^* = \pm \varepsilon^*_{HLM} dy^H \delta y^L \partial y^M \qquad (\varepsilon^*_{HLM} = \sqrt{a^*}\, \mathscr{E}_{0HLM})\,. \quad (56.5)$$

Furthermore, up to irrelevant infinitesimals, dC is the volume of the spatial parallelepiped $(x^\rho, d\overset{+}{x}^\rho, \delta \overset{+}{x}^\rho, \partial \overset{+}{x}^\rho)$, with $dx^\rho = x^\rho{}_L dy^L, \ldots, \partial x^\rho = x^\rho{}_L \partial y^L$. Then, by the consequence $\overset{+}{\varepsilon}_{\alpha\beta\gamma} = \varepsilon_{\rho\alpha\beta\gamma} (\partial x^\rho / \partial t) Dt/Ds$ of $(20.4)_1$, we have—cf. $(16.5)_1$

$$\begin{aligned} dC &= \pm \overset{+}{\varepsilon}_{\alpha\beta\gamma} dx^\alpha \delta x^\beta \partial x^\gamma = \pm \frac{Dt}{Ds} \left(\varepsilon_{\rho\alpha\beta\gamma} \frac{\partial x^\rho}{\partial t} x^\alpha{}_H x^\beta{}_L x^\gamma{}_M \right) dy^H \delta y^L \partial y^M \\ &= \pm \frac{Dt}{Ds} \left(\sqrt{\frac{-g}{a^*}}\, \mathscr{E}_{\rho\alpha\beta\gamma} \frac{\partial x^\rho}{\partial t} x^\alpha{}_1 x^\beta{}_2 x^\gamma{}_3 \varepsilon^*_{HLM} \right) dy^H \delta y^L \partial y^M\,. \end{aligned} \quad (56.6)$$

From (56.6), $(56.5)_1$, and $(56.4)_2$ we deduce $(56.4)_1$ $(dC > 0, dC^* > 0)$.

Incidentally let us remark that under the local conditions $(54.1)_{2,1,3,6}$ we have

$$1 = u^0 = \frac{\partial x^0}{\partial t} \frac{Dt}{Ds}, \qquad \frac{\partial x^0}{\partial y^L} = 0\,, \qquad -g = 1 = a^*\,, \quad (56.7)$$

so that $(56.4)_2$ becomes identical with the classical expression

$$\mathscr{D} = \left| \frac{\partial(x^1, x^2, x^3)}{\partial(y^1, y^2, y^3)} \right| \tag{56.8}$$

for the ratio dC/dC^*, in cartesian co-ordinates.

Furthermore (54.1) implies $(54.2)_3$, which by $(56.7)_5$ and (56.8) yields the first of the equalities

$$\begin{aligned} \mathscr{D}^2 &= \frac{1}{a^*} \det \|C_{LM}\| = \frac{-g}{a^*} \det \|\alpha^\rho{}_L\|^2 , \\ \mathscr{D} &= \frac{1}{a^*} \det \|\mathscr{D}_{LM}\| , \qquad \overset{+}{\varepsilon}_{\alpha\beta\gamma} \alpha^\alpha{}_A \alpha^\beta{}_B \alpha^\gamma{}_C = \mathscr{D} \varepsilon^*_{ABC} . \end{aligned} \tag{56.9}$$

Furthermore $(53.15)_1$ implies $(56.9)_2$ where $\det \|\alpha^\rho{}_L\|^2$ can be calculated using Binet's formula; $(56.9)_3$, follows from $(56.9)_1$ and $(54.4)_1$; By (54.1), equalities $(54.2)_{1,2}$ hold, so that the left hand side of $(56.9)_4$ equals $\partial(x^1, x^2, x^3)/\partial(y^A, y^B, y^C)$. Then (56.8) yields $(56.9)_4$. Now remark that the scalar $\det \|\alpha^\rho{}_L\|^2 g/a^*$ equals $g \det \|\alpha^\rho{}_L \alpha^{\sigma L}\|$ in all frames (x) and (y), so that it is a tensor.

Since $\alpha^\rho{}_L$, C_{LM}, ε^*_{ABC}, $\overset{+}{\varepsilon}_{\alpha\beta\gamma}$, $\mathscr{D}$, and the right hand sides of $(56.9)_{2,3}$ are tensors independent of the choice of $\hat{t}(x)$, equalities (56.9) hold for every choice of the functions $(52.2)_1$ which represent $\mathscr{M}$, in all frames (x) and (y).

By $(56.4)_1$ and $(56.9)_{1,3}$, the coefficient $\lambda^{(c)}$ of cubic dilation—at x^ρ or (t, y)—has, among others, the expressions

$$\delta^{(c)} = \frac{dC - dC^*}{dC^*} = \mathscr{D} - 1 = \left(\frac{\det \|C_{LM}\|}{a^*} \right)^{1/2} - 1 = \frac{1}{a^*} \det \|\mathscr{D}_{LM}\| - 1 . \tag{56.10}$$

Now we consider the oriented material parallelogram represented by the vector

$$d\sigma^*_L = {}^*n_L |d\sigma^*| = \pm \varepsilon^*_{LRS} dy^R \delta y^S \tag{56.11}$$

where the upper or lower sign holds according to whether (y) is right handed or not.

When the material point y^L (or P^*) is at the event point x^ρ, this parallelogram occupies (up to irrelevant infinitesimals) the spatial parallelogram $(x^\rho, d\overset{+}{x}{}^\rho, \delta \overset{+}{x}{}^\rho)$ with $dx^\rho = x^\rho{}_L dy^L$ and $\delta x^\rho = x^\rho{}_L \delta y^L$, which is represented by the spatial vector $d\sigma_\lambda$:

$$d\sigma_\lambda = n_\lambda |d\sigma| = \pm \overset{+}{\varepsilon}_{\lambda\rho\sigma} dx^\rho \delta x^\sigma = \pm \overset{+}{\varepsilon}_{\lambda\rho\sigma} \alpha^\rho{}_R \alpha^\sigma{}_S dy^R \delta y^S , \tag{56.12}$$

where the upper or lower sign holds according to whether (x) is right handed or not.

Since (56.11) and an analogue of $(20.5)_1$ imply $2dy^{[R}\delta y^{S]} = \pm\varepsilon^{*LRS}d\sigma^*_L$, from (56.12) we deduce

$$d\sigma_\lambda = \gamma_\lambda{}^L d\sigma^*_L \quad \text{with} \quad \gamma_\lambda{}^L = \pm\frac{1}{2}\overset{\perp}{\varepsilon}_{\lambda\rho\sigma}\varepsilon^{*LRS}\alpha^\rho{}_R\alpha^\sigma{}_S \tag{56.13}$$

where, as well as in the remainder of this section, the upper or lower signs hold according to whether the frames (x) and (y) have concordant or discordant orientations. However these orientations may always be supposed to be concordant, more right-handed, in the sequel.

The quantity $\gamma_\rho{}^L$ defined by $(56.13)_2$—where $\alpha^\rho{}_L$ can be replaced with $x^\rho{}_L$—is an absolute (non-oriented) double tensor, which is independent of the choice of $\hat{f}(x)$ and determines the correspondence $(56.13)_1$ of surface elements.

Now we assume the local conditions (54.1), whence $(54.2)_1$. Then $(56.13)_2$ yields

$$\begin{aligned}\gamma_r{}^L\alpha^s{}_L &= \pm\frac{1}{2}\mathscr{E}_{0rab}\mathscr{E}_0{}^{LAB}x^s{}_L x^a{}_A x^b{}_B = \pm\frac{1}{2}\mathscr{E}_{0rab}\frac{\partial(x^s,x^a,x^b)}{\partial(y^1,y^2,y^3)}\\ &= \pm\frac{\partial(x^s,x^{r+1},x^{r+2})}{\partial(y^1,y^2,y^3)} \geqslant 0 \qquad (x^{i+3}=x^i)\,.\end{aligned} \tag{56.14}$$

Furthermore (56.8) and the relations $\gamma_0{}^L = 0 = \alpha^0{}_L$ hold—cf. $(56.13)_2$, $(54.2)_2$. Then we have the first two of the equalities

$$\gamma_\rho{}^L\alpha^\sigma{}_L = \mathscr{D}\overset{\perp}{g}_\rho{}^\sigma\,, \qquad u^\rho\gamma_\rho{}^L = 0\,, \qquad \gamma_\rho{}^L\alpha^\rho{}_M = \mathscr{D}a^{*L}{}_M\,. \tag{56.15}$$

Relations $(56.15)_{1,2}$ say that $\overset{-1}{\alpha}{}_r^L$ equals $\mathscr{D}^{-1}\gamma_r{}^L$; thence $(56.15)_3$, easily follows. It is easy to see that (56.15) holds even if none of the assumptions (54.1) do.

For the sake of completeness we also prove the relations

$$\begin{aligned}&\mathscr{D}^2\overset{-1}{C}{}^{LM} = \gamma_\rho{}^L\gamma^{\rho M}\,, \qquad \alpha^\rho{}_L\alpha^\sigma{}_M\overset{-1}{C}{}^{LM} = \overset{\perp}{g}{}^{\rho\sigma}\,, \qquad \overset{-1}{\mathscr{D}}{}_L^A\,\mathscr{D}_{\overset{-1}{A}M} = C_{\overset{-1}{L}M}\,,\\ &\mathscr{D}_{\overset{-1}{[LM]}} = 0\,, \qquad \Big(C_{HA}\overset{-1}{C}{}^{AM} = a^{*M}_H = \mathscr{D}_{HA}\overset{-1}{\mathscr{D}}{}^{AM}\Big).\end{aligned} \tag{56.16}$$

By $(53.15)_1$ and $(56.15)_{1,3}$ we have

$$\gamma_\rho{}^L\gamma^{\rho H}C_{HA} = \gamma_\rho{}^L\gamma^{\rho H}\alpha^\sigma{}_H\alpha_{\sigma A} = \mathscr{D}\overset{\perp}{g}{}^{\rho\sigma}\gamma_\rho{}^L\alpha_{\sigma A} = \mathscr{D}^2a^{*L}{}_A$$

which by definition $(56.16)_5$ yields $(56.16)_1$. Furthermore $(56.16)_2$ follows from $(56.16)_1$ and $(56.15)_1$, $(56.16)_3$ from definitions $(56.16)_{5,6}$ and $(54.4)_1$, and $(56.16)_4$ from $(54.4)_2$ and $(56.16)_6$.

By $(56.11)_1$, $(56.12)_1$, and $(56.13)_1$ $n_\lambda|d\sigma| = \gamma_\lambda{}^{L\,*}n_L|d\sigma^*|$, which yields the relation

$$n_\lambda = \frac{1}{1+\delta_{*n}{}^{(\sigma)}}\gamma_\lambda{}^{L\,*}n_L \quad \text{with} \quad \delta_{*n}{}^{(\sigma)} = \frac{|d\sigma|-|d\sigma^*|}{|d\sigma^*|} = \frac{|d\sigma|}{|d\sigma^*|} - 1 \tag{56.17}$$

between the corresponding orientations (characterized by the unit vectors) $^*n_L$ and n_λ. In order to find useful expressions for the dilation coefficient $\delta_{*n}^{(\sigma)}$ for material surface elements which have the orientation $^*n_L$, we remark that $(56.11)_1$, $(56.12)_1$, and $(56.13)_1$ imply—see the equality written above (56.17)

$$|d\sigma|^2 = d\sigma_\rho d\sigma^\rho = \gamma_\rho{}^L \gamma^{\rho M} {}^*n_L {}^*n_M |d\sigma^*|^2 \tag{56.18}$$

which together with $(56.17)_{2,3}$, $(56.16)_1$, and $(56.16)_{3,4}$ yields

$$\delta_{*n}^{(\sigma)} + 1 = (\gamma_\rho{}^L \gamma^{\rho M} {}^*n_L {}^*n_M)^{1/2} = \mathscr{D}\left(\overset{-1}{C}{}^{LM} {}^*n_L {}^*n_M\right)^{1/2} = \mathscr{D}\left(\overset{-1}{\mathscr{D}}{}^{AL} {}^*n_L \overset{-1}{\mathscr{D}}{}_A{}^M {}^*n_M\right)^{1/2} \tag{56.19}$$

respectively. From (56.19) and $(56.9)_3$ we see that $\delta_{*n}^{(\sigma)}$ is a simple algebraic function of $\mathscr{D}^{LM}$—as well as $\delta_\lambda^{(l)}$ and $\delta^{(c)}$.

Incidentally we have

$$\gamma_\rho{}^L = \mathscr{D} \overset{+}{g}_\rho{}^\sigma \frac{\partial y^L}{\partial x^\sigma} = \mathscr{D} y^L{}_{,\overset{+}{\rho}} \quad (\text{where } x^\alpha \equiv \hat{x}^\alpha[\hat{t}(x), y(x)]) \tag{56.20}$$

for, if we consider $(56.15)_{1,2}$, as equations in $\gamma_\rho{}^L$, then (56.20) affords a solution of them—cf. $(53.7)_1$—and such a solution is unique. Furthermore the relations

$$\mathscr{D}\, d\sigma_L^* = \alpha^\rho{}_L d\sigma_\rho, \qquad \mathscr{D}|d\sigma^*|\, {}^*n_L = |d\sigma| \alpha^\lambda{}_L n_\lambda, \qquad \mathscr{D}\, {}^*n_L = (1 + \delta_{*n}^{(\sigma)}) n_\lambda \alpha^\lambda{}_L \tag{56.21}$$

hold, the first by $(56.13)_1$ and $(56.15)_3$, the second by $(56.21)_1$,$(56.11)_1$, and $(56.12)_1$, and the third by $(56.21)_2$ and $(56.17)_{2,3}$.

§ 57. The Vectors V_L^* and V_*^L for V_ρ Spatial. Expressions of $\dot{\alpha}^\rho{}_L$ and $\dot{C}_{LM}$ in Terms of $u_{\rho/\sigma}$

Convention. *For every spatial vector V_ρ we denote by V_L^* and V_*^L the vectors defined by*

$$V_L^* = V_\rho \alpha^\rho{}_L, \qquad V^\rho = \alpha^\rho{}_L V_*^L \qquad (V^\rho u_\rho = 0). \tag{57.1a}$$

The covariant [contravariant] form of vector V_L^* [V_*^L] is the more natural. Then we may call this vector the *covariant [contravariant] material image* of the spatial vector V_ρ.

Incidentally, on the basis of (56.15) we easily see that (57.1a) implies

$$V_\rho = \mathscr{D}^{-1} \gamma_\rho{}^L V_L^*, \qquad V_*^L = \mathscr{D}^{-1} \gamma_\rho{}^L V^\rho \qquad (V^\rho u_\rho = 0) \tag{57.1b}$$

and, more precisely, $(57.1\,a)_{1,3}$ is equivalent to $(57.1\,b)_{1,3}$ and $(57.1\,a)_{2,3}$ to $(57.1\,b)_{2,3}$.

From $(57.1\,a)_{1,2}$, $(53.15)_1$, and $(56.16)_2$ we easily deduce

$$V_L^* = C_{LM} V_*^M, \quad C_{LM} V_*^L W_*^M = V_*^L W_L^* = V^\rho W_\rho = \overset{-1}{C}{}^{LM} V_L^* W_M^* \tag{57.2}$$

for all spatial vectors V^ρ and W^ρ.

Now we remark that $(53.13)_1$ and $(17.4)_3$ imply

$$u_{\rho|L} = u_{\rho/\sigma} \alpha^\sigma{}_L, \quad u^\rho u_{\rho|L} = 0. \tag{57.3}$$

Furthermore we assume the identity $t \equiv s$, which implies $u^\rho = \partial x^\rho(s,y)/\partial s$, and the local conditions $(53.10)_{2,3,4}$ which imply $(53.10)_1$. Thence we have $u^\rho{}_{|L} = \partial^2 x^\rho / \partial s\, \partial y^L$. Then by $(53.7)_1$ and $(53.10)_2$

$$\grave{\alpha}^\rho{}_L = \overset{\perp}{g}{}^\rho{}_\sigma \frac{\partial^2 x^\sigma}{\partial s\, \partial y^L} + (u^\rho A_\sigma + A^\rho u_\sigma) x^\sigma{}_L \quad \left(\frac{\partial^2 x^\sigma}{\partial s\, \partial y^L} = u^\sigma{}_{|L}\right). \tag{57.4}$$

Since we have $u_L^\dagger \equiv u_\sigma x^\sigma{}_L = 0$ (hence $x^\sigma{}_L = \alpha^\sigma{}_L$), the first of the equalities

$$\grave{\alpha}^\rho{}_L = u^\rho{}_{|L} + u^\rho A_L^* = (u^\rho{}_{/\sigma} + u^\rho A_\sigma) \alpha^\sigma{}_L \tag{57.5}$$

holds by (57.4), $(57.3)_2$, and $(57.1\,a)_1$, and the second by $(57.3)_1$ and $(57.1\,a)_1$.

From $(57.5)_1$, $(53.15)_1$, and $(57.3)_1$ respectively we deduce the relations

$$\overset{\perp}{g}{}^\rho{}_\sigma \grave{\alpha}^\sigma{}_L = u^\rho{}_{|L}, \quad \grave{C}_{LM} = u_{\rho|L} \alpha^\rho{}_M + \alpha^\rho{}_L u_{\rho|M} = 2 u_{(\rho/\sigma)} \alpha^\rho{}_L \alpha^\sigma{}_M \tag{57.6}$$

which obviously hold, as well as (57.5), for every admissible choice of (x), (y), and $\hat{t}(x)$. From $(57.6)_{2,3}$ we deduce

$$\grave{\varepsilon}_{LM} = u_{(\rho/\sigma)} \alpha^\rho{}_L \alpha^\sigma{}_M \quad \text{where} \quad C_{LM} = a_{LM}^* + 2\varepsilon_{LM}. \tag{57.7}$$

§ 58. New Determination of the General Solution for the Continuity Equation. Connection of $D^c V_\rho$ and $D_c V_\rho$ with $D V_*^L$ and $D V_L^*$ for V_ρ Spatial, and Lagrangian Expression for the Electromagnetic Work $d_3 \lambda$

From $(56.9)_1$, $(57.6)_{2,3}$, and $(56.16)_2$ we deduce

$$2\mathscr{D}\grave{\mathscr{D}} = \mathscr{D}^2 \overset{-1}{C}{}^{LM} \grave{C}_{LM} = 2\mathscr{D}^2 \overset{-1}{C}{}^{LM} \alpha^\rho{}_L \alpha^\sigma{}_M u_{(\rho/\sigma)} = 2\mathscr{D}^2 u^\rho{}_{/\overset{\perp}{\rho}} \tag{58.1}$$

which by $(17.20)_2$ yields

$$\grave{\mathscr{D}} = \mathscr{D} u^\rho{}_{/\rho}, \quad \frac{D}{Ds} \log \mathscr{D} = u^\rho{}_{/\rho} \tag{58.2}$$

Let us remember that k^* was defined in § 21 by the condition that $k^* dC^*$ should be the proper gravitational mass of $d\mathscr{C}$ in the reference state. Now this state is characterized by $\mathscr{P}^*$ and S_3^*. Then k^* is a function $k^*(y)$ of y^L. Hence $\grave{k}^*=0$. Furthermore, by combining $(21.2)_1$ with $(56.4)_1$ we have

$$k^* = k\mathscr{D} \quad \text{with} \quad k^* = k^*(y^1, y^2, y^3), \qquad \grave{k}^* = 0. \tag{58.3}$$

The continuity equation $(21.3)_1$—*or* $D\log k/Ds = -u^\alpha{}_{/\alpha}$—*follows from* (58.3) *and* $(58.2)_2$. The same holds if we do not take the physical meanings of k^* and k into account and if we consider function $(58.3)_2$ as given arbitrarily. Of course we consider the motion (52.2) as given.

Now we conversely assume the continuity equation $(21.3)_1$, so that by $(58.2)_2$ $D(\log\mathscr{D} + \log k)/Ds = 0$. Then $k\mathscr{D}$ is constant along world lines, so that $k\mathscr{D}$ is a function of y^L.

We conclude that $k = \mathscr{D}^{-1} k^*(y^1, y^2, y^3)$ is the general solution of the continuity equation.

The remainder of this section is basic for electromagnetostriction and will not be applied to other purposes.

The identities $(57.1\text{a})_{2,3}$ and (57.5) imply

$$\overset{\perp}{g}{}^\rho{}_\sigma \grave{V}^\sigma = \overset{\perp}{g}{}^\rho{}_\sigma \grave{\alpha}^\sigma{}_L V_*^L + \alpha^\rho{}_L \grave{V}_*^L = u^\rho{}_{/\sigma} \alpha^\sigma{}_L V_*^L + \alpha^\rho{}_L \grave{V}_*^L. \tag{58.4}$$

Thence by $(57.1\text{a})_2$ and $(22.1)_1$ the first of the basic relations

$$\frac{D^c V^\rho}{Ds} = \alpha^\rho{}_L \frac{DV_*^L}{Ds}, \qquad \frac{DV_L^*}{Ds} = \frac{D_c V_\rho}{Ds} \alpha^\rho{}_L \qquad (\text{for } V_\rho u^\rho = 0) \tag{58.5}$$

follows. Now let W^ρ be an arbitrary spatial vector defined along W_{p*}, as well as V^ρ. Then convention $(57.1\text{a})_1$ and $(58.5)_1$ imply

$$W_L^* D V_*^L = W_\rho \alpha^\rho{}_L D V_*^L = W_\rho D^c V^\rho. \tag{58.6}$$

Furthermore $(57.2)_3$ and $(22.5)_1$ imply

$$V_*^L D W_L^* + W_L^* D V_*^L = D(V^\rho W_\rho) = V^\rho D_c W_\rho + W_\rho D^c V^\rho. \tag{58.7}$$

From (58.6), (58.7), and $(57.1\text{a})_2$ we deduce

$$V_*^L D W_L^* = V^\rho D_c W_\rho = V_*^L \alpha^\rho{}_L D_c W_\rho \tag{58.8}$$

which, by the obvious arbitrariness of V_*^L, implies $D W_L^* = \alpha^\rho{}_L D_c W_\rho$. Hence $(58.5)_2$ holds under condition $(58.5)_3$.

The fields E_ρ, H_ρ, π^ρ, and μ^ρ are spatial—cf. $(34.1)_{3,4}$, (34.2). Then by (convention (57.1 a) and) theorem $(57.2)_3$ the quantity W'—cf. $(46.4)_1$—can be given the expression

$$W' = E_L^* \pi_*^L + H_L^* \mu_*^L \qquad (W' = E_\rho \pi^\rho + H_\rho \mu^\rho). \tag{58.9}$$

Furthermore, by theorem (58.6), the elementary work $d_3\lambda$ of the electromagnetic field on dipoles, per unit proper mass—cf. $(40.16)_2$—has the Lagrangian expressions

$$d_3\lambda = E_L^* D\pi_*^L + H_L^* D\mu_*^* = DW' - \pi_*^L DE_L^* - \mu_*^L DH_L^*. \tag{58.10}$$

§ 59. The First and Second Piola-Kirchhoff Stress Tensors $K^{\rho M}$ and Y^{LM}, and Lagrangian Expressions for $dl^{(i)}$

Using $\mathscr{D}$ and $\gamma_\rho{}^L$—see (56.4) and $(56.13)_2$—we define the quantities $K^{\rho M}$ and Y^{LM}, classical analogues of which were introduced by Piola in 1833—see Truesdell and Toupin [1960, pp. 553—554]:

$$K^{\rho M} = X^{\rho\sigma} \gamma_\sigma{}^M, \qquad Y^{LM} = \frac{1}{\mathscr{D}} \gamma_\rho{}^L K^{\rho M} = \frac{1}{\mathscr{D}} \gamma_\rho{}^L \gamma_\sigma{}^M X^{\rho\sigma}. \tag{59.1}$$

By (56.15) these relations involving the spatial double tensors $X^{\rho\sigma}$ and $K^{\rho M}$ are obviously equivalent to

$$X^{\rho\sigma} = \frac{1}{\mathscr{D}} K^{\rho M} \alpha^\sigma{}_M = \frac{1}{\mathscr{D}} \alpha^\rho{}_L \alpha^\sigma{}_M Y^{LM}, \qquad K^{\rho M} = \alpha^\rho{}_L Y^{LM}. \tag{59.2}$$

From (59.1) and $(24.2)_{3,4}$ and from $(59.2)_{1,2}$ we deduce

$$u_\rho K^{\rho M} = 0; \qquad Y_{[LM]} = 0 \quad \text{if and only if} \quad X_{[\rho\sigma]} = 0. \tag{59.3}$$

Let us remember that the oriented surface element $(x^\rho, dx^\rho, \delta x^\rho)$ is represented by $d\sigma^\lambda$—see $(56.12)_{1,2}$—and that the resultant of the contact forces through this element is the spatial vector dR^ρ expressed by $(24.2)_1$. Then by $(59.1)_1$ and $(56.13)_1$

$$dR^\rho = K^{\rho M} d\sigma_M^*, \qquad \frac{dR^\rho}{|d\sigma^*|} = K^{\rho M} n_M^* \qquad (d\sigma_M^* = |d\sigma^*| n_M^*) \tag{59.4}$$

Furthermore from $(59.4)_1$, $(59.1)_2$, $(57.1\,\mathrm{b})_2$ (and $(57.1\,\mathrm{a})_2$) we deduce

$$dR_*^L = Y^{LM} d\sigma_M^* \qquad \left(dR_*^L = \frac{1}{\mathscr{D}} \gamma_\rho{}^L dR^\rho, \; dR^\rho = \alpha^\rho{}_L dR_*^L \right). \tag{59.5}$$

From (59.4) and (59.5)—and the analogue $dR_*^L / |d\sigma^*| = Y^{LM} n_M^*$ of $(59.4)_2$—we see the physical meanings of $K^{\rho M}$ and Y^{LM}.

By $dl^{(i)}/Ds$ we denote the power of internal contact forces per unit (actual) proper volume [§ 24]. Let $d^*l^{(i)}/Ds$ be the power of the same forces per unit reference (proper) volume, so that $dC^* d^* l^{(i)} = dC\, dl^{(i)}$. Then $(56.4)_1$ implies the first of the equalities

$$d^* l^{(i)} = \mathscr{D}\, dl^{(i)} = Y^{(LM)} D\varepsilon_{LM} = \tfrac{1}{2} Y^{LM} DC_{LM} \quad \text{for } X^{\rho\sigma} = X^{\sigma\rho}. \tag{59.6}$$

By $(28.1)_3$ and $(59.2)_{1,2}$ we have $\mathscr{D}\, dl^{(i)} = \alpha^\rho{}_L \alpha^\sigma{}_M Y^{LM} u_{(\rho/\sigma)}$ which by $(57.7)_1$ implies $(59.6)_2$. Thence $(56.6)_3$ follows by $(57.7)_2$.

It is natural to consider Y^{LM}, C^{LM}, and ε^{LM} as functions of t and y^L. Then we have e.g. $D\varepsilon_{LM} = (\partial \varepsilon_{LM}/\partial t)\, dt$.

By $(53.15)_1$ $Y^{LM} \dot{C}_{LM} = 2 Y^{(LM)} \alpha^\rho{}_L \dot{\alpha}_{\rho M}$, so that $(59.2)_3$ and (59.6) imply that *for* Y^{LM} *symmetric (and* $c\tau \equiv s$*)*

$$d^* l^{(i)} = K_\rho{}^L D\alpha^\rho{}_L, \qquad \Pi^{*(i)} = c K_\rho{}^L \dot{\alpha}^\rho{}_L \qquad \left(\text{for } \Pi^{*(i)} = \frac{d^* l^{(i)}}{D\tau},\ Y_{[LM]} = 0\right). \tag{59.7}$$

By $(59.3)_1$ and $(20.6\text{b})_4$ $K^{\rho L} x^\sigma{}_M D\overset{\perp}{g}_{\rho\sigma} = 0$ holds for $u_L^\dagger \equiv u_\sigma x^\sigma{}_L = 0$. Then by $(53.7)_1$ relations $(59.7)_{1,2}$ take the usual classical form for $u_L^\dagger = 0$ and $ct \equiv s$:

$$d^* l^{(i)} = K_\rho{}^L D \frac{\partial x^\rho}{\partial y^L}, \qquad \Pi^{*(i)} = K_\rho{}^L \frac{\partial^2 x^\rho}{\partial t\, \partial y^L} \qquad (u_L^\dagger = 0 = g_{\rho\sigma,\alpha} = X^{[\rho\sigma]},\ ct \equiv s). \tag{59.8}$$

§ 60. Connection Between $X^{\rho\sigma}{}_{/\overset{\perp}{\sigma}}$ and $K^{\rho M}{}_{|M}$

We assume[12)] $a^*_{LM} = \delta_{LM}$ and $t \equiv s$. Then $(56.4)_2$ becomes

$$\mathscr{D} = \pm \sqrt{-g}\, \frac{\partial(x^0, \dots, x^3)}{\partial(y^0, \dots, y^3)} \qquad (a^*_{LM} = \delta_{LM},\ t \equiv y^0 \equiv s). \tag{60.1}$$

It is well known that

$$X^{\rho\sigma}{}_{/\sigma} = \frac{1}{\sqrt{-g}} \sum_\sigma \frac{\partial \sqrt{-g}\, X^{\rho\sigma}}{\partial x^\sigma} + \begin{Bmatrix} \rho \\ \lambda\sigma \end{Bmatrix} X^{\lambda\sigma}. \tag{60.2}$$

Furthermore from $(53.8)_1$—i.e. $\alpha^\sigma{}_L = x^\sigma{}_L + u^\sigma u_L^\dagger$—and $(59.2)_1$ we deduce

$$\sqrt{-g}\, X^{\rho\sigma} = \frac{\sqrt{-g}}{\mathscr{D}} (x^\sigma{}_L + u^\sigma u_L^\dagger) K^{\rho L} \qquad (u_L^\dagger \equiv u_\sigma x^\sigma{}_L). \tag{60.3}$$

By (60.1) and the classical theorems on Jacobians $(60.4)_2$ below we have

$$\frac{\partial}{\partial x^\sigma}\left(\frac{\sqrt{-g}}{\mathscr{D}} x^\sigma{}_M\right) = \pm \frac{\partial}{\partial x^\sigma}\left[\frac{\partial(y^0, \dots, y^3)}{\partial(x^0, \dots, x^3)} x^\sigma{}_M\right] = 0. \tag{60.4}$$

Theorem $(60.4)_2$ is the identity of Euler, Piola, and Jacobi. In the degree of generality stated here, it was first proved by Piola in 1825—see Truesdell and Toupin [1960, p. 246]. We now prove $(60.4)_2$ for the ease of the reader. By the usual formula for differentiating a determinant and well known algebraic properties of Jacobians

$$\frac{\partial j}{\partial x^\sigma}=j\frac{\partial x^\rho}{\partial y^\Delta}\frac{\partial^2 y^\Delta}{\partial x^\rho \partial x^\sigma} \quad \text{where} \quad j=\frac{\partial(y^0,\ldots,y^3)}{\partial(x^0,\ldots,x^3)}. \tag{*}$$

Then

$$\frac{\partial j x^\sigma{}_K}{\partial x^\sigma}=j(x^\rho{}_{,\Delta}y^\Delta{}_{,\rho\sigma}x^\sigma{}_{,K}+x^\sigma{}_{,K\Delta}y^\Delta{}_{,\sigma}). \tag{**}$$

Furthermore by differentiating the identity $x^\rho{}_{,\Delta}y^\Delta{}_{,\sigma}=\delta^\rho{}_\sigma$ with respect to x^ρ we obtain

$$x^\rho{}_{,\Delta}y^\Delta{}_{,\rho\sigma}+x^\rho{}_{,\Delta\Sigma}y^\Sigma{}_{,\rho}y^\Delta{}_{,\sigma}=0. \tag{***}$$

The right-hand side of (**) equals the right-hand side of (***) composed with $jx^\sigma{}_{,K}$; hence $\partial(jx^\sigma{}_K)/\partial x^\sigma=0$. By $(*)_2$ and $(60.1)_1$ this yield $(60.4)_2$.

When x^ρ are taken as the independent variables and $u^\dagger_L=0$ is assumed locally, (60.3) and (60.4) imply

$$\sum_\sigma \frac{\partial\sqrt{-g}\,X^{\rho\sigma}}{\partial x^\sigma}=\frac{\sqrt{-g}}{\mathscr{D}}\left[x^\sigma{}_M\frac{\partial K^{\rho M}}{\partial x^\sigma}+K^{\rho L}u^\sigma\frac{\partial u^\dagger_L}{\partial x^\sigma}\right]. \tag{60.5}$$

Now we assume, besides $(60.1)_{2,3,4}$ all of the local conditions (54.1) on (x), (y), and $\mathfrak{t}(x)$, so that, considering y^L and $t\equiv s$ as the independent variables, we have (at $\mathscr{E}$)

$$x^\rho{}_L=\alpha^\rho{}_L\,,\quad \{\rho\sigma,\lambda\}=0\,,\quad \frac{Dx^\rho{}_L}{Ds}=u^\rho{}_{,L}\,,\quad u^\dagger_L=0=g+1\,. \tag{60.6}$$

From (60.2), (60.5), and (60.6) we deduce

$$X^{\rho\sigma}{}_{/\sigma}=\frac{1}{\mathscr{D}}\left[\sum_M x^\sigma{}_M\frac{\partial K^{\rho M}}{\partial x^\sigma}+K^{\rho L}u^\sigma u^\dagger_{L,\sigma}\right]. \tag{60.7a}$$

From $(17.3)_1$, $(60.3)_2$, $(60.6)_{3,4}$, $(17.4)_2$, and convention $(57.1\,\text{a})_1$, respectively, we deduce (at $\mathscr{E}$)

$$u^\sigma u^\dagger_{L,\sigma}=\frac{D}{Ds}u^\dagger_L=A_\sigma x^\sigma{}_L+u_\sigma\frac{Dx^\sigma{}_L}{Ds}=A_\sigma x^\sigma{}_L+u_\sigma u^\sigma{}_{,L}=A_\sigma\alpha^\sigma{}_L=A^*_L\,. \tag{60.8}$$

Hence we can turn (60.7a) into

$$X^{\rho\sigma}{}_{/\sigma}=\sum_{\sigma}\frac{\partial\sqrt{-g}\,X^{\rho\sigma}}{\partial x^{\sigma}}=\frac{1}{\mathcal{D}}\left[\sum_{M}\frac{\partial K^{\rho M}}{\partial y^{M}}+A_{\sigma}\alpha^{\sigma}{}_{L}K^{\rho L}\right]. \tag{60.7b}$$

We know that assumption (54.1) implies $(53.10)_1$, hence $K^{\rho M}{}_{|M}=K^{\rho M}{}_{,M}[\equiv K^{\rho M}(t,y)_{,M}]$. Then (17.17), $(59.2)_1$ and (60.7b) yield the basic formula

$$X^{\rho\sigma}{}_{/\overset{\perp}{\sigma}}=X^{\rho\sigma}{}_{/\sigma}-X^{\rho\sigma}A_{\sigma}=\frac{1}{\mathcal{D}}K^{\rho M}{}_{|M}\,. \tag{60.9}$$

which obviously holds for an arbitrary choice of (x), (y), and $\hat{t}(x)$.

§ 61. On $\alpha^{\rho}{}_{LM}$ and the Lagrangian Expression of $\overset{\perp}{g}{}^{\rho}{}_{\lambda}\,X^{\lambda\sigma}{}_{/\overset{\perp}{\sigma}}$

By the equality $\alpha^{\rho}{}_{L}=x^{\rho}{}_{L}+u^{\rho}u^{\dagger}_{L}$—cf. (53.8)—and $(17.6)_3$ we can develop the definition $(53.16)_1$ of $\alpha^{\rho}{}_{LM}$ into

$$\alpha^{\rho}{}_{LM}=\overset{\perp}{g}{}^{\rho}{}_{\sigma}\left(\frac{\partial x^{\sigma}}{\partial y^{L}}\right)_{|M}+u^{\dagger}_{L}u^{\rho}{}_{|M}\qquad(u^{\dagger}_{L}\equiv u_{\rho}x^{\rho}{}_{L})\,. \tag{61.1}$$

We assume the local conditions (54.1) on (x), (y), and $\hat{t}(x)$, which imply $(53.10)_1$. Then (61.1) becomes

$$\alpha^{r}{}_{LM}=\frac{\partial^2 x^{r}(t,y)}{\partial y^{L}\partial y^{M}},\qquad \alpha^{0}{}_{LM}=0\qquad(g_{\rho\sigma,\lambda}=0=a^{*}_{LM,A},\ u^{\dagger}_{L}=0,\ g_{\alpha\beta}=\delta'_{\alpha\beta})\,. \tag{61.2}$$

The special expression $(61.2)_1$ of $\alpha^{r}{}_{LM}$ coincides with the classical expression for the second position gradient in Cartesian co-ordinates. Incidentally (61.2) implies the first of the symmetry conditions

$$\alpha^{\rho}{}_{LM}=\alpha^{\rho}{}_{ML}\,,\qquad C_{BLM}=C_{BML}\,. \tag{61.3}$$

The second follows from $(61.3)_1$ and $(53.16)_2$. The validity of (61.3) is obviously independent of assumptions (54.1).[13] From $(57.7)_2$, $(53.15)_1$, $(53.7)_1$, and $(53.16)_1$ we have

$$2\varepsilon_{LM|A}=C_{LM|A}=\alpha_{\rho L}\alpha^{\rho}{}_{MA}+\alpha_{\rho M}\alpha^{\rho}{}_{MA}\,. \tag{61.4}$$

In order to compare classical and relativistic results carefully, especially in connection with elastic waves, we introduce the classical deformation system

$\varepsilon^{(c)}_{LM}$ in relativity and we prove that *if we identify the independent variables with t and y^L, and if we assume the local conditions* (54.1) *on* $(x), (y)$, *and* $\hat{t}(x)$, *then we have*

$$\frac{\partial \varepsilon_{LM}}{\partial y^A} = \varepsilon_{LM|A} = \frac{\partial \varepsilon^{(c)}_{LM}}{\partial y^A} \quad \text{where} \quad \delta_{LM} + 2\varepsilon^{(c)}_{LM} = \delta_{rs} x^r{}_L x^s{}_M \,. \tag{61.5}$$

Indeed (54.1) implies $(53.10)_1$ whence $(61.5)_1$ follows. Furthermore $(\alpha^0{}_L = 0)$ (61.2), (61.4), and $(61.5)_3$ imply $(61.5)_2$.

By $(53.16)_1$ and $(59.2)_3$ the basic relation (60.9) implies

$$\mathscr{D} \overset{\perp}{g}{}^{\rho}{}_{\lambda} X^{\lambda\sigma}{}_{/\overset{\perp}{\sigma}} = \alpha^{\rho}{}_{LM} Y^{LM} + \alpha^{\rho}{}_L Y^{LM}{}_{|M} \,. \tag{61.6}$$

Then under the local conditions (54.1) on $(x), (y)$, and $\hat{t}(x)$—which imply $(53.10)_1$ and $\alpha^r{}_L = x^r{}_L$—we have by (61.2)

$$\mathscr{D} X^{rs}{}_{/s} = \mathscr{D} \overset{\perp}{g}{}^r{}_{\rho} X^{\rho\sigma}{}_{/\overset{\perp}{\sigma}} = \sum_{LM} \left(\frac{\partial^2 x^r}{\partial y^L \partial y^M} Y^{LM} + \frac{\partial x^r}{\partial y^L} \frac{\partial Y^{LM}}{\partial y^M} \right) = \sum_{LM} \frac{\partial}{\partial x^M} \left(\frac{\partial x^r}{\partial y^L} Y^{LM} \right). \tag{61.7}$$

§ 62. Explicit Form in Co-Moving Co-Ordinates for Some of the Preceding Lagrangian Formulas

In this section—which is not essential for the sequel—we assume that (x) is a co-moving frame. Hence we can set $y^r \equiv x^r$. Then, since $ds^2 = -g_{00}(dx^0)^2$ holds along world lines, from $(17.1)_1$ and $(17.6)_1$ (and $g_{\rho}{}^{\sigma} = \delta_{\rho}{}^{\sigma}$) we deduce

$$u^{\rho} = \frac{\delta^{\rho}{}_0}{\sqrt{-g_{00}}}, \qquad u_{\rho} = \frac{g_{0\rho}}{\sqrt{-g_{00}}}, \qquad \overset{\perp}{g}_{\rho\sigma} = g_{\rho\sigma} - \frac{g_{0\rho} g_{0\sigma}}{g_{00}}, \tag{62.1}$$

$$\overset{\perp}{g}_{0\sigma} = 0; \quad \overset{\perp}{g}{}^{rs} = g^{rs}, \quad \overset{\perp}{g}{}^{0\sigma} = g^{0\sigma} - \delta_0{}^{\sigma}/g_{00} \tag{62.2}$$

and

$$\overset{\perp}{g}{}_r{}^s = \delta_r{}^s, \quad \overset{\perp}{g}{}_0{}^{\sigma} = 0, \quad \overset{\perp}{g}{}^0{}_s = -g_{0s}/g_{00} \,. \tag{62.3}$$

Since $x^r{}_L = \delta^r{}_L$, formula (53.7) becomes—see $(62.3)_{1,3}$

$$\alpha^{\rho}{}_L = \overset{\perp}{g}{}^{\rho}{}_L, \quad \text{i.e.} \quad \alpha^r{}_L = \delta^r{}_L, \quad \alpha^0{}_L = -g_{0L}/g_{00} \quad (x^r \equiv y^r) \,. \tag{62.4}$$

From $(62.4)_1$, $(53.15)_1$, and $(17.6)_4$, and from $(62.1)_3$ we obtain

$$C_{LM} = \overset{\perp}{g}_{LM} = g_{LM} - \frac{g_{0L} g_{0M}}{g_{00}} \qquad (x^r \equiv y^L). \tag{62.5}$$

Of course we have e.g. $C^{LM} = \overset{\perp}{g}_{AB} a^{*AL} a^{*BM}$ (hence $C^{LM} \neq \overset{\perp}{g}^{LM}$ generally). For the sake of completeness let us add that $(54.8)_1$, $(62.4)_1$, and $(62.3)_{1,3}$ imply

$$\begin{gathered} C'^{\rho\sigma} = \alpha^\rho{}_A \alpha^{\sigma A} = \overset{\perp}{g}^\rho{}_A \overset{\perp}{g}^\sigma{}_B a^{*AB}, \quad \text{i.e.} \\ C'^{rs} = a^{*rs}, \qquad C'^{0s} = -\frac{g_{0A}}{g_{00}} a^{*As}, \qquad C'^{00} = \frac{g_{0A} g_{0B}}{g_{00}{}^2} a^{*AB} \qquad (x^r \equiv y^r). \end{gathered} \tag{62.6}$$

Assuming also $y^0 \equiv t \equiv x^0$, we reduce $(56.4)_2$ to

$$\mathscr{D} = \sqrt{\frac{-g}{a^*}}\, u^0 = \sqrt{\frac{g}{a^* g_{00}}} \qquad (x^\lambda \equiv y^\lambda). \tag{62.7}$$

By $(62.3)_{1,2}$ and (62.4) the equations $(56.15)_{1,2}$ in $\gamma_\rho{}^L$ become

$$\gamma_r{}^s \equiv \gamma_r{}^L \overset{\perp}{g}{}^s{}_L = \mathscr{D}\, \delta_r{}^s, \qquad \gamma_0{}^L = 0 \qquad (x^\lambda \equiv y^\lambda) \tag{62.8}$$

so that their unique solution is

$$\gamma_\alpha{}^L = \mathscr{D}\, \delta_\alpha{}^L = \sqrt{\frac{g}{a^* g_{00}}}\, \delta_\alpha{}^L \qquad (x^\lambda \equiv y^\lambda). \tag{62.9}$$

Now let us remark that by $(59.1)_{2,3}$ and (62.9)

$$Y^{LM} = K^{LM} = \mathscr{D}\, X^{LM} = \sqrt{\frac{g}{a^* g_{00}}}\, X^{LM}. \tag{62.10}$$

Furthermore by $(62.4)_1$ we can turn (59.2) into

$$X^{\rho\sigma} = \frac{1}{\mathscr{D}} K^{\rho M} \overset{\perp}{g}{}^\sigma{}_M = \frac{1}{\mathscr{D}} \overset{\perp}{g}{}^\rho{}_L \overset{\perp}{g}{}^\sigma{}_M Y^{LM}, \qquad K^{\rho M} = \overset{\perp}{g}{}^\rho{}_L Y^{LM} \qquad (x^r \equiv y^r). \tag{62.11}$$

Let us take (62.3) and (62.7) into account. Then for $\rho, \sigma = 1,2,3$ (62.11) yields $(62.10)_{1,2}$. Instead the following consequences of (62.11) and (62.3)—where $X^{\rho\sigma}$ and Y^{LM} may be non-symmetric—are new:

$$\begin{aligned} X^{0s} &= \frac{1}{\mathscr{D}} K^{0M} \delta_M{}^s = \sqrt{\frac{a^* g_{00}}{g}}\, K^{0s}, \\ X^{\rho 0} &= -\frac{1}{\mathscr{D}} \frac{g_{0M}}{g_{00}} K^{\rho M} = \sqrt{\frac{a^*}{g g_{00}}}\, g_{0M} K^{\rho M}, \end{aligned} \tag{62.12}$$

$$X^{0s} = -\frac{g_{0L}}{\mathscr{D} g_{00}} Y^{Ls} = \sqrt{\frac{a^*}{g g_{00}}} g_{0L} Y^{Ls},$$

$$X^{r0} = \sqrt{\frac{a^*}{g g_{00}}} g_{0L} Y^{rL}, \tag{62.13}$$

$$X^{00} = \frac{1}{\mathscr{D}} \frac{g_{0L} g_{0M}}{(g_{00})^2} Y^{LM} = \sqrt{\frac{a^*}{g g_{00}}} \frac{g_{0L} g_{0M}}{-g_{00}} Y^{LM}, \tag{62.14}$$

$$K^{0M} = -\frac{g_{0L}}{g_{00}} Y^{LM}. \tag{62.15}$$

Theorem 62.1. *If $\mathscr{R}_{L_1 \dots L_n}$ is a material tensor, then there is exactly one spatial tensor $T_{\lambda_1 \dots \lambda_n}$ such that for every choice of the frames (x) and (y) satisfying the condition $x^r \equiv y^r$,*

$$T_{\lambda_1 \dots \lambda_n} = \delta_{\lambda_1}{}^{L_1} \dots \delta_{\lambda_n}{}^{L_n} \mathscr{R}_{L_1 \dots L_n} (= \mathscr{D}^{-1} \gamma_{\lambda_1}{}^{L_1} \dots \gamma_{\lambda_n}{}^{L_n} \mathscr{R}_{L_1 \dots L_n}) \quad (x^r \equiv y^r) \tag{62.16}$$

The converse also holds.

Indeed, given the material tensor $\mathscr{R}_{L_1 \dots L_n}$, the right hand side of $(62.16)_2$ is a spatial tensor; furthermore (62.8) implies $(62.16)_2$. Hence there is exactly one spatial tensor $T_{\lambda_1 \dots \lambda_n}$ with the required properties.

Conversely, given the spatial tensor $T_{\lambda_1 \dots \lambda_n}$ and two particular frames (x) and (y) with $x^r \equiv y^r$, $(62.1)_1$ holds, so that $T_{\lambda_1 \dots \lambda_n} = 0$ if some λ_i vanishes. Then there is a material tensor $\mathscr{R}_{L_1 \dots L_n}$ satisfying (62.16) in those frames. In addition, by (62.8) again, (62.16) holds for every choice of (x) and (y) such that $x^r \equiv y^r$.

Formulas (62.1) to (62.5) and (62.7) to (62.15) above—taken from Bressan [1963b, sect. 11]—afford us a mechanical procedure useful to check e.g. the equivalence of the theory presented in this book, or Bressan [1963b] and [1964a], with other relativistic theories of materials presented directly in co-moving co-ordinates, in connection with the common part of the fields to which these theories can be applied. For instance by $(57.7)_2$ and $(62.5)_1$

$$2\varepsilon_{LM} = \overset{\perp}{g}_{LM} - a^*_{LM} \quad (x^r \equiv y^r) \tag{62.17}$$

and it is easy to see that ε_{LM} (or $2\varepsilon_{LM}$) or the spatial space-time tensor associated to ε_{LM} (or $2\varepsilon_{LM}$) according to Theorem 62.1 coincides with the deformation tensor used e.g. by Rayner [1963], by Schöpf [1964a] and [1967, (9.1)], and by Cattaneo [1973, (3.4)].

Likewise if we want the expression in co-moving co-ordinates of the total energy tensor $U^{\alpha\beta} = k(c^2 + w) u^\alpha u^\beta + X^{\alpha\beta}$ for elastic materials that are capable of only adiabatic transformations, it suffices to use, first, the expressions $(63.4)_{2,3}$

for Y^{LM} and w in ε_{LM}, k^*, and η, then the relations $(63.1)_2$—i.e. $k^* = k\mathscr{D}$—, $(62.7)_1$, and $(62.11)_{1,2}$. One obtains

$$U^{\alpha\beta} = k(c^2 + w)u^\alpha u^\beta + X^{\alpha\beta} = k^* \sqrt{\frac{a^*}{-g}} \left[(c^2 + w)u^\alpha u^\beta + \overset{\perp}{g}{}^\alpha{}_L \overset{\perp}{g}{}^\beta{}_M \frac{\partial w}{\partial \varepsilon_{LM}} \right], \qquad (62.18)$$

where the entropy density can be regarded as a constant, so that it can be cancelled. For instance, this expression of $\mathscr{U}^{\alpha\beta}$ coincides with Cattaneo's [1973 (6.3)] and constitutes a basis of Schöpf [1964a].

Footnotes to Chapter 6

[1] Sections 65 and 66 in Möller [1952] are devoted to linear elasticity in special relativity.

[2] Let us remember that people's attention to relativistic (non linear) elasticity and electromagnetostriction was kept alive by de Donder and Dupont [1932—33] and [1936—37]. Unfortunately these two series of papers are unsatisfactory from the physical point of view and also from the mathematical one, in harmony with footnote 14 in Chapter 1.

[3] Synge [1959, p. 82] says: "In classical elasticity the strain (or deformation) of an elastic body is measured relative to a natural (unstrained) state and the basic stress-strain relation is a linear equation (Hooke's law) connecting stress and strain. These ideas are not easy to carry over into Riemann space-time... it is hard to see how a 'natural' state can exist, since gravity is always operative... though the concept of strain escapes us, the concept of rate of strain is easy to put into mathematical form." Therefore Synge considers a linear (isotropic) relation between the rate of strain and the rate of stress in general relativity. He cannot deal with equilibrium problems. However he can determine the speed of propagation for elastic acceleration waves.

Rayner [1963] is able to state a linear relation between stress and strain in general relativity for possibly anisotropic materials. Cf. footnote 1 in Chapter 7.

[4] Bennoun [1964a, b] studies from Synge's point of view elastic and more general materials; furthermore he introduces thermodynamic quantities. More in particular thermodynamic functions are introduced at each point of the world-tube $W_{\mathscr{C}}$ of $\mathscr{C}$ in S_4, but the connection between those of them which are introduced at the points of the world-line of a same material point, P^*, is not taken into account, so that e.g. the problem how to deal with the internal energy of P^* is not considered.

[5] Among other things, Lagrangian expressions for stresses and the work of internal contact forces that are quite similar with their classical analogues are derived.

[6] This theory of double tensors is employed by Truesdell and Toupin [1960] within the classical theory of continuous media. In Bressan [1964b] it is extended to general relativity also taking into account the kinematics for materials of any order n—see footnotes 15 in Chapter 1.

[7] (a) Ehlers [1961] is concerned, among other things, with Eulerian kinematics; e.g. he separates the velocity gradient into distortion and expansion tensor.

(b) Edelen [1963] proves that if the field equations are obtained from a homogeneous variational principle, then the resulting Einstein theory of general relativity receives certain intrinsic limitations. E.g. irreversible processes are disallowed.

(c) Edelen [1964] is concerned with the momentum-energy tensor and the deformation of the metric tensor in an n-dimensional Riemann-Einstein space.

(d) Schmutzer [1964] introduces the entropy 4-vector and discusses Onsager's relations.

[8] Berdichewski [1966] is in agreement with Schöpf [1964a] as far as ordinary (first-order) elasticity (with finite deformations) is concerned. This author also considers second-order elastic materials (in the adiabatic case). These materials include polar materials, i.e. those capable of couple stress (however no characterization of these materials is provided) and the classical limit of Berdichewski's theory seems to be different from common classical theories on polar materials, such as G. Grioli [1960] or Toupin [1962] and [1964]—cf. footnote 4 in Chapter 10.

9) Greek indices run from 0 to 3 except otherwise noted.

10) In accord with § 17 we say e.g. that the space-time index ρ of the double tensor $T_{\rho\ldots}{}^{L\ldots}$ is *spatial* if $u^\rho T_{\rho\ldots}{}^{L\ldots}=0$, and that $T_{\rho\ldots}{}^{L\ldots}$ is *spatial* if all its space-time indices are spatial. We mean $T_{\stackrel{\perp}{\rho}\ldots}{}^{L\ldots}$, $\stackrel{\perp}{T}_{\rho\ldots}{}^{L\ldots}$, and $(T_{\ldots}{}^{\ldots})^\perp$ in the obvious way in accordance with (17.9) and (17.9′).

11) The transport law (54.2) was given originally by Fermi in 1922. It was extended by Walker in 1932 to possibly non-spatial vectors. The resulting law defines what Synge [1960, p. 13] calls Fermi-Walker transport.

12) In this section and in § 61 we follow Bressan [1964a, §§ 6, 7], while for all other kinematical topics concerning ordinary relativistic elasticity—i.e. the one without couple stress—are substantially taken from Bressan [1963b].

13) Conditions (61.3) are immediate consequences of the commutation theorem (90.12) for Lagrangian derivatives, which will not be applied in our theory of piezo-elasticity.

Chapter 7

Elasticity, Acceleration Waves, and Variational Principles for Simple Materials

§ 63. Foundations of Elasticity

In accord with what was said in § 51, this chapter is substantially based on Bressan [1963b, d] and [1964a], except §§ 69, 70 which deal with the variational approach and are based on Schöpf [1964a]. Of course this subject is presented using the mathematical tools developed above.[1)]

The interest of the exact theory of elasticity for finite deformations in general relativity is now witnessed, first, by many papers stating physically equivalent versions of this theory, written by authors in part ignoring the previous similar works on the subject; in any case the variety of the formal treatment of that theory is more or less enriched. Second, this exact theory is being applied to astrophysical problems such as the detection of gravitational radiation by its interaction with elastic solid matter and the investigation of the properties of neutron stars that concern the mechanical behaviour of their solid crusts; the use of this theory for these purposes is considered as necessary by astrophysics —cf. Carter and Quintana [1972], [1975], who calculate the deviation from perfect fluid behaviour of a stationary rotating partly solid neutron star when it undergoes an elastic deformation due to a change of angular momentum.

In this chapter we assume the absence of electromagnetic field. Then the consequence $(28.1)_1$ of the first and second principle of thermodynamics holds. By $(58.3)_1$ and $(59.6)_2$ inequality $(28.1)_1$ can be written in the Lagrangian form

$$k^* D\mathscr{F} + k^* \eta D T + Y^{LM} D\varepsilon_{LM} \leqslant 0 \quad (k^* = k\mathscr{D}). \tag{63.1}$$

Like $(28.1)_1$, inequality $(63.1)_1$ holds as an equality if and only if, the second principle—i. e. $(25.1)_1$ for $r=0$—does.

We say that the body $\mathscr{C}$ is *elastic* at its material point P^* (or y^L) if the following two conditions hold at P^*:

a) *the specific internal energy* w *and the specific entropy* η*—hence the specific free energy* $\mathscr{F}(=w-T\eta)$ *also—are twice continuously differentiable functions of* T *and* ε_{LM}—cf. $(57.7)_2$; *furthermore the same holds for* Y^{LM}.

b) *constraints are absent, in the (narrow) sense that admissible values of* T *and* ε_{LM} *are physically compatible with arbitrary values of* DT *and* $D\varepsilon_{LM}(=D\varepsilon_{ML})$.

Now let $\mathscr{C}$ be elastic (at P^*). Then (63.1) implies the validity of

$$k^*\left(\frac{\partial \mathscr{F}}{\partial T}+\eta\right)DT+\left(k^*\frac{\partial \mathscr{F}}{\partial \varepsilon_{LM}}+Y^{LM}\right)D\varepsilon_{LM}\leqslant 0 \tag{63.1'}$$

for arbitrary values of DT and $D\varepsilon_{LM}$; hence

$$\eta=-\frac{\partial \mathscr{F}}{\partial T}, \qquad Y^{LM}=-k^*\frac{\partial \mathscr{F}}{\partial \varepsilon_{LM}}, \qquad \mathscr{F}=\mathscr{F}(T,\varepsilon_{LM}). \tag{63.2}$$

We conclude that *elastic materials have constitutive equations of the form* (63.2), *so that* $(63.1)_1$ *and consequently the second principle* $\eta\, DT \geqslant dQ$ *must hold as equalities.* In other words *elastic materials are capable of only reversible processes.*

The Helmholtz postulate [§ 28] implies $\partial^2\mathscr{F}/\partial T^2<0$, so that $(63.2)_1$ defines T as an implicit function of η and ε_{LM}. Then w and $\mathscr{F}$ are also functions of η and ε_{LM}. By $(28.1)_2$ we can turn (63.1) into

$$k^*Dw+Y^{LM}D\varepsilon_{LM}-k^*T\,D\eta\leqslant 0; \tag{63.3}$$

thence it is easy to derive the constitutive equations

$$T=\frac{\partial\tilde{w}}{\partial\eta}, \qquad Y^{LM}=-k^*\frac{\partial\tilde{w}}{\partial\varepsilon_{LM}} \qquad \text{where} \qquad w=\tilde{w}(\eta,\varepsilon_{LM}) \tag{63.4}$$

by a reasoning like the deduction of (63.2) above.

Now let us deduce from $(53.15)_1$ and $(57.7)_2$ that for every scalar (or tensor-valued) function ϑ of $C_{LM}\,(=a^*_{LM}+2\varepsilon_{LM})$ or of ε_{LM} we have—cf. Appendix A, § A 3

$$\begin{gathered}\frac{\partial\vartheta}{\partial\alpha_{\rho M}}=\frac{\partial\vartheta}{\partial C_{MB}}\alpha^{\rho}{}_{B}+\frac{\partial\vartheta}{\partial C_{AM}}\alpha^{\rho}{}_{A}=2\frac{\partial\vartheta}{\partial C_{LM}}\alpha^{\rho}{}_{L}=\frac{\partial\vartheta}{\partial\varepsilon_{LM}}\alpha^{\rho}{}_{L},\\ \vartheta=f(C_{LM})=f(a^*_{LM}+2\varepsilon_{LM}).\end{gathered} \tag{63.5}$$

Thus, in particular, $(63.2)_{2,3}$, $(63.4)_{2,3}$ and $(59.2)_3$ imply

$$K^{\rho M}=-k^*\frac{\partial\mathscr{F}}{\partial\alpha_{\rho M}}=-k^*\frac{\partial\tilde{w}}{\partial\alpha_{\rho M}} \qquad \left(\varepsilon_{LM}=\frac{C_{LM}-a^*_{LM}}{2}\right). \tag{63.6}$$

Furthermore by (63.6), $(63.1)_2$, and $(59.2)_1$ we have

$$X^{\rho\sigma}=-k\frac{\partial\mathscr{F}}{\partial\alpha_{\rho M}}\alpha^{\rho}{}_{M}=-k\frac{\partial\tilde{w}}{\partial\alpha_{\rho M}}\alpha^{\sigma}{}_{M}. \tag{63.7}$$

Incidentally the consequence $\alpha_{\sigma M}=\overset{\perp}{g}_{\rho\sigma}x^{\rho}{}_{M}$ of $(53.7)_1$ implies

$$\frac{\partial\vartheta}{\partial x^{\rho}{}_{M}}=\overset{\perp}{g}_{\rho\sigma}\frac{\partial\tilde{\vartheta}}{\partial\alpha_{\sigma M}} \qquad \text{for} \qquad \vartheta=\tilde{\vartheta}(\alpha_{\rho M},g_{\rho\sigma},u_{\rho}), \tag{63.8}$$

so that (63.6) and $(59.3)_1$ yield

$$K_\rho{}^M = -k^* \frac{\partial \mathscr{F}}{\partial x^\rho{}_M} = -k^* \frac{\partial \tilde{w}(\eta, \varepsilon_{LM}, y^L)}{\partial x^\rho{}_M} \quad (a^*_{LM} + 2\varepsilon_{LM} = \overset{\perp}{g}_{\rho\sigma} x^\rho{}_L x^\sigma{}_M). \tag{63.9}$$

The analogue of (63.7) for $x^\rho{}_L$ does not hold in general. Indeed, for instance the tensor $-k x^\sigma{}_M \partial\mathscr{F}/\partial x^\rho{}_M$ depends on the choice of $\hat{t}(x)$ and equals $X_\rho{}^\sigma$ only under the local condition $u^\dagger_L \equiv u_\rho x^\rho{}_L = 0$. Incidentally we now set

$$x^\rho{}_0 = \frac{\partial x^\rho}{\partial t}, \quad \text{whence} \quad x^\rho{}_0 = u^\rho \quad \text{for} \quad t \equiv s \tag{63.10}$$

and we assume $t \equiv s$ for the sake of simplicity. Then by $(17.6)_1$ and (53.15) [(53.7)] a function of $C_{LM}\,[\alpha^\rho{}_M]$, $g_{\rho\sigma}$, and u^ρ is a function of $g_{\rho\sigma}$ and $x^L{}_\Sigma$. Then in our relativistic elasticity $\mathfrak{F}$ and w can be thought of as functions of T (or η), $g_{\rho\sigma}$, and $x^\rho{}_\Sigma$. To consider $\partial\mathscr{F}/\partial x^\rho{}_0$ or $\partial\tilde{w}/\partial x^\rho{}_0$ is fairly in accordance with the practice of some authors—e. g. De Donder—even if they did not consider the dependence of $\mathscr{F}$ or w on (something like) $x^\rho{}_\Sigma$ through C_{LM} as we are doing. In order to compare our views on relativistic elasticity with the one followed by these authors, let us incidentally remark that for instance

$$-k^* \frac{\partial \tilde{w}(\eta, C_{LM})}{\partial u^\rho} = Y^{LM} x^\rho{}_L u^\dagger_M \qquad [C_{LM} = (g_{\rho\sigma} + u_\rho u_\sigma) x^\rho{}_L x^\sigma{}_M, t \equiv s] \tag{63.11}$$

holds—cf. Appendix A, § A 3—*under condition* $(63.11)_3$, i.e. $-g_{\rho\sigma} x^\rho{}_0 x^\sigma{}_0 \equiv 1$. Indeed from (53.15) and $(17.6)_1$ we deduce $(63.11)_2$; and by $(63.11)_2$, $(57.7)_2$, and the definition $u^\dagger_L = u_\rho x^\rho{}_L$, *for every function* $\vartheta = \vartheta(a^*_{LM} + 2\varepsilon_{LM})$ *of* $x^\rho{}_L$, u^ρ, *and* $g_{\rho\sigma}$ *we have*

$$\frac{\partial\vartheta}{\partial u^\rho} = \frac{\partial\vartheta}{\partial C_{LM}} \frac{\partial C_{LM}}{\partial u^\rho} = \frac{\partial\vartheta}{\partial C_{LM}} (x^\rho{}_L u^\dagger_M + u^\dagger_L \alpha^\rho{}_M) = \frac{\partial\vartheta}{\partial \varepsilon_{LM}} x^\rho{}_L u^\dagger_M. \tag{63.12}$$

Now $(63.4)_{2,3}$ clearly imply (63.11).

§ 64. Some Theorems on Elastic Materials

Each of the tensors $\mathscr{D}_{LM}, C_{LM}$, and ε_{LM} determines the others—cf. (54.4), $(57.7)_2$. Furthermore the polar decomposition (54.5) holds. Then, when $R^\rho{}_L$ is known, each of those tensors determines $\alpha^\rho{}_L$ (and conversely). Furthermore by (56.15) the spatial inverse $\overset{-1}{\alpha}{}^L{}_\rho$ of $\alpha^\rho{}_L$ is $\mathscr{D}^{-1}\gamma_\rho{}^L$ and is determined by the first two of the equations

$$\overset{-1}{\alpha}{}^A{}_\rho \alpha^\rho{}_B = a^*{}^A{}_B, \quad \overset{-1}{\alpha}{}^A{}_\rho u^\rho = 0, \quad \alpha^\rho{}_A \overset{-1}{\alpha}{}^A{}_\sigma = \overset{\perp}{g}{}^\rho{}_\sigma, \quad \overset{-1}{\alpha}{}^L{}_\rho = \mathscr{D}^{-1}\gamma_\rho{}^L. \tag{64.1}$$

It is also determined by $(64.1)_{2,3}$. Indeed under the local conditions (54.1) on (x), (y), and $\hat{t}(x)$ we have $\alpha^r{}_L = x^r{}_L$, $\alpha^0{}_L = 0$, hence $\overset{-1}{\alpha}{}^L{}_r = \partial y^L/\partial x^r$. Now the equivalence of $(64.1)_{1,2}$ to $(64.1)_{2,3}$ is obvious. By $(64.1)_3$, $(53.15)_1$, and $(53.16)_1$ yield

$$\alpha^\rho{}_L = \overset{-1}{\alpha}{}^{A\rho} C_{AL}, \qquad \alpha^\rho{}_{L_1 \ldots L_n} = \overset{-1}{\alpha}{}^{A\rho} C_{AL_1 \ldots L_n}. \tag{64.2}$$

We conclude that the variables

$$R^\rho{}_L, \qquad \varepsilon_{LM} \text{ (or } C_{LM}, \text{ or } \mathscr{D}_{LM}), \qquad C_{AL_1 \ldots L_i} \quad (i = 2, \ldots, n) \tag{64.3}$$

are in one-to-one correspondence with the variables

$$\alpha^\rho{}_{L_1, \ldots,} \; \alpha^\rho{}_{L_1 \ldots L_n}. \tag{64.4}$$

Now we consider elastic (and visco-elastic) materials from a more general point of view than we did in § 63.

Theorem 64.1. *We assume, for $\mathscr{C}$ at P^*, that*

(a) (*for some* $n > 0$) w *and* η—*hence* $\mathscr{F}$ *also—are functions of* T *and the variables* (64.3) *or* (64.4), *and that*

(b) *physical non-kinematical restrictions—we mean constraint relations—involving the time derivatives of these variables are absent.*

Then the following two theses hold: (i) *at* P^* *we have*

$$\eta = -\frac{\partial \mathscr{F}}{\partial T}, \qquad \mathscr{F} = \mathscr{F}(T, \varepsilon_{LM}), \qquad w = w(T, \varepsilon_{LM}), \tag{64.5}$$

and

$$\mathscr{D} X^{\rho\sigma}_{\text{irr}} u_{(\rho/\sigma)} \equiv Y^{LM}_{\text{irr}} \dot{\varepsilon}_{LM} \leqslant 0 \qquad \left(X^{\rho\sigma}_{\text{irr}} = \frac{1}{\mathscr{D}} \alpha^\rho{}_L \alpha^\sigma{}_M Y^{LM}_{\text{irr}} \right) \tag{64.6}$$

where, besides $(28.3)_1$ *the definitions*

$$Y^{LM}_{\text{rev}} = -k^* \frac{\partial \mathscr{F}}{\partial \varepsilon_{LM}}, \qquad Y^{LM} = Y^{LM}_{\text{rev}} + Y^{LM}_{\text{irr}}, \qquad X^{\rho\sigma}_{\text{rev}} = \frac{1}{\mathscr{D}} \alpha^\rho{}_L \alpha^\sigma{}_M Y^{LM}_{\text{rev}} \tag{64.7}$$

of Y^{LM}_{rev}, Y^{LM}_{irr}, *and* $X^{\rho\sigma}_{\text{rev}}$ *are understood.*

(ii) *Under the additional assumption that (at P^*) $X^{\rho\sigma}$ is a function of T and the variables* (64.3) *or* (64.4), $g_{\rho\sigma}$, *and* u^ρ, *we can assert that $\mathscr{C}$ is elastic (at P^*), so that it has the constitutive equation* (63.2).

Proof. Since η and $\mathscr{F}$ are functions of T, $R^\rho{}_L$, ε_{LM}, and $C_{AL_1 \ldots L_i}$ $(i = 2, \ldots, n)$, by definitions $(64.7)_{1,2}$ we can turn inequality (63.1) into

$$\left(\frac{\partial \mathscr{F}}{\partial T} + \eta \right) DT + \frac{\partial \mathscr{F}}{\partial R^\rho{}_B} DR^\rho{}_B + \sum_{i=2}^{n} \frac{\partial \mathscr{F}}{\partial C_{AL_1 \ldots L_i}} DC_{AL_1 \ldots L_i} + \frac{1}{k^*} Y^{LM}_{\text{irr}} D\varepsilon_{LM} \leqslant 0, \tag{64.8}$$

which holds for arbitrary values of DT, $DR^\rho{}_L,\ldots,D\varepsilon_{LM}$ (compatible with the mathematical conditions fulfilled by $R^\rho{}_L,\ldots,\varepsilon_{LM}$ by definition). Then (64.8) for $D\varepsilon_{LM}=0$ yields $(64.5)_{1,2}$—cf. Appendix A, §§ A 3, A 4. Since $w=\mathscr{F}+T\eta$, this implies $(64.5)_3$.

From $(64.5)_{1,2}$ and (64.8) we deduce $(64.6)_2$. Furthermore $(59.2)_{1,2}$ and definitions $(64.7)_{2,3}$ and $(28.3)_1$ imply $(64.6)_3$. Thence $(64.6)_1$ follows by $(57.7)_1$.

We now assume that $X^{\rho\sigma}$ is (at P^*) a function of T and the variables (64.3). Then the same holds for Y^{LM} by $(59.1)_{2,3}$ and for Y^{LM}_{irr} by $(64.7)_{1,2}$. As a consequence $(64.6)_{2,3}$ imply $Y^{LM}_{\text{irr}}\equiv 0\equiv X^{\rho\sigma}_{\text{irr}}$. Then $(64.7)_{1,2}$ yield $(63.2)_2$. By $(64.5)_{1,2}$, $(63.2)_{1,3}$ hold. Then $\mathscr{C}$ ist elastic at P*. q.e.d.

Let $\mathscr{C}$ be a non-viscous fluid at P^* [§ 29]. Then the quantities w, η, and $X^{\rho\sigma}$ are functions of T and k (and of $g_{\rho\sigma}$ and u^ρ), so that they are functions of T and the variables (64.3), i.e. assumption (a) in Theorem 64.1 holds. Indeed, more in detail, by $(56.4)_{2,3}$ $\mathscr{D}>0$, so that by $(56.9)_1$ and (58.3) k is a function of C_{LM}. Then by $(57.7)_2$ k is a function of ε_{LM}.

Since no constraints impose any restrictions on T, k, $\dot{T}$, and $u_{\rho/\sigma}$, no restrictions are imposed on $\dot{T}$ or $\dot{C}_{LM}$—cf. $(57.6)_{2,3}$—i. e. assumption (b) in Theorem 64.1 also holds.

Then by Theorem 64.1, thesis (ii), we conclude that *non-viscous fluids are elastic materials.*

In order to relate our relativistic elasticity with the views of some other authors, we now incidentally remark that by (53.8), $(63.10)_1$, and $(17.1)_1$ we respectively have

$$x^\rho{}_L=\alpha^\rho{}_L-u^\rho u^\dagger_L\,,\qquad \frac{Ds}{Dt}=g_{\rho\sigma}x^\rho{}_0x^\sigma{}_0\,,\qquad u^\rho=x^\rho{}_0\Big/\frac{Ds}{Dt}\,, \tag{64.9}$$

so that the variables $x^\rho{}_\Sigma$ are in a one-to-one correspondence with $\alpha^\rho{}_L$, Ds/Dt, u^ρ, and $u^\dagger_L\equiv u_\rho x^\rho{}_L$. Furthermore we assume that (at P^*) w, η, and $X^{\rho\sigma}$ are functions of T, $x^\rho{}_\Sigma$, and $g_{\rho\sigma}$. Then they are functions of $\alpha^\rho{}_L$, $u^\dagger_L$, Ds/Dt, $g_{\rho\sigma}$, and u^ρ.

Since $u^\dagger_L$ and Ds/Dt depend on the choice of $\hat{t}(x)$ and in every physical situation they can locally equal 0 and 1 respectively, the afore-mentioned dependence on them cannot be effective. Then, by Theorem 64.1 thesis (ii), $\mathscr{C}$ is obviously elastic at P^*, if constraints are absent.

§ 65. On Discontinuity Surfaces in Space-Time

In §§ 65 to 68, where acceleration waves in general elastic materials are dealt with (within special or general relativity) we substantially follow Bressan [1963d], especially in that we use the polarization and inertial-mass quadrics. These quadrics are useful to compare relativistic and classical results in the general case, and in particular in the case of principal waves travelling in isotropic elastic materials.[2)]

Let us mention that Cattaneo [1971b] deals with small motions of an elastic body around a configuration of stable equilibrium in general relativity—cf. also Cattaneo [1973].

We first state the main concepts and formulas concerning discontinuity surfaces quickly by using the corresponding classical theory. In the second part of this section we shall restate some of them more carefully and directly in (general) relativity.

Let σ_3 be a time-like hypersurface represented by

$$\bar{f}(t,x^1,x^2,x^3)=f(x^0,\dots,x^3)=0 \quad \text{with} \quad s=ct, \quad g^{\rho\sigma}f_{,\rho}f_{,\sigma}\geqslant 0. \tag{65.1}$$

Furthermore let functions $(52.2)_1$—which represent the motion $\mathcal{M}$ of $\mathcal{C}$—be twice continuously differentiable in a neighborhood of σ_3 except that the second derivatives $\partial^2 x^\rho/\partial y^\Lambda \partial y^\Sigma$ may have a discontinuity of the first kind on σ_3. Lastly let $g_{\alpha\beta}$ be twice continuously differentiable in the same neighborhood (and thrice nearly everywhere).

Equations $(52.2)_1$ transform σ_3 into a moving material surface σ_2^* to be called the image of σ_3 in S_3^*. Then by the assumption $x^0\equiv s\equiv ct$, the equations of the motion of σ_2^* are

$$f^*(t,y_1,y^2,y^3)\equiv f[x(t,y)]=0 \quad (x^0\equiv s\equiv ct). \tag{65.2}$$

We fix a point, x^α, of σ_3 and assume the validity at x^α of the local conditions (54.1) on (x), (y), and $\hat{t}(x)$. Then, disregarding infinitesimals of appropriately high order and using essentially the same words as in classical physics (in connection with Cartesian co-ordinates) we can define the normal unit vector $\mathcal{N}_L^*$ for σ_2^* at (t,y), the spatial one N_ρ for σ_3 at x^α, the speed of propagation V of σ_3 at x^α (which is the speed of displacement of σ_3 with respect to matter at x^α), and its Lagrangian correspondent V^* (to be identified with the speed of displacement of σ_2^* at (t,y) in S_3^*; in addition we can prove that *for* $x^0\equiv s\equiv ct$ *and under the local conditions* (54.1), N_ρ, $\mathcal{N}_L^*$, V^*, *and* V *have the expressions*

$$N_r=(\delta^{rs}\bar{f}_{,r}\bar{f}_{,s})^{-1/2}\bar{f}_{,r}=(\delta^{rs}f_{,r}f_{,s})^{-1/2}f_{,r}, \quad N_0=0, \tag{65.3}$$

$$\mathcal{N}_L^*=(\delta^{LM}f_{,L}^*f_{,M}^*)^{-1/2}f_{,L}^*, \quad V^*=(\delta^{LM}f_{,L}^*f_{,M}^*)^{-1/2}\left|\frac{\partial f^*}{\partial t}\right|, \tag{65.4}$$

$$V=(\delta^{rs}\bar{f}_{,r}\bar{f}_{,s})^{-1/2}\left|\frac{\partial\bar{f}}{\partial t}\right|=c(\delta^{rs}f_{,r}f_{,s})^{-1/2}|f_{,0}| \quad (x^0\equiv s\equiv ct). \tag{65.5}$$

The physical meaning of these formulas—which have been introduced briefly—will be considered more extensively in the second part of this section. Now let us first prove that the equalities

$$N_\rho=(\gamma_\rho{}^L\gamma^{\rho M}\mathcal{N}_L^*\mathcal{N}_M^*)^{-1/2}\gamma_\rho{}^H\mathcal{N}_H^*, \quad V=\left(\overset{-1}{C}{}^{LM}\mathcal{N}_L^*\mathcal{N}_M^*\right)^{-1/2}V^* \tag{65.6}$$

hold for every choice of (x), (y), and $\hat{t}(x)$. To this end we first assume the local conditions (54.1), hence $x^r{}_L = \alpha^r{}_L$, $\alpha^0{}_L = 0$—cf. $(54.2)_{1,2}$. Then $(65.2)_1$ yields

$$f^*_{,L} = f_{,r} x^r{}_L = f_{,\sigma} \alpha^\sigma{}_L . \tag{65.7}$$

By $(56.15)_1$ we have $\mathscr{D}\overset{\perp}{g}_\rho{}^\sigma f_{,\sigma} = \gamma_\rho{}^L \alpha^\sigma{}_L f_{,\sigma}$, which by (65.7) yields

$$\mathscr{D} f_{,\overset{\perp}{\rho}} = \gamma_\rho{}^L f^*_{,L}, \qquad \text{hence} \qquad f_{,r} = \frac{1}{\mathscr{D}} \gamma_r{}^L f^*_{,L} . \tag{65.8}$$

By $(65.3)_3$ N_ρ is a spatial unit vector and by (65.3), $(65.4)_1$, and $(65.8)_2$ N_ρ is parallel with $\gamma_\rho{}^L \mathscr{N}^*_L$, which implies $(65.6)_1$.

Now we remark that (65.8) and $(56.16)_1$ imply, respectively,

$$\delta^{rs} f_{,r} f_{,s} = \mathscr{D}^{-2} \gamma_\rho{}^L \gamma^{\rho M} f^*_{,L} f^*_{,M} = \overset{-1}{C}{}^{LM} f^*_{,L} f^*_{,M} . \tag{65.9a}$$

From $(54.1)_2$ and $(65.2)_{1,3,4}$ we deduce $\partial x^\rho / \partial t = c u^\rho = c \delta^\rho{}_0$ and $\partial f^* / \partial t = c f_{,0}$. Then $(65.4)_2$, $(65.5)_{1,2}$, and (65.9a) imply

$$V^{*2} / V^2 = \overset{-1}{C}{}^{LM} f^*_{,L} f^*_{,M} / \delta^{LM} f^*_{,L} f^*_{,M} \tag{65.9b}$$

which by $(65.4)_1$ yields $(65.6)_2$.

Let us remark that by (65.6) *a wave, σ_3, travelling at x^ρ along N_ρ with the speed of propagation V corresponds to a wave travelling in the reference space S^*_3 at (t, y) along $\mathscr{N}^*_L$ with the speed of displacement V^*, where V, V^*, N_ρ, and $\mathscr{N}^*_L$ are related by* (65.6).

Remembering the assumption that the derivatives $\partial^2 x^\rho / \partial y^A \partial y^\Sigma$ of functions $(52.2)_1$ have a discontinuity of the first kind across σ_3, and using the local conditions (54.1) on (x), (y), and $\hat{t}(x)$, we can prove the Hugoniot-Hadamard theorem substantially in the same way as in classical physics.

Thus using the symbol Δ to denote discontinuity across σ_3 (at $x^\rho = \hat{x}^\rho(t, y)$), *we may assert the existence of a non-zero spatial vector λ^ρ which satisfies the Hugoniot-Hadamard conditions*

$$\Delta \frac{\partial^2 x^s}{\partial t^2} = \lambda^s V^{*2}, \qquad \Delta \frac{\partial^2 x^s}{\partial y^L \partial y^M} = \lambda^s \mathscr{N}^*_L \mathscr{N}^*_M \qquad (\lambda^r = \lambda_r, \lambda^0 = 0) . \tag{65.10}$$

Now we wish to consider an explicit relativistic definition for the speed of propagation V of σ_3 (and a more careful characterization of N_ρ). To this end let l be an *orthogonal space-time ray associated with σ_3*, i.e. let l be a time-like line that belongs to σ_3 and is orthogonal to the locally spatial sections of σ_3; consequently l can be represented by equations of the form

$$x^\rho = x^\rho(\tau') , \tag{65.11}$$

where

$$f[x(\tau')]=0, \qquad c\,d\tau' = -u_\rho dx^\rho, \tag{65.12}$$

and where

$$\frac{dx^\rho}{d\tau'}\delta\overset{\perp}{x}_\rho=0 \quad \text{for every } \delta x^\rho \text{ with } \quad f_{,\overset{\perp}{\rho}}\delta\overset{\perp}{x}^\rho=0. \tag{65.13}$$

By $(65.12)_2$ $d\tau'$ is called by Cattaneo the *standard time* between the event points x^ρ and $x^\rho+dx^\rho$—cf. $(65.11)_1$—with respect to an observer joined with matter. The physical meaning of this time—which was studied extensively by C. Cattaneo [1958] or [1960] and hinted at in Sect. 18—can be grasped by using a locally natural and proper frame, say (x). Indeed from (17.7) we have $u_0=-1$ in such a frame, hence $c\,d\tau'=dx^0$. In special relativity (x) can be identified with a Minkowskian frame, so that $d\tau'$ appears to be the increment of the natural time $c^{-1}x^0$ for an (inertial) observer who is locally stationary with respect to matter. In general relativity the analogue holds for a locally natural frame and in addition τ' can be identified (along W_{p*}) with a time co-ordinate, x^0, which plays the role of local natural time for an observer joined with matter.

Hence τ' seems to be the best relativistic local analogue of classical time, in connection with phenomena inside matter, such as elastic waves.

Condition (65.13) is equivalent to the requirement that $dx^\rho/d\tau'$ be orthogonal to every spatial linear element $\delta\overset{\perp}{x}^\rho$ tangent to σ_3. From (65.13) and (18.1) we deduce

$$\overset{\perp}{g}{}^\rho{}_\sigma\frac{dx^\sigma}{d\tau'}=h\overset{\perp}{g}{}^{\rho\sigma}f_{,\sigma} \quad \text{with} \quad h^2=\frac{d\overset{\perp}{\sigma}{}^2}{d\tau'^2}\Big/ f_{/\overset{\perp}{\rho}}f^{/\rho} \tag{65.14}$$

where $d\overset{\perp}{\sigma}$ is the spatial (or standard) length $((d\overset{\perp}{x}{}^\rho d\overset{\perp}{x}_\rho)^{1/2}\geqslant 0)$ of dl.

Now let us consider the arc of l whose end points are $x^\alpha(0)$ and $x^\alpha(\tau')$, and the spatial length $\lambda(\tau')$ of this arc. We define V as the limit of the ratio $\lambda(\tau')/\tau'$ as τ' approaches zero:

$$V=\lim_{\tau'\to 0}\frac{1}{\tau'}\int_0^{\tau'}\left(\overset{\perp}{g}_{\rho\sigma}\frac{dx^\rho}{d\tau'}\frac{dx^\sigma}{d\tau'}\right)^{1/2}d\tau'=\left(\frac{d\overset{\perp}{x}{}^\rho}{d\tau'}\frac{d\overset{\perp}{x}_\rho}{d\tau'}\right)^{1/2}=\frac{d\overset{\perp}{\sigma}}{d\tau'}. \tag{65.15}$$

By $(65.12)_1$ we have $f_{/\rho}dx^\rho/d\tau'=0$ which by $(17.6)_1$ and $(65.12)_2$ yields

$$f_{/\overset{\perp}{\rho}}\frac{dx^\rho}{d\tau'}=f_{/\rho}u^\rho u_\sigma\frac{dx^\sigma}{d\tau'}=-cf_{/\rho}u^\rho. \tag{65.16}$$

By (65.16) and $(65.14)_1$, and by $(65.14)_2$, respectively, we have $-cf_{/\rho}u^\rho= hf_{/\overset{\perp}{\rho}}f^{/\rho}=\pm(f_{/\overset{\perp}{\rho}}f^{/\rho})^{1/2}d\overset{\perp}{\sigma}/d\tau'$, which by (65.15) yields

$$V=c(f_{/\overset{\perp}{\rho}}f^{/\rho})^{-1/2}|f_{/\rho}u^\rho| \tag{65.17}$$

In locally proper and natural co-ordinates (65.17) obviously becomes (65.5) —see (65.1).

Furthermore by (65.16) $f_{/\dot{p}}dx^p/d\tau'>0$ if and only if, $f_{/\rho}u^\rho<0$. Then the vector $f_{/\dot{p}}$ has the orientation of the propagation of σ_3 if and only if, $f_{/\rho}u^\rho<0$. Hence we can uniquely define the spatial normal unit vector N_ρ of σ_3, oriented in the sense of the propagation of σ_3, as follows

$$N_\rho = -f_{/\sigma}u^\sigma |f_{/\sigma}u^\sigma|^{-1}(f_{/\dot{p}} f^{/p})^{-1/2} f_{/\dot{p}}. \tag{65.18}$$

In the above special co-ordinates, (65.18) is equivalent to (65.3) if and only if, the sign of f is so chosen that $f_{/0}$ is negative.

§ 66. Dynamic Equations of Elastic Acceleration Waves

We aim at dealing with elastic waves in a way by which a careful comparison with the corresponding classical results may appear.

We assume $s \equiv ct$ and the local conditions (54.1) on (x), (y), and $\hat{t}(x)$, so that (24.9) and $(61.7)_{1,2}$ hold. Furthermore we consider the adiabatic case $(q^\rho \equiv 0)$. Then

$$\left[k\left(1+\frac{w}{c^2}\right)\delta^{rs}+\frac{1}{c^2}X^{rs}\right]a_s = -\frac{1}{\mathscr{D}}(Y^{LM}x^r{}_{,LM}+Y^{LM}{}_{,M}x^r{}_L)\left(a^r=\frac{\partial^2 x^r}{\partial t^2}\right), \tag{66.1}$$

where $(66.1)_2$ follows from our assumptions above and $(24.10)_1$. Since $\mathscr{D}k=k^*$ —cf. (58.3)—and $\alpha^\rho{}_L = x^\rho{}_L$—cf. $(54.2)_1$—, (66.1) and $(59.2)_{1,2}$ yield

$$\left[k^*\left(1+\frac{w}{c^2}\right)\delta^r{}_s+\frac{\delta_{ls}}{c^2}x^r{}_L x^l{}_M Y^{LM}\right]\frac{\partial^2 x^s}{\partial t^2} = -Y^{LM}\frac{\partial^2 x^r}{\partial y^L \partial y^M} - Y^{LM}{}_{,M}x^r{}_L. \tag{66.2}$$

We also assume that $\mathscr{C}$ is elastic at every material point [§ 63], so that $(63.4)_{2,3}$ hold. Then, considering y^L and $x^r{}_A$ $(x^r{}_A = x_{,rA})$ as twelve independent arguments, we have the first of the equalities—where "$_{,M}$" and "$/\partial y_M$" are used for total and partial derivation respectively, like in $\tilde{w}(y^L,\varepsilon_{AB})_{,M} \equiv \partial\tilde{w}/\partial y_M + (\partial\tilde{w}/\partial\varepsilon_{AB})\partial\varepsilon_{AB}/\partial y_M$

$$\begin{aligned} -Y^{LM}{}_{,M}\frac{x^r{}_L}{k^*} - \frac{\partial}{\partial y^M}\left(\frac{\partial k^*\tilde{w}}{\partial\varepsilon_{LM}}\right)\frac{x^r{}_L}{k^*} &= \frac{\partial^2\tilde{w}}{\partial\varepsilon_{LM}\partial\varepsilon_{AB}}\frac{\partial\varepsilon_{AB}}{\partial y^M}x^r{}_L \\ = \frac{\partial^2\tilde{w}}{\partial\varepsilon_{LA}\partial\varepsilon_{MB}}\frac{\partial\varepsilon^{(c)}_{MB}}{\partial y^A}x^r{}_L &= \frac{\partial^2\tilde{w}}{\partial\varepsilon_{LA}\partial\varepsilon_{MB}}x^s{}_{,MA}x_{sB}x^r{}_L = \frac{\partial^2\tilde{w}}{\partial\varepsilon_{LA}\partial\varepsilon_{MB}}x^r{}_A x_{sB}x^s{}_{,LM}. \end{aligned} \tag{66.3}$$

The second follows from the consequence $(61.5)_{1,2}$ of our local assumptions (54.1), the third from $(61.5)_3$ and the relation $\varepsilon_{[MB]}=0$, the fourth from the same relation.

By (66.3) and $(63.4)_2$ equation (66.2) becomes $(\alpha^\rho{}_L = x^\rho{}_L,\ k^* = k^*(y^L))$

$$q_{rs}\frac{\partial^2 x^s}{\partial t^2} = p_{rs}{}^{LM}\frac{\partial^2 x^s}{\partial y^L \partial y^M} + \frac{1}{k^*}\frac{\partial}{\partial y^M}\left(k^*\frac{\partial \tilde{w}}{\partial \varepsilon_{LM}}\right)x^r{}_L, \tag{66.4}$$

where y^L and $x^r{}_A$ are considered again as independent arguments and where the following definitions are understood:

$$q^{rs} = \left(1+\frac{w}{c^2}\right)\delta^{rs} + \frac{X^{rs}}{c^2 k} = \left(1+\frac{w}{c^2}\right)\delta^{rs} - \frac{1}{c^2}x^r{}_L x^s{}_M\frac{\partial \tilde{w}}{\partial \varepsilon_{LM}} = q^{sr},$$

$$p^{rsLM} = \frac{\partial \tilde{w}}{\partial \varepsilon_{LM}}\delta^{rs} + \frac{\partial^2 \tilde{w}}{\partial \varepsilon_{LA}\partial \varepsilon_{MB}}x^r{}_A x^s{}_B = p^{srLM} = p^{rsML}. \tag{66.5}$$

Let us incidentally remark that we have

$$p_{rs}{}^{LM} = \frac{\partial^2 w^{(c)}}{\partial x^r{}_L \partial x^s{}_M} \qquad \text{where} \qquad w^{(c)}(\eta, x^r{}_L, y^L) = \tilde{w}(\eta, \varepsilon^{(c)}_{LM}, y^L) \tag{66.6}$$

because $(66.6)_2$ and $(61.5)_3$ imply

$$\frac{\partial w^{(c)}}{\partial x^r{}_L} = \frac{\partial \tilde{w}}{\partial \varepsilon_{LM}}x^r{}_M \qquad (\delta_{LM} + 2\varepsilon^{(c)}_{LM} = \delta_{rs}x^r{}_L x^s{}_M),$$

which by $(66.5)_4$ yields $(66.6)_1$.

Formulas (66.4) to (66.6) coincide with their classical analogues up to the terms in c^{-2} in the expression $(66.5)_1$ of q^{rs}.

Now we consider again the hypersurface σ_3 where $\partial^2 x^r/\partial y^A \partial y^\Sigma$ have a discontinuity of the first kind [§ 65]. From the physical point of view this can occur along an adiabatic phenomenon; we know that we can write $\eta = \text{const}$, $\Delta\eta = 0$.

We assume that the body $\mathscr{C}$ is elastic and regular in the reference configuration; more precisely let the functions $k^*(y)$ and $\tilde{w}(\eta, \varepsilon_{LM}, y^L)$ be twice continuously differentiable with respect to ε_{LM} and y^L. Then the last term in (66.4) is continuous across σ_3, as well as q_{rs} and $p_{rs}{}^{LM}$, whereas the discontinuities of $\partial^2 x^r/\partial t^2$ and $x^r{}_{,LM}$ are given by (65.10). As a consequence, by writing (66.4) at both sides of σ_3 and by subtracting the resulting equalities from one another we obtain

$$q_{rs}\lambda^s V^{*2} = p_{rs}\lambda^s \qquad \text{where} \qquad p_{rs} = p_{rs}{}^{LM}\mathcal{N}^*_L\mathcal{N}^*_M = p_{sr}. \tag{66.7}$$

Obviously, an effective discontinuity $(\lambda_1{}^2 + \lambda_2{}^2 + \lambda_3{}^2 > 0)$ is possible if and only if, at least one solution of the equation

$$\Delta = \Delta(x) = \det \|p_{rs} - x\,q_{rs}\| = 0 \tag{66.8}$$

in $x\ (=V^{*2})$ is positive or vanishing (stationary waves).

Incidentally let us remark that, understanding (24.3) with ($q^\rho \equiv 0$, hence) $Q_{\alpha\beta} \equiv 0$, the dynamic discontinuity equations (66.7)—based on (66.1)—can be written $\overset{\perp}{g}_\alpha{}^\rho \varDelta \mathscr{U}_{\rho\beta}{}^{/\beta} = 0$ where—in contrast with (36.8)—by $\varDelta$ we denote the discontinuity across σ_3. Of course $\varDelta \mathscr{U}_{\alpha\beta}{}^{/\beta} = 0$ also holds on σ_3; however we have $\varDelta \mathscr{U}_{\alpha\beta/\gamma} \neq 0$ on σ_3, in general, because of the assumed discontinuity of $x^r{}_{,LM}$. Then the gravitation equations $(23.1)_1$ imply $\varDelta A_{\alpha\beta/\gamma} \neq 0$ on σ_3, whence by a well known expression for the Ricci tensor $A_{\alpha\beta}$ we generally have $g_{\alpha\beta,\rho\sigma\gamma} \neq 0$ on σ_3.

Dautcourt [1963] gives a suitable expression in $f_{/\rho}$—cf. $(65.2)_2$—for the equation $\varDelta \mathscr{U}_{\alpha\beta}{}^{/\beta} = 0$. This expression is used by Schöpf [1965a] as a starting point of his relativistic theory of elastic waves.

§ 67. Polarization and Inertial-Mass Quadrics. Acoustic Axes

Every material so far encountered in experiment abundantly fulfills the following condition, which forbids extremely large normal tensions:

$$X^{\rho\sigma} N_\rho N_\sigma > -k(c^2 + w) \equiv -\rho \quad \textit{for every spatial unit vector } N_\rho. \tag{67.1}$$

This condition becomes $p > -\rho$ *in a pure pressure state* ($X^{\rho\sigma} = p\overset{\perp}{g}{}^{\rho\sigma}$).

By $(66.5)_1$ inequality $(67.1)_1$ is equivalent to $q_{rs} N_r N_s > 0$, hence to the condition that the quadric which belongs to the spatial 3-space $S_3{}^{(t)}$ tangent to S_4 at x^ρ and is represented by the second of the equations

$$p_{rs}\xi^r \xi^s = \pm 1, \qquad q_{rs}\xi^r \xi^s = 1 \tag{67.2}$$

should be an ellipsoid. Then the characteristic equation $(66.8)_3$ has three (possibly coinciding) real solutions $V^{*2}_{(l)}$ $(l = 1, 2, 3)$.

We call the union of (the real parts of) the quadrics $(67.2)_1$ *polarization quadric* —in accordance with the classical theory—because of its properties to be presented shortly—, and the quadric $(67.2)_2$ *inertial-mass quadric*, because $k q_{rs}$ is the inertial-mass tensor, i. e. the quantity which constitutes the coefficient of intrinsic acceleration in the dynamic equations—cf. e. g. (66.1) or more generally (24.9).

An axis through the common center of the afore-mentioned quadrics, which meets them in points where the tangent planes are parallel, will be called *acoustic axis*, relative to the propagation direction $N_\rho (= \gamma_\rho{}^L \mathscr{N}^*_L |d\sigma^*|/|d\sigma|)$—cf. (56.17).

The spatial unit vector λ^ρ is parallel with an acoustic axis (relative to N_ρ) if and only if, λ^ρ solves equation $(66.7)_1$ for a (possibly negative) value $V^{*2}_{(l)}$ of V^{*2}, that is a solution of the characteristic equation $(66.8)_3$. An acoustic axis which is connected with the solution $V^{*2}_{(l)}$ of $(66.8)_3$ in this way, can be said to be *relative to the squared (Lagrangian) speed of propagation* $V^{*2}_{(l)}$—in accordance with Truesdell [1961, p. 280]. Let $a^\rho{}_{(l)}$ be the (spatial) unit vectors of three independent (generally non-orthogonal) acoustic axes.

By a locally spatial linear transformation we can change (x) into the frame $(\bar{x})$ which is locally proper and geodesic and whose co-ordinate lines are locally tangent to u^ρ and $a^\rho{}_{(l)}$. Then, denoting by $\bar{p}_{rs}$ and $\bar{q}_{rs}$ the components of p_{rs} and q_{rs} in the frame $(\bar{x})$ and by $V^*_{(r)}$ the solutions of $(66.8)_3$, we have[3)]

$$V^{*2}_{(r)} = p_{(r)}/q_{(r)}, \qquad \bar{p}_{rs} = p_{(r)}\delta_{rs}, \qquad \bar{q}_{rs} = q_{(r)}\delta_{rs}. \tag{67.3}$$

Under the obvious condition (67.1)—by which $(67.2)_2$ represents an ellipsoid—$q_{(r)} > 0$ holds, so that by (67.3) *elastic waves can propagate in the arbitrary spatial direction* N_ρ *with a suitable [non-zero] speed if and only if, at least one of* $p_{(1)}$ *to* $p_{(3)}$ *is* $\geqslant 0$ $[>0]$, *i.e. if and only if the quadratic form* $p_{rs}\xi^r\xi^s$ *is positive [strictly positive] definite.*

In accordance with stability properties usually considered in classical physics —cf. Coleman and Noll [1959 (8.3)]—we assume that under the local conditions (54.1) on (x), (y), and $\hat{t}(x)$ we have—cf. $(66.5)_4$—

$$0 < p_{rs}{}^{LM}\delta x^r{}_L \delta x^s{}_M \qquad (\delta x^r{}_L = x^l{}_L \delta \bar{x}^r{}_l,\ \delta \bar{x}^{(r}{}_{l)} \neq 0) \tag{67.4}$$

for every non-rigid virtual increment $\delta x^r{}_L$ *of* $x^r{}_L$—defined by $(67.4)_{2,3}$.

Incidentally it is well known that, given the possible motion $x^r = x^r(t,y)$ within classical physics and in Cartesian co-ordinates, we have $\partial x^r{}_L/\partial t = v^r{}_{/l}x^l{}_L$, hence $dx^r{}_L = x^l{}_L v^r{}_{/l}dt$. Then $(67.4)_2$ holds for $\delta x^r{}_L = dx^r{}_L$ and $\delta\bar{x}^r{}_l = v^r{}_{/l}dt$; furthermore the local non-rigidity condition $v_{(r/l)} \neq 0$ is equivalent to $(67.4)_3$.

By $(66.7)_2$ *assumption* (67.4) *yields the following condition of strong ellipticity:*

$$p^{rs}\lambda_r\lambda_s > 0 \qquad (p^{rs} = p^{rsLM}\mathcal{N}^*_L\mathcal{N}^*_M) \tag{67.5}$$

for every spatial unit vector λ_ρ *at* x^ρ, *and every unit vector* $\mathcal{N}^*_L$ *at* y^L.

Indeed let us fix the afore-mentioned unit vectors λ_ρ and $\mathcal{N}^*_L$. We assume, besides (54.1), that $(67.4)_1$ holds for every $\delta x^\rho{}_L$ of the form $(67.4)_{2,3}$. Lastly we define N_ρ and $\delta\bar{x}^s{}_L$ as follows:

$$N_r x^r{}_L |d\sigma| = \mathcal{D}\mathcal{N}^*_L |d\sigma^*|, \qquad N_0 = 0, \qquad \delta\bar{x}^r{}_l = \bar{\varepsilon}\lambda^r N_l \qquad (\bar{\varepsilon} \neq 0), \tag{67.6}$$

so that by $(56.21)_2$ N_ρ is the spatial normal unit-vector of the (spatial) surface corresponding to the material surface represented by $|d\sigma^*|\mathcal{N}^*_L$.

Then the denial of $(67.4)_3$ implies the following absurd result:

$$0 = 2\delta\bar{x}^{(l}{}_{r)}\lambda^r N_l/\bar{\varepsilon} = (\lambda_r N_l + N_r\lambda_l)\lambda^r N^l = 1 + (\lambda_r N^r)^2.$$

Hence $(67.4)_3$ holds. Then, after setting $(67.4)_2$—i.e. $\delta x^r{}_L = \bar{\varepsilon}\lambda^r N_l x^l{}_L$—and remembering assumption (67.4), from $(67.4)_{2,3}$ we both deduce $(67.4)_1$ and see that we can put it into the form

$$\bar{\varepsilon}^2 p_{rs}{}^{LM} x^l{}_L x^m{}_M N_l N_m \lambda^r\lambda^s > 0. \tag{67.7}$$

From (67.7), $(67.6)_1$, and $(67.5)_2$ we deduce $(67.5)_1$. q.e.d.

Furthermore (67.5) and $(67.3)_2$ imply $p_{(r)}>0$. Thence we conclude that *if for the elastic body $\mathscr{C}$ conditions (67.1) and (67.3) hold at x^ρ in connection with the unit vector $\mathscr{N}^*_L$ at y^L (and with given values for η and ε_{LM}), then for $l=1,2,3$ the acoustic axis $a^\rho_{(l)}$ is effective, that is an elastic acceleration wave can travel through x^ρ, with the discontinuity vector parallel to $a^\rho_{(l)}$ and with the Lagrangian speed of propagation $V^*_{(l)}$, so that its speed of propagation equals the value $V_{(l)}$ of V given by $(65.6)_2$ for $V^*=V^*_{(l)}$.*

§ 68. Pure Pressure States. Isotropic Elastic Materials. Comparison with the Classical Theory

Let us use locally natural and proper co-ordinates again and let us consider a pure pressure state $(X^{rs}=p\,\delta^{rs})$, so that by (21.4) equality $(66.5)_1$ becomes

$$q_{rs}=\left(1+\frac{k\,w+p}{k\,c^2}\right)\delta_{rs}=\frac{\rho+p}{k\,c^2}\,\delta_{rs} \quad \text{for} \quad X^{\rho\sigma}=p\,\overset{\perp}{g}{}^{\rho\sigma}(p\gtrless 0)\,. \tag{68.1}$$

Hence (in these states) q_{rs} is a (spatially) isotropic tensor as well as its classical analogue, which is δ_{rs}. Incidentally this is the reason by which this tensor is never explicitly considered in dealing with fluids—which are necessarily in such states.

In a pure pressure state—cf. (68.1)—the quadric $(67.2)_2$ is a sphere because $q_{(r)}=(\rho+p)k^{-1}c^{-2}$ by (68.1) and $(67.3)_3$, so that the relativistic *acoustic axes are the symmetry axis of the polarization quadric* $(67.2)_1$, as well as in the general classical case, and in particular they are orthogonal.

The comparison of the relativistic propagation speeds $V^*_{(l)}$ and $V_{(l)}$ for elastic waves with their respective classical analogues $V^{*(c)}_{(l)}$ and $V_{(l)}{}^{(c)}$ is interesting for $w=0$—i.e. $\rho=k\,c^2$—and in particular in case the reference configuration is identified with the present one. The classical analogues of $(67.3)_1$ and $(65.6)_2$ are

$$V^{*(c)}_{(r)}=p_{(r)}\,, \quad V_{(r)}{}^{(c)}=\left(\overset{-1}{C}{}^{LM}\,\mathscr{N}^*_L\,\mathscr{N}^*_M\right)^{-1/2}\,V^{*(c)}_{(r)}\,. \tag{68.2}$$

By $(68.1)_{1,2}$ for $\rho=k\,c^2$, (68.2) and $(65.6)_2$ imply

$$V^*_{(r)}=V^{*(c)}_{(r)}\left(1+\frac{p}{\rho}\right)^{-1/2}, \quad V_{(r)}=V_{(r)}{}^{(c)}\left(1+\frac{p}{\rho}\right)^{-1/2} \quad (w=0)\,. \tag{68.3}$$

The cases $p\leqslant 0$ are also interesting, for we are dealing with solids. By the remark below condition (67.1), this condition implies $1+p/\rho>0$ in the case $(68.1)_3$.

We say that the elastic body $\mathscr{C}$ is *isotropic in the reference state* $(\mathscr{P}^*, S^*_3)$ being considered if $w=\tilde{w}(\eta,\varepsilon_{LM}\,y^L)$ is an isotropic function of ε_{LM}. Then by $(63.4)_2$ and an analogue of Theorem 38.4 Y^{LM} also is such a function.

Let $\mathscr{C}$ be elastic and isotropic in $(\mathscr{P}^*, S^*_3)$. Then w depends on ε_{LM} through the principal stretches $\varepsilon_{(1)},\varepsilon_{(2)},\varepsilon_{(3)}$. Furthermore we know from the classical theory that $(63.4)_2$ implies the following: *Every principal axis of strain*—i.e. [§ 54] *every symmetry axis of the quadratic* $\varepsilon_{LM}\eta^L\eta^M=1$ [4)]*—is a principal material axis of tension,* i.e. *a symmetry axis of the quadric* $Y^{LM}\xi_L\xi_M=1$.

Indeed let us assume that the locally cartesian frame (y) characterizes a principal triad of strain [§ 54]. Furthermore let us change the frame (y) into the frame (y'): $y'^3=y^3$, $y'^L=-y^L$ $(L=1,2)$. Then, on the one hand $Y'^{L3}=-Y'^{L3}$ $(L=1,2)$; on the other hand all principal stretches $\varepsilon_{(r)}$ are unchanged, so that by the assumed isotropy $Y'^{L3}=Y^{L3}$ $(L=1,2)$. We conclude that $Y^{L3}=0$ $(L=1,2)$. Likewise we can prove $Y^{LM}=0$ for $L\neq M$. Then the frame (y) characterizes a triad of principal material axes of tension. q.e.d.

We remember [§ 54] that by $\lambda^L{}_{(A)}$ we denote the unit vector of the A-th axis of a principal triad of strain, and that $\alpha^\rho{}_L\lambda^L{}_{(A)}$ are three orthogonal vectors (which characterize what may be called the spatial image of this triad). These vectors are parallel with $\gamma^\rho{}_L\lambda^L{}_{(A)}$, because from $(56.15)_3$ and $(54.11)_2$ we deduce

$$\gamma_\rho{}^L\lambda_{(r)L}\alpha^\rho{}_M\lambda^M{}_{(s)}=\mathscr{D}a^*_{LM}\lambda^L{}_{(r)}\lambda^M{}_{(s)}=\mathscr{D}\delta_{rs}\,.$$

Furthermore, since $\mathscr{C}$ is assumed to be elastic and isotropic in $(\mathscr{P}^*,S_3^*)$, *the vectors* $\alpha^\rho{}_L\lambda^L{}_{(A)}$ *characterize the principal directions of tension.*

Indeed, since we are considering the elastic case, $Y_{LM}\lambda^M{}_{(B)}$ is parallel with $\lambda^L{}_{(A)}$ by an assertion above; hence $Y_{LM}\lambda^L{}_{(A)}\lambda^M{}_{(B)}=0$ for $A\neq B$. Then $(59.2)_{1,2}$ imply $\mathscr{D}X_{\rho\sigma}\alpha^\rho{}_L\lambda^L{}_{(A)}\alpha^\sigma{}_M\lambda^M{}_{(B)}=Y_{LM}\lambda^L{}_{(A)}\lambda^M{}_{(B)}=0$ for $A\neq B$. q.e.d.

Now we consider a *principal wave*, i.e. a wave travelling along a principal axis of strain, so that the unit vector $\mathscr{N}^*_L$ of the propagation direction in the body coincides with $\lambda_{(r)L}$ for some r.

Incidentally the analogue of $\mathscr{N}^*_L$ in space time is the spatial unit vector $N_\rho=|d\sigma|^{-1}|d\sigma^*|\gamma_\rho{}^L\mathscr{N}^*_L$—cf. (56.17); it is such that $|d\sigma|N_\rho$ and $|d\sigma^*|\mathscr{N}^*_L$ represent corresponding surface elements; furthermore by the parallelism of $\gamma^\rho{}_L\lambda^L{}_{(A)}$ with $\alpha^\rho{}_L\lambda^L{}_{(A)}$ asserted above, *for* $\mathscr{N}^*_L=\lambda_{(r)L}$ *we have*

$$N^\rho=l^\rho{}_{(r)}\qquad \textit{where}\qquad l^\rho{}_{(r)}=(C_{LM}\lambda^L{}_{(r)}\lambda^M{}_{(r)})^{-1/2}\alpha^\rho{}_H\lambda^H{}_{(r)}\,. \tag{68.4}$$

The following analogue of a known classical theorem—cf. Truesdell [1961, p. 275]—can be proved as in classical physics: *if $\mathscr{C}$ is isotropic, then the spatial images of the principal axes of strain are symmetry axes of the quadric* $X_{rs}\xi^r\xi^s=1$, *i.e. principal axes of space tension, and symmetry axes of the polarization quadric* $(67.2)_1$.

For the ease of the reader we write a proof of the assertion above.

Besides the afore-mentioned isotropy of $\mathscr{C}$ we assume that the locally geodesic frames (y) and (x) characterize the principal triad of strain at y^L and its spatial image at x^ρ respectively. Thus we have $\alpha^\rho{}_L=0$ for $\rho\neq L$ and $\alpha^r{}_L=v_{(r)}$ for $L=r$, where $v_{(r)}$ is the r-th principal stretch:

$$1+2\varepsilon_{(r)}=v_{(r)}{}^2\qquad (v_{(r)}>0)\,. \tag{a}$$

Furthermore for a principal wave, e. g. a wave travelling along $\lambda^L{}_{(H)}$, we have $\mathcal{N}^*_L=\lambda_{(H)L}=\delta_{HL}$. Then from $(66.7)_2$ and $(66.5)_4$, respectively, we deduce

$$p_{rs}=p_{rs}{}^{HH}=\frac{\partial\tilde{w}}{\partial\varepsilon_{HH}}\delta_{rs}+\frac{\partial^2\tilde{w}}{\partial\varepsilon_{Hr}\partial\varepsilon_{Hs}}v_{(r)}v_{(s)}\,. \tag{b}$$

Now we consider the transformation

$$y'^1=-y^1\,,\quad y'^2=-y^2\,,\quad y'^3=y^3\,. \tag{c}$$

Let (c) turn ε_{LM} into ε'_{LM} and $\tilde{w}(\varepsilon_{LM},\ldots)$ into $\tilde{w}'(\varepsilon'_{LM},\ldots)$. Then, since $\partial^2\tilde{w}/\partial\varepsilon_{LM}\partial\varepsilon_{AB}$ is a tensor—see Appendix A 3

$$\frac{\partial^2\tilde{w}'}{\partial\varepsilon'_{12}\partial\varepsilon'_{13}}=-\frac{\partial^2\tilde{w}}{\partial\varepsilon_{12}\partial\varepsilon_{13}}\,,\qquad \frac{\partial^2\tilde{w}'}{\partial\varepsilon'_{13}\partial\varepsilon'_{33}}=-\frac{\partial^2\tilde{w}}{\partial\varepsilon_{13}\partial\varepsilon_{33}}\,. \tag{d}$$

Furthermore, since $\tilde{w}(\varepsilon_{LM},\ldots)$ is an isotropic function of ε_{LM},

$$\tilde{w}'(\varepsilon'_{LM},\ldots)=\tilde{w}(\varepsilon_{LM},\ldots)\quad\text{and}\quad \frac{\partial^2\tilde{w}'}{\partial\varepsilon'_{LM}\partial\varepsilon'_{AB}}=\frac{\partial^2\tilde{w}}{\partial\varepsilon_{LM}\partial\varepsilon_{AB}}\quad\text{for}\quad \varepsilon'_{LM}=\varepsilon_{LM}\,. \tag{e}$$

Since the principal frame of strain is being referred to, we have $\varepsilon_{LM}=0$ for $L\neq M$. By (c) this obviously implies $(e)_3$, hence $(e)_2$. By $(e)_2$ and (d) the right hand sides of $(d)_1$ and $(d)_2$ vanish.

Likewise we can prove $\partial^2\tilde{w}/\partial\varepsilon_{Hr}\partial\varepsilon_{Hs}=0$ for $r\neq s$. This result and (b) imply $p_{rs}=0$ for $r\neq s$. Then the spatial image of a principal axis of strain is a symmetry axis of the polarization quadric $(67.2)_1$. q. e. d.

It must be remarked that in classical physics, *for every wave in a possibly anisotropic elastic body, the symmetry axes of the polarization quadric are the acoustic axes*—i. e. the axes which are parallel with the solutions of the classical analogue ($c=\infty, q_{rs}=\delta_{rs}$) of equation $(66.7)_1$ in the discontinuity vector λ^ρ (of the elastic acceleration wave). *The same does not occur generally in relativity.* However we are going to show that it occurs in the particular case being considered.

Let the co-ordinate lines x^r be locally tangent to the principal axes of tensions. Then in terms of the principal tensions $-p_{(r)}$, $(66.5)_1$ becomes

$$q_{rs}=\left(1+\frac{w+p_{(r)}}{c^2k}\right)\delta_{rs}\qquad (X_{rr}=p_{(r)},\ X_{rs}=0\ \text{for}\ r\neq s)\,. \tag{68.5}$$

Thus $q_{rs}=0$ for $r\neq s$. Hence *the principal axes of tension are the symmetry axis of the inertial-mass quadric* $(67.2)_2$.

The two italicized assertions above imply that in *an isotropic elastic body $\mathscr{C}$ the acoustic axes for the principal waves are the spatial images of the principal axes of strain (and that they are principal axes of tension)* just as in classical physics.

We denote by $V_{(l,r)}$ the speed of a wave which locally has the discontinuity vector parallel with the spatial image x^l of a principal axis of strain and is travelling along x^r, and by $V^*_{(l,r)}$ we denote its Lagrangian counterpart. Let $V_{(l,r)}{}^{(c)}$ and $V^*_{(l,r)}{}^{(c)}$ be the classical analogues of $V_{(l,r)}$ and $V^*_{(l,r)}$ respectively. Then, as we already said in dealing with a pure pressure state, $V_{(l,r)}{}^{(c)}=p_{(r)}$ holds. Furthermore we have the analogue of (68.2) for $V^*_{(l,r)}{}^{(c)}$. Now we can deduce the following analogue of (68.3) for principal waves in an isotropic body (considered in an arbitrary state) provided w be so chosen that it vanishes in the reference configuration:

$$V^*_{(l,r)}=V^*_{(l,r)}{}^{(c)}\left(1+\frac{p_{(r)}}{\rho}\right)^{-1/2}, \qquad V_{(l,r)}=V_{(l,r)}{}^{(c)}\left(1+\frac{p_{(r)}}{\rho}\right)^{-1/2} \qquad (w=0). \tag{68.6}$$

In § 30, where relativistic elastic waves in non-viscous fluids were dealt with, we stressed a point of view in contrast with the customary one. According to the former [latter] point of view the fact that a non-viscous fluid can be considered as a particular elastic material, is [is not] taken into account; then the classical mass density (entering the constitutive equations of the fluid) is regarded as a kinematic [dynamic] quantity, so that it is relativized into k [$c^{-2}\rho$]; this implies $V\neq V^{(c)}$ by (30.12) [$V=V^{(c)}$ after $(30.11)_2$]. (The difference in these results is purely formal.)

No (natural) analogue of the latter point of view holds for (68.6) in the general case ($p_{(r)}\neq p_{(s)}$ for $r\neq s$). The fact that (68.3) is a particular case of (68.6) and that $(68.3)_2$ coincides with the result $(30.12)_1$ obtained for fluids in accord with our point of view, motivates our considering this point of view.

§ 69. A Principle Concerning the Variation of the Metric Tensor of Riemannian Space-Time in the Adiabatic Elastic Case

We consider an elastic body, $\mathscr{C}$, with a vanishing coefficient of thermal conduction, so that the constitutive equations (63.4) hold for $\mathscr{C}$ and the entropy η can be assumed to be constant in $W_{\mathscr{C}}$.

We have a view to stating some variational principles for $\mathscr{C}$ in general relativity [§§ 69, 70], where variations of space-time metric and world-lines are considered. They are due to Schöpf [1964a] who generalized to general elastic bodies such as $\mathscr{C}$ the variational principles stated for ideal fluids by Taub [1954]—see also Fock [1964, sect. 47, 48] who improved them.

Let the motion (52.2) of $\mathscr{C}$ be assigned and the metric tensor $g_{\alpha\beta}=\hat{g}_{\alpha\beta}(x^\rho)$ also. Then we can calculate e.g. u^α, C_{LM}, and $w=\hat{w}(C_{LM}, y^L)$ both as function of $g_{\alpha\beta}$ and y^Σ—i.e. $g_{\alpha\beta}$, y^L, and t—and as functions of $g_{\alpha\beta}$ and x^ρ.

The functions $\hat{w}$, $\hat{g}_{\alpha\beta}$, and $\hat{x}^\rho$—cf. (52.2)—are assumed to be in $C^{(2)}$, i.e. to be twice continuously differentiable.

We consider a variation, $\delta g_{\alpha\beta}$, of the metric tensor in S_4 and we denote by $\delta_g f$ the corresponding variation of any physical magnitude f. Thus in particular we have

$$\delta_g\left(\frac{Ds}{Dt}\right)^2 = -x^\rho{}_0 x^\sigma{}_0 \delta g_{\rho\sigma} \qquad \left[\left(\frac{Ds}{Dt}\right)^2 = -x^\rho{}_0 x^\sigma{}_0 g_{\rho\sigma},\ x^\rho{}_0 = \frac{Dx^\rho}{Dt}\right]. \tag{69.1}$$

Since $u^\alpha = x^\alpha{}_0 Dt/Ds$ and $Dt/Ds>0$, (69.1) implies

$$\delta_g u^\alpha = \frac{x^\alpha{}_0}{2}\left(\frac{Ds}{Dt}\right)^{-3} x^\rho{}_0 x^\sigma{}_0 \delta g_{\rho\sigma} = \tfrac{1}{2} u^\alpha u^\rho u^\sigma \delta g_{\rho\sigma} \tag{69.2a}$$

and

$$\delta_g u_\alpha = \delta_g(g_{\alpha\sigma} u^\sigma) = u^\sigma \delta g_{\alpha\sigma} + \tfrac{1}{2} u_\alpha u^\rho u^\sigma \delta g_{\rho\sigma}. \tag{69.2b}$$

By (69.2b) and $(17.6)_1$ we have

$$\delta_g \overset{\perp}{g}_{\alpha\beta} = \delta g_{\alpha\beta} + u^\sigma u_\beta \delta g_{\alpha\sigma} + u_\alpha u^\rho \delta g_{\rho\beta} + u_\alpha u_\beta u^\rho u^\sigma \delta g_{\rho\sigma}, \tag{69.3}$$

which easily yields

$$\delta_g \overset{\perp}{g}_{\alpha\beta} = \overset{\perp}{g}_\alpha{}^\rho \overset{\perp}{g}_\beta{}^\sigma \delta g_{\rho\sigma}. \tag{69.4}$$

By (53.15) and $(53.7)_1$, (69.4) yields

$$\delta_g C_{LM} = x^\rho{}_L x^\sigma{}_M \delta \overset{\perp}{g}_{\rho\sigma} = \alpha^\rho{}_L \alpha^\sigma{}_M \delta g_{\rho\sigma} \qquad (C_{LM} = \overset{\perp}{g}_{\alpha\beta} x^\alpha{}_L x^\beta{}_M). \tag{69.5}$$

By $(56.9)_1$, $(69.5)_{1,2}$, and $(56.16)_2$, respectively, we have

$$2\mathscr{D}^{-1}\delta_g \mathscr{D} = \overset{-1}{C}{}^{LM} \delta_g C_{LM} = \overset{-1}{C}{}^{LM} \alpha^\rho{}_L \alpha^\sigma{}_M \delta g_{\rho\sigma} = \overset{\perp}{g}{}^{\rho\sigma} \delta g_{\rho\sigma}. \tag{69.6}$$

From (58.3) we deduce $\mathscr{D}\delta_g k + k\delta_g \mathscr{D} = 0$, which by (69.6) yields

$$\frac{1}{k}\delta_g k = -\frac{1}{\mathscr{D}}\delta_g \mathscr{D} = -\tfrac{1}{2}\overset{\perp}{g}{}^{\rho\sigma}\delta g_{\rho\sigma} \qquad (k^* = k\mathscr{D}). \tag{69.7}$$

Furthermore, since $g = \det \|g_{\rho\sigma}\|$, we have

$$\delta_g g = g g^{\rho\sigma} \delta g_{\rho\sigma}, \qquad \delta_g \sqrt{-g} = \frac{-g}{2\sqrt{-g}} g^{\rho\sigma} \delta g_{\rho\sigma} = \frac{\sqrt{-g}}{2} g^{\rho\sigma} \delta g_{\rho\sigma}. \tag{69.8}$$

Let us set

$$w_c = c^2 + w, \qquad \text{so that} \qquad \rho = k w_c. \tag{69.9}$$

Furthermore $(57.7)_2$ yields $\delta_g C_{LM} = 2\delta_g \varepsilon_{LM}$. Then from $(69.7)_3$ and $(63.4)_3$, from $(63.4)_2$ and $(69.5)_{1,2}$, and from $(59.2)_{1,2}$, respectively, we deduce

$$k\,\delta_g w_c = \frac{k^*}{2\mathscr{D}}\,\frac{\partial w_c}{\partial \varepsilon_{LM}}\,\delta_g C_{LM} = -\frac{1}{2\mathscr{D}}\,Y^{LM}\,\alpha^\rho{}_L\,\alpha^\sigma{}_M\,\delta g_{\rho\sigma} = -\tfrac{1}{2}X^{\rho\sigma}\,\delta g_{\rho\sigma}\,. \qquad (69.10\text{a})$$

Equalities $(69.8)_{2,3}$ and $(69.7)_{1,2}$, and $(69.9)_2$ and $(17.6)_1$, respectively, imply that

$$\delta_g(\sqrt{-g}\,k)\,w_c = \frac{\sqrt{-g}}{2}\,k\,w_c(g^{\rho\sigma} - \overset{\perp}{g}{}^{\rho\sigma})\,\delta g_{\rho\sigma} = -\frac{\sqrt{-g}}{2}\,\rho\,u^\rho u^\sigma\,\delta g_{\rho\sigma}\,. \qquad (69.10\text{b})$$

From $(69.9)_2$, (69.10a), and (69.10b) we deduce

$$\delta_g(\sqrt{-g}\,\rho) = -\frac{\sqrt{-g}}{2}(\rho\,u^\rho u^\sigma + X^{\rho\sigma})\,\delta g_{\rho\sigma} \qquad (\rho = k\,w_c)\,. \qquad (69.11)$$

Let C_4 be a bounded domain—see footnote 5 to Chapter 2—belonging to $W_{\mathscr{C}}$ and let its boundary (or frontier) $\mathscr{F}C_4$ oriented outwards, be smooth enough to allow us to use the Green formulas. Then, as is well known, by definition $(16.10)_1$

$$\delta_g \int_{C_4} R\sqrt{-g}\,dx = -\int_{C_4} A^{\alpha\beta}\,\delta g_{\alpha\beta}\sqrt{-g}\,dx \quad (R_{\alpha\beta} = R_{\alpha\rho}{}^\rho{}_\beta,\ R = R^{\alpha\beta}g_{\alpha\beta},\ dx = dx^0\ldots dx^3) \qquad (69.12)$$

holds for every twice continuously differentiable variation $\delta g_{\alpha\beta}$ *of* $g_{\alpha\beta}$ *which vanishes on* $\mathscr{F}C_4$ *together with* $\delta g_{\alpha\beta,\gamma}$, *hence with* $\delta g_{\alpha\beta/\gamma}$.

Indeed $(69.8)_{2,3}$ and $(69.12)_3$ imply

$$\delta_g \int_{C_4} R\sqrt{-g}\,dx = \int_{C_4} \sqrt{-g}\left(\frac{R}{2}\,g^{\alpha\beta}\,\delta g_{\alpha\beta} + R_{\alpha\beta}\,\delta g^{\alpha\beta}\right)dx + \int_{C_4} \sqrt{-g}\,g^{\alpha\beta}\,\delta_g R_{\alpha\beta}\,dx\,. \qquad (\text{a})$$

From (16.8) and $(69.12)_2$ we deduce in locally geodesic co-ordinates

$$R_{\alpha\beta} = 2\left\{{\rho \atop \alpha\ [\rho}\right\}_{,\beta]}, \quad \text{hence} \quad 2\delta_g\left\{{\rho \atop \alpha\ [\rho}\right\}_{,\beta]} = \delta_g R_{\alpha\beta} = 2\left(\delta_g\left\{{\rho \atop \alpha\ [\rho}\right\}\right)_{/\beta]}. \qquad (\text{b})$$

Since $D_\alpha{}^\rho{}_\sigma = \delta_g\left\{{\rho \atop \alpha\ \sigma}\right\}$ is a tensor, $(\text{b})_3$ holds in every frame. Then

$$g^{\alpha\beta}\,\delta_g R_{\alpha\beta} = g^{\alpha\beta}(D_\alpha{}^\rho{}_{\rho/\beta} - D_\alpha{}^\rho{}_{\beta/\rho}) = v^\gamma{}_{/\gamma} \qquad \text{for} \qquad v^\gamma = D^{\gamma\rho}{}_\rho - D^{\alpha\gamma}{}_\alpha\,. \qquad (\text{c})$$

On $\mathscr{F}C_4$ we have $\delta g_{\alpha\beta} = 0 = \delta g_{\alpha\beta,\gamma}$, hence $\delta g_{\alpha\beta/\gamma} = 0$ and $D_\alpha{}^\rho{}_\sigma = 0 = v^\gamma$. Then by the divergence theorem the last integral in (a) is zero.

Furthermore, the equality $g_{\rho\sigma}g^{\sigma\beta}=\delta_\rho^{\ \beta}$ yields $g_{\rho\sigma}\delta g^{\sigma\beta}=-g^{\sigma\beta}\delta g_{\rho\sigma}$; hence $\delta g^{\alpha\beta}=-g^{\alpha\rho}g^{\beta\sigma}\delta g_{\rho\sigma}$, so that $R_{\alpha\beta}\delta g^{\alpha\beta}=-R^{\alpha\beta}\delta g_{\alpha\beta}$. Then by $(69.12)_{2,3}$ and $(16.10)_1$ $2^{-1}Rg^{\alpha\beta}\delta g_{\alpha\beta}+R_{\alpha\beta}\delta g^{\alpha\beta}=-A^{\alpha\beta}\delta g_{\alpha\beta}$. Now we see—cf. (a)—that $(69.12)_1$ holds.

From (69.11) and (69.12) we deduce

$$\delta_g \int_{C_4} \sqrt{-g}\left(R+\frac{16\pi h}{c^4}\rho\right)dx$$

$$=-\int_{C_4}\sqrt{-g}\left(A^{\alpha\beta}+\frac{8\pi h}{c^4}\mathscr{U}^{\alpha\beta}\right)\delta g_{\alpha\beta}dx \quad (\mathscr{U}^{\alpha\beta}=\rho u^\alpha u^\beta+X^{\alpha\beta}) \tag{69.13}$$

for the above variation $\delta g_{\alpha\beta}$. Then *gravitation equations* (23.1) *hold for* $\mathscr{C}$ *in* C_4 *under definition* $(69.13)_2$ *if and only if*,

$$\delta_g \int_{C_4}\sqrt{-g}\left(R+\frac{16\pi h}{c^4}\rho\right)dx=0 \quad [\rho=k(c^2+w)=kw_c] \tag{69.14}$$

holds for all variations $\delta g_{\alpha\beta}$ *of* $g_{\alpha\beta}$ *which are in* $C^{(2)}$ *and vanish on* $\mathscr{F}C_4$ *together with* $\delta g_{\alpha\beta,\gamma}$.

As a consequence the last italicized variational condition can be used as a principle for $\mathscr{C}$, instead of gravitation equations.

The extension of the above result to polar and more general materials is performed in §§ 96 to 99.

§ 70. Variation of World Lines in the Adiabatic Elastic Case

We keep the assumptions on $\mathscr{C}$ made in § 69 and we consider again the field $g_{\alpha\beta}$ and the motion $\mathscr{M}$ of $\mathscr{C}$ in S_4—represented by the functions $x^\alpha=\hat{x}^\alpha(y^\Sigma)$—as given. But now we have in mind a variation δx^α of $\mathscr{M}$.

We consider a motion $\mathscr{M}_\lambda$ depending on the real parameter λ:

$$z^\alpha={}^*z^\alpha(\lambda,y^\Sigma) \quad \text{with} \quad {}^*z^\alpha(0,y^\Sigma)=\hat{x}^\alpha(y^\Sigma), \tag{70.1}$$

and we assume that the motions $(70.1)_1$ are twice continuously differentiable.

We also define the functions $z^\alpha(\lambda,x^\rho)$ by means of the condition

$$z^\alpha=z^\alpha[\lambda,\hat{x}^\rho(y^\Sigma)]={}^*z^\alpha(\lambda,y^\Sigma), \quad \text{whence} \quad z^\alpha(0,x^\rho)=x^\alpha, \tag{70.2}$$

and the functions $\zeta^\rho(x)$ and ${}^*\zeta^\rho(y^\Sigma)$ by means of

$$\zeta^\rho=\zeta^\rho(x)=\left[\frac{\partial z^\rho(\lambda,x)}{\partial\lambda}\right]_{\lambda=0}, \quad {}^*\zeta^\rho(y^\Sigma)=\zeta^\rho[\hat{x}(y^\Sigma)]=\left(\frac{\partial^* z^\rho}{\partial\lambda}\right)_{\lambda=0}. \tag{70.3}$$

Every component $\chi^{\cdots}_{\cdots}$ of a tensor which is a function of the event point z^α and the material point y^L can be expressed as a function, $\phi^*_\lambda(y^\Sigma)$ of y^Σ by using transformation $(70.1)_1$, and as a function

$$\chi^{\cdots}_{\cdots}=\phi_\lambda(z)=\phi^*_\lambda[y^\Sigma(\lambda,z)] \tag{70.4}$$

of z by using the converse $y^\Sigma=y^\Sigma(\lambda,z)$ of $(70.1)_1$.

We wish to evaluate the increment $\Delta\chi^{\cdots}_{\cdots}$ of $\chi^{\cdots}_{\cdots}$ in the correspondence $\mathscr{M}_0\to\mathscr{M}_\lambda$ at a given event point, say x^ρ (we write e.g. "x" for "$x_0,\ldots,x_3$") and for a given value of λ. To this end we associate with x and λ the quantity z such that for some y^Σ

$$z={}^*z(\lambda,y^\Sigma)\qquad\text{where}\qquad x^\rho=\hat{x}^\rho(y^\Sigma)\qquad(\text{hence } y^\Sigma=y^\Sigma(\lambda,z))\,. \tag{70.1b}$$

Then, by $(70.4)_2$, $\phi_\lambda(z)=\phi^*_\lambda(y^\Sigma)$. Since $x^\alpha={}^*z^\alpha(0,y^\Sigma)$, the analogue of this equality for $\lambda=0$ is $\phi_0(x)=\phi^*_0(y)$. Now we can write

$$\begin{aligned}\Delta\chi^{\cdots}_{\cdots}&=\phi_\lambda(x)-\phi_0(x)=\phi_\lambda(x)-\phi_\lambda(z)+\phi_\lambda(z)-\phi_0(x)\\&=-[\phi_\lambda(z)-\phi_\lambda(x)]+\phi^*_\lambda(y^\Sigma)-\phi^*_0(y^\Sigma)\\&=-\left[\frac{\partial\phi_\lambda(z')}{\partial z'^\rho}\right]_{z'^\sigma=x^\sigma+\vartheta^{(\sigma)}(z^\sigma-x^\sigma)}\frac{z^\rho-x^\rho}{\lambda}\lambda+\lambda\left[\frac{\partial\phi^*}{\partial\lambda'}\right]_{\lambda'=\vartheta'\lambda}\qquad(|\vartheta|<1>|\vartheta'|)\,,\end{aligned} \tag{70.5}$$

where $\vartheta^{(\sigma)}$ and ϑ' obviously are suitable real numbers depending on λ and fulfilling $(70.5)_{5,6}$. By taking $(70.3)_2$ and $(70.2)_3$ into account and by assuming that we are using locally geodesic co-ordinates at x^ρ, we deduce from (70.5)

$$\lim_{\lambda\to0}\frac{\Delta\chi^{\cdots}_{\cdots}}{\lambda}=-\frac{\partial\phi_0}{\partial x^\rho}\zeta^\rho+\left[\frac{\partial\phi^*_\lambda}{\partial\lambda}\right]_{\lambda=0}\qquad\left(\frac{\partial\phi_0}{\partial x^\rho}=\chi^{\cdots}_{\cdots/\rho},\ \frac{\partial\phi^*_\lambda}{\partial\lambda}=\frac{D\phi^*_\lambda}{D\lambda}\right). \tag{70.6a}$$

Since x^ρ is fixed, $\Delta\chi^{\cdots}_{\cdots}$ is a function of λ and by (70.5) its differential $\delta_{\mathscr{M}}\chi^{\cdots}_{\cdots}$ for $\lambda=0$ has the expression

$$\begin{gathered}\delta_{\mathscr{M}}\chi^{\cdots}_{\cdots}=-\chi^{\cdots}_{\cdots/\rho}\zeta^\rho d\lambda+\delta^*\chi^{\cdots}_{\cdots}\\\left(\delta_{\mathscr{M}}\chi^{\cdots}_{\cdots}=\left[\frac{\partial\Delta\chi^{\cdots}_{\cdots}}{\partial\lambda}\right]_{\lambda=0}d\lambda,\quad\delta^*\chi^{\cdots}_{\cdots}=\left[\frac{D\phi^*_\lambda}{D\lambda}\right]_{\lambda=0}d\lambda\right)\end{gathered} \tag{70.6b}$$

which obviously holds in every frame. The differential $\delta^*\chi^{\cdots}_{\cdots}$ is the Lagrangian or material contribution to $\delta\chi^{\cdots}_{\cdots}$ in that it is the first order increment of $\chi^{\cdots}_{\cdots}$ in the correspondence $\mathscr{M}_0\to\mathscr{M}_{d\lambda}$ at (the fixed) material point y^L (and at the fixed instant $t=y^0$). Furthermore the vector $\zeta^\rho d\lambda$—cf. $(70.3)_1$—is the displacement of the material point y^L at t in $\mathscr{M}_0\rightarrowtail\mathscr{M}_{d\lambda}$.

Since the field $g_{\alpha\beta}$ is unaffected by the correspondence $\mathscr{M}_0\to\mathscr{M}_\lambda$, we have $\delta_{\mathscr{M}}g_{\alpha\beta}=0$, By $(70.6\text{b})_1$ this yields

$$\delta^*g_{\alpha\beta}=g_{\alpha\beta/\gamma}\zeta^\gamma d\lambda=0\quad\text{besides}\quad\delta_{\mathscr{M}}\chi_{\alpha\beta\ldots}=g_{\alpha\rho}\delta_{\mathscr{M}}\chi^\rho{}_{\beta\ldots}\,. \tag{70.7}$$

From $(69.5)_3$, $(70.7)_{1,2}$, and $(17.6)_1$ we have

$$\delta^* C_{LM} = \overset{\perp}{g}_{\rho\sigma}(x^\rho{}_L \delta^* x^\sigma{}_M + x^\sigma{}_M \delta^* x^\rho{}_L) + x^\rho{}_L x^\sigma{}_M (u_\rho \delta^* u_\sigma + u_\sigma \delta^* u_\rho). \tag{70.7b}$$

We have a view to proving the general theorem 70.2 below[5] which is useful to reduce variations of world-lines to variations of the metric tensor in the present case and in more general situations—cf. sections 97—99.[6] Therefore we assume that $\mathscr{C}$ cannot conduct heat as well as in section 69, but that its specific internal energy w is a function of y^L, $C_{BL_1} (=2\varepsilon_{LM})$, ..., $C_{BL_1 \ldots L_n}$.

In connection with the motion $\mathscr{M}$ of $\mathscr{C}$ and an event point, $\mathscr{E}$, in C_4 we often have to evaluate a (double) tensor valued magnitude $T_{\rho \ldots L \ldots}{}^{\sigma \cdots M \cdots}$ $(= T_{\ldots}^{\ldots})$ such as ρ, k, or $\alpha^\rho{}_{L_1 \ldots L_n}$. Given a material frame (y) and a space-time frame (x), this double tensor has an expression in terms of a representation $\hat{x}^\rho$ of $\mathscr{M}$ and those $\hat{g}_{\alpha\beta}$ and $\hat{a}_{LM}$ of the fields $\hat{\mathbf{g}}$ and $\hat{\mathbf{a}}$ of the metric tensors in S_4 and the reference configuration C^* respectively. This expression is *tensorial*, i.e. it behaves in a certain well known way under changes of the frames (x) and (y). Furthermore it is *$\hat{t}$-invariant*, i.e. $T_{\ldots}^{\ldots}$ is unaffected by changes of the time parameter $t = \hat{t}(x)$ —cf. (52.5)—up to whose choice the representation $\hat{x}^\rho$ of $\mathscr{M}$ is determined.

Now let $U^{\rho \cdots}{}_{\sigma \cdots}$ and $V^{L \cdots}{}_{M \cdots}$ be two arbitrary affine tensors at x^ρ and y^L respectively, that are independent of $\mathscr{M}$, $\hat{\mathbf{g}}$, and $\hat{\mathbf{a}}$; and set

$$\psi = U^{\rho \cdots}{}_{\sigma \cdots} T_{\rho \cdots}{}^{\sigma \cdots}{}_{L \cdots}{}^{M \cdots} V^{L \cdots}{}_{M \cdots} \qquad (\delta^* U_{\ldots}^{\ldots} = \delta_g U_{\ldots}^{\ldots} = 0 = \delta^* V = \delta_g V). \tag{70.8}$$

Then ψ is a scalar $\hat{t}$-independent (tensorial) functional of $\mathscr{M}$, $\hat{\mathbf{g}}$, and $\hat{\mathbf{a}}$. In the sequel $\hat{\mathbf{a}}$ and (y) will be kept fixed. Therefore we can represent ψ in the form

$$\psi = \psi[\mathscr{E}, \mathscr{M}, \hat{\mathbf{g}}] = \psi_{(x)}[x^\rho, \hat{x}^\rho, \hat{g}_{\alpha\beta}]. \tag{70.9}$$

Incidentally, if e.g. ψ is w, to evaluate ψ we have a function of the form $w = \hat{w}(y^L, C_{BL_1}, \ldots, C_{BL_1 \ldots L_n})$. We find y^L and t by solving equations $(52.2)_1$, i.e. by applying the inverse of the transformation $\hat{x}^\rho$ to x^ρ. Then we evaluate $C_{BL_1 \ldots L_i}$ by use of $\hat{x}^\rho{}_{,\lambda_1 \ldots \lambda_i}(t, y^L)$ and $\hat{g}_{\alpha\beta, \lambda_1 \ldots \lambda_i}(x^\rho)$. At last we obtain $\psi = w$ by means of the function $\tilde{w}$ above.

Theorem 70.1. *Let the double tensor* $T_{\rho \ldots}{}^{\sigma \cdots}{}_{L \ldots}{}^{M \cdots} (= T_{\ldots}^{\ldots}) = T_{\ldots}^{\ldots}(\mathscr{E}, \mathscr{M}, \hat{\mathbf{g}})$ *have a $\hat{t}$-invariant tensorial expression in* x^ρ, $\hat{x}^\rho$, $\hat{g}_{\alpha\beta}$, *and* $\hat{a}_{LM}$. *Then for every variation* δx^ρ *of* $\mathscr{M}$ *we have*—cf. (70.3), (70.6b)

$$\delta_{\mathscr{M}} T_{\rho \ldots}{}^{\sigma \cdots}{}_{L \cdots}{}^{M \cdots} = \delta_* T_{\rho \ldots}{}^{\sigma \cdots}{}_{L \cdots}{}^{M \cdots}, \quad \textit{where} \quad \delta_* = \delta_g \quad \textit{for} \quad \delta g_{\alpha\beta} = 2\zeta_{(\alpha/\beta)} d\lambda. \tag{70.10}$$

Indeed in any case we can reduce our problem to the analogue for a scalar by means of definition (70.8), in that by our assumptions on $T_{\ldots}^{\ldots}$, $U_{\ldots}^{\ldots}$, and $V_{\ldots}^{\ldots}$, ψ turns out to have a $\hat{t}$-invariant tensorial expression.

Now consider the above family $\mathcal{M}_\lambda$ of varied motions, represented by (70.1), or by the equations $z^\alpha = z^\alpha(\lambda, x)$—cf. (70.2). Let us fix λ; more precisely set $\lambda = d\lambda$. Then on the one hand those equations represent a one to one correspondence $\mathscr{E} \to \mathscr{E}'$ in space time, which transforms $\mathcal{M}(=\mathcal{M}_0)$ into $\mathcal{M}_{d\lambda}$: $\mathscr{E}'$ is the event point of co-ordinates $z^\alpha = z^\alpha(d\lambda, x)$ in (x). On the other hand the equations $z^\alpha = z^\alpha(d\lambda, x)$ determine a new frame, say (z). Remark that $\mathscr{E}$ and $\mathscr{E}'$ have the co-ordinates z^ρ in (z) and (x) respectively.

We denote e.g. by $\hat{g}^z_{\alpha\beta}(z^\rho)$ the components in (z) of the tensor $\mathbf{g}$ at $\mathscr{E}$, and we call $\gamma = \hat{\gamma}(\mathscr{E})$ the tensor at $\mathscr{E}$ whose components $\hat{\gamma}^z_{\alpha\beta}(z^\gamma)$ in (z) equal those of $\mathbf{g}(\mathscr{E}')$ in (x):

$$\hat{\gamma}^z_{\alpha\beta}(z^\rho) = \hat{g}_{\alpha\beta}(z^\rho) \text{ (for all } z^\rho), \quad \text{i.e.} \quad \hat{\gamma}^z_{\alpha\beta} = \hat{g}_{\alpha\beta}\,. \tag{70.11}$$

Thus γ is symmetric, so that it can be regarded as a metric tensor. Remembering the expression of Christoffel's symbols

$$2\zeta_{(\alpha/\beta)} = \zeta_{\beta,\alpha} + \zeta_{\alpha,\beta} - (g_{\alpha\tau,\beta} + g_{\tau\beta,\alpha} - g_{\alpha\beta,\tau})\zeta^\tau = g_{\rho\beta}\zeta^\rho{}_{,\alpha} + g_{\alpha\sigma}\zeta^\sigma{}_{,\beta} + g_{\alpha\beta,\tau}\zeta^\tau\,. \tag{70.12}$$

By $(70.3)_2$ $z^\alpha(d\lambda, x) \approx x^\alpha + \zeta^\alpha d\lambda$, and $z^\rho{}_{,\alpha} = \dfrac{\partial z^\rho(d\lambda, x)}{\partial x^\alpha} \approx \delta^\rho{}_\alpha + \zeta^\rho{}_{,\alpha} d\lambda$.

Then, by (70.11) we have at $\mathscr{E}$ (for $\lambda = d\lambda$ and $(d\lambda)^2$ negligible)

$$\begin{aligned}\hat{\gamma}_{\alpha\beta}(x^\gamma) &\approx z^\rho{}_{,\alpha} z^\sigma{}_{,\beta}\, \hat{g}_{\rho\sigma}(x^\gamma + \zeta^\gamma d\lambda) \approx (\delta^\rho{}_\alpha + \zeta^\rho{}_{,\alpha} d\lambda)(\delta^\sigma{}_\beta + \zeta^\sigma{}_{,\beta} d\lambda)(g_{\rho\sigma} + g_{\rho\sigma,\tau}\zeta^\tau d\lambda) \\ &\approx \hat{g}_{\alpha\beta}(x^\gamma) + (g_{\rho\beta}\zeta^\rho{}_{,\alpha} + g_{\alpha\sigma}\zeta^\sigma{}_{,\beta} + g_{\rho\beta}\zeta^\rho{}_{,\alpha} + g_{\alpha\beta,\tau}\zeta^\tau) d\lambda\,. \end{aligned} \tag{70.13}$$

We define $\delta g_{\alpha\beta}$ by means of $(70.10)_3$—cf. (70.3). Then (70.12) and (70.13) yield

$$\delta g_{\alpha\beta} = 2\zeta_{(\alpha/\beta)} d\lambda = \hat{\gamma}_{\alpha\beta}(x^\gamma) - \hat{g}_{\alpha\beta}(x^\gamma)\,. \tag{70.14}$$

The analogue of (70.9) for (x), $\mathscr{E}'$, and $\mathcal{M}_{d\lambda}$ [for (z), $\mathscr{E}$, and $\mathcal{M}$] holds. This yields the first [last] of the equalities

$$\psi[\mathscr{E}', \mathcal{M}_{d\lambda}, \hat{\mathbf{g}}] = \psi_{(x)}[z^\rho, \hat{z}^\rho, \hat{g}_{\alpha\beta}] = \psi_{(z)}[z^\rho, \hat{z}^\rho, \hat{\gamma}^z_{\alpha\beta}] = \psi(\mathscr{E}, \mathcal{M}, \hat{\gamma})\,. \tag{70.15}$$

Given the numbers z^ρ and functions $\hat{z}^\rho$ and $\hat{g}_{\alpha\beta}$, it is obvious that the choice of (x) does not affect the scalar $\psi_{(x)}[z^\rho, \hat{z}^\rho, \hat{g}_{\alpha\beta}]$. Therefore this equals $\psi_{(z)}[z^\rho, \hat{z}^\rho, \hat{g}_{\alpha\beta}]$. Hence $(70.11)_2$ implies $(70.15)_2$. From (70.15), (70.14), and $(70.10)_{2,3}$ we deduce

$$\delta_{\mathcal{M}}\psi = \psi[\mathscr{E}, \mathcal{M}_{d\lambda}, \hat{\mathbf{g}}] - \psi[\mathscr{E}, \mathcal{M}, \hat{\mathbf{g}}] = \psi[\mathscr{E}, \mathcal{M}, \hat{\gamma}] - \psi[\mathscr{E}, \mathcal{M}, \hat{\mathbf{g}}] = \delta_*\psi\,. \tag{70.16}$$

By $(70.8)_{1-5}$ this yields $(70.10)_1$. q.e.d.

Theorem 70.2. (See footnotes 5 and 6 in Chapter 7.) (a) *Let the specific internal energy w of $\mathscr{C}$ be a function $\hat{w}$ of $y^L, C_{BL_1}, \ldots, C_{BL_1 \ldots L_n}$. Then*

$$\delta_{\mathcal{M}} A = \delta_{\mathcal{M}} \int_{C_4} \rho\sqrt{-g}\, dx = \delta_* \int_{C_4} \rho\sqrt{-g}\, dx = \delta_* A \quad \textit{with} \quad A = \int_{C_4} \left(\frac{c^4}{16\pi h} R + \rho\right)\sqrt{-g}\, dx\,. \tag{70.17}$$

(b) *In addition let* $\mathscr{U}^{\alpha\beta}$ *be a symmetric tensor having a* $\hat{t}$*-independent tensorial expression in the motion* $\mathscr{M}$ *of* $\mathscr{C}$ *and the metric tensor field* $\hat{\mathbf{g}}$. *Furthermore let this tensor be in* $C^{(1)}$ *when it is regarded as a function of* x^ρ *defined on* C_4. *At last let*

$$2\delta_g \int_{C_4} \rho\sqrt{-g}\,dx = -\int_{C_4} \mathscr{U}^{\alpha\beta}\,\delta g_{\alpha\beta}\sqrt{-g}\,dx \tag{70.18}$$

hold for every variation $\delta g_{\alpha\beta}$ *of* $g_{\alpha\beta}$ *such that it is in the class* $C^{(n-1)}$ *and*

$$(\delta g_{\alpha\beta})_{,\lambda_1\ldots\lambda_i} = 0 \quad on \quad \mathscr{F}C_4 \quad (i=0,\ldots,n-2). \tag{70.19}$$

Then we have—cf. (70.3) *and* (70.6b)$_2$

$$\delta_{\mathscr{M}} A = (d\lambda)\int_{C_4} \mathscr{U}^{\rho\sigma}{}_{/\sigma}\zeta_\rho\sqrt{-g}\,dx \qquad (d\lambda = \text{const}) \tag{70.20}$$

for every variation $\zeta^\rho d\lambda$ *of the motion* $\mathscr{M}$ *of* $\mathscr{C}$ *such that* ζ^ρ *is in* $C^{(n)}$ *and*

$$\zeta^\rho{}_{,\lambda_1\ldots\lambda_i} = 0 \quad on \quad \mathscr{F}C_4 \quad (i=0,\ldots,n-1). \tag{70.21}$$

(c) *Under the assumptions above the conservations equations* $\mathscr{U}^{\rho\sigma}{}_{/\sigma}=0$ *hold in* C_4 *if and only if* $\delta_{\mathscr{M}}A=0$ *for every variation* $\delta\mathscr{M}$ *of* $\mathscr{M}$ *by which the boundary conditions* (70.21) *are satisfied.*

Proof. Since R is unaffected by $\delta\mathscr{M}$, (70.17)$_4$ yields (70.17)$_1$. Furthermore by (70.10) $\delta_{\mathscr{M}}(\rho\sqrt{-g}) = \delta_*(\rho\sqrt{-g})$ and $\delta_{\mathscr{M}}A = \delta_* A$. Hence (70.17)$_{2,3}$ hold.

To prove (b) assume (70.21) and that ζ^ρ is in $C^{(n)}$, which by the definition (70.10)$_3$ of $\delta g_{\alpha\beta}$ yields (70.19) and that $\delta g_{\alpha\beta}$ is in $C^{(n-1)}$. Then, by an assumption, (70.18) holds. Thence, by (70.10)$_{2,3}$ and (70.17)$_3$, we obtain

$$\delta_* A = -(d\lambda)\int_{C_4} \mathscr{U}^{\rho\sigma}\zeta_{(\rho/\sigma)}\sqrt{-g}\,dx\,.$$

Furthermore $\mathscr{U}_{[\alpha\beta]}=0$ by an hypothesis. Hence Green's lemma and (70.17) yield (17.20)$_1$.

Part (c) follows immediately from the validity of (70.20) for all $\delta\mathscr{M}$ that satisfy (70.21) under definition (70.3). q.e.d.

Theorem 70.3. *Let* $\mathscr{C}$ *not be capable of conducting heat. In addition let* $\mathscr{C}$ *have the internal energy* $w=\hat{w}(y^L,\varepsilon_{LM})$; *(and let* $\hat{w}$ *and the metric field be of class* $C^{(2)}$*). Then the conservation equations for* $\mathscr{C}$

$$\mathscr{U}^{\alpha\beta}{}_{/\beta}=0 \quad with \quad \mathscr{U}^{\alpha\beta} = \rho u^\alpha u^\beta + X^{\alpha\beta} \tag{70.22}$$

hold in C_4, *if and only if*

$$\delta_{\mathscr{M}}\int_{C_4}\rho\sqrt{-g}\,dx = 0, \quad or\ equivalently \quad \delta_{\mathscr{M}}\int_{C_4}\left(R+\frac{16\pi}{c^4}\rho\right)\sqrt{-g}\,dx = 0 \tag{70.23}$$

for the variations of world lines characterized by any twice continuously differentiable field $\zeta^\rho=\zeta^\rho(x)$—cf. (70.3)—*that vanishes on* $\mathscr{F}C_4$.

Indeed the assumptions on w made in Theorem 70.2 hold for $n=1$. Furthermore under the hypothesis of our theorem equality (69.11) was deduced without using any boundary condition on $\delta g_{\alpha\beta}$. It yields (70.18) under the definition $(70.22)_2$ of $\mathscr{U}^{\alpha\beta}$. Then by Theorem 70.2(c) the conservation equations (70.22) hold in C_4 if and only if $\delta_{\mathscr{M}} A=0$ for all solutions ζ^ρ of (70.21) for $n=1$. Furthermore by $(70.17)_4$ and $(70.17)_1$ the condition $\delta_{\mathscr{M}} A=0$ is equivalent to each of the equalities $(70.23)_{1,2}$. Hence the thesis holds. q.e.d.

Incidentally, since $(17.22)_1$ is a consequence of gravitation equations $(23.1)_1$ under definition $(70.22)_2$, in the first place *equalities* (70.23) *hold for the above variations of world lines* as a consequence of the validity of $(23.1)_1$ in C_4. In the second place *the afore-mentioned validity of* $(70.23)_1$ *and* $(70.23)_2$ *is implied by the validity of equality* $(69.14)_1$ *for all variations* $\delta g_{\alpha\beta}$ *of* $g_{\alpha\beta}$ *that are in* $C^{(2)}$ *and vanish on* $\mathscr{F}C_4$ *together with* $\delta g_{\alpha\beta,\gamma}$ even in case gravitation equations $(23.1)_1$ are not postulated.

Footnotes to Chapter 7

[1] Bressan [1963b] to [1967b] and Schöpf [1964a] to [1967] are mutually compatible theories. In the first approximation they are also compatible with Rayner [1963] where Hooke's linear law is relativized by assuming that stress is a linear function of (non-linearized) strain. However, as is to be expected from the approximation character of linear theories, Schöpf [1965a, p. 350] remarks that if for conservative systems a variational principle holds, then the strain stress relation in co-moving co-ordinates cannot be linear in an exact theory.

Schöpf's remark is in accord with the following assertion proved by Bressan [1964a, p. 71] in order to relate Bressan [1964a] to Rayner [1963]—cf. Appendix D:

If (i) in co-moving co-ordinates the stress X_{rs} of $\mathscr{C}$ at P^* is a linear function of $\varepsilon_{LM}=(\overset{\downarrow}{g}_{LM}-a^*_{LM})/2$ —cf. $(62.5)_1$ and $(57.7)_2$—and (ii) (at least as far as isentropic processes are concerned) either the proper density ρ of total energy or the internal energy w per unit conventional mass is a function of ε_{LM} (and η) at P^*, then the stress must vanish identically.

[2] Acceleration waves are dealt with in Synge [1959] and Rayner [1963] in the isotropic and general case, respectively, on the basis of different relativizations of Hooke's linear law. Besides Bressan [1963d], let us mention Schöpf [1965a] on general elastic waves in general relativity.

[3] Practically all in §§ 67, 68 is taken directly from Bressan [1963d], except that we have corrected two obviously mistaken orthogonality assertions, one of which refers to the above transformation of (x) into $(\bar{x})$.

[4] By $(57.7)_2$ and (54.4) the quadrics

$$\varepsilon_{LM}\xi^L\xi^M=1, \quad C_{LM}\xi^L\xi^M=1, \quad \mathscr{D}_{LM}\xi^L\xi^M=1 \tag{*}$$

have the same symmetry axes.

[5] This is the first time that Theorem 70.2 is published. In proving it Bressan took advantage from an unpublished part of the first version of Pitteri [1975b], written to reach a completely different goal, and from Pitteri's useful criticism.

[6] For simple elastic materials Schöpf [1964a] proves a set of formulas on variations of the metric tensor and the analogous set for variations of world-lines, in order to deal with the two variational principles involving these kinds of variations. The same do Bressan [1972b] and Pitteri [1975a, b] in generalizing Schöpf's results to elastic materials capable of couple stresses and to general elastic materials of order $n\geqslant 2$ respectively. In all of these cases, Theorem 70.2 allows us to deal with the variational principles connected with world lines very quickly: in particular we can help considering the second set of variational formulas mentioned above—see Theorem 70.3, Theorem 97.1, Theorem 98.2, and Theorem 99.5.

Chapter 8

Piezo-Elasticity and Magnetoelastic Waves from the Lagrangian Point of View

§ 71. Introduction

In this chapter we consider the conservation equations $(23.3)_1$, and the gravitation equations (23.1) in general relativity, in the general elastic case, with inclusion of thermodynamics and electromagnetism. In § 72 we lay down foundations of piezo-elasticity (for finite deformations). [1)] § 72 is based on the consequence (43.8) of conservation or gravitation equations, which is the equation of the balance of energy, and on the second principle in the general form (46.2).

The remainder of this chapter is concerned with ideal conductors. In §§ 73—75 we consider some extensions of the operations $T_{\cdots} \to T^*_{\cdots}$, $T^{\cdots} \rightarrowtail T_*^{\cdots}$, D^c, and D_c, and some properties of them connected with Born-rigidity. To the latter purpose we also use a formula of second-order relativistic Lagrangian kinematics: (91.11a)[2)] §§ 73—75 serve, among other things, as preliminaries for dealing in § 76 with certain properties of magnetizable ideal conductors.[3)] In §§ 77, 78 we deal with magneto-elastic waves in piezo-elastic ideal conductors.[4)] This theory contains the one for fluids [§§ 49, 50] as a particular case.[5)]

§ 72. Foundations of Piezo-Elasticity

In this section we consider the determination ${}^7\mathscr{U}_{\alpha\beta}$ of $\mathscr{U}_{\alpha\beta}$ (and the one ${}^7X_{\alpha\beta}$ of $X_{\alpha\beta}$) which is related to the determination ${}^7E_{\alpha\beta}$ of $E_{\alpha\beta}$—cf. $(40.18)_2$—trough (43.5). In accord with this and with (43.9) we use inequality (46.2) for $n=3$, which by $(58.3)_1$, $(58.10)_1$, and $(59.6)_{1,2}$ can be developed into

$$k^*(D\mathscr{F}+\eta DT-E^*_L D\pi^L_* - H^*_L D\mu^L_*)+Y^{LM}D\varepsilon_{LM}\leqslant 0 \quad (\mathscr{F}=w-T\eta). \tag{72.1}$$

By $(46.4)_{1,2}$ we can turn (72.1) into

$$k^*(D\bar{\mathscr{F}}+\eta DT+\pi^L_* DE^*_L+\mu^L_* DH^*_L)+Y^{LM}D\varepsilon_{LM}\leqslant 0 \quad (\bar{\mathscr{F}}=\bar{w}-\eta T). \tag{72.2}$$

and equality $(72.2)_2$ follows from $(46.4)_{3,4}$. We can deduce the following immediate consequence of (72.2),

$$k^*(D\bar{w} - T D\eta + \pi_*^L D E_L^* + \mu_*^L D H_L^*) + Y^{LM} D\varepsilon_{LM} \leqslant 0 \quad (\bar{\mathscr{F}} = \bar{w} - \eta T), \tag{72.3}$$

from (46.14) in the same way as (72.2) was derived from (46.2).

We say that $\mathscr{C}$ is *piezo-elastic* or *polarizable and elastic* at P^*, if at P^* (i) the magnitudes $\bar{\mathscr{F}}$, Y^{LM}, π_*^L, and μ_*^L are functions of T, ε_{LM}, E_L^*, and H_L^*, and (ii) internal constraints are absent in the sense that $\dot{T}$, $\dot{\varepsilon}_{LM}$ (or $u_{\alpha/\beta}^{\perp}$), $\dot{\pi}_*^L$, and $\dot{\mu}_*^L$ are physically independent of one another and of T, ε_{LM}, E_L^*, and H_L^*—cf. the definition of polarizable non-viscous fluid in § 46.

Obviously, in case $\mu_*^L \equiv 0$ $[\pi_*^L \equiv 0]$ but π_*^L $[\mu_*^L]$ can be unequal to 0 at P^*, we shall say that $\mathscr{C}$ is *electrically* [*magnetically*] *piezo-elastic* at P^*.

Let $\mathscr{C}$ be piezo-elastic at P^*. Then, after $D\bar{\mathscr{F}}$ has been developed, the left hand side of $(72.2)_1$ becomes a linear form in DT, $D\varepsilon_{LM}$, DE_L^*, and DH_L^*. Since $(72.2)_1$ must hold for arbitrary values of these differentials, we have the constitutive equations

$$\begin{aligned} \eta &= -\frac{\partial \bar{\mathscr{F}}}{\partial T}, \qquad Y^{LM} = -k^* \frac{\partial \bar{\mathscr{F}}}{\partial \varepsilon_{LM}}, \\ \pi_*^L &= -\frac{\partial \bar{\mathscr{F}}}{\partial E_L^*}, \qquad \mu_*^L = -\frac{\partial \bar{\mathscr{F}}}{\partial H_L^*} \quad \text{where} \quad \bar{\mathscr{F}} = \bar{\mathscr{F}}(T, \varepsilon_{LM}, E_L^*, H_L^*). \end{aligned} \tag{72.4}$$

As a consequence inequalities (72.2), (72.1), and (46.2) hold as equalities, hence so does the second principle $T D\eta \geqslant dQ$. Now we can conclude that a *piezo-elastic material is capable of only reversible processes.*

By the Helmholtz postulate—see the last part of § 46—$(T D\eta =) dQ > 0$ for $DT > 0$ and $D\varepsilon_{LM} = DE_L^* = DH_L^* = 0$, so that by the usual reasoning, we see that $(72.4)_1$ defines T in terms of η, ε_{LM}, E_L^*, and H_L^*. Thus we can express $\bar{\mathscr{F}}$ and $\bar{w}$—see $(46.4)_{1,2}$—as such functions. Now it is easy to see that we can deduce

$$\begin{aligned} T &= \frac{\partial \bar{w}}{\partial \eta}, \qquad Y^{LM} = -k^* \frac{\partial \bar{w}}{\partial \varepsilon_{LM}}, \\ \pi_*^L &= -\frac{\partial \bar{w}}{\partial E_L^*}, \qquad \mu_*^L = -\frac{\partial \bar{w}}{\partial H_L^*}, \qquad \bar{w} = \bar{w}(\eta, \varepsilon_{LM}, E_L^*, H_L^*) \end{aligned} \tag{72.5}$$

from (72.3) in practically the same way in which (72.4) was derived from (72.2).

Incidentally, by (63.5) and $(59.2)_3$ equalities $(72.4)_2$ and $(72.5)_2$ imply the following analogues of $(63.6)_{1,2}$:

$$K_\rho{}^M = -k^* \frac{\partial \bar{\mathscr{F}}}{\partial \alpha^\rho{}_M} = -k^* \frac{\partial \bar{w}}{\partial \alpha^\rho{}_M} \quad (2\varepsilon_{LM} = \alpha^\rho{}_L \alpha_{\rho M} - a_{LM}^*). \tag{72.6}$$

We say that the body $\mathscr{C}$ (piezo-elastic at P^*) is *isotropic at* P^* *in the reference state* $(\mathscr{P}^*, \Sigma_3^*)$ if $\bar{\mathscr{F}}$ is an isotropic function of ε_{LM}, π_*^L, and μ_*^L (according to the well known analogue of Definition 38.1).

We assume this isotropy. Then by $(72.2)_2$, $(72.5)_1$, and a well known obvious analogue of Theorem 38.4, $\bar{w}$ also is an isotropic function of the variables ε_{LM}, π_*^L, and μ_*^L. By a well known classical theorem we may conclude that $\bar{\mathscr{F}}$ and $\bar{w}$ depend on those variables (12 scalars) through the principal components $\varepsilon_{(r)}$ of ε_{LM} and the components of π_*^L and μ_*^L in an arbitrary principal triad of strain.

If $\mathscr{C}$ is both electrically and magnetically polarizable, then it is reasonable to assume that relations $(72.4)_{3,4}$, $[(72.5)_{3,4}]$ define E_L^* and H_L^* as functions of T $[\eta]$, ε_{LM}, π_*^L, and μ_*^L. Then by (46.4) $\mathscr{F}$ $[w]$ is a function of the same arguments. In this case the analogue of the deduction of (72.4) [(72.5)] from (72.2) [(72.3)] can be easily set up using (72.1) [the consequence

$$k^*(Dw - TD\eta - E_L^* D\pi_*^L - H_L^* D\mu_*^L) + Y^{LM} D\varepsilon_{LM} \leqslant 0 \tag{72.7}$$

of $(72.1)_{1,2}$] instead of (72.2) [(72.3)]. E.g. by the analogue of the afore-mentioned deduction of (72.5) we derive

$$\begin{cases} T = \dfrac{\partial w}{\partial \eta}, \qquad Y^{LM} = -k^* \dfrac{\partial w}{\partial \varepsilon_{LM}}, \\ E_L^* = \dfrac{\partial w}{\partial \pi_*^L}, \qquad H_L^* = \dfrac{\partial w}{\partial \mu_*^L}, \qquad w = w(\eta, \varepsilon_{LM}, \pi_*^L, \mu_*^L). \end{cases} \tag{72.8}$$

For instance, in case $\mathscr{C}$ is magnetically but not electrically polarizable at P^*, we have $\pi_*^L \equiv 0$ and E_L^* is arbitrary. Then w can be reasonably be thought of as a function of η, ε_{LM}, and μ_L^*; and the constitutive equations $(72.8)_{1,2,4}$ can be asserted.

§ 73. Extension of the Operations $T\ldots \to T^*_{\cdot\cdot}$, $T^{\cdots} \to T_*^{\cdots}$, D^c and D_c to Tensors of Arbitrary Order

It is useful to extend conventions (57.1a) to an arbitrary tensor (or double tensor) $T_{\ldots}^{\ldots}$ (defined on $W_{\mathscr{C}}$) by setting e.g.

$$T_{L_1 \ldots L_n}^* = T_{\rho_1 \ldots \rho_n} \alpha^{\rho_1}{}_{L_1} \cdots \alpha^{\rho_n}{}_{L_n}, \qquad \alpha^{\rho_1}{}_{L_1} \cdots \alpha^{\rho_n}{}_{L_n} T_*^{L_1 \ldots L_n} = \overset{+}{T}{}^{\rho_1 \ldots \rho_n}. \tag{73.1}$$

Thence by $(53.7)_1$ and $(53.15)_1$ we easily obtain

$$T_{L_1 \ldots L_n}^* = C_{L_1 M_1} \cdots C_{L_n M_n} T_*^{M_1 \ldots M_n}. \tag{73.2}$$

Furthermore by (73.1) and (17.9)

$$\overset{\perp}{T}_{\rho_1\ldots\rho_n} U^{\rho_1\ldots\rho_n} = T^*_{L_1\ldots L_n} \underset{*}{U}^{L_1\ldots L_n}. \tag{73.3}$$

We can easily extend to general spatial tensors the definitions (22.1) and $(22.2)_1$ of D^c, D_c, and D_r, as well, as the theorems $(22.2)_2$ to (22.6). Among these extensions let us write explicitly, first, the equalities

$$\frac{D^c T^{\rho_1\ldots\rho_n}}{Ds} = \left(\frac{DT^{\cdots}}{Ds}\right)^{\perp} - \sum_{l=1}^{n} u^{\rho_l}{}_{/\overset{\perp}{\sigma}} T^{\rho_1\ldots\rho_{l-1}\sigma\rho_{l+1}\ldots\rho_n} \tag{73.4}$$

and

$$\frac{D_c T^{\rho_1\ldots\rho_n}}{Ds} = \left(\frac{DT^{\cdots}}{Ds}\right)^{\perp} + \sum_{l=1}^{n} u_\sigma{}^{/\overset{\perp}{\rho_l}} T^{\rho_1\ldots\rho_{l-1}\sigma\rho_{l+1}\ldots\rho_n}, \tag{73.5}$$

where convention (17.9′) is used. (From (73.4) and (73.5) it appears that $D^c T_{\ldots}^{\cdots}$ and $D_c T_{\ldots}^{\cdots}$ are spatial.) Then let us write

$$\frac{D}{Ds}(\overset{\perp}{T}_{\rho_1\ldots\rho_n} U^{\rho_1\ldots\rho_n}) = \left(\frac{D^c}{Ds}\overset{\perp}{T}_{\rho_1\ldots\rho_n}\right) U^{\rho_1\ldots\rho_n} + T_{\rho_1\ldots\rho_n} \frac{D_c}{Ds}\overset{\perp}{U}^{\rho_1\ldots\rho_n}, \tag{73.6}$$

$$\frac{D_c}{Ds} T^{\rho_1\ldots\rho_n} - \frac{D^c}{Ds} T^{\rho_1\ldots\rho_n} = 2\sum_{l=1}^{n} u^{(\rho_l}{}_{/\overset{\perp}{\sigma})} T^{\rho_1\ldots\rho_{l-1}\sigma\rho_{l+1}\ldots\rho_n}, \tag{73.7}$$

and

$$\frac{D_c \overset{\perp}{T}_{\rho_1\ldots\rho_n}}{Ds} \alpha^{\rho_1}{}_{L_1}\ldots\alpha^{\rho_n}{}_{L_n} = \frac{D}{Ds} T^*_{L_1\ldots L_n}, \qquad \frac{D^c \overset{\perp}{T}{}^{\rho_1\ldots\rho_n}}{Ds} = \alpha^{\rho_1}{}_{L_1}\ldots\alpha^{\rho_n}{}_{L_n} \frac{D \underset{*}{T}^{L_1\ldots L_n}}{Ds}. \tag{73.8}$$

We deduce (73.6) and (73.7) from (73.4) and (73.5) directly. To deduce (73.8) let us first remark that by (17.9), (17.9′), (73.4), and (73.5)

$$\frac{D^c}{Ds}(\overset{\perp}{T}_{\rho_1\ldots\rho_n} \overset{\perp}{W}_{\sigma_1\ldots\sigma_p}) = \left(\frac{D^c}{Ds}\overset{\perp}{T}_{\rho_1\ldots\rho_n}\right)\overset{\perp}{W}_{\sigma_1\ldots\sigma_p} + \overset{\perp}{T}_{\rho_1\ldots\rho_n}\frac{D^c}{Ds}\overset{\perp}{W}_{\sigma_1\ldots\sigma_p} \tag{73.9}$$

and that the analogue holds for D_c. Then we easily see that by $(57.1\,\mathrm{a})_{1,2}$ and (58.5), equalities $(73.8)_{1,2}$ hold in case $T^{\rho_1\ldots\rho_n}$ equals a product, $V_{(1)}{}^{\rho_1}\ldots V_{(n)}{}^{\rho_n}$, of arbitrary spatial vectors.

Now let $T\ldots$ be arbitrary and let $U^{\cdots}$ equal the above product. Then by $(73.1)_1$ and the analogue of $(73.8)_2$ for $U^{\cdots}$

$$T^*_{L_1\ldots L_n} \frac{D\underset{*}{U}^{L_1\ldots L_n}}{Ds} = T_{\rho_1\ldots\rho_n}\alpha^{\rho_1}{}_{L_1}\ldots\alpha^{\rho_n}{}_{L_n}\frac{D\underset{*}{U}^{L_1\ldots L_n}}{Ds} = T_{\rho_1\ldots\rho_n}\frac{D^c \overset{\perp}{U}{}^{\rho_1\ldots\rho_n}}{Ds}. \tag{73.10}$$

From (73.3), (73.6), and (73.10) we easily obtain

$$\left(\frac{D}{Ds}T^*_{L_1\ldots L_n}\right)U_*^{L_1\ldots L_n}=\frac{D}{Ds}(\overset{\perp}{T}_{\rho_1\ldots\rho_n}U^{\rho_1\ldots\rho_n})-T^*_{L_1\ldots L_n}\frac{D}{Ds}U_*^{L_1\ldots L_n}$$
$$=\left(\frac{D_c}{Ds}\overset{\perp}{T}_{\rho_1\ldots\rho_n}\right)U^{\rho_1\ldots\rho_n} \tag{73.11}$$
$$=\left(\frac{D_c}{Ds}\overset{\perp}{T}_{\rho_1\ldots\rho_n}\right)\alpha^{\rho_1}{}_{L_1}\ldots\alpha^{\rho_n}{}_{L_n}U_*^{L_1\ldots L_n}.$$

Since $U_*^{L_1\ldots L_n}$ is the product $V_{(1)*}^{L_1}\ldots V_{(n)*}^{L_n}$ of arbitrary material vectors, (73.11) yields $(73.8)_1$. We can deduce $(73.8)_2$ by (73.3), (73.6), $(73.8)_1$, and $(73.1)_1$.

Obviously the operations D^c and D_c commute with the one of raising indices. As to contracting tensors, e.g. by (73.1) and $(56.16)_2$ we easily deduce

$$\overset{-1}{C}{}^{AB}T^*_{ABL_3\ldots L_n}=W^*_{L_3\ldots L_n} \tag{73.12}$$

where $W_{\rho_3\ldots\rho_n}=T_\sigma{}^\sigma{}_{\rho_3\ldots\rho_n}$.

§ 74. On Rigid Motions in the Born Sense

The motion $\mathcal{M}$ of $\mathcal{C}$ in S_4 is said to be *rigid in the Born* sense if $u_{(\rho/\overset{\perp}{\sigma})}\equiv 0$ holds in $W_{\mathcal{C}}$.

By $(56.15)_{1,2}$ the relation $(57.6)_{2,3}$ determines the spatial tensor $u_{(\rho/\overset{\perp}{\sigma})}$ uniquely in terms of $\dot{C}_{LM}$ and $\alpha^\rho{}_L$. Moreover this relation implies the following

Theorem 74.1. *C_{LM} is constant along W_{P*} if and only if the Born (and Rosen) condition $u_{(\rho/\overset{\perp}{\sigma})}=0$ holds on W_{P*}.*

Incidentally by $(53.7)_1$ and (53.15)

$$|d\overset{\perp}{x}|^2=\overset{\perp}{g}_{\rho\sigma}dx^\rho dx^\sigma=C_{LM}dy^L dy^M \qquad (d\overset{\perp}{x}{}^\rho=\alpha^\rho{}_L dy^L). \tag{74.1}$$

Hence we have the following

Theorem 74.2. *The tensor C_{LM} is constant along W_{P*} if and only if, for every material linear element dy^L (at P^*) the norm $|d\overset{\perp}{x}|^2$ is constant.*

By Theorems 74.1, 74.2 the motion $\mathcal{M}$ of $\mathcal{C}$ is rigid in the Born sense if and only if the spatial length of every material line l^* of $\mathcal{C}$ does not change, or more precisely if and only if *the spatial lengths of the intersections of W_{l*} with two arbitrary space-like sections of S_4 coincide.*[6)]

Now we assume that $\mathcal{M}$ is rigid in the Born sense ($u_{(\rho/\overset{\perp}{\sigma})}\equiv 0$). Then by (73.7) the first of the equalities

$$D^c = D_c\,, \quad \acute{k} = 0 = \acute{\mathscr{D}} \quad (u_{(\rho/\overset{\perp}{\sigma})} = 0) \tag{74.2}$$

holds. Furthermore $u^\rho{}_{/\rho} = u^\rho{}_{/\overset{\perp}{\rho}} = 0$—cf. (20.12). Hence $(21.3)_3$ yields $(74.2)_2$, so that by $(58.3)_1$ we have $(74.2)_3$.

In classical physics it is well known that if at an instant the eulerian velocity field v_r is rigid ($v_{(r/s)} \equiv 0$), then it is linear ($v_{r/sh} \equiv 0$). We now prove the following corresponding theorem in general relativity.

Theorem 74.3. *If the motion $\mathscr{M}$ of $\mathscr{C}$ is rigid in the Born sense, then*—cf. (17.9′)

$$[u_{\beta/\lambda\mu} + 2u_{\beta/(\lambda}A_{\mu)} - R_{\beta\lambda\mu\delta}u^\delta]^\perp \equiv 0 \quad (u_{(\rho/\overset{\perp}{\sigma})} \equiv 0)\,. \tag{74.3 a}$$

Proof. By $(17.6)_1$ and by $(17.4)_1$ and convention (17.9′)

$$\overset{\perp}{g}_{\rho\beta/\gamma} = u_{\rho/\gamma}u_\beta + u_\rho u_{\beta/\gamma}\,, \qquad (u_{\alpha/\rho}\overset{\perp}{g}{}^\rho{}_{\beta/\gamma})^\perp = A_\alpha u_{\beta/\overset{\perp}{\gamma}}\,. \tag{74.4}$$

By $(74.3\,\mathrm{a})_2$ and $(74.4)_2$

$$0 = [(u_{\alpha/\overset{\perp}{\beta}} + u_{\beta/\overset{\perp}{\alpha}})_{/\gamma}]^\perp = (u_{\alpha/\beta\gamma} + u_{\beta/\alpha\gamma})^\perp + A_\alpha u_{\beta/\overset{\perp}{\gamma}} + A_\beta u_{\alpha/\overset{\perp}{\gamma}}\,. \tag{74.5}$$

Furthermore by the cyclic property of Riemann's tensor

$$\varepsilon^{\rho\alpha\beta\gamma}u_{\alpha/\beta\gamma} = \varepsilon^{\rho\alpha\beta\gamma}u_{\alpha/[\beta\gamma]} = \varepsilon^{\rho\alpha\beta\gamma}R_{\delta\alpha\beta\gamma}u^\delta = 0\,. \tag{74.6}$$

Now let us use locally natural and proper co-ordinates. Furthermore let Σ' denote summation over the even permutations (a,b,c) of $(1,2,3)$. As a consequence of (74.6)

$$0 = \tfrac{1}{2}\varepsilon^{0\alpha\beta\gamma}u_{\alpha/\beta\gamma} = \Sigma' u_{a/bc} = u_{a/bc} + (u_{b/c} + u_{c/b})_{/a} + 2u_{c/[ab]}\,. \tag{74.7}$$

Hence by (74.5) and (16.9), and by $(74.3\,\mathrm{a})_2$ ($u_{\iota/a} = -u_{a/\iota}$), respectively, we have

$$0 = u_{a/bc} - 2A_{(b}u_{c)/a} + u^\delta R_{\delta cab} = u_{a/bc} + 2u_{a/(c}A_{b)} - R_{abc0} \tag{74.8}$$

which is equivalent to the tensor formula $(74.3\,\mathrm{a})_1$. q.e.d.

Incidentally, in classical physics homographic motions can be characterized by the condition $v_{r/sh} \equiv 0$. A straightforward relativization of its is—cf. (17.9′)

$$(u_{\beta/(\lambda\mu)})^\perp \equiv 0 \tag{74.3 b}$$

which differs from $(74.3\mathrm{a})_1$ by very little terms. However among conditions $(74.3\mathrm{a})_1$ and (74.3b) only the former complies with the important requirement of being implied by Born rigidity. Hence only the former can characterize (a sort of) *relativistic homographic motions.*

In the theory of second order kinematics formula (91.11a) on C_{BLM} is proved—cf. $(53.16)_2$ and footnote 2 in Chapter 8.

By $(57.7)_2$ and convention (91.1) below, that formula is equivalent to

$$\grave{C}_{BLM}=\grave{C}_{BR}\overset{-1}{C}{}^{RS}C_{SLM}+(u_{\beta/\lambda\mu}+2u_{\beta/(\lambda}A_{\mu)}-R_{\beta\lambda\mu\delta}u^{\delta})\alpha^{\beta}{}_{B}\alpha^{\lambda}{}_{L}\alpha^{\mu}{}_{M}\,. \tag{74.9}$$

Hence by Theorems 74.1—74.3, *Born rigidity implies the constancy of* C_{BLM} *along world-lines:*

$$\grave{C}_{BLM}=0 \quad \text{for} \quad u_{(\alpha/\overset{\perp}{\beta})}\equiv 0 \quad (\grave{f}\equiv Df/Ds)\,. \tag{74.10}$$

§ 75. Born Rigidity and Stationary Tensors

We shall say that the tensor $T_{\dots}^{\dots}$ defined on $W_{\mathscr{C}}$ (or on W_{P*}) is *stationary with respect to* $\mathscr{C}$ *if* $D^c T_{\dots}^{\dots}/Ds\equiv 0$.

Theorem 75.1. *Let the motion* $\mathscr{M}$ *of* $\mathscr{C}$ *be rigid in the Born sense. Then the following theses hold:*

(a) *The following four conditions are mutually equivalent:*

$$\frac{D^c T_{\dots}^{\dots}}{Ds}=0=\frac{D_c T_{\dots}^{\dots}}{Ds}\,,\qquad \frac{DT^{*}_{\dots}}{Ds}=0=\frac{DT_{*}^{\dots}}{Ds}\,. \tag{75.1}$$

(b) *If* $T_{\rho_1\dots\rho_n}$ *and* $U_{\sigma_1\dots\sigma_p}$ *are spatial tensor fields (in* $W_{\mathscr{C}}$*) that are stationary (with respect to* $\mathscr{C}$*), then such are* $T_{\rho_1\dots\rho_n}U_{\sigma_1\dots\sigma_p}$ *and (for* $n\geqslant 2$*)* $T_{\sigma}{}^{\sigma}{}_{\rho_3\dots\rho_n}$.

(c) *If* $T_{\dots}^{\dots}$ *is a stationary spatial field, then such is* $(T_{\dots}^{\dots}{}_{/\sigma})^{\perp}$.

(d) *The spatial Ricci tensor* $\overset{\perp}{\varepsilon}_{\alpha\beta\gamma}$ *is stationary.*

Proof. We proved that $(74.2)_4$ implies $(74.2)_1$; by (73.8) this implies thesis (a).

The part of thesis (b) concerning $T_{\rho_1\dots\rho_n}U_{\sigma_1\dots\sigma_p}$ follows from (73.9); furthermore by (73.12), by Theorem 74.1, and by the equivalence of conditions $(75.1)_{1,3}$ we easily deduce the part of thesis (b) concerning $T_{\sigma}{}^{\sigma}{}_{\rho_3\dots\rho_n}$.

To prove thesis (c) we assume $D^c T_{\dots}^{\dots}/Ds\equiv 0$. Then we obtain $DT^{*}_{\dots}/Ds\equiv 0$ by thesis (a). Furthermore we consider $T^{*}_{\dots}$ $[T_{\dots}]$ as a function of y^L and t [of x^{α}] and we identify t with s. Then by $(53.13)_2$

$$\frac{D}{Ds}T^{*}_{\dots|M}\equiv\frac{D}{Ds}T^{*}_{\dots|M}\equiv 0 \quad \left(\text{besides } \frac{DT^{*}_{\dots}}{Ds}\equiv 0\right). \tag{75.2}$$

By $(73.1)_1$ and $(53.13)_1$

$$T_{\rho_1\dots\rho_n|M}\,\alpha^{\rho_1}{}_{L_1}\dots\alpha^{\rho_n}{}_{L_n}=V^{*}_{L_1\dots L_nM} \quad \text{for} \quad V_{\rho_1\dots\rho_n\sigma}=T_{\rho_1\dots\rho_n|\sigma}\,. \tag{75.3}$$

By $(53.16)_{2,3}$ and $(56.16)_2$

$$\overset{\perp}{g}{}^{\rho}{}_{\sigma}\alpha^{\sigma}{}_{L_i|M}=\alpha^{\rho}{}_{H}\overset{-1}{C}{}^{HA}C_{AL_iM}\,. \tag{75.4}$$

Then by $(73.1)_1$ and $(75.3)_2$

$$T^*_{L_1\dots L_n|M}=V^*_{L_1\dots L_nM}+\sum_{i=1}^{n}T^*_{L_1\dots L_{i-1}HL_{i+1}\dots L_n}\overset{-1}{C}{}^{HA}C_{AL_iM}\qquad (T_{\dots}=\overset{\perp}{T}_{\dots})\,. \tag{75.5}$$

By (75.2), Theorem 74.1, and (74.10) $T^*_{L_1\dots L_n|M}$, $\overset{-1}{C}{}^{HA}$, and C_{AL_iM} are constant along world-lines. Hence by (75.5) the same holds for $V^*_{L_1\dots L_nM}$. Then $D^cV_{\rho_1\dots\rho_n0}/Ds\equiv 0$ by the equivalence of $(75.1)_{1,3}$. We conclude that thesis (c) holds.

To prove thesis (d) we remark that by $(56.9)_4$

$$T^*_{ABC}=\mathscr{D}\varepsilon^*_{ABC}\quad\text{for}\quad T_{\alpha\beta\gamma}=\overset{\perp}{\varepsilon}_{\alpha\beta\gamma}\,. \tag{75.6}$$

Furthermore the Born rigidity of $\mathscr{M}$ yields $(74.2)_3$. Then $DT^*_{ABC}/Ds\equiv 0$, so that $D^c\overset{\perp}{\varepsilon}_{\alpha\beta\gamma}/Ds\equiv 0$ by $(75.6)_2$ and thesis (a). q.e.d.

In spite of its not being used in the remainder of the book, let us consider the following theorem.

Theorem 75.2. (a) *Born rigidity is equivalent to the independence of* $\overset{\perp}{g}_{\alpha\beta}$ *from* x^0 *in a co-moving frame* (x).

(b) *We can choose* (x), $\hat{t}(x)$, *and the reference state* $(\mathscr{P}^*, S_3^*)$ [§ 52] *in such a way that*

$$x^r\equiv y^r\,,\qquad \hat{t}(x)\equiv x^0\,,\qquad a^*_{LM}(x^r)\equiv\overset{\perp}{g}_{LM}(x^r,0)\qquad (\text{in } W_{\mathscr{C}})\,. \tag{75.7}$$

By such a choice, for $u_{(\alpha/\beta)}\equiv 0$ *(in* $W_{\mathscr{C}}$*) we have*

$$C_{rs}=\overset{\perp}{g}_{rs}(x)=a^*_{rs}(x^r)\,,\qquad \overset{\perp}{g}_{0\sigma}=0\,,\qquad \overset{\perp}{g}{}^{rs}=a^{*rs}\qquad (\text{in } W_{\mathscr{C}}) \tag{75.8}$$

and

$$T_{L_1\dots}{}^{M_1\dots}=T^*_{L_1\dots}{}^{M_1\dots}=T_{*L_1\dots}{}^{M_1\dots}\qquad (\alpha^r{}_L=\delta^r{}_L)\qquad \textit{for}\qquad T^{\dots}_{\dots}=\overset{\perp}{T}{}^{\dots}_{\dots}\,. \tag{75.9}$$

(c) *Assuming Born rigidity, the spatial field* $T^{\dots}_{\dots}=T^{\dots}_{\dots}(x)$ *(defined in* $W_{\mathscr{C}}$*) is stationary (with respect to* $\mathscr{C}$*) if and only if it is independent of* x^0 *in co-moving co-ordinates.*

To prove Theorem 75.2 we choose (x), $\hat{t}(x)$, and $(\mathscr{P}^*, \Sigma_3^*)$ in such a way that (75.7) holds. By $(75.7)_1$ we have $(62.5)_1$ and $(62.2)_1$, i.e. $(75.8)_{1,3}$ respectively.

By Theorem 74.1 $u_{(\alpha/\overset{\perp}{\beta})}\equiv 0$ (in $W_{\mathscr{C}}$) if and only if $\grave{C}_{LM}=0$. By $(75.7)_{1,2}$ and $(75.8)_1$ the last condition holds if and only if $\overset{\perp}{g}_{\alpha\beta,0}=0$. We conclude that part (a) holds.

Now let $u_{(\alpha/\overset{+}{\beta})} \equiv 0$ (hold in $W_{\mathscr{C}}$). Then $\overset{+}{g}_{\alpha\beta,0} \equiv 0$, so that $(75.7)_3$ yields $(75.8)_2$. Since by $(75.8)_3$

$$\overset{\perp}{g}_{rh} \overset{\perp}{g}^{sh} = \delta_r{}^s = a^*_{rh} a^{*sh},$$

$(75.8)_2$ yields $(75.8)_4$.

By $(75.7)_1$ we have $(62.4)_2$, i.e. $(75.9)_3$. Furthermore $(75.9)_{3,4}$, $(75.8)_{2,4}$, and (73.1) yield $(75.9)_{1,2}$. Thus we have proved part (b).

To prove part (c) we also assume $(75.9)_4$. Then by Theorem 75.1(a) $D^c T_{\cdot\cdot\cdot}^{\cdot\cdot\cdot}/Ds \equiv 0$ if and only if $DT^{*\cdot\cdot\cdot}_{\cdot\cdot\cdot}/Ds \equiv 0$, which by $(75.9)_1$ and $(75.7)_{1,2}$ is equivalent to $T_{\cdot\cdot\cdot}^{\cdot\cdot\cdot}{}_{,0} \equiv 0$. Thus we have also proved part (c). q.e.d.

§ 76. Some Invariance Properties of Ideal Conductors

We consider an ideal conductor, $\mathscr{C}$, so that (48.2) and (48.3) hold. First we put the invariance condition $(48.3)_2$ into the Lagrangian forms (76.2) and (76.3) below in the case where the motion $\mathscr{M}$ of $\mathscr{C}$ is arbitrary but $\mathscr{C}$ is piezo-elastic, so that (72.5) hold—cf. footnote 1 to Chapter 8. After this we consider some invariance properties holding for an arbitrary ideal conductor in case $u_{(\alpha/\overset{+}{\beta})} \equiv 0$—cf. footnote 4 in Chapter 8.

Remembering conventions $(57.1\text{a})_{1,2}$ on spatial vectors, from $(34.1)_{2,4}$ and $(57.2)_1$ we deduce

$$b^L_* = h^L_* + \mu^L_* = \overset{-1}{C}{}^{LA} h^*_A + \mu^L_* \qquad \text{where} \qquad B^\rho = k b^\rho, \; H^\rho = k h^\rho. \tag{76.1}$$

By $(76.1)_3$ and $(48.3)_2$ $D^c b^\rho/Ds = 0$, so that by $(58.5)_1$ $\alpha^\rho{}_L D b^L_*/Ds = 0$. Hence $b^L_* = \text{const}$ along world-lines, which by $(76.1)_{1,2}$ and $(72.5)_4$ yields

$$\overset{-1}{C}{}^{LA} h^*_A - \frac{\partial \bar{w}}{\partial H^*_L} \equiv b^L_* = \text{const} \qquad \text{along world-lines.} \tag{76.2}$$

Since $k^* = k\mathscr{D}$ by $(58.3)_1$ and $k h^*_L = H^*_L$ by $(76.1)_4$ and (57.1a), we may write (76.2) in the form

$$\mathscr{D}\overset{-1}{C}{}^{LA} H^*_A - k^* \frac{\partial \bar{w}}{\partial H^*_L} = k^* b^L_* = \text{const} \qquad \text{along world-lines.} \tag{76.3}$$

Theorem 76.1. *Let the motion $\mathscr{M}$ of the ideal conductor $\mathscr{C}$ be locally rigid in the Born sense. Then*

$$\frac{D B_\alpha B^\alpha}{Ds} = 0 = \frac{D I^\alpha I_\alpha}{Ds}, \qquad \frac{D B^\alpha I_\alpha}{Ds} = 0 \qquad (u_{(\alpha/\overset{+}{\beta})} \equiv 0), \tag{76.4}$$

where

$$I_\alpha = j''_\alpha - c\overset{\perp}{\varepsilon}_\alpha{}^{\beta\gamma} A_\beta B_\gamma = j''_\alpha - \frac{1}{c}\varepsilon_\alpha{}^{\beta\gamma} a_\alpha B_\beta \quad (\simeq j''_\alpha). \tag{76.5}$$

Proof. Let $(76.4)_4$ hold. Then $u^\alpha{}_{/\alpha} \equiv 0$ by $(17.20)_2$, so that by $(48.4)_1$ and $(22.1)_1$ B_α is stationary ($D^c B_\alpha / Ds = 0$). Furthermore $(76.5)_1$ and $(48.3)_4$ imply

$$I_\alpha = c\overset{\perp}{\varepsilon}_\alpha{}^{\beta\gamma} B_{\beta/\gamma}. \tag{76.6}$$

Then by Theorem 75.1 (c), (d), (b) I_α is also stationary.

Now $(76.4)_{1,2,3}$ follow by Theorem 75.1 (b) and (73.4) for $n=0$. q.e.d.

§ 77. Dynamic Equations for Piezo-Elastic Ideal Conductors

In this section—cf. footnote 1 to Chapter 8—we consider an ideal conductor with zero thermal conductivity, so that (48.2), (48.3), (48.5), and (76.1—3) hold. By (48.2) and $(40.18)_2$

$$^7E_{\alpha\beta} = W(u_\alpha u_\beta + \overset{\perp}{g}_{\alpha\beta}) - H_\alpha H_\beta \qquad (2W = H_\alpha H^\alpha,\ E_\alpha \equiv 0 \equiv q^\alpha). \tag{77.1}$$

Having in mind to deal with acceleration waves, we also assume that $\mathscr{C}$ is piezo-elastic so that (72.5) holds, and that $\mathscr{C}$ is undergoing an adiabatic process. Then we may assume $\eta = \text{const}$ with respect to t and y^L.

By (77.1) and (43.5) the total energy tensor $^7\mathscr{U}_{\alpha\beta}$ becomes

$$^7\mathscr{U}_{\alpha\beta} = k(w + w^{(H)} + c^2) u_\alpha u_\beta + {}^7\overset{\perp}{E}_{\alpha\beta} + {}^7X_{\alpha\beta} \quad \text{with} \quad w^{(H)} = \frac{W}{k} = \frac{H^\alpha H_\alpha}{2k}. \tag{77.2}$$

Since—cf. $(17.9)_3$—the tensor $^7\overset{\perp}{E}{}^{\rho\sigma}$ is spatial, the analogues of definition $(59.1)_{2,3}$ and relation $(59.2)_{1,2}$ involving $X^{\rho\sigma}$ can be asserted for it:

$$W\overset{\perp}{g}{}^{\rho\sigma} - H^\rho H^\sigma = \overset{\perp}{E}{}^{\rho\sigma} = \frac{1}{\mathscr{D}} \alpha^\rho{}_L \alpha^\sigma{}_M Z^{LM} \quad \text{where} \quad Z^{LM} = \frac{1}{\mathscr{D}} \gamma_\rho{}^L \gamma_\sigma{}^M \overset{\perp}{E}{}^{\rho\sigma}; \tag{77.3}$$

so that Z^{LM} is the analogue of Y^{LM} for $\overset{\perp}{E}{}^{\rho\sigma}$.

From $(77.1)_2$ and $(57.2)_4$ we deduce the first two of the equalities

$$2W = H_\rho H^\rho = \overset{-1}{C}{}^{PQ} H^*_P H^*_Q, \quad \frac{1}{\mathscr{D}} \gamma_\rho{}^L \gamma_\sigma{}^M W \overset{\perp}{g}{}^{\rho\sigma} = \frac{\mathscr{D}}{2} \overset{-1}{C}{}^{PQ} H^*_P H^*_Q \overset{-1}{C}{}^{LM}. \tag{77.4}$$

Thence $(77.4)_3$ follows by $(56.16)_1$. Furthermore $(57.1\text{b})_1$ and $(56.16)_1$ yield

$$\mathscr{D} H_\rho = \gamma_\rho{}^P H^*_P, \quad H^\rho \gamma_\rho{}^L = \mathscr{D} \overset{-1}{C}{}^{LP} H^*_P. \tag{77.5}$$

From $(77.3)_{3,1}$, $(77.4)_3$, and $(77.5)_2$ we deduce

$$Z^{LM}=\mathscr{D}\left(\tfrac{1}{2}\overset{-1}{C}{}^{PQ}\overset{-1}{C}{}^{LM}-\overset{-1}{C}{}^{LP}\overset{-1}{C}{}^{MQ}\right)H_P^*H_Q^*=Z^{ML}. \tag{77.6}$$

We can put (77.6) into the form

$$\mathscr{D}^{-1}Z^{LM}=\left(\begin{vmatrix}\overset{-1}{C}{}^{PQ} & \overset{-1}{C}{}^{PL}\\ \overset{-1}{C}{}^{MQ} & \overset{-1}{C}{}^{ML}\end{vmatrix}-\tfrac{1}{2}\overset{-1}{C}{}^{LM}\overset{-1}{C}{}^{PQ}\right)H_P^*H_Q^*; \tag{77.7}$$

thence by well known properties of determinants and by $(56.9)_1$ we have

$$\begin{aligned}\mathscr{D}^{-1}Z^{LM}&=\left(\varepsilon^{*PLR}\varepsilon^{*QMS}C_{RS}-\tfrac{1}{2}\overset{-1}{C}{}^{LM}\overset{-1}{C}{}^{PQ}\right)H_P^*H_Q^*\\&=\left(\frac{\partial^2\mathscr{D}^2}{\partial C_{PQ}\partial C_{LM}}-\tfrac{1}{2}\overset{-1}{C}{}^{LM}\overset{-1}{C}{}^{PQ}\right)H_P^*H_Q^*.\end{aligned} \tag{77.8}$$

Equation (66.2) was deduced for $u_L=0$ in any locally natural (x) [geodesic (y)] from the dynamic equations $(23.4)_2$ for $\mathscr{U}_{\alpha\beta}={}^7\mathscr{U}_{\alpha\beta}$—cf. (77.2)—in case $w^{(H)}\equiv 0$, $q_\alpha\equiv 0$, and $\overset{\perp}{E}_{\alpha\beta}\equiv 0$. In the case $q^\alpha\equiv 0$, $w^{(H)}\not\equiv 0$, and $\overset{\perp}{E}_{\alpha\beta}\not\equiv 0$ we obviously obtain (in the same frames and for $u_L^\dagger=0$), by the analogous deduction

$$k^*q^r{}_s\frac{\partial^2x^s}{\partial t^2}=-(Y^{LM}+Z^{LM})x^r{}_{,LM}-(Y^{LM}+Z^{LM})_{,M}x^r{}_L\quad(X_{\alpha\beta}\equiv{}^7X_{\alpha\beta},\ldots) \tag{77.9}$$

where—cf. $(66.5)_1$, (77.3)—the first of the equalities

$$q^{rs}=\left(1+\frac{w+w^{(H)}}{c^2}\right)\delta^{rs}+\frac{X^{rs}+\overset{\perp}{E}{}^{rs}}{kc^2}=\left(1+\frac{w+k^{-1}H_lH^l}{c^2}\right)\delta^{rs}+\frac{X^{rs}-H^rH^s}{kc^2}=q^{sr} \tag{77.10}$$

holds. By $(77.2)_{2,3}$ and $(77.3)_1$ we have $(77.10)_{2,3}$. Since the relations $k^*=k\mathscr{D}$ —cf. $(58.3)_1$—, $\alpha^\rho{}_L=x^\rho{}_L$, $(59.2)_{1,2}$, $(72.5)_{2,5}$, and $(77.3)_1$ hold, $(77.10)_1$ becomes

$$q^{rs}=\left(1+\frac{w+w^{(H)}}{c^2}\right)\delta^{rs}+\frac{x^r{}_Lx^s{}_M}{c^2}\left(\frac{Z^{LM}}{k^*}-\frac{\partial\bar{w}}{\partial\varepsilon_{LM}}\right)\quad[\bar{w}=\bar{w}(\eta,y^L,\varepsilon_{LM},H_L^*)]. \tag{77.11}$$

In the present case Y^{LM} depends on η, y^L, ε_{LM} (and η is constant as well as in the elastic case [§ 66]), and also on H_L^*—cf. $(72.5)_{2,5}$. Then the present expression of $-Y^{LM}{}_{,M}$ can be obtained from the implicit expression (66.3) of its, by replacing w with $\bar{w}$ and by adding the terms in $H^*_{I,M}$ due to the fact that H_L^* is an argument of the expression (77.11) of $\bar{w}$:

$$-Y^{LM}{}_{,M}=k^*\frac{\partial^2\bar{w}}{\partial\varepsilon_{LM}\partial\varepsilon_{AB}}\delta_{rs}x^r{}_Ax^s{}_{,BM}+\frac{\partial}{\partial y^M}\left(k^*\frac{\partial\bar{w}}{\partial\varepsilon_{LM}}\right)+k^*\frac{\partial^2\bar{w}}{\partial\varepsilon_{LM}\partial H_I^*}H^*_{I,M} \tag{77.12}$$

where—as well as in the sequel—"$_{,A}$" or "$_{|A}$" ["$\partial/\partial y^A$"] means total [partial] derivative with respect to y^A.

By (77.6) or (77.8) we can consider Z^{LM} as a function of ε_{LM}, through $\mathscr{D}$ and $\overset{-1}{C}{}^{AB}$, and of H^*_A. Then—see $(61.5)_3$

$$Z^{LM}{}_{,M} = \frac{\partial Z^{LM}}{\partial \varepsilon_{AB}} \delta_{rs} x^r{}_A x^s{}_{,BM} + \frac{\partial Z^{LM}}{\partial H^*_I} H^*_{I,M} \tag{77.13}$$

where we have

$$\frac{\partial Z^{LM}}{\partial H^*_I} = \mathscr{D}\left(\overset{-1}{C}{}^{LM} \overset{-1}{C}{}^{PI} - 2\,\overset{-1}{C}{}^{(LP} \overset{-1}{C}{}^{M)I}\right) H^*_P\,, \tag{77.14}$$

$$\frac{\partial \mathscr{D}}{\partial \varepsilon_{AB}} = \mathscr{D}\,\overset{-1}{C}{}^{AB}\,, \tag{77.15}$$

and

$$\frac{\partial Z^{LM}}{\partial \varepsilon_{AB}} = \mathscr{D}\left[2\varepsilon^{*PM(A}\varepsilon^{*QLB)} + \overset{-1}{C}{}^{L(A}\overset{-1}{C}{}^{B)M}\overset{-1}{C}{}^{PQ} + \overset{-1}{C}{}^{LM}\overset{-1}{C}{}^{P(A}\overset{-1}{C}{}^{B)Q}\right] H^*_P H^*_Q + Z^{LM}\overset{-1}{C}{}^{AB}\,. \tag{77.16}$$

Indeed relation (77.14) is an immediate consequence of (77.6).

From $(56.9)_1$ and $(56.5)_2$, and from $(57.7)_2$ [Appendix A 3], respectively, we deduce

(a) $$6\mathscr{D}^2 = \varepsilon^{*L_1L_2L_3}\varepsilon^{*M_1M_2M_3} C_{L_1M_1} C_{L_2M_2} C_{L_3M_3}\,, \qquad \frac{\partial C_{HK}}{\partial \varepsilon_{AB}} = 2\,\delta_H{}^{(A}\delta_K{}^{B)}\,;$$

thence we easily obtain

(b) $$\frac{\partial \mathscr{D}^2}{\partial \varepsilon_{AB}} = \varepsilon^{*AL_1L_2}\varepsilon^{*BM_1M_2} C_{L_1M_1} C_{L_2M_2} = 2\mathscr{D}^2\,\overset{-1}{C}{}^{AB}$$

which yields (77.15). In order to prove (77.16) we remark that

(c) $$C_{HK}\overset{-1}{C}{}^{KM} = \delta_H{}^M\,, \quad \text{hence} \quad \frac{\partial C_{HK}}{\partial \varepsilon_{AB}}\overset{-1}{C}{}^{KM} = -C_{HK}\frac{\partial \overset{-1}{C}{}^{KM}}{\partial \varepsilon_{AB}}\,.$$

From $(c)_2$ and $(a)_2$ [Appendix A 3] we have

(d) $$\frac{\partial \overset{-1}{C}{}^{LM}}{\partial \varepsilon_{LM}} = -\overset{-1}{C}{}^{LH}\frac{\partial C_{HK}}{\partial \varepsilon_{AB}}\overset{-1}{C}{}^{KM} = -2\,\overset{-1}{C}{}^{L(A}\overset{-1}{C}{}^{B)M}\,.$$

From $(77.8)_1$, (77.15), and (d) we easily obtain (77.16).

Equalities (77.8), (77.13), (77.14), and (77.16) allow us to develop the divergence $Z^{LM}{}_{,M}$, present in the dynamic equations (77.9) for piezo-elastic ideal conductors. This and (77.8) allow us to develop all terms in the same equations which are connected with the magnetic field—cf. (77.1) and (77.3).

§ 78. Magneto-Elastic Acceleration Waves in Piezo-Elastic Ideal Conductors

We consider a magneto-elastic acceleration wave, σ_3, in the piezo-elastic ideal conductor $\mathscr{C}$ with zero thermal conductivity. We obviously assume that $\mathscr{C}$ is regular and more precisely that the functions $k^*=k^*(y)$ and $\bar{w}=\bar{w}(y^L,\eta,\varepsilon_{LM},H_L^*)$ are twice continuously differentiable.

We mean that σ_3 is a time-like surface such that (i) $\partial^2 x^\rho/\partial t^2$ (with $ct\equiv s$) has a discontinuity of the first kind through σ_3, (ii) the functions $x^\rho=x^\rho(y^0,\dots,y^3)$ —which represent $\mathscr{M}$—, $\partial x^\rho/\partial y^\Lambda$, the fields $g_{\alpha\beta}$, $g_{\alpha\beta,\gamma}$, and H_ρ are continuous in a neighbourhood, $\mathscr{N}(\sigma_3)$, of σ_3, and (iii) the same functions and fields are continuously differentiable in $\mathscr{N}(\sigma_3)-\sigma_3$.

As a consequence $\mathscr{D}$, k, C_{LA}, and ε_{LA} are continuous in $\mathscr{N}(\sigma_3)$ and continuously differentiable in $\mathscr{N}(\sigma_3)-\sigma_3$ as well as k^*; furthermore the same holds for H_L^*, μ_L^*—cf. $(72.5)_4$—, B_L^*, b_L^*, and B_ρ. In particular $(76.2)_2$ says that b_*^L *is a function of* y^L *(and not of* t*)* in spite of the presence of the discontinuity surface σ_3.

Since acceleration waves occur only in adiabatic processes, we may assume that η is constant in $\mathscr{N}(\sigma_3)$.

Consider the event point $\mathscr{E}$ on σ_3 and let (x), (y), and $\hat{t}(x)$ fulfill the local conditions (54.1) at $\mathscr{E}$, so that all of equations (77.9) to (77.16) hold at $\mathscr{E}$.

The Lagrangian spatial differentiation of (76.3) yields

$$\left(\mathscr{D}\overset{-1}{C}{}^{AB}-k\frac{\partial^2\bar{w}}{\partial H_A^*\partial H_B^*}\right)H^*_{B|M}=\left(-H_B^*\frac{\partial\big(\mathscr{D}\overset{-1}{C}{}^{AB}\big)}{\partial\varepsilon_{HK}}+k^*\frac{\partial^2\bar{w}}{\partial H_A^*\partial\varepsilon_{HK}}\right)\varepsilon_{HK|M}+\cdots \tag{78.1}$$

where the dots stand for terms which are continuous across σ_3.

Furthermore by (77.15), $(57.7)_2$, and $(56.9)_1$, respectively, we have

$$\frac{\partial\mathscr{D}\overset{-1}{C}{}^{AB}}{\partial\varepsilon_{HK}}=\frac{\partial^2\mathscr{D}}{\partial\varepsilon_{HK}\partial\varepsilon_{AB}}=\frac{2}{\mathscr{D}}\frac{\partial^2\mathscr{D}^2}{\partial C_{HK}\partial C_{AB}}=\frac{2}{\mathscr{D}}\varepsilon^{*HAR}\varepsilon^{*KBS}C_{RS}\,. \tag{78.2}$$

By writing each of equations (78.1) and (77.9)—and also (77.12) to (77.14) and (77.16)—at both sides of σ_3 (at $\mathscr{E}$) and by subtracting the two results from one another, we obtain conditions on the discontinuities of $\partial^2 x^\rho/\partial y^\Lambda\partial y^\Sigma$ and $H^*_{L,M}$.

We denote by $\mathscr{V}_{IA}$ the inverse of the tensor coefficient of $H^*_{B,M}$ in (78.1):

$$\mathscr{V}_{IA}\left(\mathscr{D}\overset{-1}{C}{}^{AB}-k^*\frac{\partial^2\bar{w}}{\partial H_A^*\partial H_B^*}\right)=\delta_I{}^B\,. \tag{78.3a}$$

If $\mathscr{C}$ is not magnetizable, we have

$$\mathscr{V}_{IA}=\mathscr{D}^{-1}C_{IA} \qquad \text{for} \qquad \bar{w}=\bar{w}(\eta,\varepsilon_{LM},y^{L}). \tag{78.3b}$$

Hence equation (78.3a) in $\mathscr{V}_{IA}$ has a solution in case μ_{*}^{L}—cf. $(72.5)_4$—does not vary too fast. *We now consider the case when the tensor $\mathscr{V}_{IA}$ fulfilling* (78.3a) *exists.*

We denote discontinuity across σ_3 by Δ, as well as in § 65. By (78.3a) and (78.2) the condition on $\Delta H^*_{B,M}$ and $\Delta\varepsilon_{HK|M}$ obtained from (78.1) in the way explained above can be put into the form

$$\Delta H^*_{I|M}=\mathscr{V}_{IA}\left(k^*\frac{\partial^2\bar{w}}{\partial H^*_A\,\partial\varepsilon_{HK}}-\frac{2}{\mathscr{D}}\varepsilon^{*AHR}\varepsilon^{*BKS}C_{RS}H^*_B\right)\Delta\varepsilon_{HK|M}\,. \tag{78.4}$$

Our present goal is to write the dynamic equations in the discontinuity vector λ^{ρ} for $\partial^2 x^{\rho}/\partial y^{\Lambda}\partial y^{\Sigma}$. To attain it we use the local conditions (54.1) on (x), (y), and $\hat{t}(x)$, and we consider equation (78.4) and the discontinuity equations obtained from (77.9), (77.12), and (77.13):

$$\begin{cases} k^* q^r{}_s\,\Delta\dfrac{\partial^2 x^s}{\partial t^2}=-(Y^{LM}+Z^{LM})\Delta x^r{}_{,LM}-x^r{}_B(\Delta Y^{BM}{}_{,M}+\Delta Z^{BM}{}_{,M}),\\[2ex] -\Delta Y^{BM}{}_{,M}=k^*\dfrac{\partial^2\bar{w}}{\partial\varepsilon_{BM}\,\partial\varepsilon_{AL}}\delta_{hs}x^h{}_A\,\Delta x^s{}_{,LM}+k^*\dfrac{\partial^2\bar{w}}{\partial\varepsilon_{BM}\,\partial H^*_I}\Delta H^*_{I,M}\,,\\[2ex] \Delta Z^{BM}{}_{,M}=\dfrac{\partial Z^{BM}}{\partial\varepsilon_{AL}}\delta_{hs}x^h{}_A\,\Delta x^s{}_{,LM}+\dfrac{\partial Z^{BM}}{\partial H^*_I}\Delta H^*_{I,M}\,,\end{cases} \tag{78.5}$$

where (77.14) and (77.16) hold. By $(72.8)_2$ and $(78.5)_{2,3}$ we can turn $(78.5)_1$ into

$$q_{rs}\,\Delta\frac{\partial^2 x^s}{\partial t^2}=p'_{rs}{}^{LM}\,\Delta x^s{}_{,LM}+p''_r{}^{IM}\,\Delta H^*_{I|M} \tag{78.6}$$

where $(\Delta x^s{}_{,[LM]}=0$ and)

$$\begin{aligned} p'^{rsLM}&=\left(\frac{\partial\bar{w}}{\partial\varepsilon_{LM}}-\frac{Z^{LM}}{k^*}\right)\delta^{rs}+x^r{}_B\,x^s{}_A\left(\frac{\partial^2\bar{w}}{\partial\varepsilon_{A(L}\partial\varepsilon_{BM)}}-\frac{1}{k^*}\frac{\partial Z^{BM)}}{\partial\varepsilon_{A(L}}\right),\\ p''^{rIM}&=x^r{}_K\left(\frac{\partial^2\bar{w}}{\partial\varepsilon_{KM}\,\partial H^*_I}-\frac{1}{k^*}\frac{\partial Z^{KM}}{\partial H^*_I}\right). \end{aligned} \tag{78.7}$$

By (78.4), the relation $\Delta\varepsilon_{HL|M}=\delta_{rs}x^r{}_{(H}\Delta x^s{}_{,L)M}$—see (61.5), $(57.7)_2$—, and (78.7) we can turn (78.6) into

$$q_{rs}\,\Delta\frac{\partial^2 x^s}{\partial t^2}=p_{rs}{}^{LM}\,\Delta x^s{}_{,LM} \tag{78.8}$$

where

$$p_r{}^{sLM}=p'_r{}^{sLM}+p''_r{}^{IM}\mathscr{V}_{IA}\left(k^*\frac{\partial^2\bar{w}}{\partial H^*_A\,\partial\varepsilon_{HL}}-\frac{2}{\mathscr{D}}\varepsilon^{*A(HR}\varepsilon^{*BL)S}C_{RS}H^*_B\right)x^s{}_H\,. \tag{78.9}$$

We can develop (78.8) completely by taking (78.9), (78.7), (77.14), (77.16), and (78.3a) into account.

Now we consider the unit vector N_ρ of the spatial normal to σ_3 at $\mathscr{E}$, its Lagrangian analogue $\mathscr{N}^*_L$, and the propagation speeds V and V^*—cf. (65.5), (65.6). Then there exists a spatial vector λ^ρ which fulfills the Hugoniot-Hadamard conditions (65.10). As a consequence (78.8) becomes

$$(p_{rs} - V^{*2} q_{rs})\lambda^s = 0 \quad \text{where} \quad p_{rs} = p_{rs}{}^{LM} \mathscr{N}^*_L \mathscr{N}^*_M . \tag{78.10}$$

The acceleration wave σ_3 has an effective discontinuity at $\mathscr{E}$ if and only if, equation (78.10)—cf. $(66.7)_{1,2}$—*has a non-zero solution λ^s.* This in turn occurs if and only if, the equation

$$\Delta(x) = \det \|p_{rs} - x q_{rs}\| = 0 \tag{78.11}$$

—cf. (66.8)—in x has a solution V^{*2} with V^* real (and $V^* \geqslant 0$). Furthermore, if this is the case, then the Eulerian speed V is determined by $(65.6)_2$.

In the purely elastic case ($E_\alpha \equiv H_\alpha \equiv 0$) $w^{(H)} \equiv 0$ holds by $(77.2)_{2,3}$, $w \equiv \bar{w}$ by $(46.4)_{1,3}$, and $Z^{LM} \equiv 0$ by (77.6). Consequently the double tensors q_{rs}, $p_{rs}{}^{LM}$, and p_{rs} given by (77.11), by (78.9) and (78.7), and by $(78.10)_2$, become those denoted in the same way in dealing with elastic acceleration waves—cf. $(66.5)_{1,2,4}$, $(66.7)_2$. Then in the present case we may call $(67.2)_1$ the *generalized inertial mass quadric* and $(67.2)_2$ the *(generalized) polarization quadric* relative to the Lagrangian propagation direction $\mathscr{N}^*_L$. An axis that contains the common center of these quadrics and cuts them in points with parallel tangent planes may be called *generalized acoustic axis.*

In fluids Alfvén waves can propagate with a speed whose classical limit is a function of magnetic quantities, mass density, and the propagation direction N_α—cf. (65.3). No equally simple result holds for elastic materials, even in case they are isotropic and B_α is parallel to a principal direction of strain, except in very particular cases.

Footnotes to Chapter 8

1) § 72 is based on Bressan [1966e] and the remainder of the chapter on Bressan [1972a]. For historical hints concerning this chapter see also § 44.

2) At the present point the reader can understand the proof of formula (91.11a) by simply reading §§ 90, 91.

3) In the case of non-magnetizable materials ($B_\alpha \equiv H_\alpha$) the results in § 76 coincides with those due to Schöpf [1965b] who extended to general relativity some results obtained by Carstiou [1963] in classical physics.

4) § 72 and the part of § 76 up to formula (76.3) constitute the preliminary part of this chapter which enable the reader to understand §§ 77, 78 on magneto-elastic waves.

5) Another work on waves in solids in the presence of the electromagnetic field is due to Pichon [1965]—see footnote 2 in Chapter 5.

6) The spatial length of the (regular) line $x^\rho = x^\rho(\lambda)$ $(0 \leqslant \lambda \leqslant 1)$ in S_4 is $\int_0^1 \left(\overset{+}{g}_{\rho\sigma} \frac{dx^\rho}{d\lambda} \frac{dx^\sigma}{d\lambda} \right)^{1/2} d\lambda$.

Chapter 9

Materials with Memory and Axiomatic Foundations

§ 79. Introduction to a Relativistic Theory of Materials with Memory

In this and the next three sections we state the foundations of a theory of materials with memory in special or general relativity and in § 83 to § 85 we briefly show how the relativistic theory which has already been developed in this tract can be presented in an axiomatic way.[1]

Materials with memory have been studied in classical physics for several decades, and in particular Volterra's contributions [1909] and [1912] are well known. Recently a new theory for these materials was set up from a general point of view, mainly by Noll and Coleman. More in particular Noll [1958], besides presenting the mathematical tools fit to work out the theory from this point of view, stated the principle of material (frame) indifference (which had been enunciated rather intuitively by Zaremba in 1903 and Jaumann in 1906 and 1911) in an explicit and general form—cf. Truesdell and Noll [1965, Sects. 19, 19A, 79].[2]

Other principles for materials with memory have been stated by B. Coleman, partly in collaboration with W. Noll; one of these is the principle of fading memory—cf. Truesdell and Noll [1965, Sect. 38].

Among the afore-mentioned principles only the one of material indifference will be explicitly stated in (general and special) relativity [§ 81], because, on the one hand, the others can be stated naturally and directly in a completely Lagrangian form, and in this same form they can also be stated in general relativity, practically with the same words; on the other hand the physical meaning of the principle of material indifference is expressed in classical physics by a condition involving space-time. This condition is proved to be equivalent to a condition in a (completely) Lagrangian form, relativization of which is nearly trivial. In order to take the physical meaning of the principle into account, following substantially Bressan [1964b] we shall postulate the relativization of the former condition, for materials of order n, and shall deduce the relativization of the latter, in analogy with what is done in classical physics.[3]

Bragg [1965] states his principle of non sentient response, which is based on an apparent metric (i. e. on apparent times and lengths), in special relativity for simple (first order) purely mechanical materials. He proves that in this case his principle is equivalent to Bressan's relativistic version of the objectivity principle. Roughly speaking, in connection with a fixed material point P^* of the body $\mathscr{C}$

being considered, Bressan's formulation of the principle is similar to the classical analogue as far as a neighbourhood of P^*, of order n, is concerned. For Bragg's formulation this similarity holds in a stronger way, namely as far as the localization of the motion of $\mathscr{C}$ at P^*—cf. Truesdell and Noll [1965, Sect. 22]—is concerned. Söderholm [1970] does not use Bragg's apparent metric, but follows him in considering localizations of motions of $\mathscr{C}$ at P^*. This allows him to state a relativistic principle of objectivity in flat spaces, for the most general purely mechanical materials.[4)]

The extension of Söderholm's principle to thermic phenomena appears not to be (completely) true even in special relativity—cf. the Fourier-Eckart law. Furthermore even in the purely mechanical case its extension to general relativity seems easy on the basis of the frame-work developed so far, but poor as far as its physical content is concerned. No such extension has yet appeared.

Conceptually Bragg's principle is different from Bressan's, so that the equivalence proved by Bragg is not obvious. I believe this especially for in my opinion a natural extension of Bragg's principle to higher order materials is possible, and it differs in some cases from Bressan's principle, but only by terms of order c^{-n} for some $n \geqslant 2$. Even if this is a mere conjecture, in connection with it, it is worthwhile remarking that the choice of one among several relativistic objectivity principles that differ from one another by such small terms is a matter of taste. The question which of them is true has little physical ground, for we cannot expect the principle of objectivity to be true with such a degree of accuracy—cf. e. g. Bressan [1969], [1972c].[5)] In accordance with this Müller [1969] on relativistic thermodynamics, does not comply with the objectivity principle; and Müller [1970] remarks that not even the non-relativistic limit $(c \to \infty)$ of [1969] does. The author only regrets this.

There are other alternative relativistic versions of (some parts of the) principle of objectivity—cf. Maugin [1972] and Lianis [1966], [1973].[6)] Let us add that some attempts at dealing somehow with phenomena involving memory, in several relativity, were written much earlier than the papers mentioned above.[7)]

A little discussion of the objectivity principle can be found in the second part of § 81. Some additional remarks can be found in Bressan [1974 c].

§ 80. Intrinsic Kinematic Histories. Total Geodesic Derivatives

The concept of history is essential to theories of materials with memory. Therefore we now consider a real function, $\alpha(t)$, defined for every real number t, to be interpreted as the proper time $c^{-1}s$ of a given material point, P^*, of $\mathscr{C}$. Furthermore let $\alpha^t(\tau)$ be the function defined (only) for $\tau \geqslant 0$ by the condition

$$\alpha^t(\tau) = \alpha(t-\tau) \quad \text{for} \quad \tau \geqslant 0, \quad \text{hence} \quad \alpha^t(0) = \alpha(t). \tag{80.1}$$

We call α^t *the history of* α *up to the instant* t.

Now let us associate with P^* an orthonormal triad $\lambda^{*L}{}_{(A)}$, to be thought of as a twice continuously differentiable function of the co-ordinates y^L of P^*.

Let $x^\rho = x^\rho(t, y^L)$ be any point in the world tube $W_{\mathscr{C}}$ of $\mathscr{C}$ and let the spatial orthonormal triad $f^\rho_{(r)} = f^\rho_{(r)}(t', t, y)$ associated with the event point $x^\rho(t, y^L)$ and depending on $t' = c^{-1}s'$ be that Fermi triad for the material point y^L, which coincides with $R^\rho{}_L \lambda^{*L}{}_{(r)}$ for $t' = t$—see § 55, (55.3). Thus along the world line of P^* we have

$$\frac{Df^{(\rho)}_{(r)}}{Dt'} = c f^\sigma_{(r)} A_\sigma u^\rho, \qquad f^\rho_{(r)} f_{(s)\rho} = \delta_{rs},$$
$$f^\rho_{(r)} u_\rho = 0, \qquad f^\rho_{(r)}(t, x) = R^\rho{}_M \lambda^{*\,M}_{(r)}. \tag{80.2}$$

Let $T_{\rho_1 \dots \rho_n}{}^{L\dots} = T_{\rho_1 \dots \rho_n}{}^{L\dots}(x)$ be a spatial double tensor field defined on $W_{\mathscr{C}}$. By (52.2) $T_{\rho_1 \dots \rho_n}{}^{L\dots}$ can be considered as a function $\hat{T}_{\rho_1 \dots \rho_n}{}^{L\dots}(t, y^L)$ of t and y^L. For every t' let $T_{(r_1)\dots(r_n)}{}^{L\dots}$, or e. g. $T^{(r_1)\dots(r_n)L\dots}$ be the intrinsic components of $\hat{T}_{\rho_1 \dots \rho_n}{}^{L\dots}(t', y^L)$ in the triad $f^\rho_{(r)}(t', t, y^L)$, being understood that, for the sake of simplicity, within every processes considered in §§ 80—82 *the time parameter t is assumed to be so chosen, that* $ct = s$ *holds along the world line of* y^L*:*

$$T_{(r_1)\dots(r_n)}{}^{L\dots} = \hat{T}_{\rho_1 \dots \rho_n}{}^{L\dots}(t', t, y^L)\, f^{\rho_1}_{(r_1)} \cdots f^{\rho_n}_{(r_n)}. \tag{80.3}$$

Considering the point $x^\rho = x^\rho(t, y^L)$ as being fixed, $T_{(r_1)\dots(r_n)}{}^{L\dots}$ is a scalar function of t'. We shall express its history up to the value t of t' $(= c^{-1}s)$ by

$$\{T_{(r_1)\dots(r_n)}{}^{L\dots}\}_{\lambda^*_{(r)}, x} = \{T_{(r_1)\dots(r_n)}{}^{L\dots}\}_{\lambda^*_{(r)}, P^*, t} = \{T_{(r_1)\dots(r_n)}{}^{L\dots}\}_x. \tag{80.4}$$

Similar notations will be used for any double tensor $T_{\rho\dots}{}^{\sigma\dots}{}_{L\dots}{}^{M\dots}$.

Now let us consider in $W_{\mathscr{C}}$ the fields A_ρ, $\alpha^\rho{}_{L_1}, \dots, \alpha^\rho{}_{L_1 \dots L_n}$—cf. (53.7), $(53.16)_1$. We shall call the set

$$\{c^2 A_{(r)}\}_{\lambda^*_{(r)}, P^*, t}, \qquad \{\alpha^{(r)}{}_{L_1}\}_{\lambda^*_{(r)}, P^*, t}, \dots, \qquad \{\alpha^{(r)}{}_{L_1 \dots L_n}\}_{\lambda^{*L}{}_{(r)}, P^*, t} \tag{80.5}$$

intrinsic kinematic history of order n, at x^ρ*—or for* P^* *and up to t—relative to the material triad,* $\lambda^{*L}{}_{(r)}$. Explicit reference to $\lambda^{*L}{}_{(r)}$ in formulas such as (80.5) will be omitted sometimes.

History (80.5) is related to a Fermi triad, which is the relativistic analogue of a classical co-moving triad with an invariant orientation. Hence, if so to speak, we consider a neighbourhood $\mathscr{N}$ of P^*, of order n, then the classical analogue of the history (80.5) determines the motion of $\mathscr{N}$ up to the instant t with an exception for the final values of the position, velocity, and orientation of $\mathscr{N}$. In other words, knowing this classical analogue is equivalent to knowing that the motion of $\mathscr{N}$ has certain equations with respect to an inertial frame, without specifying this frame.

Incidentally the functions forming the classical analogue $\{a_r\}^*_{y,t}$, $\{\alpha^r{}_{L_1}\}^*_{y,t}, \dots, \{\alpha^r{}_{L_1 \dots L_n}\}^*_{y,t}$ of history (80.5) can be given arbitrarily to a large extent provided the symmetry condition with respect to L_i and L_j $(i,j=1\ldots n)$ and some qualitative conditions such as $\det \|\alpha^r{}_L\|>0$ be fulfilled. The same holds in special relativity. In general relativity $\alpha^\rho{}_{L_1\ldots L_h}$ is not symmetric with respect to the L's for $h>2$. However the completely symmetric part (for $h=1,\ldots,n$) can still be given arbitrarily, to a large extent, at least for small values of n.

By $(80.2)_4$ and $(54.7)_1$, for $t'=t$ we have $R_{(r)L}=R_{\rho L}f^\rho_{(r)}=R_{\rho L}R^\rho{}_M\lambda^{*}_{(r)}{}^M=a^*_{LM}\lambda^*_{(r)}{}^M$. Furthermore by (80.3) and $(54.5)_1$ $\alpha^{(r)}{}_L=R^{(r)}{}_L\mathscr{D}^L{}_M$. Hence

$$R_{(r)L}=\lambda^*_{(r)L} \quad \text{and} \quad \alpha^{(r)}{}_M=\lambda^{*(r)}{}_L\mathscr{D}^L{}_M \quad \text{in} \quad (t,y). \tag{80.6}$$

Let us add that by (80.3) and $(80.2)_3$

$$\frac{D}{Dt'}T_{(r_1)\ldots(r_n)}{}^{L\ldots}=c\frac{DT_{\rho_1\ldots\rho_n}{}^{L\ldots}}{Ds'}f^{\rho_1}_{(r_1)}\ldots f^{\rho_n}_{(r_n)} \quad (\text{for } T_{\ldots}^{\ldots}=\overset{\perp}{T}{}_{\ldots}^{\ldots},\ s'=ct'). \tag{80.7}$$

Hence by $(57.6)_1$

$$\frac{1}{c}\frac{D\alpha^{(r)}{}_L}{Dt'}=\frac{D\alpha_{\rho L}}{Ds'}f^\rho_{(r)}=u_{\rho|L}f^\rho_{(r)} \quad (s'=ct'). \tag{80.8}$$

Let us now incidentally consider an event point x^ρ in $W_{\mathscr{C}}$ and the totally geodesic hypersurface $S_x{}^{(t,g)}$ through x^ρ, orthogonal to u^ρ at x^ρ, i. e. the union of the geodesics through x^ρ which are orthogonal to u^ρ at x^ρ. We can choose $t=\hat{t}(x)$ in such a way that for $t=0$ equations $(52.2)_1$ represent $S_x{}^{(t,g)}$ when y^1, y^2, y^3 are thought of as parameters. Then we call the r-th total covariant derivative of any double tensor $T_{\rho\ldots}{}^{A\ldots}$ with respect to the mapping $(52.2)_1$ of S^*_3 into $S_x{}^{(t,g)}$—cf. Appendix A2—the r-th *total (time-orthogonal) geodesic derivative* of $T_{\rho\ldots}{}^{A\ldots}$ at x^ρ and we denote it by

$$T_{\rho\ldots}{}^{A\ldots}{}_{\|L_1\ldots L_r}. \tag{80.9}$$

Bressan [1964b] used, instead of (80.5), the history obtained from (80.5) by replacing $\alpha^\rho{}_{L_1\ldots L_n}$ with $x^\rho{}_{\|L_1\ldots L_n}$. To know the intrinsic kinematic history obtained in this way is equivalent to knowing the history (80.5). Hence the use of (80.5) raises no question.

However let us note that, as was shown by A. Grioli [1971] who investigated total geodesic derivatives, for $n>2$ (and only for $h>2$). $x^\rho{}_{\|L_1\ldots L_n}$ is generally different from $\alpha^\rho{}_{L_1\ldots L_n}$, unlike $\alpha^\rho{}_{L_1\ldots L_n}$, it is independent of the velocity distribution, and in special relativity it has a simpler expression than $\alpha^\rho{}_{L_1\ldots L_n}$. Therefore it is certainly preferable to $\alpha^\rho{}_{L_1\ldots L_n}$ for constructing a relativistic theory of elasticity for materials of order $n>2$—cf. footnote 15 in Chapter 1.[8)]

§ 81. A Relativistic Version of the Principle of Material (Frame) Indifference

At the event point x^ρ of $W_{\mathscr{C}}$, and in connection with the material orthonormal triad $\lambda^{*L}{}_{(A)}$ at the material point y^L (which passes through x^ρ) we consider the histories

$$\{c^2 A_{(r)}\}_x, \{\alpha^{(r)}{}_{L_1}\}_x, \ldots, \{\alpha^{(r)}{}_{L_1 \ldots L_n}\}_x, \{T\}_x, \{\theta_{(r)}\}_x, \{\bar{\rho}\}_x, \{E_{(r)}\}_x, \{H_{(r)}\}_x, \quad (81.1)$$

where T is the absolute temperature, $\theta_\rho = T_{/\rho}^{\perp} + TA_\rho$, $\bar{\rho}$ is the proper density of electric charge, and E_ρ and H_ρ are the electric and magnetic fields respectively. The histories (81.1) form a set $\mathfrak{H}$ to be called the *(local) thermo-electromagneto-kinematic history of order n, at* x^ρ—or *up to the instant t, for the material point* y^L.

The history above is determined by the fields

$$g_{\alpha\beta}(x), \quad x^\alpha(t,y), \quad T(x), \quad \bar{\rho}(x), \quad E^\alpha(x), \quad H^\alpha(x), \quad (81.1')$$

whose physical meanings and regularity properties have already been specified. We are referring to the actual process of the universe, or to any process that can be described in the above way and can (ideally) take place; hence the stated physical equations—Einstein's, Maxwell's (and constitutive equations)—are supposed to hold. In dealing with histories it is sometimes useful to regard the collection $\mathscr{P}$ of fields (81.1') as having the afore-mentioned regularity properties but as possibly failing to solve the physical equations. In this case we shall briefly call $\mathscr{P}$ an *admissible (thermo-electromagnetokinematic) process*—cf. Definition 85.1. Let $\tilde{\mathscr{P}}$:

$$\tilde{g}_{\alpha\beta}(x), \quad \tilde{x}^\alpha(t,y), \ldots, \tilde{H}^\alpha(x)$$

be another of these processes (possibly coinciding with $\mathscr{P}$ up to a change of the reference space-time frame). We shall use a wave sign to denote quantities connected with $\tilde{\mathscr{P}}$. Let $\|\omega_{rs}\| = \|\omega_{rs}(t')\|$ be an arbitrary proper orthonormal matrix of order 3, which is a continuously differentiable function of t' ($t' \leqslant t$) and fulfills the condition $\omega_{rs}(t) = \delta_{rs}$.

Consider the thermo-electromagnetokinematic history $\tilde{\mathfrak{H}}$ of $\tilde{\mathscr{P}}$ at $\tilde{x} = \tilde{x}(t,y)$ and assume

$$\{\tilde{\alpha}^{(r)}{}_L\}_{\tilde{x}} = \{\omega^r{}_s \alpha^{(s)}{}_{L_1}\}_x, \ldots, \{\tilde{\alpha}^{(r)}{}_{L_1 \ldots L_n}\}_{\tilde{x}} = \{\omega^r{}_s \alpha^{(s)}{}_{L_1 \ldots L_n}\}_x, \{\tilde{\mathscr{V}}_{(r)}\}_{\tilde{x}} = \{\omega_r{}^s \mathscr{V}_{(s)}\}_x \quad (81.2\text{a})$$

where $\mathscr{V}_\rho$ can be θ_ρ, E_ρ, or H_ρ (e.g. $\omega_{rs} = \omega_r{}^s$), and

$$\{\tilde{T}\}_{\tilde{x}} = \{T\}_x, \quad \{\tilde{\bar{\rho}}\}_{\tilde{x}} = \{\bar{\rho}\}_x \quad (\omega_{rs}(t) = \delta_{rs}, \omega_{lr}\omega^l{}_s = \delta_{rs}\ ct' = s' \text{ for } t' \leqslant t)\,. (81.2\text{b})$$

Then the histories $\mathfrak{H}$ and $\tilde{\mathfrak{H}}$ (of order n) are said to be *equivalent*. They are said to be *rotationally [translationally] equivalent* if, besides (81.2a), we have

$$\{\tilde{A}^{(r)}\}_{\tilde{x}} = \{\omega_{rs} A^{(s)}\}_x \quad [\omega_{rs}(t') = \delta_{rs} \text{ for } t' \leqslant t]\,. \quad (81.2\text{c})$$

Obviously the relations above between the histories $\mathfrak{H}$ and $\tilde{\mathfrak{H}}$ are equivalence relations and induce corresponding equivalence relations that hold for the processes $\mathscr{P}$ and $\tilde{\mathscr{P}}$ where these processes may belong to different determinations of space-time: If the histories $\mathfrak{H}$ and $\tilde{\mathfrak{H}}$ of order n, are equivalent or rotationally [translationally] equivalent, then $\mathscr{P}$ and $\tilde{\mathscr{P}}$ can be said to be *n-equivalent* or *rotationally [translationally] n-equivalent* respectively at P^*, the material point y^L, and at the instant t—or at (t, y).

Now let $\mathscr{P}$ fulfill the physical equations. Then, especially for small values of n, some other processes $\tilde{\mathscr{P}}$ that are n-equivalent to $\mathscr{P}$ at (t, y) and fulfill the physical equations are expected to exist. This existence is essential in order that the principle of material indifference stated in section 8.2 may have any satisfactory physical content.

Incidentally, to illustrate the physical meaning of the notions introduced above, let $E_\rho \equiv H_\rho \equiv 0$ for the sake of simplicity, so that (81.1) reduces to a thermokinematic history, and let us consider, in classical physics, a thermomechanical process $\mathscr{P} = (\mathscr{F}, \mathfrak{D})$ described in the Euclidean space-time frame $\mathscr{F}$ [§ 19] by the collection $\mathfrak{D}$ of functions. Let $\tilde{\mathscr{P}} = (\tilde{\mathscr{F}}, \mathfrak{D})$ be the (admissible) process described in another Euclidean space-time frame $\tilde{\mathscr{F}}$ by the collection $\mathfrak{D}$ again. Furthermore we denote by τ_0 [$\tilde{\tau}_0$] the time origin of $\mathscr{F}$ [$\tilde{\mathscr{F}}$]. Then the thermokinematic history (of P^*) in the process $\mathscr{P}$ up to $\tau_0 + \tau$ is equivalent to the one in $\tilde{\mathscr{P}}$ up to $\tilde{\tau}_0 + \tau$ for every real τ (and for some, hence for every choice of P^*)—see Truesdell and Noll [1965, Sect. 18]. The converse also holds.

Likewise the relativistic processes $\mathscr{P}$ and $\tilde{\mathscr{P}}$ are n-equivalent at (t, y) if and only if they appear locally in the same way (i.e. have the same description) with respect to two suitable observers moving along the world lines of P^* in $\mathscr{P}$ and $\tilde{\mathscr{P}}$ respectively. By "locally" we mean "as far as a neighbourhood of P^* or order n is concerned" in the sense suggested by the definition (81.1) of $\mathfrak{H}$. More precisely, if (81.2a, b) holds then one of those observers can be identified with the Fermi frame $f_{(r)}{}^\rho$ considered in $\mathscr{P}$, the other with the frame $\tilde{\phi}_{(r)}{}^\rho = \omega_r{}^s \tilde{f}_{(s)}{}^\rho$; hence $\alpha^{(r)}{}_{L_1 \dots L_a} \tilde{\phi}_{(r)}{}^\rho = \tilde{\alpha}^{(s)}{}_{L_1 \dots L_a} \tilde{f}_{(r)}{}^\rho$ and $\mathscr{V}^{(r)} \tilde{\phi}_{(r)}{}^\rho = \tilde{\mathscr{V}}^{(s)} \tilde{f}_{(s)}{}^\rho$, where $\mathscr{V}_\rho$ can be θ_ρ, E_ρ or H_ρ. Thus (81.2a, b) implies that the same collection of functions $\mathfrak{H}$ can be regarded as the *relative history of order n* of P^* both in $\mathscr{P}$, with respect to the frame $f_{(r)}{}^\rho$, and in $\tilde{\mathscr{P}}$ with respect to the frame $\tilde{\phi}_{(r)}{}^\rho$. The coverse also holds. The direct and converse assertions can be made explicit, in a slightly more general form: *Let $ct = s$ hold along the world lines of y^L in the admissible processes $\mathscr{P}$ and $\tilde{\mathscr{P}}$; then $\mathscr{P}$ and $\tilde{\mathscr{P}}$ are n-equivalent at (t, y) if and only if, the spatial orthonormal frames $\phi_{(r)}(t')$ and $\tilde{\phi}_{(r)}(t')$ can be so chosen at $x(y, t')$ and $\tilde{x}(y, t')$ respectively, for every $t' \leqslant t$, that* (i) *they turn out to be continuous functions of t', and* (ii) *for $t' \leqslant t$ any of the magnitudes $\alpha^\rho{}_{L_1}, \dots, \alpha^\rho{}_{L_1 \dots L_n}$, T, θ_ρ, $\bar{\rho}$, E_ρ, and H_ρ takes values in $\mathscr{P}$ and $\tilde{\mathscr{P}}$, at t', that have coinciding corresponding components in the frames $\phi_{(r)}{}^\rho(t')$ and $\phi_{(r)}{}^\rho$ respectively.*

In the relativistic theory developed in the foregoing chapters the actual values at x^ρ of the quantities

$$\rho, \quad X^{\rho\sigma}, \quad q^\rho, \quad j^\rho, \quad P^\rho, \quad M^\rho \tag{81.3}$$

were often regarded as given by constitutive functions of $\alpha^\rho{}_L$, T, $\theta_\rho = T_{/\rho}^{\perp} + TA_\rho$, $\bar{\rho}$, E_ρ, and H_ρ (and of the co-ordinates y^L of P^*). Now, in order to deal with materials with memory, we assume that, if we fix an orthonormal triad $\lambda^{*L}_{(r)} = \hat{\lambda}^{*L}_{(r)}(y)$, then the actual components of the magnitudes (81.3) at x^ρ in the triad $\lambda^\rho{}_{(r)} = R^\rho{}_L \lambda^{*L}{}_{(r)}$ are the values of given constitutive functionals of the actual thermo-electromagnetokinematic history at x^ρ—cf. (81.1):

$$\begin{aligned}&\rho = G_{(1)}(\mathfrak{H}, y^L), \qquad X^{(r)(s)} = G^{rs}_{(2)}(\mathfrak{H}, y^L), \qquad q^{(r)} = G^r_{(3)}(\mathfrak{H}, y^L),\\ &j^{(r)} = G^r_{(4)}(\mathfrak{H}, y^L), \qquad P^{(r)} = G^r_{(5)}(\mathfrak{H}, y^L), \qquad M^{(r)} = G^r_{(6)}(\mathfrak{H}, y^L),\end{aligned} \tag{81.4}$$

where $\mathfrak{H}$ is the set (81.1) of histories (principle of local action).

If the Cattaneo-Vernotte law of heat conduction is accepted, then the coefficient $\bar{\kappa}^{\rho\sigma}$ is to be included in (81.3) and the corresponding functional in (81.4). In this connection it is useful to take into account Lianis [1974], hinted at in § 26.

Principle of Material indifference. *Every one of the functionals* (81.4) *takes the same value for equivalent admissible thermo-electromagnetokinematic histories.*

The replacement of "equivalent" with "rotationally equivalent" or "translationally equivalent" in the principle above turns it into its *rotational* or *translational* part respectively. In dealing with certain phenomena that are in experimental disagreement with the full principle, it is convenient to postulate only one of these parts—cf. Bressan [1972c] and Lianis [1973].

Remembering (53.15), (53.16), and convention $(57.1)_1$, and writing $\{\ \}^*_{t,y}$ for $\{\ \}_{x(t,y)}$, we see that the history (81.1) determines the set

$$\{C_{AL_1}\}^*_{t,y}, \dots, \{C_{AL_1 \dots L_n}\}^*_{t,y}, \quad \{T\}^*_{t,y}, \quad \{\theta^*_L\}^*_{t,y}, \quad \{\bar{\rho}\}^*_{t,y}, \quad \{E^*_L\}^*_{t,y}, \quad \{H^*_L\}^*_{t,y} \tag{81.5}$$

which will be denoted by $\mathfrak{H}^*$ and called the *Lagrangian thermo-electromagnetokinematic history at* (t, y). We shall also say that $\mathfrak{H}^*$ is the *(full) Lagrangian counterpart* of $\mathfrak{H}$. Let us add $\{c^2 A^*_L\}^*_{t,y}$, i.e. $\{c^2 A_\rho \alpha^\rho{}_L\}^*_{t,y}$ $[\{u^*_{[LM]}\}^*_{t,y}$, i.e. $\{u_{[\rho/\sigma]} \alpha^\rho{}_L \alpha^\sigma{}_M\}^*_{t,y}]$ to $\mathfrak{H}^*$. Then the result will be called the *translational [rotational] Lagrangian counterpart of* $\mathfrak{H}$.

Truesdell and Toupin [1960] explain (in Section 7) within classical physics that "the field equations and jump conditions express the general principles of mechanics, thermodynamics, and electromagnetism, while constitutive equations define *ideal materials*, which are mathematical models of particular classes of materials encountered in nature". Furthermore these authors list (at pp. 700—704) the mathematical principles that a theorist "may call to his aid when he attempts to formulate definitive constitutive equations." They add that no ideal material is known for which all these principles have been demonstrated to hold, although for the simpler classical theories it is generally believed that they do.

The afore-mentioned mathematical principles include the one of material indifference, which incidentally is sometimes briefly called the *objectivity principle.* We have postulated a relativistic version of it in accordance with the point of view above. Of course the principle, as well as classical physics itself, is expected to hold only within a certain degree of approximation which perhaps is higher for mechanical than (thermodynamic or) electromagnetic constitutive equations.

In harmony with the considerations above let us first assume that the constitutive functional $G^{(r)}_{(\mathfrak{A})}$ contains the history $\{\theta_{(r)}\}_x$ effectively—cf. $(25.2)_2$. In this case, which certainly occurs for $\mathfrak{A}=3$, the principle of material indifference is fulfilled in a rather ficticious way. More precisely, let $\mathfrak{K}$ be the history obtained from $\mathfrak{H}$, e.i. the collection (81.1), by replacing $\{\theta_{(r)}\}_x$ with $\{T_{/(r)}\}$; and let $\tilde{\mathfrak{K}}$ be an history equivalent to $\mathfrak{K}$, i. e. let $\tilde{\mathfrak{K}}$ be related to $\mathfrak{K}$ by an obvious analogue of (81.2a, b). Then one might object that a relativization of the objectivity principle which is closer than ours to the classical sense of the principle would be the following:

$$G^{r}_{(\mathfrak{A})}(\mathfrak{K}, y^L) = G^{r}_{(\mathfrak{A})}(\tilde{\mathfrak{K}}, y^L) \tag{81.6}$$

holds for an arbitrary choice of the equivalent histories $\mathfrak{K}$ *and* $\tilde{\mathfrak{K}}$.

Given $\mathfrak{K}$, $\tilde{\mathfrak{K}}$ is determined by the above functions $A^{(r)}=A^{(r)}(t)$ and $\omega_{rs}=\omega_{rs}(t')$ $(t'\leqslant t)$—cf. (81.2b). Our objectivity principle implies (81.6) for all $\mathfrak{K}$ and $\omega_{rs}{}^{(t)}$, only in the cases when $(81.2\text{c})_1$ holds, i. e. if and only if $\mathfrak{K}$ and $\tilde{\mathfrak{K}}$ are rotationally equivalent.

Hence, one might object that, if $G^{r}_{(\mathfrak{A})}(\mathfrak{K}, y^L)$ depends on $\{\theta_{(r)}\}_x$ effectively, then it is truly objective only in a reduced form: the value of $G_{(r)}(\mathfrak{K}, y^L)$ is independent (in a suitable sense) of the "rotation history" $\{R^{(r)}{}_L\}_x$ but may depend on the acceleration history $\{\omega_{rs} A^{(s)}\}_x$; more precisely only the rotational part of the objectivity principle (81.6) holds.

When heat conduction is taken into account—cf. (81.4) for $\mathfrak{A}=3$—θ_ρ cannot be replaced by $T_{/\rho}^{\perp}$ even in C. Eckart's very simple version of the Fourier law—cf. (25.2) for $\kappa^{\alpha\beta}=\kappa \overset{\perp}{g}{}^{\alpha\beta}$.

It is true that the definitive form of the Fourier law is yet unknown, but in thermomechanical equilibrium the relation $\theta_\rho \equiv 0$ (and not $T_{/\rho}^{\perp}\equiv 0$) must hold as a consequence of the Maxwell equations, conservation equations, and Stefan-Boltzmann laws (in finite terms), in accord with the fact that in relativity energy weights (and weight is equivalent to an inertial force) [§ 45]. This prevents any relativistic version of the Fourier law from being truly objective in the sense above connected with the history $\mathfrak{K}$, but is compatible with our objectivity principle. On the other hand this also explains why in classical physics the preceding difficulties concerning the objectivity principle in thermodynamics do not arise.

Our relativistic objectivity principle is a modification of its version (81.6), not a reduction, and in particular it does not follow from the afore-mentioned reduced validity of (81.6). Furthermore experiment cannot now detect whether a local equation holds for θ_ρ or for $T_{/\rho}^{\perp}$; this contributes to justifying our use of $\theta_{(r)}$ (instead of $T_{/(r)}$) in the principle.

Let us now remember that Müller [1969] and [1970] on relativistic thermodynamics does not comply with the objectivity principle even in the non-relativistic limit $(c\to\infty)$ admittedly by very small terms.

Furthermore some considerations in classical physics show that even the classical constitutive equations of non-linear elasticity should be corrected by very small terms which are not objective—cf. footnote 5 in Chapter 9. This confirms the very good acceptability of the objectivity principle, but also specifies a limit to its validity in classical physics. Analogous conclusions are reached by C. A. Grioli [1972] on gyromagnetic phenomena.[9)]

We can add—cf. e. g. Bressan [1972c]—that the Barnet experiment on gyromagnetic penomena is incompatible with the objectivity of $G^r_{(6)}$—cf. (81.4); more precisely an increasing angular velocity parallel to the magnetic field slightly influences the magnetic polarization. Hence for $\mathfrak{A}=6$ we can assert at most that

$$G^r_{(\mathfrak{A})}(\mathfrak{H}, y^L) = G^r_{(\mathfrak{A})}(\tilde{\mathfrak{H}}, y^L) \tag{81.7}$$

holds for an arbitrary choice of the histories $\mathfrak{H}$ and $\tilde{\mathfrak{H}}$ that have the forms (81.1) and (81.2a, b) respectively for $\omega_{rs}(t') \equiv \delta_{rs}$ $(t' \leqslant t)$.

Hence for $G^r_{(6)}$ at most the translational (part of the) objectivity principle holds: *the value of $G^r_{(\mathfrak{A})}(\mathfrak{H}, y^L)$ is independent of the history of (intrinsic) acceleration but may depend on the rotation history $\{R^{(r)}{}_L\}_x$.*

Some claims of restrictions for the validity of the rotational part of the objectivity principle in classical physics are based on the kinetic theory of gases. Chapman and Enskog's expression of Boltzmann equation in terms of a certain small parameter affords a method for deriving constitutive equations. The second approximation gives—cf. Chapman and Cawling [1970, Chap. 15]—(the specialization to ideal gases of) Burnett's heat conduction equations:

$$q_r = -\kappa T_{/r} + \frac{b}{T} \varepsilon_r{}^{ih} \omega_i T_{/h} \tag{81.8}$$

where the constant b is positive and very little. That the last term in (81.8) contrasts with the objectivity principle was emphazized by Müller [1972], Edelen and McLennan [1973], and Söderholm [1976]. Among other things, Söderholm concluded that for the material equations of a gas the objectivity principle is unnecessary up to the first order and incorrect in the second order". Truesdell [1976] strongly criticizes the claims above as based on results of formal processes of "approximation" in the kinetic theory—such as (81.8)—wrongly regarded as constitutive equations.

Experiment is now unable to test relativistic corrections to classical theories in connection with local phenomena observed from a (nearly) co-moving frame. Hence one of the main results now achieved by the relativistic theory of materials is a rather conceptual one: logical compatibility with electromagnetism and gravitation theory, and an agreement with actual experiments at least as good as the one of the corresponding classical theory.

In most relativistic problems our full version of the principle of material indifference is quite acceptable; for instance its acceptance does not destroy the advantages characteristic of relativistic treatments in describing very fast motions or certain non-local phenomena. This is in accord with the known great importance of the objectivity principle as a general classical principle and its usefulness in describing at least the gross (intrinsic) behavior of every material element.

If one is willing to take into account e. g. the afore-mentioned slight deviations from the objectivity principle,—cf. Bressan [1972c]—then the preceding treatment can immediately afford him the means to do this. For instance one may assert in compliance with Truesdell's principle of equipresence—cf. Truesdell and Toupin [1960, p. 703]—that $G^r_{(\mathfrak{A})}(\mathfrak{H}, y^L)$ is objective fully or translationally, according to whether its dependence on $\{A_{(r)}\}_x$—cf. (81.1)—is effective or not.

§ 82. Some Consequences of the Principle of Material Indifference

As a preliminary we prove the following:

Theorem 82.1. *Two thermo-electromagnetokinematic histories are equivalent if and only if, they have the same Lagrangian counterpart.*

Proof. We first assume that history (81.1) is equivalent to

$$\{c^2\tilde{A}_{(r)}\}_x, \quad \{\tilde{\alpha}^{(r)}{}_{L_1}\}_x, \dots, \{\tilde{\alpha}^{(r)}{}_{L_1\dots L_n}\}_x, \quad \{\tilde{T}\}_x,$$
$$\{\tilde{\theta}_{(r)}\}_x, \quad \{\tilde{\bar{\rho}}\}_x, \quad \{\tilde{E}_{(r)}\}_x, \quad \{\tilde{H}_{(r)}\}_x; \tag{82.1}$$

hence (81.2a, b) holds, so that, for $t'\leqslant t$, $\tilde{\alpha}^{(r)}{}_A\tilde{\alpha}_{(r)L_1\dots L_K}=\omega^r{}_h\alpha^{(h)}{}_A\omega_{rs}\alpha^{(s)}{}_{L_1\dots L_K}$ where $\omega^r{}_h\omega_{rs}=\delta_{hs}$. Then by $(53.15)_1$ and $(53.16)_2$

$$\tilde{C}_{AL_1\dots L_K}=\omega^r{}_h\alpha^{(h)}{}_A\omega_{rs}\alpha^{(s)}{}_{L_1\dots L_K}=\delta_{hs}\alpha^{(h)}{}_A\alpha^{(s)}{}_{L_1\dots L_K}=C_{AL_1\dots L_K} \qquad (t'\leqslant t). \tag{82.2}$$

Furthermore, by convention $(57.1\,\text{a})_1$,

$$\tilde{\mathscr{V}}^*_L=\tilde{\mathscr{V}}_{(r)}\alpha^{(r)}{}_L=\omega_r{}^h\mathscr{V}_{(h)}\omega^r{}_s\alpha^{(s)}{}_L=\delta^h{}_s\mathscr{V}_{(h)}\alpha^{(s)}{}_L=\mathscr{V}^*_L \tag{82.3}$$

in (t',y) for $t'\leqslant t$, where $\mathscr{V}_\rho$ can be θ_ρ, E_ρ, or H_ρ. We conclude that histories (81.1) and (82.1) have the same Lagrangian counterpart (81.5).

Now we conversely assume the above conclusion, hence (82.2), and we want to prove the equivalence of histories (81.1) and (82.1). To this end we remember that there is only one tensor $\mathscr{D}^L{}_M$ for which $\det\|\mathscr{D}_{LM}\|>0$ and (54.4) holds. Furthermore there are two double tensors $R^\rho{}_A$ and $\tilde{R}^\rho{}_A$ fulfilling conditions $(54.5)_1$ and (54.7) and their analogues for $\tilde{R}^\rho{}_A$, which are

$$\tilde{\alpha}^\rho{}_L=\tilde{R}^\rho{}_A\mathscr{D}^A{}_L, \quad \tilde{R}^\rho{}_L\tilde{R}_{\rho M}=\alpha^*_{LM}, \quad \tilde{R}^\rho{}_A\tilde{R}^{\sigma A}=\overset{\perp}{\tilde{g}}{}^{\rho\sigma}, \quad \tilde{u}_\rho\tilde{R}^\rho{}_L=0. \tag{82.4}$$

We have $(55.4)_{1,2}$ for $\lambda^L{}_{(r)}=\lambda^{*L}_{(r)}$ and its analogue for $\tilde{R}^\rho{}_A$:

$$\tilde{R}_{(r)(A)}=\tilde{f}_{(r)\rho}\tilde{l}^\rho{}_{(A)} \qquad (\tilde{l}^\rho{}_{(A)}=\tilde{R}^\rho{}_L\lambda^{*L}{}_{(A)}) \tag{82.5}$$

where $\tilde{f}_{(r)\rho}$ is the determination at the event point $\tilde{x}^\rho=\tilde{x}^\rho(t',y)$ [§ 80] on the world line of y^L in $\tilde{\mathscr{P}}$ for $t'\leqslant t$, of the Fermi triad defined by (80.2). These formulas say that $R_{(r)(A)}$, $[\tilde{R}_{(r)(A)}]$ are the components of $R^\rho{}_A$ $[\tilde{R}^\rho{}_A]$ in the orthonormal frames $f_{(r)}{}^\rho$ $[\tilde{f}_{(r)}{}^\rho]$ and $\lambda^{*L}_{(A)}$. This easily yields—see (54.7)

$$R_{(r)(A)}R_{(s)}{}^{(A)}=\delta_{rs}=\tilde{R}_{(r)(A)}\tilde{R}_{(s)}{}^{(A)}, \quad R_{(r)(A)}R^{(r)}{}_{(B)}=\delta_{AB}=\tilde{R}_{(r)(A)}\tilde{R}^{(r)}{}_{(B)}. \tag{82.6}$$

We now deduce that

$$\tilde{\alpha}^{(r)}{}_L=\omega^r{}_s\alpha^{(s)}{}_L \quad \text{for} \quad \omega_{rs}=\tilde{R}_{(r)(B)}R_{(s)}{}^{(B)}=\tilde{R}_{(r)L}R_{(s)}{}^L \tag{82.7a}$$

and equalities $(81.2\,\text{b})_{3,4}$ hold

Indeed $(54.5)_1$ implies $\alpha^{(s)}{}_L = R^{(s)}{}_{(A)} \mathscr{D}^{(A)}{}_L$ which by $(82.6)_3$ yields $R_{(s)}{}^{(B)} \alpha^{(s)}{}_L = \delta^B{}_A \mathscr{D}^{(A)}{}_L = \mathscr{D}^{(B)}{}_L$. Then from the equality $\tilde{\alpha}^{(r)}{}_L = \tilde{R}^{(r)}{}_{(B)} \mathscr{D}^{(B)}{}_L$—cf. its analogue for $\alpha^{(s)}{}_L$ above—we deduce (82.7a).

By $(80.6)_1$ and its analogue for $\tilde{R}^{\rho}{}_L$, $(82.7\text{a})_{2,3}$ yields, for $t'=t$, $\omega_{rs} = \lambda^*_{(r)L} \lambda^{*L}_{(s)} = \delta_{rs}$, i. e. $(81.2\text{b})_3$ holds.

By $(82.7\text{a})_2$, $(82.6)_3$, and $(82.6)_2$ we respectively have

$$\omega_{rl}\,\omega_s{}^l = \tilde{R}_{(r)(B)} R_{(l)}{}^{(B)} \tilde{R}_{(s)(A)} R^{(l)(A)} = \tilde{R}_{(r)(B)} \tilde{R}_{(s)(A)} \delta^{AB} = \delta_{rs};$$

hence $(81.2\text{b})_4$ holds. By $(81.2\text{b})_4$ and $(82.7\text{a})_1$

$$\delta^A{}_B = \overset{-1}{\tilde{\alpha}}_{(r)}{}^A \delta^r{}_s \alpha^{(s)}{}_B = \overset{-1}{\tilde{\alpha}}_{(r)}{}^A \omega_l{}^r \omega^l{}_s \alpha^{(s)}{}_B = \omega_l{}^r \overset{-1}{\tilde{\alpha}}_{(r)}{}^A \tilde{\alpha}^{(l)}{}_B \tag{82.7b}$$

in (t',y) for $t' \leqslant t$, and

$$\overset{-1}{\tilde{\alpha}}_{(l)}{}^A = \omega_l{}^r \overset{-1}{\alpha}_{(r)}{}^A \tag{82.7c}$$

in (t',y) for $t' \leqslant t$.

Since the histories (81.1) and (82.1) are assumed to have the same Lagrangian counterpart, we obtain the equalities

$$\tilde{\alpha}_{(r)A} \tilde{\alpha}^{(r)}{}_{L_1 \ldots L_K} = C_{AL_1 \ldots L_K} = \alpha_{(r)A} \alpha^{(r)}{}_{L_1 \ldots L_K}$$

in (t',y) for $t' \leqslant t$, which imply $(82.8)_{1,3}$ below respectively in (t',y) for $t' \leqslant t$, while $(82.8)_2$ follows from (82.7c)

$$\tilde{\alpha}^{(r)}{}_{L_1 \ldots L_K} = \delta^{rl} \overset{-1}{\tilde{\alpha}}_{(l)}{}^A C_{AL_1 \ldots L_K} = \delta^{rl} \omega_l{}^s \overset{-1}{\alpha}_{(s)}{}^A C_{AL_1 \ldots L_K} = \omega^r{}_h \alpha^{(h)}{}_{L_1 \ldots L_K} \quad (K=1,\ldots,n). \tag{82.8}$$

Furthermore, by convention $(57.1\text{a})_1$, (82.7c), and the present assumption $\{\tilde{\mathscr{V}}^*_L\}^*_{t,y} = \{\mathscr{V}^*_L\}^*_{t,y}$, we easily deduce

$$\tilde{\mathscr{V}}_{(r)} = \overset{-1}{\tilde{\alpha}}{}^L{}_{(r)} \mathscr{V}^*_L = \omega_r{}^s \overset{-1}{\alpha}_{(s)}{}^L \mathscr{V}^*_L = \omega_r{}^s \mathscr{V}_{(s)} \tag{82.9}$$

in (t',y) for $t' \leqslant t$, where $\mathscr{V}_\rho$ can be θ_ρ, E_ρ, or H_ρ. At this point we easily see that the history (82.1) satisfies conditions (81.2a, b) and hence is equivalent to the history (81.1). q.e.d.

The theorem above stated in Bressan [1964b] (up to the use of totally geodesic derivatives) can be extended into the following

Theorem 82.2. *The thermo-electromagnetokinematic histories* $\mathfrak{H}$ *and* $\tilde{\mathfrak{H}}$—cf. (81.1) *and* (82.2)—*are rotationally [translationally] equivalent if and only if they have the same translational [rotational] Lagrangian counterpart.*

Proof. We assume that $\mathfrak{H}$ and $\tilde{\mathfrak{H}}$ are equivalent and $\tilde{\mathfrak{H}}^* = \mathfrak{H}^*$, which is not restrictive for by Theorem 82.1 this is implied by anyone of the conditions whose equivalence is to be proved. Now remark that $\mathfrak{H}$ and $\tilde{\mathfrak{H}}$ are rotationally equivalent if and only if

(a) $\tilde{A}_{(r)} = \omega_{rs} A^{(s)}$ at (t', y) for $t' \leqslant t$.

This is the analogue of $(81.2\text{a})_3$ for A_ρ. Under condition (82.7a), which is now valid by our initial assumption, $(81.2\text{a})_3$ was substantially proved to be equivalent to $\tilde{\mathscr{V}}^*_L = \mathscr{V}^*_L$ in (t', y) for $t' \leqslant t$—cf. (82.3) and (82.9). Thus (a) holds, if and only if

$$\tilde{A}^*_L = A^*_L \quad \text{in } (t', y) \text{ for } t' \leqslant t, \text{ i. e. if and only if} \quad \{\tilde{A}^*_L\}^*_{t,y} = \{A^*_L\}^*_{t,y}.$$

By our first hypothesis this is equivalent to the coincidence of the translational Lagrangian counterparts of $\mathfrak{H}$ and $\tilde{\mathfrak{H}}$.

It remains to prove the part of the theorem on the rotational Lagrangian counterparts of $\mathfrak{H}$ and $\tilde{\mathfrak{H}}$. They coincide if and only if

(b) $\{\tilde{u}^*_{[LM]}\}^*_{t,y} = \{u^*_{[LM]}\}^*_{t,y}$ where $u^*_{LM} = u_{\rho/\sigma} \alpha^\rho{}_L \alpha^\sigma{}_M$.

We have $\{\tilde{C}_{LM}\}^*_{t,y} = \{C_{LM}\}^*_{t,y}$, hence $\{D\tilde{C}_{LM}/Ds\}^*_{t,y} = \{D C_{LM}/Ds\}^*_{t,y}$.

By (57.7) this yields $\{\tilde{u}^*_{(LM)}\}^*_{t,y} = \{u^*_{(LM)}\}^*_{t,y}$. We conclude that (b) is equivalent to

(c) $\{\tilde{u}^*_{LM}\}^*_{t,y} = \{u^*_{LM}\}^*_{t,y}$.

By (80.8) and $(\text{b})_2$

(d) $$\delta_{rs} \alpha^{(s)}{}_M \frac{D\alpha^{(r)}{}_L}{Ds'} = u_{\rho|L} f_{(s)}{}^\rho \alpha^s{}_M = u^*_{LM}.$$

Since $\det \|\alpha^{(r)}{}_L\| \neq 0$, (d) is equivalent to

(e) $$\frac{D\alpha^{(r)}{}_L}{Dt'} = c u^*_{LM} \overset{-1}{\alpha}{}_{(s)}{}^M \delta^{sr} \quad \text{where } \delta_{rs} \alpha^{(r)}{}_L \overset{-1}{\alpha}{}^{(s)}{}_M = \delta_{LM}.$$

We know that $\overset{-1}{\alpha}{}^{(s)}{}_M$ is a function of $\alpha^{(r)}{}_L$, that along the world-line of P^* u^*_{LM} is a function of t', and that the function $x^L = x^L(t', y)$ can be assumed to be smooth enough to assure the validity of the uniqueness theorem for the system (e) of nine ordinary differential equations in the nine unknown functions $\alpha^r{}_L$ of t'. Furthermore the initial conditions $\alpha^{(r)}{}_L(t, y) = \mathscr{D}_L{}^M \lambda^*_{(r)M}$ hold—cf. $(80.6)_2$. Let us add that the functions $\tilde{\alpha}^{(r)}{}_L(t', y)$ of t' satisfy the analogue of equations (e) for $\tilde{u}^*_{LM}$ and the same initial conditions for $t' = t$ (because $\tilde{\mathfrak{H}}^* = \mathfrak{H}^*$ yields $\tilde{C}_{LM} = C_{LM}$, hence $\tilde{\mathscr{D}}_L{}^M = \mathscr{D}_L{}^M$, in (t, y)). We conclude that (c) holds if and only if $\alpha^r{}_L(t') = \tilde{\alpha}^r{}_L(t')$ for $t' \leqslant t$, i. e. if and only if $\omega_{rs}(t') = \delta_{rs}$ for $t' \leqslant t$.

Thus we can further conclude that the rotational Lagrangian counterparts of $\mathfrak{H}$ and $\tilde{\mathfrak{H}}$ coincide if and only if $\mathfrak{H}$ and $\tilde{\mathfrak{H}}$ are translationally equivalent. q.e.d.

Let us show that by convention $(57.1\text{a})_2$ the components $q^{(r)}$ of q^ρ and $X^{(r)(s)}$ of $X^{\rho\sigma}$ in the triad $R^\rho{}_A \lambda^{*A}_{(r)}$ fulfill the conditions

$$q^{(r)} \lambda^{*A}_{(r)} = q^\rho R_\rho{}^A = \mathscr{D}^A{}_L q^L_* \quad (\lambda^{*A}_{(r)} \lambda^{*(r)}{}_B = \delta^A{}_B) \tag{82.10}$$

and

$$X^{(r)(s)} \lambda^{*A}_{(r)} \lambda^{*B}_{(s)} = X^{\rho\sigma} R_\rho{}^A R_\sigma{}^B = \frac{1}{\mathscr{D}} \mathscr{D}^A{}_L \mathscr{D}^B{}_M Y^{LM} \tag{82.11}$$

respectively. Indeed by the definition of $q^{(r)}$, by $(82.10)_3$, by $(57.1\text{a})_2$ and $(54.5)_1$, and by $(54.7)_1$, respectively, we deduce

$$q^{(r)} \lambda^{*A}_{(r)} = q^\rho R_\rho{}^B \lambda^{*(r)}{}_B \lambda^{*A}_{(r)} = q^\rho R_\rho{}^A = R^\rho{}_B \mathscr{D}^B{}_L q^L_* R_\rho{}^A = \mathscr{D}^A{}_L q^L_* .$$

We can deduce (82.11) by the same procedure except that $(57.1\text{a})_2$ has to be replaced with $(59.2)_{1,2}$. The relations (82.10) and (82.11), just proved yield

$$q^L_* = \overset{-1}{\mathscr{D}}{}^L{}_A \lambda^{*A}_{(r)} q^{(r)}, \qquad Y^{LM} = \mathscr{D} \overset{-1}{\mathscr{D}}{}^L{}_A \overset{-1}{\mathscr{D}}{}^M{}_B \lambda^{*A}_{(r)} \lambda^{*B}_{(s)} X^{(r)(s)} \tag{82.12}$$

and the analogues of $(82.12)_1$ hold for j^L_*, P^L_*, and M^L_*.

Now suppose that we know the actual Lagrangian history $\mathfrak{H}^*$ expressed by (81.5) for the material point y^L and the instant t. This history is independent of the above orthonormal triad $\lambda^{*L}_{(A)}$ at y^L. Then we can fix this triad arbitrarily. After having done this we can determine the Fermi triad $f_{(r)}{}^\rho(t')$ considered in § 80 for $t' \leqslant t$, and we can construct a thermo-electromagnetokinematic history, $\tilde{\mathfrak{H}}$, whose Lagrangian counterpart is $\mathfrak{H}^*$. By Theorem 82.1 $\tilde{\mathfrak{H}}$ is equivalent to the actual thermo-electromagnetokinematic history $\mathfrak{H}$ at the event point $x^\rho = x^\rho(t, y^L)$. By the principle of material indifference the values of the constitutive functionals $G_{(1)}$ to $G_{(6)}$ for $\tilde{\mathfrak{H}}$, $\lambda^{*L}_{(A)}$, and the tensor $a^*_{LM} = \hat{a}^*_{LM}(y)$ which is also supposed to be known on $D^*_3 = \phi(\mathscr{C})$, are the values of $G_{(1)}$ to $G_{(6)}$ for $\mathfrak{H}$, $\lambda^{*L}_{(A)}$, and a^*_{LM}; hence they are the components of the actual values of the magnitudes (81.3) at x^ρ in the triad $\lambda^\rho{}_{(r)} = R^\rho{}_L \lambda^{*L}_{(r)}$ (where $R^\rho{}_L$ need not be known). Then by using (82.12) and the analogues of $(82.12)_1$ for j^L_*, P^L_*, and M^L_*, we can calculate the present values of

$$\rho, \quad Y^{LM}, \quad q^L_*, \quad j^L_*, \quad P^L_*, \quad M^L_* . \tag{82.13}$$

These values are independent of the triad $\lambda^{*L}_{(A)}$; hence they are functions of $\mathfrak{H}^*$, y^L, and a^*_{LM} alone. Thus we have proved the following theorem:

Theorem 82.3. *The present values at* $x^\rho = \hat{x}^\rho(t, y)$ *of the magnitudes* (82.13) *are given by six constitutive functionals* $G^*_{(1)}$ *to* $G^*_{(6)}$ *of the Lagrangian thermo-electromagnetokinematic history* $\mathfrak{H}^*$, *that also depend on* y^L *(explicitly) and on* $\hat{a}^*_{LM}$ *(implicitly):*

$$\begin{aligned} &\rho = G^*_{(1)}(\mathfrak{H}^*, y), &&Y^{LM} = G^{*LM}_{(2)}(\mathfrak{H}^*, y), &&q^L_* = G^{*L}_{(3)}(\mathfrak{H}^*, y), \\ &j^L_* = G^{*L}_{(4)}(\mathfrak{H}^*, y), &&P^L_* = G^{*L}_{(5)}(\mathfrak{H}^*, y), &&M^L_* = G^{*L}_{(6)}(\mathfrak{H}^*, y). \end{aligned} \tag{82.14}$$

This theorem allows us to carry over to special and general relativity most of the principles stated in classical physics for constitutive functionals of materials with memory—e.g. the one of the fading memory—and most of their consequences concerning the form of constitutive equations.

§ 83. On the Axiomatic Foundations of the Preceding Theory. Primitive Notions and First Axioms

Usually, books on special or general relativity do not undertake the task of an axiomatic presentation of their subjects, even in case they are well known and rather recent.[10)] However people are becoming more and more interested in axiomatic presentations. Therefore we now briefly speak of an axiomatic presentation of general relativity which substantially is the one considered in Bressan [1963b] and (especially) [1964b]. For the sake of simplicity we shall not take into account the possibility of ruptures, unlike Bressan [1964b]; furthermore we shall not be much interested in regularity conditions.

The complex character of our axiomatic theory is strongly connected with its possibility of dealing with materials with memory. Its reduction to elastic bodies is rather obvious and involves considerable simplifications.

The axiomatization that we have in mind is informal and physical in that the primitive concepts are to be characterized in it before stating the axioms[11)], and in an operational way, e.g. like Synge [1960], [1965]—cf. § 15.

Among the primitive notions we include those of *event point* and *material point*. Of course we identify the set of event points with S_4 and the set of material points with the body $\mathscr{C}$; furthermore we have a view to supposing suitable conditions on these sets.

The following primitive notions are relations and it is useful to present them in the form of matrices:[12)]

a) *The one-to-one mapping ϕ of S_4 onto the set of 4-tuples of real numbers is an admissible frame for S_4.*

b) *The real numbers $g_{\alpha\beta}=\tilde{g}_{\alpha\beta}(\phi,x)$, regarded as functions of the admissible frame ϕ and the real numbers x^ρ, are the components in ϕ of the chronometric tensor.*

c) *l is the world line of the material point P^*.*

By material point we mean an element of a 3-dimensional continuous body. In this theory matter can be defined as the set of material points. Bodies are portions of matter having certain structural properties that can be expressed after the remaining primitive concepts are listed and the axioms are stated.

We accept, among others, the assertion (A_1) to (A_5) in § 15—where "being suitably smooth" is understood in the way specified after assertion (A_2)—as axioms on the primitive notions (a) to (c), event point, and material point. As a consequence of (A_1) and (A_2) S_4, the admissible frames, and the chronometric tensor determine a Riemann space $(S_4, K, \mathfrak{g})$—see Appendix A2.

Let us remember that the notions (a) and (b) have an operational physical characterization [§ 15] which is very complex and requires the use of some auxiliary notions such as those of past and future, particles, clocks, and proper time. Incidentally the notions (a) and (b) are intimately connected to one another as it appears from the assertions (A_1) and (A_2) and from their intuitive characterization [§ 15].

Let us emphasize that to consider some properties of the primitive notions for characterizing them intuitively is quite different from asserting some properties of the same notions by means of axioms. This difference is analogous to the one between the uses of the metalanguage and the object language in logic.

By the afore-mentioned difference, to assert by an axiom (or a theorem) a property of a primitive notion, after having considered the same property for characterizing the notion intuitively, is not only non-circular but may be compulsory for constructing a rigorous and efficient axiomatic theory. In particular the property (c) in § 15 was used to characterize the chronometry, and says that $ds^2>0$ along the possible world-lines of material points. This use is not sufficient to imply within the axiomatic theory that $ds^2>0$ along world lines of material points. The axioms must be framed in such a way that this fact is deducible in the axiomatic theory—see axiom (A_4) in § 15.

In accordance with the foregoing considerations, as we remembered above, the notions of proper time, past, and future on the one hand had an auxiliary role in the characterization of the notions (a) and (b); on the other hand they had to be reintroduced in the axiomatic theory; they were substantially defined in terms of event point and the notions (a) and (b) [§ 15].

Let us add that from a strictly logical point of view it must be considered that our way of presenting primitive notions such as (a) and (b) contains implicitly some axioms stating elementary (structural) properties of the same notions.

More precisely from a strictly logical point of view it is better to replace the matrices (a) and (b) above with

(a′) *ϕ is an admissible frame,*

(b′) *$\tilde{g}$ is the field of the covariant components of the chronometric tensor in the admissible frame ϕ,*

and to add the following two axioms:

(A_1^*) *If ϕ is an admissible frame, then it is a one-to-one mapping of S_4 onto the set $R^{(4)}$ of four-tuples of real numbers,*

(A_2^*) *If $\tilde{g}$ is as in* (b′), *then it is a function whose counterdomain is the Cartesian product of the set of admissible frames with $R^{(4)}$.*

We now consider other primitive matrices:

(d) *T is the absolute temperature at the event-point $\mathscr{E}$ (belonging to a world-line).*

(e) *$\bar{\rho}$ is the proper density of electric charge (in given units) at the event point $\mathscr{E}$ (belonging to a world-line)* [§ 34].

(f) [(h)] E_ρ $[H_\rho]$ *is the* ρ*-th covariant component in the admissible frame* ϕ, *of the electric [magnetic] field at* $\mathscr{E}$ *with respect to an observer of 4-velocity* w^ρ.

Experiments to measure $T, \bar{\rho}, E_\rho$, and H_ρ are well known. In particular Stratton [1941, § 23e] points out a procedure for measuring E_ρ and D_ρ [§ 34] in case $P_\rho = D_\rho - E_\rho \neq 0$.[13)] Incidentally in the present theory D_ρ, P_ρ, B_ρ, and M_ρ, as well as j^α, ρ, and other magnitudes, will be introduced by means of constitutive equations, so that they are not primitive concepts and in particular no intuitive physical characterization is wanted for them.

Of course for the primitive notions (d) to (h) some structural axioms are also understood.

From a strictly logical point of view the following structural axioms have to be stated.

(A_3^*) *Matrix* (d) [(e)] *implies that* $\mathscr{E}$ *is an event point and that* $T[\bar{\rho}]$ *is a real number*

(A_4^*) *Matrix* (f) [(h)] *implies that* $\mathscr{E}$ *is an event point,* ϕ *is an admissible frame,* $E_\rho[H_\rho]$ *is a real number as well as* w^ρ, *and* $g_{\alpha\beta} w^\alpha w^\beta = -1$ *where* $g_{\alpha\beta} = \tilde{g}_{\alpha\beta}[\phi, \phi^0(\mathscr{E}), \ldots, \phi^3(\mathscr{E})]$ *(and matrix* (b) *or* (b′) *holds).*

Of course the notions (d) to (h) are not necessary for a relativistic theory of kinematics and gravitation.

§ 84. On Kinematic Axioms and the Notion of Physical Possibility

Axioms (A_1), (A_2), (A_4) and (A_5) [§ 15] imply the existence of a one-to-one mapping ϕ_* of $\mathscr{C}$ into a non-empty domain $D^*_{(3)}$ in $R^{(3)}$—cf. footnote 5 in Chapter 2—and the existence of a real valued function, $a^*_{rs} = \hat{a}^*_{rs}(y^1, y^2, y^3)$, with domain $D^*_{(3)}$ for which it is physically possible that there is an admissible frame ϕ such that, for every $P^* \in \mathscr{C}$,

(i) $\phi^L_*(P^*)$ equals the L-th co-ordinate y^L in ϕ for the intersection of the world-line of P^* with the hypersurface $x^0 = 0$, and

(ii) equality $(52.1)_1$ holds (in the obvious sense).

Then ϕ_* can be called a *(regular) material frame and* $\hat{a}^*_{rs}$ *the field of the components in* ϕ_* *of the material metric tensor.* We shall say that $D^*_{(3)}$, ϕ_*, *and* $\hat{a}^*_{LM}$ *represent a reference configuration* [§ 52].

Let $(52.2)_1$ be a one-to one mapping of $R^{(1)} \times D^*_{(3)}$ into $R^{(4)}$, where $\times$ denotes cartesian product. We say that this mapping *represents the (actual) motion* $\mathscr{M}$ *of* $\mathscr{C}$ *in the admissible frame* ϕ *and the material frame* ϕ_* if (i) $\partial x^0/\partial t > 0$ in $R^{(1)} \times D^*_{(3)}$ and (ii) for $P^* \in \mathscr{C}$ and $y^L = \phi^L_*(P^*)$ the event point $\phi^{-1}[\hat{x}^\alpha(t, y^1, y^2, y^3)]$ describes the world-line of P^* when t describes $R^{(1)}$.

The present tract is in accordance with the following regularity axiom of kinematics:

(A_6) *If* ϕ_* *is a (regular) material frame, then we can choose the admissible frame* ϕ *and the functions* (52.2) *in such a way that* $\hat{x}^0(t, y^L) \equiv t$, *these functions represent the motion of* $\mathscr{C}$ *in* ϕ *and* ϕ_*, *and for them the regularity assumptions* (i) *to* (iii) *considered in* § 52 *hold.*

Of course (A_6) and (A_2) [§ 15] imply that the regularity assumptions (i) to (iii) in § 49 hold in connection with every admissible frame ϕ.

Let us now remark that axiom (A_6) involves the notion of possibility through the material frame ϕ_*. Furthermore all axioms above are understood to hold *necessarily*, i.e. in every physically possible process of the whole matter, or briefly in every (physically) possible world.[14)]

From a strictly logical point of view, by the considerations above the present axiomatic theory, which is very similar to the one considered in Bressan [1964b], is to be based on a theory of modal logic (which just deals with possibility); furthermore, at least in the past decades, people usually disliked axiomatizations which are not based on extensional logic, and this is mainly due to the fact that, as it appears from Hughes and Cresswell [1968], at least the theories of modal logic published before 1968 are not developed enough.[15)] Therefore in Bressan [1964b] it is pointed out that to the axiomatic relativistic theory presented in that paper we can apply the procedure introduced in Bressan [1962] to translate into an extensional language the ordinary modal language used by Painlevé [1922] and Signorini [1954, Chapt. X] in an axiomatization of classical mechanics based on some ideas of E. Mach (and P. Painlevé).[16)]

Incidentally a feature of the procedure presented in Bressan [1962] which favours its applicability to the present axiomatic theory is that in both works the physical possibility of a proposition p is understood as a property of p to be tested experimentally as well as e.g. in the theory of modal logic presented by Burks [1951]—and it is not understood e.g. as the logical compatibility of p with all physical axioms.[17)]

If we apply to our theory the afore-mentioned procedure presented in Bressan [1962], then on the one hand the concept of event point is used as an absolute concept and usual theories of general relativity do not determine any privileged absolute concept of event point [§ 15] except the few theories where Fock's conjecture [§ 10] or something like that is accepted; furthermore the resulting language would be an unusual extensional language.

Let us consider the present relativistic axiomatic theory from a strictly logical point of view. It has the afore-mentioned two defects if we base it on Bressan [1962]; but they disappear if we understand it as based on Bressan [1972d] where a semantical and syntactical theory is constructed from a general point of view[18)] and where the object language, denoted by ML^ν, is translated into an extensional language, so that the use of ML^ν is about as unobjectionable as the one of extensional languages.[19)]

§ 85. Conservation Equations and Maxwell Equations in Our Axiomatic Theory

We complete our axiomatic basis by means of one fundamental axiom, (A_7) below, which introduces the conservation equations and the Maxwell equations in our theory which is fit to treat materials with memory (and which can be reduced in the elastic case by obvious simplifications). After axiom (A_7) the principle of material indifference can be stated practically as in § 81.

We shall not be completely explicit about regularity conditions, among other things, because they depend on the problem being considered.

Our relativistic axioms will not imply the validity of the requirement of just setting mentioned by Truesdell and Toupin [1960, p. 701]. According to it constitutive equations connecting a given set of variables should be such that, when combined with all conservation principles (affecting the same variables), there should result a unique solution corresponding to initial and boundary data, and a solution that depends continuously on that data.

However our axioms seem compatible with this requirement, which unfortunately can rarely be used even in classical physics because it requires much mathematical labor even in the simplest classes of boundary value problems. The situation with this requirement in relativity is even worse and we shall not consider it.

We now define *admissible* and *actual (thermo-electromagnetokinematic) processes.*

Definition 85.1. *Let $D^*_{(3)}$, ϕ_*, and $\hat{a}^*_{LM}$ represent a reference configuration* [§ 84]. *We shall say that the functions $\hat{g}_{\rho\sigma}$, $\hat{x}^\rho$, $\hat{T}$, $\hat{\bar{\rho}}$, $\hat{E}_\rho$, and $\hat{H}_\rho$ represent or constitute an admissible (thermo-electromagnetokinematic) process with respect to ϕ_* and $\hat{a}^*_{LM}$ if conditions* (a) *to* (c) *below hold.*

(a) *The functions $E_\rho = \hat{E}_\rho(x)$, $H_\rho = \hat{H}(x)$, $g_{\rho\sigma} = \hat{g}_{\rho\sigma}(x)$, $g_{\rho\sigma,\alpha}$, and $g_{\rho\sigma,\alpha\beta}$ are defined on $R^{(4)}$ and are continuously differentiable in $R^{(4)}$ except that $E_{\rho,\alpha}$, $H_{\rho,\alpha}$, and $g_{\rho\sigma,\alpha\beta\gamma}$ may have a discontinuity of the first kind on some hypersurfaces.*

(b) *The functions $\hat{x}^\rho$*—i.e. $(52.2)_1$—*represent a transformation of $R^{(1)} \times D^*_{(3)}$ into a domain, say D_4, in $R^{(4)}$*—cf. *footnote* 5 in Chapter 2—*and they fulfill the regularity conditions* (i) *to* (iii) *assumed below* (52.2).

(c) *The functions $T = \hat{T}(x)$ and $\bar{\rho} = \hat{\bar{\rho}}(x)$ are defined on D_4*—*see* (b); *furthermore $\bar{\rho}$, T, $T_{,\alpha}$, and $T_{,\alpha\beta}$ are continuous in $R^{(4)}$ except that $\bar{\rho}$ and $T_{,\alpha\beta}$ may have a discontinuity of the first kind on some hypersurfaces.*

Definition 85.2. *Let the functions $\hat{g}_{\rho\sigma}$, $\hat{x}^\rho$, $\hat{T}$, $\hat{\bar{\rho}}$, $\hat{E}_\rho$, and $\hat{H}_\rho$ represent an admissible thermo-electromagnetokinematic process with respect to ϕ_* and $\hat{a}^*_{LM}$; furthermore let $x^r(t, y) \equiv y^r$. We say that the functions $\hat{g}_{\rho\sigma}$ to $\hat{H}_\rho$ above represent the actual thermomagnetokinematic process with respect to ϕ_* and $\hat{a}^*_{LM}$, if there is an admissible frame ϕ which fulfills the following conditions:*

(a) *$\hat{g}_{\rho\sigma}$ is the field of the components of the chronometric tensor in ϕ.*

(b) *The functions $\hat{x}^\rho$ represent the actual motion of matter with respect to ϕ_* and ϕ.*

(c) *For* $x \in R^{(4)}$, $\mathscr{E} = \phi^{-1}(x)$, $T = \hat{T}(x)$, $\bar{\rho} = \hat{\bar{\rho}}(x)$, $E_\rho = \hat{E}_\rho(x)$, *and* $H_\rho = \hat{H}_\rho(x)$, *the quantities* T, $\bar{\rho}$, E_ρ, *and* H_ρ *have the physical meanings expressed by the matrices* (d) *to* (h) *in* § 83 *(*i.e. *these matrices hold) for* $w^\rho = \delta^\rho{}_0$.[20)]

We now define *admissible systems of constitutive functionals*, where for the sake of simplicity the order n of the history $\mathfrak{H}$ being referred to is assumed to equal 1.

Definition 85.3. *Let* $D^*_{(3)}$, ϕ_*, *and* $\hat{a}^*_{LM}$ *represent a reference configuration and let* $\lambda^{*L}{}_{(A)} = \hat{\lambda}^{*L}{}_{(A)}(y)$ $(y \in D^*_{(3)})$ *be an orthonormal triad which is twice continuously differentiable with respect to* y^L. *We say that the functionals* $G_{(1)}$ *to* $G_{(6)}$ *constitute an admissible system of constitutive functionals of order* $n=1$ *(for the magnitudes* ρ, $X^{\rho\sigma}$, q^ρ, j^ρ, P^ρ, *and* M^ρ*) if they have the form* (81.4) *for* $y \in D^*_{(3)}$ *and* $n=1$, *and if they are suitably regular (*e.g. *they are twice continuously differentiable with respect to all arguments where continuity and differentiability with respect to the history* $\mathfrak{H}$ *of order* $n=1$ *is to be related to a suitable Hilbert space).*

Let us remark that the admissible entities introduced by definitions 85.1 and 85.3 are purely mathematical entities; they are absolute, i.e. independent of phenomena and in particular of notions such as admissible frame, chronometric tensor, and so on.

In order to simplify our basic dynamical axiom (A_7) below we introduce another mathematical entity, the notion of *world process* corresponding to an admissible system of constitutive functionals, $G_{(1)}$ to $G_{(6)}$, to an admissible thermo-electromagnetokinematic process, and to certain reference entities.

Definition 85.4. *We assume that* $G_{(1)}, \ldots, G_{(6)}$ *is an admissible system of functionals of order* $n=1$ *relative to* ϕ_*, $\hat{a}^*_{LM}$, *and* $\hat{\lambda}^{*L}{}_{(A)}$, *that* $D^*_{(3)} = \phi_*(\mathscr{C})$, *and that* $\hat{g}_{\rho\sigma}$, $\hat{x}^\rho$, $\hat{T}$, $\hat{\bar{\rho}}$, $\hat{E}_\rho$, *and* $\hat{H}_\rho$ *constitute an admissible thermo-electromagnetokinematic process, say* $\mathscr{P}$, *relative to* ϕ_*.

We say that the functions $\hat{g}_{\rho\sigma}$, $\hat{F}_{\rho\sigma}$, $\hat{f}_{\rho\sigma}$, $\hat{J}_\alpha$, *and* $\hat{\mathscr{U}}_{\alpha\beta}$ *constitute the world process corresponding to* $\mathscr{P}$, ϕ_*, $\hat{a}^*_{LM}$, $\hat{\lambda}^{*L}_{(A)}$, $\hat{k}^*$, *and* $G_{(1)}$ *to* $G_{(6)}$ *if for every* $x \in R^{(4)}$ *we have*

$$\hat{F}_{\alpha\beta}(x) = F_{\alpha\beta}\,, \quad \hat{f}_{\alpha\beta}(x) = f_{\alpha\beta}\,, \quad \hat{\mathscr{U}}_{\alpha\beta}(x) = \mathscr{U}_{\alpha\beta}\,, \quad \hat{J}_\alpha(x) = J_\alpha \tag{85.1}$$

where either of the following cases hold:

Case 1. $x \in D_4$, *where* D_4 *is the counterdomain of the transformation* $x^\rho = \hat{x}^\rho(t, y)$ $[(t, y) \in R^{(1)} \times D^*_{(3)}]$, *and moreover if* $\mathfrak{H}$ *is the admissible thermo-electromagnetokinematic history at* x^ρ *constructed according to* (81.1) *starting out from* $\hat{g}_{\rho\sigma}, \hat{x}^\rho, \ldots, \hat{H}_\rho$, *if* ρ, $X^{(r)(s)}$, $q^{(r)}$, $j^{(r)}$, $P^{(r)}$, *and* $M^{(r)}$ *are the values of the functionals* $G_{(1)}$ *to* $G_{(6)}$ *for* $\mathfrak{H}$, y^L, $\lambda^{*L}_{(A)} = \hat{\lambda}^{*L}_{(A)}(y)$, *and* $\hat{a}^*_{LM}$ *according to* $(81.4)_1$, *if* $R^\rho{}_L$ *is calculated by means of* $(53.7)_1$, $(53.15)_1$, (54.4), *and* $(54.5)_2$ *from* $\hat{g}_{\rho\sigma}$ *and* $\hat{x}^\rho$, *and if* $X^{\rho\sigma}$ *and* q^ρ *are determined by*

$$X^{\rho\sigma} = R^\rho{}_L R^\sigma{}_M \lambda^{*L}_{(r)} \lambda^{*M}_{(s)} X^{(r)(s)}\,, \qquad q^\rho = R^\rho{}_L \lambda^{*L}_{(r)} q^{(r)} \tag{85.2}$$

and j^ρ, P^ρ, *and* M^ρ *are determined by obvious analogues of* $(85.2)_2$, *then among the right hand sides of equalities* (85.1) *the quantities* $F_{\alpha\beta}$ *and* $f_{\alpha\beta}$ *are given by* (34.7) *and* $(34.8)_{2,3}$, ${}^7E_{\alpha\beta}$ *by* $(40.18)_2$, $\mathscr{U}_{\alpha\beta}$ *by* (43.5) *for* $s=3$, *and* J_α *by*

$$J_\alpha = j_\alpha + \bar{\rho} u_\alpha . \tag{85.3}$$

Case 2. $x \notin D_4$ *and if we set* $D_\rho = E_\rho = \hat{E}_\rho(x)$, $B_\rho = H_\rho = \hat{H}_\rho(x)$, *and* $P_\rho = M_\rho = 0$, *then the quantities* $F_{\alpha\beta}$ *and* $f_{\alpha\beta}$ *are given by* (34.7) *and* $(34.8)_{2,3}$ *(as in* Case 1*) and in addition we have* $\mathscr{U}_{\alpha\beta} = {}^7E_{\alpha\beta}$ *and* $J_\alpha = 0$, *where* $(40.18)_2$ *holds.*

We are now able to state the following basic dynamical axiom:

(A_7) *There is a reference configuration represented by* $D^*_{(3)}$, ϕ_*, *and* $\hat{a}^*_{LM}$ *(with* $D^*_{(3)} = \phi_*(\mathscr{C})$*), a twice continuously differentiable field* $\hat{\lambda}^{*L}_{(A)}$ *of orthonormal triads (with respect to* $\hat{a}^*_{LM}$*), and an admissible system* $G_{(1)}$ *to* $G_{(6)}$ *of constitutive functionals of order* $n=1$, *such that necessarily (*i.e. *in every physically possible world) we can choose an admissible frame* ϕ *for which*

(i) *the actual thermo-electromagnetokinematic process* $\mathscr{P} = \{\hat{g}_{\rho\sigma}, \hat{x}^\rho, \hat{T}, \hat{\bar{\rho}}, \hat{E}_\rho, \hat{H}_\rho\}$ [Definition 85.1] *with respect to* ϕ, ϕ_*, *and* $\hat{a}^*_{LM}$ *exists, and*

(ii) *the world process* $\hat{g}_{\alpha\beta}$, $\hat{F}_{\alpha\beta}$, $\hat{f}_{\alpha\beta}$, $\hat{J}_\alpha$, $\hat{\mathscr{U}}_{\alpha\beta}$, [Definition 4] *corresponding to* $\mathscr{P}$, ϕ_*, $\hat{a}^*_{LM}$, *and* $\hat{\lambda}^{*L}_{(A)}$ *fulfills the Einstein gravitation equations* $(23.1)_1$ *and the Maxwell equations* (35.1).

An admissible system $G_{(1)}$ to $G_{(6)}$ of constitutive functionals such as the one considered in axiom (A_7) can be simply called *system of constitutive functionals.* Now we can assert the principle of material indifference [§ 81] on the system of constitutive functionals (81.4).

Incidentally axiom (A_7) holds also in connection with the physically possible process by which the reference configuration is determined. For this process $\rho = G_{(1)}(\mathfrak{H}, y)$ equals $k^* = k^*(y)$.

Let us further remark that in order to prove the uniqueness of the admissible set of functionals fulfilling axiom (A_7), suitable possibility assumptions should be postulated. Reasonably they should assert the physical possibility of suitable "local processes" for the typical material point. However, especially because of the general kind of the materials being considered, this topic, which appears to be intimately connected with the requirement of just setting, need further investigation, as does the same requirement in classical physics.

If $\mathscr{C}$ is assumed to be elastic [§ 63], then certain processes are physically possible by definition and the uniqueness of the functionals $G_{(1)}$ to $G_{(6)}$ appears provable.

In spite of the lack of the proof of uniqueness above in the general case, our axioms (A_1) to (A_7) provide a sufficient axiomatic basis for many actual treatments of general relativity and for the one presented in this tract so far, in connection with materials of order 1—cf. footnote 15 in Chapter 1.

Footnotes to Chapter 9

1) For both topics we substantially follow Bressan [1964b]. However Theorem 82.2, an extension of Theorem 82.1, is new.
2) The importance of Noll's rigorous version of the principle of material (frame) indifference appears, among other things, from the fact that—as is remarked in Truesdell and Noll [1965, 19A]—at least two of the persons who ought have known that principle better, Truesdell and Zaremba, propose non-invariant theories, which they corrected later on the basis of Noll's version of the principle.
3) The slight differences between the versions of the objectivity principles presented here and in Bressan [1964b] are due in part to a change in Bressan's view about relativistic thermodynamics —cf. Bressan [1967a]—and in part to reasons of uniformity: the relativistic n-th position gradient used in Bressan [1964b], $\tilde{\bar{x}}^{\rho}{}_{L_1 \ldots L_n}$, is calculated on the totally geodesic hypersurface Σ_{tg} orthogonal to the actual 4-velocity. In this book it is replaced by $\alpha^{\rho}{}_{L_1 \ldots L_n}$—cf. the second part of § 80. A study of the $\tilde{\bar{x}}^{\rho}{}_{L_1 \ldots L_n}$'s and their relations to the $\alpha^{\rho}{}_{L_1 \ldots L_n}$'s is performed by A. Grioli [1971a].
4) That Bragg's and Söderholm's formulations are not completely similar with the classical formulation of the objectivity principle has this counterpart. The latter formulation is independent of the principle of local action—cf. Truesdell and Noll [1965, pp. 44, 56]—while the former formulations depend on this principle.

Söderholm [1970] considers constitutive functionals that are functions of the localization of the motion of $\mathscr{C}$ at P^*. The (general) corresponding Lagrangian form is proved by the author to be equivalent to the original one and to coincide with Bressan's in the cases dealt with by both theories.
5) Bressan [1969] considers certain 1-dimensional equilibrium problems of a heavy elastic crystal, $\mathscr{C}$, parallel with the gravity acceleration g_r. He shows that, when $\mathscr{C}$ is regarded as a continuum body, the stretch u is a function $\alpha(\tau, g)$ of the tension τ and the modulus g of g_r. In particular $u = \alpha(\tau, 0)$ is the corresponding constitutive equation according to ordinary elasticity. The following rough estimate for the very slight influence of gravity on elasticity is reached:

$$\alpha(\tau, g) - \alpha(\tau, 0) \simeq \tfrac{1}{12} \frac{\partial^2 \alpha(\tau, 0)}{\partial \tau^2} k^{4/3} M^{2/3} N^{-2/3} g^2 \tag{$*$}$$

where k is the 3-dimensional mass density, M is the ionic weight, and N is Avogadro's number. In c.g.s. units and for $g = 980$, the right hand side of ($*$) roughly equals $10^{-11} k^{4/3} M^{2/3} \partial^2 \alpha(\tau, 0)/\partial \tau^2$.

The constitutive equation $u = \alpha(\tau, g)$ is in slight disagreement with the principle of material indifference.

In terms of classical physics and in connection with constitutive equations Bressan [1972c] considers a partial version of the local equivalence principle, briefly PLE-principle (which is basic for the construction of general relativity) and the objectivity principle divided into a translational and a rotational part, in order to present some well known theories from a more general point of view and to construct some generalizations of these theories.

Among other things the opportunity of combining the PLE-principle with only one of the aforementioned parts of the objectivity principle is considered in cases where the other part is in slight disagreement with experiment. This in the case with the Barnett effect and the Einstein- de Haas effect.
6) Lianis [1973] states a version of something like the rotational part of the objectivity principle in general relativity, apparently for simple materials, using a covariance condition. He also takes the apparent metric (substantially introduced by Bragg [1965]) into account.

Incidentally Lianis [1973] speaks of Bressan's paper [1964b] (in Italian) as if its content were included in Bressan [1967a] (in English)—cf. Bressan [1974c].
7) Finzi [1931] combined somehow the concepts of Einstein's theory of relativity with those of Volterra's classical theory of materials with memory. However he did not write any mechanic or electromagnetic constitutive equation. So to speak, among other things the gravitational equations proposed by Finzi [1931] consist of an hereditary version of Poisson's law relating mass density and gravitational potential.
8) Of course a theory of elasticity of order $n \geqslant 2$ in general relativity must be based on an expression of the total energy tensor $\mathscr{U}_{\alpha\beta}$ different from (24.3) or (43.5). Such a theory has been constructed for materials capable of couple stresses [§ 87] which are of order $n = 2$. In this case $\mathscr{U}_{\alpha\beta}$ has the expression (87.3).

9) Einstein-De Haas experiments on gyromagnetic phenomena can be described within a generalized classical theory of continuous bodies, where a density of intrinsic moment of momentum is taken into account. C.A. Grioli [1972] considers from this point of view some of the afore-mentioned experiments and some ideal experiments similar to them. He shows that some among them are unable to decide whether the objectivity principle or a certain other natural principle holds, and that some other experiments are able to do this but by very little terms.

10) Among these books let us mention Synge [1960] and [1965].

11) Kleene [1952, p. 28] speaks of *informal* or *material axiomatics* in case the primitive notions are known prior to the axioms, and of *formal* or *existential axiomatics* in case the primitive notions are defined, in a certain sense, by the axioms.

12) Here—as well in logic—by matrix we mean an expression which contains variables and becomes a proposition after substituting suitable terms for the variables.

13) A determination of D_ρ or E_ρ at the material point P^* when P^* occupies the event point $\mathscr{E}$ can be made as follows, at least in the static case.

Let ξ be an arbitrary spatial axis through $\mathscr{E}$. By removing some matter from the body let us make an infinitesimal cylindrical hollow with the center in P^*, the revolution axis coinciding with ξ, and the height h infinitesimal of a larger order than the basis σ.

Let us remember that $D_\rho = E_\rho$ in vacuo and that across the discontinuity surfaces of the dielectric tensor $\eta_{\alpha\beta}{}^{\rho\sigma}$—cf. (34.12)—the tangential component $E_\rho{}^{(t)}$ of E_ρ and the normal component $D_\rho{}^{(n)}$ of D_ρ are continuous.

Let us determine the electric field E'_ρ in the hollow, precisely at $\mathscr{E}$, by measuring the acceleration of a test particle with infinitesimal mass and charge, which initially is in $\mathscr{E}$ with zero "spatial velocity".

Since the hollow is infinitesimal, the electric field in the matter element at the intersection of the positive axis ξ with the boundary of the hollow equals, up to negligible quantities, the electric field at $\mathscr{E}$ in case the hollow does not exist. Then, by our assumptions on the infinitesimal hollow (in particular on h and σ) and by the afore-mentioned properties of D_ρ and E_ρ in vacuo and in connection with $\eta_{\alpha\beta}{}^{\rho\sigma}$, the component of $E_\rho[D_\rho]$ normal to [parallel with] ξ in the absence of the hollow coincides with the same component of E'_ρ in the presence of the hollow.

By the arbitrariness of ξ such ideal experiments can determine E_ρ and D_ρ at $\mathscr{E}$ at least in the static case.

14) As is well known, *the proposition p is possible*, briefly $\Diamond p$ [*p is necessary*, briefly Np] can be expressed in terms of necessity [possibility] as follows: $\sim N \sim p$ [$\sim \Diamond \sim p$], where "$\sim$" stands for negation.

15) Hughes and Cresswell [1968, p. 210] say that "the topics of identity and description are among the most difficult in modal logic and in the present state of the subject are still full ob obscurities and unsolved problems", and that some difficulties arise in a straightforward set theory based on modal logic.

16) Bressan [1962] substantially considers the set PPW of physically possible worlds (briefly PP-worlds); furthermore he replaces the contingent matrix *P is the position of the material point P* at the instant t* with

P is the position of P at the instant t in the PP-world w.*

Similar transformations are performed on other contingent relations. Thus every non-modal sentence A^* of the ordinary modal language used e.g. by Painlevé [1922] is transformed into a sentence $A_{(w)}$ where the letter w may occur but only free—i.e. w is not quantified in $A_{(w)}$.

Furthermore the modal sentence A^* *is physically necessary [possible]* is replaced by the non-modal sentence $A_{(w)}$ *holds for every [for a suitable]* $w \in \mathrm{PPW}$.

The application of the procedure just hinted at to our relativistic axiomatic theory is simply based on the transformation of the contingent primitive matrices (a) to (h) in § 83, which consists of adding the phrase "in the PP-world w"; thus for instance the matrix (c), i.e. *l is the world-line of the material point P**, is transformed into *l is the world-line of P* in the PP-world w.*

17) Bressan [1962] deals with the axiomatization of classical mechanics according to P. Painlevé [1922]. As a consequence the notion of possibility is essentially involved by the axioms themselves. Hence it would be circular to understand the possibility of the proposition p as its logical compatibility with all physical axioms.

18) In the modal theory for the general modal language ML^ν presented by Bressan [1972d], among other things, a solution has been given to all afore-mentioned unsolved problems hinted at by Hughes and Cresswell [1968],—cf. footnote 15 in Chapter 9—and in addition absolute concepts—such as

the privileged concept of event point considered in § 15—are studied from a purely logical point of view.

Let us add that if we base the present relativistic axiomatic theory on Bressan [1972d], then *material point* is to be understood as an absolute concept—as well as *real number* and all purely mathematical concepts; but *event point* can be treated as an extensional concept, where, following R. Carnap, we say that the property F is *extensional* if the following is the case:

If x happens to be equal to y and F holds for x, then F also holds for y (even in case x is not necessarily equal to y)—cf. Bressan [1972d, Def. 6.11].

Lastly let us say that every absolute concept determines the corresponding extensional concept, while the converse is not true.

[19] Now the opinion of people about modal logic is changing: more and more logicians are working on general theories of modal logic, and the importance of the notion of possibility in rigorous systematizations of scientific theories is gaining more and more consideration. This appears e.g. from the Symposium *Modality and the analysis of scientific propositions* held in the 1972 meeting of the philosophy of science association (PSA)—see Part 8 in PSA 1972, and in particular Suppes [1972] and Bressan [1974a, b].

Suppes [1972] says on p. 305 "I have gradually come to the position that modal concepts, especially as expressed in the use of probability concepts, are essential to standard scientific talk", cf. also Van Fraassen [1972].

[20] Since $w^\rho = \delta^\rho{}_0$, the matrices (f) and (h) in § 83 imply that at the arbitrary event point $\mathscr{E}$ the fields E_ρ and H_ρ are evaluated with respect to an observer joined to the frame ϕ. By the assumption $x^r(t, y) \equiv y^r$ this observer is joined to the body $\mathscr{C}$ for $\mathscr{E} \in W_{\mathscr{C}}$.

Chapter 10

Couple Stresses and More General Stresses

§ 86. Introduction

Let $d\sigma_\lambda$ $(u^\lambda d\sigma_\lambda = 0)$ represent the infinitesimal spatial oriented surface $d\sigma$ through the event point $\mathscr{E}$ (or x^ρ) in the world tube $W_{\mathscr{C}}$ of body $\mathscr{C}$—cf. (24.1). We consider the forces exerted by the matter elements contiguous to the negative face of $d\sigma$ on those contiguous to the positive one. In the (classical) theory of first order materials—cf. footnote 15 in Chapter 1—these forces are assumed (in first order approximation) to be equivalent to their resultant dR_α—cf. (24.2)—applied at $\mathscr{E}$. In other words the resultant $d\mathscr{M}_\alpha$ $(u^\alpha d\mathscr{M}_\alpha = 0)$ of the moments of the same forces with respect to $\mathscr{E}$, which is also called the intrinsic *torque* of these forces, is assumed to vanish.

In the last decades *polar materials*, i.e. materials for which $d\mathscr{M}_\alpha$ can be non-zero, were studied in classical physics by several authors—cf. Truesdell and Noll [1965, § 98, p. 389].[1)] In particular the first correct theories, linearized and general, were obtained by Areo and Kuvshinskii [1960] and by Grioli respectively. The latter theory (for finite deformations) is equivalent to the one of Toupin [1962], as Mindlin and Tiersten showed. Several papers on polar materials and more general materials (Cosserat continua) appeared in classical physics.[2)]

A theory of polar materials (with finite deformations) was constructed in (special or) general relativity by Bressan [1966a, b, c], [1968], and [1972b] and is mainly devoted to the thermo-elastic case. This theory has been extended by Pitteri [1975a, b] to general materials of order $n \geqslant 2$, in the adiabatic case.

Polar and more general materials are considered in general relativity by a Russian school, in particular by Sedov [1956a, b] and Berdichewski [1966]. These theories differ from Bressan's considerably.

In §§ 87 to 89 the basic relativistic equation for the case of couple stresses are dealt with from the Eulerian point of view; in §§ 90 to 94 couple stresses are dealt with from the Lagrangian point of view. In §§ 95—98 we extend to the case of couple stresses the variational principles (69.14) and (70.20) connected with the variation of space-time metric and the variation of world lines respectively, and the corresponding equivalence theorems. The same principles are extended to general materials of order $n \geqslant 2$ in § 99.

The analogue of Cauchy's fundamental theorem on dR_α (which allows us to introduce the stress tensor) holds for $d\mathcal{M}_\alpha$—cf. Truesdell and Toupin [1960, p. 543]. It says that at $\mathscr{E}$ there exists a spatial tensor $m^{\beta\lambda\mu}$ such that for every spatial surface through $\mathscr{E}$ represented by $d\sigma_\mu$ we have

$$d\mathcal{M}_\alpha = \overset{\perp}{\varepsilon}_{\alpha\beta\lambda} m^{\beta\lambda\mu} d\sigma_\mu, \qquad m^{(\beta\lambda)\mu} = 0 = m^{\beta\lambda\mu} u_\mu = m^{\beta\lambda\mu} u_\lambda. \tag{86.1}$$

Assuming that, given the oriented spatial surface $d\sigma_\mu$, we can measure $d\mathcal{M}_\alpha$, conditions (86.1) determine $m^{\beta\lambda\mu}$ uniquely. Each of conditions $(86.1)_{2,3,4}$ is essential for this.

The physical situation represented at $\mathscr{E}$ by $m^{\beta\lambda\mu}$ may be called a *couple stress* and $m^{\beta\lambda\mu}$ the couple stress tensor at $\mathscr{E}$.

§ 87. Contributions of Couple Stresses to the Expression of $\mathcal{U}_{\alpha\beta}$ and to the Equation of Energy Balance

In order to write an acceptable expression of the total energy tensor $\mathcal{U}_{\alpha\beta}$ capable of taking couple stress into account, we set

$$v^\alpha = 2m^{(\alpha\rho\sigma)} u_{\rho/\sigma} \quad \text{whence—cf. (86.1)—} v^\alpha u_\alpha = 0, \tag{87.1}$$

$$\mathcal{M}^{\alpha\beta} = m^{\alpha\lambda\beta}{}_{/\lambda} + m^{\beta\lambda\alpha}{}_{/\lambda} + v^\alpha u^\beta + u^\alpha v^\beta, \tag{87.2}$$

and

$$\mathcal{U}_{\alpha\beta} = \mathcal{U}^{(E)}_{(\alpha\beta)} + \mathcal{M}_{\alpha\beta} \quad \text{where} \quad \mathcal{U}^{(E)}_{\alpha\beta} = \rho u_\alpha u_\beta + X_{\alpha\beta} + Q_{\alpha\beta} + E_{\alpha\beta}. \tag{87.3}$$

We show that by the generalization (87.3) of (36.1) the above goal is attained. From $u_\alpha m^{\alpha\beta\gamma} = 0$—cf. $(86.1)_{2,4}$—we deduce

$$\begin{aligned} u_\alpha m^{\alpha\beta\gamma}{}_{/\beta\gamma} &= (u_\alpha m^{\alpha\beta\gamma}{}_{/\beta})_{/\gamma} - u_{\alpha/\gamma} m^{\alpha\beta\gamma}{}_{/\beta} = -(u_{\alpha/\beta} m^{\alpha\beta\gamma})_{/\gamma} - (u_{\alpha/\gamma} m^{\alpha\beta\gamma})_{/\beta} + u_{\alpha/\gamma\beta} m^{\alpha\beta\gamma} \\ &= (m^{\sigma\rho\tau} u_{\rho/\sigma} + m^{\tau\rho\sigma} u_{\rho/\sigma})_{/\tau} + u_{\alpha/\gamma\beta} m^{\alpha\beta\gamma}, \end{aligned} \tag{87.4}$$

which by $(87.1)_1$ and (16.9) yields

$$u_\alpha m^{\alpha\beta\gamma}{}_{/\beta\gamma} = v^\tau{}_{/\tau} + m^{\alpha\beta\gamma}(u_{\alpha/\beta\gamma} - u^\rho R_{\rho\alpha\beta\gamma}). \tag{87.5}$$

For any twice continuously differentiable tensor $\mathscr{R}^{\alpha\beta\gamma}$

$$\mathscr{R}^{\rho\sigma\alpha}{}_{/\rho\sigma} - \mathscr{R}^{\rho\sigma\alpha}{}_{/\sigma\rho} = \mathscr{R}^{\rho\sigma\lambda} R_\lambda{}^\alpha{}_{\rho\sigma} \tag{87.6a}$$

holds. Indeed by a well known generalization of (16.9) we have

$$\mathscr{R}^{\rho\sigma\alpha}{}_{/\rho\sigma} - \mathscr{R}^{\rho\sigma\alpha}{}_{/\sigma\rho} = \mathscr{R}^{\lambda\sigma\alpha} R_\lambda{}^\rho{}_{\rho\sigma} + \mathscr{R}^{\rho\lambda\alpha} R_\lambda{}^\sigma{}_{\rho\sigma} + \mathscr{R}^{\rho\sigma\lambda} R_\lambda{}^\alpha{}_{\rho\sigma}. \tag{87.6b}$$

Furthermore $(16.10)_2$ implies $R_\lambda{}^\rho{}_{\rho\sigma}=R_\sigma{}^\rho{}_{\rho\lambda}$, whence

$$\mathscr{R}^{\lambda\sigma\alpha}R_\lambda{}^\rho{}_{\rho\sigma}=\mathscr{R}^{\sigma\lambda\alpha}R_\lambda{}^\rho{}_{\rho\sigma}=-\mathscr{R}^{\rho\lambda\alpha}R_\lambda{}^\sigma{}_{\rho\sigma}$$

which by (87.6b) yields (87.6a).

By $(86.1)_2$, (87.6) implies

$$-2m^{\beta\gamma\alpha}{}_{/\gamma\beta}=m^{\beta\gamma\alpha}{}_{/\beta\gamma}-m^{\beta\gamma\alpha}{}_{/\gamma\beta}=m^{\beta\gamma\lambda}R_\lambda{}^\alpha{}_{\beta\gamma}=-R^\alpha{}_{\lambda\beta\gamma}m^{\beta\gamma\lambda}. \tag{87.7}$$

By $(87.1)_2$ formula $(22.9)_1$ holds both for $T'_\rho=v_\rho$ and $T''_\rho=\overset{+}{T}_{\rho\sigma}=T=0$, and for $T''_\rho=v_\rho$ and $T'_\rho=\overset{+}{T}_{\rho\sigma}=T=0$. Then

$$-u^\alpha(v_\alpha u^\beta)_{/\beta}=v_\alpha A^\alpha, \qquad -u^\alpha(u_\alpha v^\beta)_{/\beta}=v^\beta{}_{/\beta}. \tag{87.8}$$

From (87.2), (87.5), (87.7), and (87.8) we deduce

$$-u_\alpha\mathscr{M}^{\alpha\beta}{}_{/\beta}=-m^{\alpha\beta\gamma}u_{\alpha/\beta\gamma}-v^\beta{}_{/\beta}+u^\rho R_{\rho\alpha\beta\gamma}(m^{\alpha\beta\gamma}-\tfrac{1}{2}m^{\beta\gamma\alpha})+v_\alpha A^\alpha+v^\beta{}_{/\beta}. \tag{87.9}$$

By $(86.1)_2$ $R_{\rho\alpha\beta\gamma}m^{\alpha\beta\gamma}=-R_{\rho\alpha\beta\gamma}m^{\alpha\gamma\beta}=R_{\rho\alpha\beta\gamma}m^{\gamma\alpha\beta}$; thence

$$2R_{\rho\alpha\beta\gamma}(m^{\alpha\beta\gamma}-\tfrac{1}{2}m^{\beta\gamma\alpha})=R_{\rho\alpha\beta\gamma}(m^{\alpha\beta\gamma}+m^{\gamma\alpha\beta}-m^{\beta\gamma\alpha}). \tag{87.10}$$

By the cyclic properties of Riemann's tensor we have

$$R_{\rho\alpha\beta\gamma}(m^{\alpha\beta\gamma}+m^{\gamma\alpha\beta}+m^{\beta\gamma\alpha})=0 \tag{87.11}$$

which by (87.10) yields

$$R_{\rho\alpha\beta\gamma}(m^{\alpha\beta\gamma}-\tfrac{1}{2}m^{\beta\gamma\alpha})=-R_{\rho\alpha\beta\gamma}m^{\beta\gamma\alpha}=R_{\beta\gamma\alpha\rho}m^{\beta\gamma\alpha}=R_{\alpha\beta\gamma\rho}m^{\alpha\beta\gamma}. \tag{87.12}$$

We now show that the relativistic definition $(28.1)_3$ of the work $dl^{(i)}$ of internal contact forces can be generalized to the case of couple stress into

$$\frac{dl^{(i)}}{D\tau}=-c\,u_\alpha(X^{(\alpha\beta)}+\mathscr{M}^{\alpha\beta})_{/\beta} \qquad (c\,D\tau=Ds). \tag{87.13}$$

Indeed by (87.9) and (87.12) we can turn $(87.13)_1$ into

$$\frac{dl^{(i)}}{Ds}=X^{(\alpha\beta)}u_{(\alpha/\overset{+}{\beta})}-m^{\alpha\beta\gamma}(u_{\alpha/\beta\gamma}-R_{\alpha\beta\gamma\rho}u^\rho)+v^\alpha A_\alpha. \tag{87.14}$$

By $(87.1)_1$ v^α has the (magnitude) order of c^{-1}, which in comparing relativistic theories with classical theories can act as a first-order infinitesimal. Then $c\,v^\alpha A_\alpha$ has the order of c^{-2}. Furthermore $R_{\alpha\beta\gamma\delta}$ is zero in special relativity, and in general relativity it has the order of $c^{-4}\rho=c^{-2}k\,w$—cf. $(23.1)_1$ and (21.4).

Then, since $X^{\alpha\beta}$ and $m^{\alpha\beta\gamma}$ are spatial, in locally natural and proper co-ordinates (87.14) becomes, up to very small terms (at least as small as c^{-2}),

$$\frac{dl^{(i)}}{D\tau} = X^{rs} v_{(r/s)} - m^{rsi} v_{r/si} \qquad (v_{r/s} = c u_{r/s},\ v_{r/si} = c u_{r/si}). \tag{87.15}$$

The expression $(87.15)_1$ for the power density of internal contact forces is the classical one in any Galileian or Euclidean frame.[3]

For the sake of completeness we explicitly add that, by (87.13), $(24.6)_2$, and $(36.2)_2$, under the determination (87.3) of $\mathscr{U}_{\alpha\beta}$ equation $(23.4)_{3,4}$ of energy balance becomes

$$k\frac{Dw}{D\tau} + \frac{dl^{(i)}}{D\tau} = k q_{\text{ass}} + \Pi^{(e)} \qquad \left(\frac{\Pi^{(e)}}{c} = u_\alpha E^{(\alpha\beta)}{}_{/\beta},\ \frac{k}{c} q_{\text{ass}} = u_\alpha Q^{\alpha\beta}{}_{/\beta}\right), \tag{87.16}$$

where $\Pi^{(e)}$ is the proper density of the supply of energy. (In our views $E_{\alpha\beta}$ is to be identified with ${}^6E_{\alpha\beta}$ or ${}^7E_{\alpha\beta}$, whence $E_{[\alpha\beta]}=0$—cf. (40.18).)

By what was said of this equation in the case $\mathscr{M}_{\alpha\beta} \equiv 0$ [§§ 24, 36] and by what we said of $dl^{(i)}$ in the general case, (87.16) is an acceptable relativistic version of the equation of energy balance in the case of couple stress.

In § 88 we shall show that the expression (87.3) of $\mathscr{U}_{\alpha\beta}$ gives rise also to an acceptable relativistic version of the Cauchy equations of continuous media. Furthermore since $\mathscr{M}_{\alpha\beta}$ is much smaller than $\rho = k(c^2+w)$, (23.2) is an acceptable relativistic version of the Poisson equation for gravitation potential also under the definition (87.3) of $\mathscr{U}_{\alpha\beta}$. As a consequence (87.3) gives rise to acceptable versions of conservation equations and gravitation equations in the general case being considered.

Now we justify in a direct way—and in § 92 [§§ 96, 97] we shall justify more satisfactorily from a Lagrangian [variational] point of view—the presence of the terms $v^\alpha u^\beta$ and $u^\alpha v^\beta$ in (87.2). By (87.1) the term—$v^\beta{}_{/\beta}$, which is present in (87.13) through (87.9), has no classical analogue—cf. (87.15)—and it has the same magnitude order as the term $-m^{\alpha\beta\gamma} u_{\alpha\beta/\gamma}$ whose presence in (87.13)—and (87.14)—is essential. Both terms constitute the contribution of the term $m^{\alpha\lambda\beta}{}_{/\lambda}$ to $-u_\alpha \mathscr{M}^{\alpha\beta}{}_{/\beta}$—see (87.2), (87.5), and (87.9). The presence of $u^\alpha v^\beta$ in (87.2) is essential to eliminate—cf. $(87.8)_2$—the term $-v^\beta{}_{/\beta}$ in (87.9).

Furthermore the (acceptable) presence of $m^{\beta\lambda\alpha}{}_{/\lambda}$ and $v^\alpha u^\beta$ in (87.2) is a consequence of the symmetry of $\mathscr{M}_{\alpha\beta}$, which is important in general relativity because otherwise we ought to take $\mathscr{M}_{(\alpha\beta)}$ into account, in that by (23.5) the symmetric part of $\mathscr{U}_{\alpha\beta}$ must fulfill conservation equations.

For the sake of uniformity it is natural to assume $\mathscr{M}_{\alpha\beta}$ as the contribution of couple stresses to $\mathscr{U}_{\alpha\beta}$ also in special relativity.

It must be added that in this theory we can replace $\mathscr{U}_{\alpha\beta}$ with $\mathscr{U}_{\alpha\beta} + \Delta\mathscr{U}_{\alpha\beta}$ provided $\Delta\mathscr{U}^{\alpha\beta/}{}_{\beta}=0$ holds (even if $\mathscr{U}_{\alpha\beta}$ and $\mathscr{U}_{\alpha\beta}+\Delta\mathscr{U}_{\alpha\beta}$ are not interchangeable as far as gravitation equations are concerned).

Incidentally in special relativity we have $\Delta\mathscr{U}_{\alpha\beta}{}^{/\beta}=0$ for

$$\Delta\mathscr{U}^{\alpha\beta} = m^{\lambda\beta\alpha}{}_{/\lambda} \quad \text{or} \quad \Delta\mathscr{U}^{\alpha\beta} = m^{\alpha\beta\lambda}{}_{/\lambda} - m^{\alpha\lambda\beta}{}_{/\lambda}. \tag{87.17}$$

Hence by (87.2) we can accept, instead of $(87.3)_1$, either of the definitions—cf. $(87.3)_2$ and $(86.1)_2$

$$\mathscr{U}_{\alpha\beta}=\mathscr{U}^{(E)}_{\alpha\beta}+m^{\alpha\lambda\beta}{}_{/\lambda}+2v^{(\alpha}u^{\beta)}, \qquad \mathscr{U}_{\alpha\beta}=\mathscr{U}^{(E)}_{\alpha\beta}+m^{\alpha\beta\lambda}{}_{/\lambda}+2v^{(\alpha}u^{\beta)}. \tag{87.18}$$

The presence of A_α in the expression (87.14) of $dl^{(i)}$ within a very small term is not surprising. The same occurs in the expression (24.5) of kq_{ass} substantially stated by Eckart in 1940.

§ 88. The Relativistic Cauchy Equations of Continuous Media in the Case of Couple Stresses

From $(87.7)_{1,2}$ we deduce $m^{\beta\lambda\rho}{}_{/\lambda\beta}=-2^{-1}m^{\beta\gamma\lambda}R_\lambda{}^\rho{}_{\beta\gamma}$. Furthermore by $(87.1)_2$, (22.8) holds for $T=\overset{\perp}{T}{}^{\alpha\beta}=0$ and $T'_\alpha=T''_\alpha=v_\alpha$. Then (87.2) yields

$$\overset{\perp}{g}_{\alpha\rho}\mathscr{M}^{\rho\beta}{}_{/\beta}=\overset{\perp}{g}_{\alpha\rho}\left(m^{\rho\beta\gamma}{}_{/\beta\gamma}-\tfrac{1}{2}m^{\beta\gamma\delta}R_{\beta\gamma\delta}{}^{\rho}+k\frac{D}{Ds}\frac{v^\rho}{k}\right)+u_{\alpha/\beta}v^\beta. \tag{88.1}$$

Incidentally by (87.6) and equality $R_\alpha{}^\rho{}_{\beta\gamma}=R_{\beta\gamma\alpha}{}^\rho$, by equality $R_{\beta\gamma\delta}{}^\rho=-R^\rho{}_{\delta\beta\gamma}$, and by $(87.12)_{1,2}$ respectively we have

$$\begin{aligned} m^{\rho\beta\gamma}{}_{/\beta\gamma}-2^{-1}m^{\beta\gamma\delta}R_{\beta\gamma\delta}{}^\rho &= m^{\rho\beta\gamma}{}_{/\gamma\beta}+(m^{\delta\beta\gamma}-2^{-1}m^{\beta\gamma\delta})R_{\beta\gamma\delta}{}^\rho \\ &= m^{\rho\beta\gamma}{}_{/\gamma\beta}-(m^{\alpha\beta\gamma}-\tfrac{1}{2}m^{\beta\gamma\alpha})R^\rho{}_{\alpha\beta\gamma}=m^{\rho\beta\gamma}{}_{/\gamma\beta}-m^{\beta\gamma\alpha}R_{\beta\gamma\alpha}{}^\rho. \end{aligned}$$

As a consequence (88.1) is equivalent to

$$\overset{\perp}{g}_{\alpha\rho}\mathscr{M}^{\rho\beta}{}_{/\beta}=\overset{\perp}{g}_{\alpha\rho}\left(m^{\rho\beta\gamma}{}_{/\gamma\beta}-m^{\beta\gamma\delta}R_{\beta\gamma\delta}{}^{\rho}+k\frac{D}{Ds}\frac{v^\rho}{k}\right)+u_{\alpha/\beta}v^\beta. \tag{88.2}$$

In the case $E_{\alpha\beta}\equiv 0\equiv m^{\alpha\beta\gamma}$ (whence $\mathscr{M}_{\alpha\beta}\equiv 0$) the expression (87.3) of $\mathscr{U}_{\alpha\beta}$ coincides with (24.3) which gives rise to the version (24.8) of the first Cauchy equation $(23.4)_{1,2}$ (of continua media) and to the version (24.7) of the consequence $\mathscr{U}_{[\alpha\beta]}=0$ of gravitation equations. Then by (88.1) equation $(23.4)_{1,2}$ or $(23.5)_2$ yields

$$\begin{aligned} (\rho\overset{\perp}{g}_{\alpha\gamma}+X_{(\alpha\gamma)})A^\gamma=-\overset{\perp}{g}_{\alpha\rho}\left(X^{(\rho\beta)}{}_{\overset{\perp}{/\beta}}+m^{\rho\beta\gamma}{}_{/\beta\gamma}-\frac{1}{2}m^{\beta\gamma\delta}R_{\beta\gamma\delta}{}^\rho+k\frac{D}{Ds}\frac{q^\rho+v^\rho}{k}\right) \\ -u_{\alpha/\rho}(q^\rho+v^\rho)-\overset{\perp}{g}_{\alpha\rho}E^{(\rho\beta)}{}_{/\beta}. \end{aligned} \tag{88.3}$$

The last term in (88.3) can be replaced by the ponderomotive force per unit proper volume $\overset{\perp}{K}_\alpha$—cf. $(36.2)_1$—in the case $E_{[\alpha\beta]}\equiv 0$, which occurs if $E_{\alpha\beta}$ is identified with ${}^6E_{\alpha\beta}$ or ${}^7E_{\alpha\beta}$, as we proposed to do in connection with materials being dealt with from the Eulerian or Lagrangian point of view, respectively.

In classical physics Cauchy's first and second laws of motion read, in any locally proper Euclidean frame—cf. Truesdell and Toupin [1960, (205.2), (205.10)],

$$k a^r = f^r - X^{rs}{}_{/s}, \qquad m^{rsi}{}_{/i} = X^{[rs]} + h^{rs} \qquad (h^{(rs)} = 0), \tag{88.4}$$

where a^r is the acceleration (of $\mathscr{C}$ at $\mathscr{E}$), f^r the resultant per unit volume, of the forces at a distance and the dragging force, and h^{rs} represents the assigned couple at a distance—cf. Truesdell and Toupin [1960, p. 538]—per unit volume. By $(88.4)_2$ we can turn $(88.4)_1$ into—cf. Truesdell and Toupin [1960, (205.17), p. 548]

$$k a^r = f^r - X^{(rs)}{}_{/s} - m^{rsi}{}_{/is} + h^{rs}{}_{/s} = 0. \tag{88.5}$$

Now we assume that the frame being considered is locally non-rotating and freely falling. Then f^r is the ponderomotive force per unit volume and for $E_{[\alpha\beta]} \equiv 0$ it is relativized into the last term in (88.3). The case $E_{[\alpha\beta]} \not\equiv 0$ will be briefly considered after the second Cauchy equation.

Since h^{rs} is to be identified with $E^{[rs]}$ $(= \overset{\perp}{E}{}^{[rs]})$—cf. $(36.4)_1$—, in the case being considered we have $h^{rs} \equiv 0$.

We know [§§ 24, 36] that for $m^{\alpha\beta\gamma} \equiv 0$ and $h^{rs} \equiv 0$ equation (88.3) is an acceptable relativistic version of (88.5). To realize the analogue in the general case we only need to prove that $-\overset{\perp}{g}_{\alpha\gamma} \mathscr{M}^{\gamma\beta}{}_{/\beta}$—cf. (88.1)—i.e. the contribution of $m^{\alpha\beta\gamma}$ (and v^α) to (88.3), is an acceptable relativization of $m^{rsi}{}_{/is}$.

To the above end we remark that the terms $u^\alpha{}_{/\rho} v^\rho$ and $D(v^\rho/k)/Ds$ have the (magnitude) order of c^{-2} by (87.1); $R_{\beta\gamma\delta}{}^\rho$ has the same order. Hence we only need to consider $\overset{\perp}{g}_{\alpha\rho} m^{\rho\beta\gamma}{}_{/\beta\gamma}$ in locally natural co-ordinates:

$$m^{r\beta\gamma}{}_{/\beta\gamma} = m^{rbc}{}_{/bc} + m^{r0c}{}_{/0c} + m^{r\beta 0}{}_{/\beta 0}. \tag{88.6}$$

By $(86.1)_3$ we have $m^{\alpha\beta\rho}{}_{/\sigma} u_\rho + m^{\alpha\beta\rho} u_{\rho/\sigma} = 0$ whence

$$m^{\alpha\beta\rho}{}_{/\sigma\tau} u_\rho + m^{\alpha\beta\rho}{}_{/\sigma} u_{\rho/\tau} + (m^{\alpha\beta\rho} u_{\rho/\sigma})_{/\tau} = 0. \tag{88.7}$$

The last two terms on the left-hand side of (88.7) have the order of c^{-1}. Hence the same holds for the first, which for $u_\rho = -\delta_{\rho 0}$ becomes $m^{\alpha\beta 0}{}_{/\sigma\tau}$. As a consequence $m^{r0c}{}_{/0c}$ can be neglected. The possibility of neglecting $m^{r\beta 0}{}_{/\beta 0}$ follows from $(86.1)_{3,4}$ in the same way. We conclude that (88.3) is an acceptable relativization of (88.5) for $E_{[\alpha\beta]} \equiv 0$.

The second law of motion of Cauchy $(88.4)_2$ can be used as a definition of $X^{[rs]}$ and considering it is not essential for the sequel. The same holds for the remainder of this section. However for some additional (complementary) considerations let us note that either of the equalities

$$(m^{\alpha\beta\gamma}{}_{/\gamma})^\perp = X^{[\alpha\beta]} + \overset{\perp}{E}{}^{[\alpha\beta]}, \qquad (m^{\alpha\beta\gamma}{}_{\overset{\perp}{/\gamma}})^\perp = X^{[\alpha\beta]} + \overset{\perp}{E}{}^{[\alpha\beta]} \tag{88.8}$$

is an acceptable relativistic version of $(88.4)_2$. Indeed $(86.1)_3$ and (17.17) yield the first of the equalities

$$m^{\alpha\beta\gamma}{}_{/\gamma} - m^{\alpha\beta\gamma}{}_{/\overset{\perp}{\gamma}} = m^{\alpha\beta\gamma} A_\gamma = (m^{\alpha\beta\gamma}{}_{/\gamma} - m^{\alpha\beta\gamma}{}_{/\overset{\perp}{\gamma}})^{\perp} . \tag{88.9}$$

By $(86.1)_{2,3,4}$ the sides of $(88.9)_1$ are spatial, so that $(88.9)_2$ holds. Furthermore $m^{\alpha\beta\gamma} A_\gamma$ has the order of c^{-2} and for $u^\alpha = \delta_0{}^\alpha$ and $h^{rs} \equiv E^{[rs]}$ (the spatial part of) $(88.8)_2$ becomes identical with $(88.4)_2$.

Let us consider general relativity. Then $E'''_{\alpha\beta} \equiv 0$—cf. $(36.4)_2$—must hold by the definition of $E_{\alpha\beta}$. As a consequence, either of $(88.8)_{1,2}$ can have the form $\mathscr{U}_{[\alpha\beta]} = 0$—cf. $(23.3)_2$—by a suitable definition of $\mathscr{U}_{\alpha\beta}$ (even if $\overset{\perp}{E}_{[\alpha\beta]} \not\equiv 0$).

For instance we can set, instead of $(87.3)_1$, $\mathscr{U}_{\alpha\beta} = U'_{\alpha\beta}$ with

$$U'_{\alpha\beta} = \mathscr{U}^{(E)}_{\alpha\beta} + 2[u_{(\alpha} v_{\beta)} + u_{[\alpha} \lambda_{\beta]} + m_{(\alpha\gamma\beta)}{}^{/\gamma}] - m_{\alpha\beta\gamma}{}^{/\gamma} , \tag{88.10}$$

where $(87.3)_2$ holds and λ^β is defined by

$$\lambda^\beta = u_{\rho/\gamma} m^{\rho\beta\gamma} \qquad (\lambda^\beta u_\beta = 0) , \tag{88.11}$$

so that λ^β has the order of c^{-1}. The relations

$$u^\alpha \lambda^\beta - \lambda^\alpha u^\beta = m^{\alpha\beta\gamma}{}_{/\gamma} - (m^{\alpha\beta\gamma}{}_{/\gamma})^{\perp} = m^{\alpha\beta\gamma}{}_{/\overset{\perp}{\gamma}} - (m^{\alpha\beta\gamma}{}_{/\overset{\perp}{\gamma}})^{\perp} \tag{88.12}$$

will now be proved. From $\overset{\perp}{g}{}^\alpha{}_{\rho/\gamma} = u^\alpha{}_{/\gamma} u_\rho + u^\alpha u_{\rho/\gamma}$—cf. $(17.6)_1$—and $(86.1)_{2,3,4}$ we deduce

$$m^{\alpha\beta\gamma}{}_{/\gamma} = (\overset{\perp}{g}{}^\alpha{}_\rho \overset{\perp}{g}{}^\beta{}_\sigma m^{\rho\sigma\gamma})_{/\gamma} = \overset{\perp}{g}{}^\alpha{}_\rho \overset{\perp}{g}{}^\beta{}_\sigma m^{\rho\sigma\gamma}{}_{/\gamma} + u^\alpha u_{\rho/\gamma} m^{\rho\beta\gamma} + u^\beta u_{\sigma/\gamma} m^{\alpha\sigma\gamma}$$

which by $(88.11)_1$ yields $(88.12)_1$. Furthermore $(88.12)_2$, follows from (88.9).

Let us add that by (88.10) and $(86.1)_2$ $\mathscr{U}_{(\alpha\beta)}$ coincides with the right hand side of $(87.3)_1$, so that by (23.5) the conservation equations deduced under the definition $(87.3)_1$ of $\mathscr{U}_{\alpha\beta}$ hold again; furthermore by $(88.12)_1$, $(86.1)_2$, and $(24.3)_2$—i.e. $Q_{[\alpha\beta]} = 0$—, the consequence $U'_{[\alpha\beta]} = 0$—cf. $(23.3)_2$—of gravitation equations is equivalent to $(88.8)_1$.

Let us incidentally add that, by the definition $U''_{\alpha\beta} = U'_{\alpha\beta} - m_{\alpha\beta}{}^\gamma A_\gamma$, by $(88.9)_1$, (88.10), and $(86.1)_2$ the analogues of the above assertions involving $U'_{\alpha\beta}$ and $(88.8)_1$ hold for $U''_{\alpha\beta}$ and $(88.8)_2$.

Since $(88.8)_{1,2}$ may be regarded as different definitions of $X^{[\alpha\beta]}$, they are compatible with the same physical phenomena relevant for motion. Hence we have merely considered four presentations of a same physical theory [(87.3) and either of $(88.8)_{1,2}$; $\mathscr{U}_{\alpha\beta} = U'_{\alpha\beta}$; $\mathscr{U}_{\alpha\beta} = U''_{\alpha\beta}$].

Lastly let us note that we can set $\mathscr{U}_{\alpha\beta} = U'_{\alpha\beta}$ for $E_{\alpha\beta} = {}^5E_{\alpha\beta}$—cf. (40.11)—(${}^5E'''_{\alpha\beta} \equiv 0 \not\equiv {}^5E_{[\alpha\beta]}$). Then, generally, the difference $\overset{\perp}{g}_{\alpha\gamma} E^{[\gamma\beta]}{}_{/\beta}$ between the last term in (88.3) and $\overset{\perp}{K}{}^\alpha$—cf. $(36.2)_1$—is unequal to 0, and so is the difference $\overset{\perp}{g}_{\alpha\gamma} X^{[\gamma\beta]}{}_{/\beta}$ between $-\overset{\perp}{g}_{\alpha\gamma} X^{(\gamma\beta)}{}_{/\beta}$ and the resultant $-\overset{\perp}{g}_{\alpha\gamma} X^{\gamma\beta}{}_{/\beta}$ of the contact forces per unit volume. These differences can have an identically vanishing resultant only in case $m^{\alpha\beta\gamma} \equiv 0$—cf. $(88.8)_1$.

§ 89. The Non-Working Part of $m^{\alpha\beta\gamma}$

Let us set

$$m = \frac{1}{6} \overset{\perp}{\varepsilon}_{\alpha\beta\gamma} m^{\alpha\beta\gamma}, \qquad \bar{m}^{\alpha\beta\gamma} = m^{\alpha\beta\gamma} - m \overset{\perp}{\varepsilon}^{\alpha\beta\gamma}. \tag{89.1}$$

We now assume that

$$m^{\alpha\beta\gamma} = m \overset{\perp}{\varepsilon}^{\alpha\beta\gamma} \tag{89.2}$$

holds in an open region Ω of S_4. Then $(87.1)_1$ yields $v^\alpha = 0$. Furthermore (87.2) implies $\mathscr{M}_{\alpha\beta} = 0$. We conclude that the *linear invariant* m—cf. $(89.1)_1$—*has no influence on gravitation and conservation equations, under the definition* (87.3) *of* $\mathscr{U}_{\alpha\beta}$. In particular it gives no contribution to the work $dl^{(i)}/D\tau$—cf. (87.14).

Now we conversely assume that $m^{\alpha\beta\gamma}$ gives no contribution to $dl^{(i)}/D\tau$ in Ω in any virtual motion. Then $dl^{(i)}/D\tau = 0$ holds in particular for $u_{\alpha/\beta} = 0$ and for arbitrary admissible values of $u_{\alpha/(\beta\gamma)}$. By (16.9) this implies

$$m^{\alpha\beta\gamma}(u_{\alpha/(\beta\gamma)} + \tfrac{1}{2} u^\rho R_{\rho\alpha\beta\gamma} - R_{\alpha\beta\gamma}{}^\rho u_\rho) = 0 \quad \textit{for every} \quad (u_{\alpha/(\beta\gamma)})^\perp. \tag{89.3}$$

Thence we deduce $m^{\alpha(\beta\gamma)} = 0$, which together with $(86.1)_2$ yields the validity of (89.2) in Ω.

We have shown that the contribution of the field $m^{\alpha\beta\gamma}$—cf. $(86.1)_{2,3,4}$—to $dl^{(i)}/D\tau$ is zero in the open region Ω of S_4 for every virtual motion if and only if, (89.2) holds.

Hence, if $m^{\alpha\beta\gamma}$ is arbitrary, $\bar{m}^{\alpha\beta\gamma}$ can be called the *working part* of $m^{\alpha\beta\gamma}$ and $m \overset{\perp}{\varepsilon}^{\alpha\beta\gamma}$ the *non-working* or the *negligible part* of $m^{\alpha\beta\gamma}$.

It must be added that, it is true, $m \overset{\perp}{\varepsilon}^{\alpha\beta\gamma}$ does not affect gravitation or conservation equations, but in general we cannot cross it out e.g. in the second Cauchy equation $(88.8)_1$, because the formula

$$\overset{\perp}{g}{}^\alpha{}_\rho \overset{\perp}{g}{}^\beta{}_\sigma (m \overset{\perp}{\varepsilon}^{\rho\sigma\gamma})_{/\gamma} = (m_{/\gamma} + m A_\gamma) \overset{\perp}{\varepsilon}^{\alpha\beta\gamma} \tag{89.4}$$

can be easily deduced from (35.11); it suffices to remark that $\overset{\perp}{g}{}^\alpha{}_\rho \overset{\perp}{g}{}^\beta{}_\sigma u_{\lambda/\beta} \overset{\perp}{\varepsilon}^{\lambda\rho\sigma\gamma}$ vanishes because all of the covariant indices ρ, σ, λ and β are spatial.

§ 90. Some Commutation Formulas for Lagrangian Spatial Derivatives

As a preliminary for the next theorem let us prove that *if* $Dt \equiv Ds$ *and* $u_L^\dagger = 0$, *then* $[\dot{f} = Df/Ds,\ u_L^\dagger \equiv u_\rho x^\rho{}_L,\ A_L^* \equiv A_\rho \alpha^\rho{}_L]$

$$u^\rho{}_{|L} = u^\rho{}_{;L} = \dot{x}^\rho{}_L, \qquad \dot{u}_L^\dagger = A_L^* \qquad (Dt \equiv Ds,\ u_L^\dagger = 0). \tag{90.1}$$

We assume $u^{\dagger}_L = 0$. Then (53.6) implies $(90.1)_1$. In addition to $(90.1)_{4,5}$ let (54.1) hold for the sake of simplicity. Then we have $u^\rho \equiv \partial x^\rho/\partial s \equiv \dot{x}^\rho$, which by (A 2.6) and (53.5) yields

$$u^\rho{}_{;L} = \frac{\partial^2 x^\rho(s,y)}{\partial y^L \partial s} = \frac{\partial}{\partial s} x^\rho{}_L(s,y) = \dot{x}^\rho{}_L . \tag{90.2}$$

This is $(90.1)_2$, which by its tensor form holds even if the frames (x) and (y) do not fulfil $(54.1)_{3\ldots 6}$.

Formula $(90.1)_3$ holds, for it has the tensor form and is included in $(60.8)_{2,3,4}$, whose validity was deduced under conditions $(90.1)_{4,5}$. q.e.d.

Theorem 90.1. *Let the double tensor $T^\beta{}_L$ be a function of x^ρ, t, y^L, continuous together with its first and second derivatives. Then under analogous hypotheses on functions* (52.2) *(which represent the motion of matter) we have*—cf. (20.2), $(57.1)_1$

$$\frac{D}{Ds} T^\beta{}_{L|M} - (\dot{T}^\beta{}_L)_{|M} = T^\tau{}_L R_\tau{}^\beta{}_{\gamma\delta} \alpha^\gamma{}_M u^\delta + \dot{T}^\beta{}_L A^*_M , \tag{90.3a}$$

$$\frac{D}{Ds} \alpha^\rho{}_{L|M} - (\dot{\alpha}^\rho{}_L)_{|M} = -R^\rho{}_{\beta\gamma\delta} \alpha^\beta{}_L \alpha^\gamma{}_M u^\delta + \dot{\alpha}^\rho{}_L A^*_M . \tag{90.4}$$

$$\frac{D}{Ds} \alpha^\rho{}_{L|M} - u^\rho{}_{|LM} = -R^\rho{}_{\beta\gamma\delta} \alpha^\beta{}_L \alpha^\gamma{}_M u^\delta + 2u^\rho{}_{|(L} A^*_{M)} + u^\rho (A^*_L A^*_M + A^*_{L|M}) . \tag{90.5}$$

Proof. Possibly using the inverse of transformation $(52.2)_1$, we can express $T^\beta{}_L$ as a function of x^ρ. Then we have the first of the equalities

$$\dot{T}^\beta{}_L = T^\beta{}_{L/\gamma} u^\gamma, \quad T^\beta{}_{L|M} = T^\beta{}_{L;M} + \dot{T}^\beta{}_L u^\dagger_M = T^\beta{}_{L/\gamma} x^\gamma{}_M + T^\beta{}_{L/M} + \dot{T}^\beta{}_L u^\dagger_M ; \tag{90.6}$$

the second follows from $(53.6)_1$, and $(90.6)_3$ from (A 2.6).

We assume $u^\dagger_L = 0$ (locally). Then, on the one hand, $(90.6)_{2,3}$ and (17.2) imply

$$\frac{D}{Ds} T^\beta{}_{L|M} = (T^\beta{}_{L/\gamma\delta} x^\gamma{}_M + T^\beta{}_{L/M\delta}) u^\delta + T^\beta{}_{L/\gamma} \dot{x}^\gamma{}_M + \frac{D}{Ds} (\dot{T}^\beta{}_L u^\dagger_M) . \tag{90.7}$$

On the other hand (53.6), $(90.6)_1$, and (A 2.6) yield $(u^\dagger_L = 0)$

$$(\dot{T}^\beta{}_L)_{|M} = (\dot{T}^\beta{}_L)_{;M} = (T^\beta{}_{L/\gamma\delta} x^\delta{}_M + T^\beta{}_{L/\gamma M}) u^\gamma + T^\beta{}_{L/\gamma} u^\gamma{}_{|M} . \tag{90.8}$$

We have $T^\beta{}_{L/M\delta} = T^\beta{}_{L/\delta M}$—cf. (A 2.3), (A 2.4). Furthermore in addition to $u^\dagger_L = 0$ we assume $Dt \equiv Ds$, so that $(90.1)_{1,2,3}$ hold. Then from (90.7) and (90.8) we deduce

$$\frac{D}{Ds} T^\beta{}_{L|M} - (\dot{T}^\beta{}_L)_{|M} = (T^\beta{}_{L/\gamma\delta} - T^\beta{}_{L/\delta\gamma}) x^\gamma{}_M u^\delta + \dot{T}^\beta{}_L A^*_M . \tag{90.9}$$

By (53.8), $(90.1)_5$ yields $x^\gamma{}_M = \alpha^\gamma{}_M$, so that by (16.9) we can turn (90.9) into formula (90.3a), whose validity is obviously independent of assumptions $(90.1)_{4,5}$ [and (54.1)].

We deduce (90.4) from (90.3a) by setting $T^\beta{}_L = \alpha^\beta{}_L$.

Lastly $(57.5)_1$ implies

$$(\dot{\alpha}^\rho{}_L)_{|M} = u^\rho{}_{|LM} + u^\rho{}_{|M} A^*_L + u^\rho A^*_{L|M}, \tag{90.10}$$

which by (90.4) and $(57.5)_1$ again yields (90.5). q.e.d.

It is obvious how to prove the following extension of (90.3) to the tensor $T_{L_1 \dots L_a}{}^{\beta_1 \dots \beta_b}$:

$$\frac{D}{Ds} T_{L_1 \dots L_a}{}^{\beta_1 \dots \beta_b}{}_{|M} - (\dot{T}_{L_1 \dots L_a}{}^{\beta_1 \dots \beta_b})_{|M}$$

$$= \sum_{i=1}^{b} T_{L_1 \dots L_a}{}^{\beta_1 \dots \beta_{i-1} \tau \beta_{i+1} \dots \beta_b} R_\tau{}^{\beta_i}{}_{\gamma\delta} \alpha^\gamma{}_M u^\delta + \dot{T}_{L_1 \dots L_a}{}^{\beta_1 \dots \beta_b} A^*_M. \tag{90.3b}$$

The following theorem will not be used in connection with couple stresses. However it may be interesting in connection with non-simple materials—cf. footnote 15 in Chapter 1.

Theorem 90.2. *Under the hypotheses of Theorem 90.1 we have*

$$\alpha^\rho{}_{A|B} - \alpha^\rho{}_{B|A} = 2 u^\rho u^*_{[AB]} \quad \textit{where} \quad u^*_{AB} = u_{\rho|B} \alpha^\rho{}_A = u_{\rho/\sigma} \alpha^\rho{}_A \alpha^\sigma{}_B \tag{90.11}$$

and

$$T^\beta{}_{L|AB} - T^\beta{}_{L|BA} = T^\gamma{}_L R_\gamma{}^\beta{}_{\rho\sigma} \alpha^\rho{}_A \alpha^\sigma{}_B + T^\beta{}_M R^{*M}{}_{LAB} + 2 \dot{T}^\beta{}_L u^*_{[AB]}. \tag{90.12}$$

Before proving the theorem let us remark that the extension of (90.12)—as well as the one of (90.3a)—to the arbitrary double tensor $T_{L \dots}{}^{M \dots}{}_{\alpha \dots}{}^{\beta \dots}$ is obvious.

Proof. From $(57.3)_1$ we have $(90.11)_3$.

Sinse $u_\lambda \alpha^\lambda{}_L = 0$—cf. $(53.7)_1$—, $u_{\sigma|B} \alpha^\sigma{}_A + u_\sigma \alpha^\sigma{}_{A|B} = 0$, which by $(90.11)_2$ yields the first of the equalities

$$u^*_{[AB]} = -u_\sigma \alpha^\sigma{}_{[A|B]}, \qquad \alpha^\rho{}_{A|B} = \alpha^\rho{}_{AB} - u^\rho u_\sigma \alpha^\sigma{}_{A|B}. \tag{90.13}$$

The second follows from $(53.16)_1$ and $(17.6)_1$. By $(61.3)_1$ $\alpha^\rho{}_{[AB]} = 0$ holds, so that (90.13) yields $(90.11)_1$.

In order to prove (90.12), we assume for the moment $u^\dagger_L = 0$. Then from $(90.6)_2$ and (53.6) we deduce

$$T^\beta{}_{L|AB} = T^\beta{}_{L;AB} + (\dot{T}^\beta{}_L u^\dagger_A)_{|B} \tag{90.14}$$

which yields

$$T^\beta{}_{L|[AB]} = T^\beta{}_{L;[AB]} + \dot{T}^\beta{}_L u^\dagger_{[A|B]} \qquad (u^\dagger_L = 0). \tag{90.15}$$

By (A 2.6), (A 2.9), theorem (16.9), and its analogue for R^*_{ABCD} we have

$$2T^{\beta}{}_{L;[AB]}=T^{\alpha}{}_{L}R_{\alpha}{}^{\beta}{}_{\gamma\delta}x^{\gamma}{}_{A}x^{\delta}{}_{B}+T^{\beta}{}_{M}R^{*M}{}_{LAB}\,. \tag{90.16}$$

Definitions $(52.8)_2$ and $(90.11)_2$ imply the first two of the equalities

$$u^{\dagger}_{A|B}=u_{\rho|B}x^{\rho}{}_{A}+u_{\rho}x^{\rho}{}_{A|B}=u^{*}_{AB}+u_{\rho}x^{\rho}{}_{A|B}\,,\qquad u^{\dagger}_{[A|B]}=u^{*}_{[AB]}\qquad (u^{\dagger}_{L}=0)\,. \tag{90.17}$$

Furthermore (53.6) and (A 2.9) yield $x^{\rho}{}_{A|B}=x^{\rho}{}_{B|A}$ for $u^{\dagger}_{L}=0$. Then $(90.17)_3$ holds. From (90.15), (90.16), and $(90.17)_3$ we deduce (90.12), whose validity is obviously independent of assumption $(90.15)_2$. q.e.d.

§ 91. A Useful Expression for $\grave{C}_{LAB}$

The following notations are useful:

$$\alpha^{\lambda_1\ldots\lambda_n}_{L_1\ldots L_n}=\alpha^{\lambda_1}_{L_1}\ldots\alpha^{\lambda_n}_{L_n}\qquad (=x^{\lambda_1}_{L_1}\ldots x^{\lambda_n}_{L_n}\text{ for } u^{\dagger}_{L}=0)\,. \tag{91.1}$$

Since $(17.4)_3$ yields $u_{\rho/\beta}=\overset{+}{g}_{\rho}{}^{\sigma}u_{\sigma/\beta}$, and $(53.16)_{2,3}$ include $C_{BLM}=\alpha_{\rho B}\alpha^{\rho}{}_{L|M}$, from $(57.5)_1$ we deduce

$$\grave{C}_{BLM}=u_{\rho|B}\alpha^{\rho}{}_{LM}+\alpha_{\rho B}\frac{D}{Ds}\alpha^{\rho}{}_{L|M}+A^{*}_{B}u_{\rho}\alpha^{\rho}{}_{L|M}\,. \tag{91.2}$$

By $(90.11)_2$ we have $(u_{\rho}\alpha^{\rho}{}_{L}=0)$

$$\alpha_{\rho B}u^{\rho}{}_{|(L}A^{*}_{M)}=u^{*}_{B(L}A^{*}_{M)}\,,\qquad u_{\rho}\alpha^{\rho}{}_{L|M}=-u_{\rho|M}\alpha^{\rho}{}_{L}=-u^{*}_{LM}\,. \tag{91.3}$$

Hence by (90.5) and $(91.1)_1$ we can turn (91.2) into

$$\grave{C}_{BLM}=u_{\rho|B}\alpha^{\rho}{}_{LM}+\alpha_{\rho B}u^{\rho}{}_{|LM}-R_{\beta\lambda\mu\delta}u^{\delta}\alpha^{\beta\lambda\mu}_{BLM}+u^{*}_{BL}A^{*}_{M}+u^{*}_{BM}A^{*}_{L}-u^{*}_{LM}A^{*}_{B}\,. \tag{91.4}$$

By $(17.20)_1$ and $(53.16)_1$, and by $(91.3)_{2,3}$, respectively, we have

$$u^{\rho}{}_{/\lambda}\alpha^{\lambda}{}_{L|M}=u^{\rho}{}_{/\lambda}\alpha^{\lambda}{}_{LM}-A^{\rho}u_{\lambda}\alpha^{\lambda}{}_{L|M}=u^{\rho}{}_{/\lambda}\alpha^{\lambda}{}_{LM}+A^{\rho}u^{*}_{LM}\,. \tag{91.5}$$

From $(57.3)_1$ and (91.5) we deduce

$$u^{\rho}{}_{|LM}=u^{\rho}{}_{/\lambda}\alpha^{\lambda}{}_{LM}+u^{\rho}{}_{/\lambda\mu}\alpha^{\lambda}{}_{L}\alpha^{\mu}{}_{M}+A^{\rho}u^{*}_{LM}\,. \tag{91.6}$$

Furthermore (57.3) yields

$$u_{\rho|B}\alpha^{\rho}{}_{LM}+\alpha_{\rho B}u^{\rho}{}_{/\lambda}\alpha^{\lambda}{}_{LM}=\alpha^{\lambda}{}_{B}(u_{\rho/\lambda}+u_{\lambda/\rho})\alpha^{\rho}{}_{LM}=2\alpha^{\lambda}{}_{B}u_{(\lambda/\rho)}\overset{+}{g}{}^{\rho}{}_{\sigma}\alpha^{\sigma}{}_{LM}\,. \tag{91.7}$$

From $(56.16)_2$ and $(53.16)_2$, respectively, we deduce

$$\overset{\perp}{g}{}^{\rho}{}_{\sigma}\alpha^{\sigma}{}_{LM}=\alpha^{\rho}{}_{R}\overset{-1}{C}{}^{RS}\alpha_{\sigma S}\alpha^{\sigma}{}_{LM}=\alpha^{\rho}{}_{R}\overset{-1}{C}{}^{RS}C_{SLM} \tag{91.8}$$

which together with $(57.7)_1$ implies

$$2\alpha^{\lambda}{}_{B}u_{(\lambda/\rho)}\overset{\perp}{g}{}^{\rho}{}_{\sigma}\alpha^{\sigma}{}_{LM}=2\grave{\varepsilon}_{BR}\overset{-1}{C}{}^{RS}C_{SLM}\,. \tag{91.9}$$

From (91.6) and (91.1) and from (91.7), (91.9), and convention $(57.1\text{a})_1$, respectively, we deduce

$$\begin{aligned} u_{\rho|B}\alpha^{\rho}{}_{LM}+\alpha_{\rho B}u^{\rho}{}_{|LM} &= (u_{\rho|B}\alpha^{\rho}{}_{LM}+\alpha_{\rho B}u^{\rho}{}_{/\lambda}\alpha^{\lambda}{}_{LM})+u_{\rho/\lambda\mu}\alpha^{\rho\lambda\mu}_{BLM}+\alpha_{\rho B}A^{\rho}u^{*}_{LM} \\ &= 2\grave{\varepsilon}_{BR}\overset{-1}{C}{}^{RS}C_{SLM}+u_{\rho/\lambda\mu}\alpha^{\rho\lambda\mu}_{BLM}+A^{*}_{B}u^{*}_{LM}\,. \end{aligned} \tag{91.10}$$

By combining (91.4) with (91.10) and by taking $(90.11)_{2,3}$, convention $(57.1\text{a})_1$, and (91.1) into account, we deduce

$$\grave{C}_{BLM}=2\grave{\varepsilon}_{BR}\overset{-1}{C}{}^{RS}C_{SLM}+(u_{\beta/\lambda\mu}+2u_{\beta/(\lambda}A_{\mu)}-R_{\beta\lambda\mu\delta}u^{\delta})\alpha^{\beta\lambda\mu}_{BLM}\,. \tag{91.11a}$$

Incidentally the last term in the right hand side of (91.11a) is symmetric with respect to L and M because so are the remaining terms in (91.11a). We can see this symmetry in a direct way by remarking that we have

$$U_{\alpha[\beta\gamma]}=0 \quad \text{for} \quad U_{\alpha\beta\gamma}=u_{\alpha/\beta\gamma}-R_{\alpha\beta\gamma\delta}u^{\delta} \tag{91.12}$$

because $(91.12)_2$ and the cyclic properties of $R_{\alpha\beta\gamma\delta}$ imply

$$2U_{\alpha[\beta\gamma]}=u^{\delta}(R_{\delta\alpha\beta\gamma}-R_{\alpha\beta\gamma\delta}+R_{\alpha\gamma\beta\delta})=-u^{\delta}(R_{\beta\gamma\alpha\delta}+R_{\alpha\beta\gamma\delta}+R_{\gamma\alpha\beta\delta})=0.$$

We know that A_μ and $R_{\beta\lambda\mu\delta}$ are very small, of the order of c^{-2}. Furthermore $(91.1)_2$ holds for $u^{\dagger}_{L}=0$. Hence if we choose (x), (y), and $\hat{t}(x)$ in such a way that (54.1) holds, then, up to terms having the order of c^{-2}, (91.11a) is equivalent to the corresponding classical formula (in Cartesian co-ordinates)

$$\frac{d}{dt}C_{BLM}=2\frac{d\varepsilon_{BR}}{dt}\overset{-1}{C}{}^{RS}C_{SLM}+v_{r/lm}x^{r}{}_{B}x^{\lambda}{}_{L}x^{m}{}_{M}\,. \tag{91.11b}$$

The classical definitions of the quantities C_{LM}, ε_{LM}, and C_{BLM}, present in (91.11b), coincide with the relativistic expressions of the same quantities for the above choice of (x), (y), and $\hat{t}(x)$:

$$C_{LM}=\delta_{rs}x^{r}{}_{L}x^{s}{}_{M}=\delta_{LM}+2\varepsilon_{LM}\,, \qquad C_{SLM}=\delta_{sr}x^{s}{}_{S}x^{r}{}_{,LM}\,. \tag{91.13}$$

because $(54.2)_3$ and $(57.7)_2$ imply $(91.13)_{1,2}$ while $(53.16)_2$, $(54.2)_{1,2}$, and $(61.2)_1$ yield $(91.13)_3$.

§ 92. A Lagrangian Expression for the Work of Stress and Couple Stress in Special or General Relativity

We introduce the Lagrangian analogue $\mathfrak{m}_*^{BLM}$ of $m^{\beta\lambda\mu}$ by the first of the equalities—cf. $(56.4)_2$, $(56.13)_2$

$$\mathfrak{m}_*^{BLM} = \mathscr{D}^{-2} \gamma_\rho^B \gamma_\lambda^L \gamma_\mu^M m^{\beta\lambda\mu} = -\mathfrak{m}_*^{LBM}, \qquad m^{\beta\lambda\mu} = \mathscr{D}^{-1} \alpha_{BLM}^{\beta\lambda\mu} \mathfrak{m}_*^{BLM}. \tag{92.1}$$

The second follows from $(86.1)_2$, and $(92.1)_3$ from $(92.1)_1$, (91.1), $(56.15)_1$, and the spatial character of $m^{\beta\lambda\mu}$—cf. $(86.1)_{2,3,4}$.

From (91.11a) multiplied by $\mathscr{D}^{-1}\mathfrak{m}_*^{BLM}$ and $(92.1)_3$ we deduce

$$u_{\beta/\lambda\mu} m^{\beta\lambda\mu} = \mathscr{D}^{-1} \Big(\dot{C}_{[BL]M} - 2\dot{\varepsilon}_{BR} \overset{-1}{C}{}^{RS} C_{SLM} \Big) \mathfrak{m}_*^{BLM} + (R_{\beta\lambda\mu\delta} u^\delta - 2 u_{\beta/(\lambda} A_{\mu)}) m^{\beta\lambda\mu}. \tag{92.2}$$

By $(86.1)_2$ and $(87.1)_1$ we have

$$v^\mu A_\mu = -2 m^{\beta(\mu\lambda)} u_{\beta/\lambda} A_\mu = -2 u_{\beta/(\lambda} A_{\mu)} m^{\beta\lambda\mu}, \tag{92.3}$$

which together with (92.2) implies

$$m^{\alpha\beta\gamma}(u_{\alpha/\beta\gamma} - R_{\alpha\beta\gamma\delta} u^\delta) - v^\alpha A_\alpha = \mathscr{D}^{-1} \mathfrak{m}_*^{BLM} \Big(\dot{C}_{[BL]M} - 2\dot{\varepsilon}_{BR} \overset{-1}{C}{}^{RS} C_{SLM} \Big). \tag{92.4}$$

By $(57.7)_1$, $(59.2)_{1,2}$, and (92.4) the Lagrangian work $d^* l^{(i)} = \mathscr{D}\, dl^{(i)}$—cf. $(59.6)_1$, (87.14)—of stress and couple stress has the following basic expression:

$$d^* l^{(i)} = \mathscr{D}\, dl^{(i)} = \Big(Y^{(BR)} + 2\mathfrak{m}_*^{(BLM} \overset{-1}{C}{}^{R)S} C_{SLM} \Big) D\varepsilon_{BR} - \mathfrak{m}_*^{BLM} D C_{[BL]M}. \tag{92.5}$$

A classical expression for $d^* l^{(i)}$—in Grioli's or Toupin's theory—coincides exactly with our relativistic one (92.5). The analogue does not hold for the Eulerian work $dl^{(i)}$—cf. (87.14). The afore-mentioned exact coincidence and the equivalence of $(92.5)_2$ to (87.14) is a good justification for the relativistic expression (87.14) of $dl^{(i)}$ and in particular for the term $v^\alpha A_\alpha$ which appears in (87.14). Hence this exact coincidence is a good justification for the expression (87.3) of $\mathscr{U}_{\alpha\beta}$, and in particular for the very small terms in v^α which appear in it. An additional justification of (87.3) and (87.14), that perhaps is more complete, will be given in §§ 96, 97 from the variational point of view.

§ 93. Elasticity with Couple Stress

We accept the gravitation equations (23.1) under the definition (87.3) of $\mathcal{U}_{\alpha\beta}$. Then by comparing (87.3) with $(23.1)_2$ and by remembering $(24.5)_1$, $(24.6)_2$, and (87.13), we can put equation $(23.4)_{3,4}$ of energy balance into the form

$$k\dot{w}+\frac{dl^{(i)}}{Ds}=\frac{k}{c}q_{\text{ass}}+u_\alpha E^{\alpha\beta}{}_{/\beta}\qquad\left(\frac{k}{c}q_{\text{ass}}=-q^\rho{}_{/\rho}-q^\rho A_\rho\right). \tag{93.1}$$

We also accept the second principle (25.1).

Let us first consider the case $E_{\alpha\beta}\equiv 0$ $(E_\alpha\equiv H_\alpha\equiv 0)$. In this case (25.1) for $r=0$ and (93.1) imply $(28.1)_1$ which by (92.5) can be developed into

$$k^*D\mathcal{F}\leqslant -k^*\eta DT-(Y^{(BR)}+2\mathfrak{m}_*^{BLM}\overset{-1}{C}{}^{RS}C_{SLM})D\varepsilon_{BR}+\mathfrak{m}_*^{BLM}DC_{[BL]M}\,. \tag{93.2}$$

We say that the body $\mathscr{C}$, capable of couple stress, is *elastic* at its material point P^* (or y^L) if the following three conditions hold at P^*:

a) *the specific internal energy w and the specific entropy* η*—hence the specific free energy* $\mathcal{F}$ $(=w-T\eta)$ *also—are functions of* T, ε_{LM}, *and* C_{BLM}—cf. $(57.7)_2$,

b) Y^{LM} *is a function of* T, ε_{LM}, C_{BLM}, *and possibly of other parameters* $p_1,\dots,p_n$ *which (as well as* $R^\rho{}_L$ *and* $C_{BL_1\dots L_r}$—cf. $(54.5)_2$, $(53.16)_2$*—and unlike* $u_{\rho/\sigma}$*) take (only) values that are logically (i.e. "definitionally") compatible with arbitrary values of T and* $\dot{\varepsilon}_{LM}$ $(=\dot{\varepsilon}_{ML})$,

c) *constraints are absent in the (narrow) sense that admissible values of T,* ε_{LM}, *and* C_{BLM} *are physically compatible with arbitrary values of* DT, $D\varepsilon_{LM}$ $(=D\varepsilon_{ML})$, *and* DC_{BLM} $(=DC_{BML})$*—which in particular implies the physical independence of these differentials.*

Let now $\mathscr{C}$ be elastic (at P^*). Then we can turn (93.2) (at P^*) into

$$\begin{aligned}&k^*\left(\frac{\partial\mathcal{F}}{\partial T}+\eta\right)DT+\left(k^*\frac{\partial\mathcal{F}}{\partial\varepsilon_{AB}}+Y^{(AB)}+2\mathfrak{m}_*^{ALM}\overset{-1}{C}{}^{BS}C_{SLM}\right)D\varepsilon_{AB}\\&\qquad+\left(k^*\frac{\partial\mathcal{F}}{\partial C_{[BL]M}}-\mathfrak{m}_*^{BLM}\right)DC_{[BL]M}+k^*\frac{\partial\mathcal{F}}{\partial C_{(BL)M}}DC_{(BL)M}\leqslant 0\,.\end{aligned} \tag{93.3}$$

We can replace the tensor C_{BLM} by the two

$$E_{BLM}=C_{[BL]M}\,,\qquad S_{BLM}=C_{(BL)M}\,. \tag{93.4}$$

The only identity fulfilled by C_{BLM} by its definition $(53.16)_2$ is $C_{B[LM]}=0$—cf. $(61.3)_2$. Hence the only identities on DE_{BLM} are

$$DE_{(BL)M}=0\,,\qquad \varepsilon^{*BLM}DE_{BLM}=0. \tag{93.5}$$

Indeed from $(93.4)_1$ we deduce $(93.5)_1$, and by $(61.3)_2$ also $(93.5)_2$.

Conversely we now assume (93.5) and prove that the system

$$DE_{BLM}=DC_{[BL]M}, \quad DC_{B[LM]}=0 \tag{a}$$

in DC_{BLM} is compatible. Indeed, in case B, L, and M are distinct, $(93.5)_2$ implies

$$DE_{BLM}+DE_{LMB}+DE_{MBL}=0. \tag{b}$$

In the remaining case (b) follows from $(93.5)_1$. Furthermore, setting

$$DC_{BLM}=\tfrac{4}{3}DE_{B(LM)}, \tag{c}$$

condition $(a)_2$ is obviously fulfilled, and condition $(a)_1$ also because (c), $(93.5)_1$, and (b) respectively imply

$$\begin{aligned} 3DC_{[BL]M}&=DE_{BLM}+DE_{BML}-DE_{LBM}-DE_{LMB}\\ &=DE_{BLM}-DE_{MBL}+DE_{BLM}-DE_{LMB}=3DE_{BLM}. \end{aligned} \tag{d}$$

The left hand side of (93.3) is linear in DT, $D\varepsilon_{AB}$, and $DC_{[BL]M}$. Furthermore inequality (93.3) holds for arbitrary values of DT, $D\varepsilon_{AB}(=D\varepsilon_{BA})$, $DE_{BLM}=-DE_{LBM}$, and $DS_{BLM}=DS_{LBM}$ fulfilling (93.5). Then there exist a function λ of T, ε_{AB}, and C_{BLM} for which

$$\begin{aligned} &\eta=-\frac{\partial\mathscr{F}}{\partial T}, \qquad \mathfrak{m}_*^{BLM}=k^*\frac{\partial\mathscr{F}}{\partial C_{[BL]M}}+\lambda\varepsilon^{*BLM},\\ &Y^{(AB)}=-k^*\frac{\partial\mathscr{F}}{\partial\varepsilon_{AB}}-2\mathfrak{m}_*^{(ALM}\overset{-1}{C}{}^{B)S}C_{SLM}, \qquad \frac{\partial\mathscr{F}}{\partial C_{(BL)M}}=0. \end{aligned} \tag{93.6}$$

Incidentally $(61.3)_2$ implies $\varepsilon^{*ABC}C_{ABC}=0$, so that by $(93.6)_2$ we can turn $(93.6)_3$ into

$$Y^{(AB)}=-k^*\frac{\partial\mathscr{F}}{\partial\varepsilon_{AB}}-k^*\left(\frac{\partial\mathscr{F}}{\partial C_{[AL]M}}\overset{-1}{C}{}^{BS}+\frac{\partial\mathscr{F}}{\partial C_{[BL]M}}\overset{-1}{C}{}^{AS}\right)C_{SLM}. \tag{93.7}$$

By $(93.6)_{4,1}$ $\mathscr{F}$ and η are functions of T, ε_{LM}, and $C_{[BL]M}$. The same holds for w by $(28.1)_2$.

The constitutive equation for $Y^{[AB]}$ is not needed because $X^{[\rho\sigma]}$ is determined by Cauchy's second equation of motion $(88.8)_1$ for $E_{\alpha\beta}\equiv 0$.

Let us now assume that $\mathfrak{m}_*^{BLM}$ is a function of T, ε_{AB}, and $C_{[BL]M}$. Hence the same holds for λ. Furthermore $\varepsilon^{*BLM}C_{[BL]M}\equiv 0$. Then, in every physical situation the definition of the function $\mathscr{F}=\mathscr{F}(T,\varepsilon_{LM},C_{[BL]M})$ can be such that $(93.6)_2$ holds for $\lambda\equiv 0$—cf. Appendix A, §A3.

Let us briefly remark that by Helmholtz's postulate [§ 28] we can deduce —cf. (63.3), (63.4)—$(63.4)_1$ and the equalities obtained from $(93.6)_{2,3,4}$ and (93.7) by the substitution of w for $\mathscr{F}$, where w is regarded as a function of η, ε_{AB}, and $C_{[BL]M}$.

By (91.1) and an obvious analogue of $(56.9)_4$

$$\mathscr{D}^{-1}\alpha^{\beta\lambda\mu}_{BLM}\lambda\varepsilon^{*BLM} = \lambda\overset{\perp}{\varepsilon}^{\beta\lambda\mu}, \tag{93.8}$$

so that *the contribution of* λ *to* $m^{\beta\lambda\mu}$ *is* $\lambda\overset{\perp}{\varepsilon}^{\beta\lambda\mu}$.

We know [§ 89] that such a contribution does not affect gravitation or conservation equations. It only appears in the second Cauchy equation $(88.8)_1$ according to (89.4).

Let us now remark that $(53.15)_1$ and $(53.16)_{1,2}$ imply the first of the equalities

$$C_{MB|L} = C_{MBL} + C_{BML} = C_{MBL} + C_{BLM}, \qquad C_{M[B|L]} = C_{[BL]M}. \tag{93.9}$$

From $(61.3)_2$ we deduce $(93.9)_2$ and, by $(93.9)_{1,2}$, also $(93.9)_3$. Relations $(93.9)_3$ are a straightforward relativization of a result of Toupin [1962]. They enable us to replace the argument C_{BLM} with $C_{M[B|L]}$ in our expressions of $\mathscr{F}$, w, η, Y^{LM}, and $\mathfrak{m}^{BLM}_{*}$.

Let now the following two assumptions hold at P^*, for the body $\mathscr{C}$ capable of couple stress:

a′) w, η, *and* Y^{LM} *are functions of* T, ε_{LM}, C_{BLM}, *and the parameters* $p_1, \ldots, p_n$ *(such as* $R^{\rho}{}_{L}$ *and* $C_{BL_1\ldots L_r}$*) whose values are logically compatible with arbitrary values of* DT, $D\varepsilon_{LM}\,(=D\varepsilon_{ML})$. $DC_{BLM}\,(=DC_{BML})$ *and* $Dp_1, \ldots, Dp_n$*—so that in particular the mutual independence of these differentials is understood,*

b′) *constraints are absent in the (narrow) sense that all admissible values of* T, ε_{LM}, C_{BLM}, *and* $p_1, \ldots, p_n$ *are physically compatible with arbitrary values of the differentials* $DT, \ldots, Dp_n$ *mentioned in* b′).

In analogy with some considerations developed in § 64 it is easy to deduce from a′) and b′), on the basis of (93.2), that $\mathscr{F}\,(=w-\eta T)$ is independent of $p_1, \ldots, p_n$ and that $\mathscr{C}$ is elastic at P^*, and in particular (93.6) and (93.7) hold.

§ 94. Hints at Non-Viscous Fluids Capable of Couple Stress and at Electromagnetoelasticity with Couple Stress

Let $\mathscr{C}$ be elastic and capable of couple stress. We say that $\mathscr{C}$ is a *non-viscous fluid (possibly capable of couple stress)* if for every event point $\mathscr{E}$ in $W_{\mathscr{C}}$ the power $dl^{(i)}/Ds$ of stress and couple stress—cf. (87.13)—is zero in every virtual motion that is isochoric at $\mathscr{E}$ ($u^{\alpha}{}_{/\alpha} \equiv u^{\alpha}{}_{/\overset{\perp}{\alpha}} = 0$ but $u_{\alpha/(\beta\gamma)}$ arbitrary).

Theorem 94.1. *Let* $\mathscr{C}$ *be an elastic body possibly capable of couple stress. Then* $\mathscr{C}$ *is a non-viscous fluid if and only if,* $\mathscr{C}$ *has the constitutive equations*

$$m^{\alpha\beta\gamma} = m\overset{\perp}{\varepsilon}^{\alpha\beta\gamma}, \qquad X^{(\alpha\beta)} = p\overset{\perp}{g}^{\alpha\beta} \tag{94.1}$$

where m *is a function of* T, ε_{LM}, *and* C_{BLM}.

Proof. We first assume that $\mathscr{C}$ is a non-viscous fluid. Then $dl^{(i)}/Ds=0$ (at the arbitrary event point $\mathscr{E}$ in $W_{\mathscr{C}}$) for all admissible values of $u_{\alpha/\beta}$ and $(u_{\alpha/\beta\gamma})^{\perp}$ such that $u^{\alpha}{}_{/\alpha}=0$ holds. Since by (87.14) $dl^{(i)}/Ds$ is a linear function of $u_{\alpha/\beta}$ and $u_{\alpha/\beta\gamma}$, and $(u_{\alpha/(\beta\gamma)})^{\perp}$ is arbitrary, we have $m^{\alpha(\beta\gamma)}=0$ which by $(86.1)_2$ yields $(94.1)_1$. Then $X^{(\alpha\beta)}u_{\alpha/\beta}=dl^{(i)}/Ds=0$ for every $u_{\alpha/\beta}$ with $\overset{\perp}{g}{}^{\alpha\beta}u_{\alpha/\beta}=0$. Hence there is a Lagrangian multiplier p for which $(94.1)_2$ holds.

Conversely we now assume that (94.1) holds identically for our elastic body $\mathscr{C}$. Then $m^{\alpha\beta\gamma}$ does not affect $dl^{(i)}/Ds$ [§ 89], so that by $(94.1)_2$ $dl^{(i)}/Ds=0$ for $u^{\alpha}{}_{/\overset{\perp}{\alpha}}=0$. Hence $\mathscr{C}$ is a non-viscous fluid capable of couple stress. q.e.d.

Theorem 94.2. *Let $\mathscr{C}$ be an elastic body possibly capable of couple stress. Then $\mathscr{C}$ is a non-viscous fluid if and only if, $Dl^{(i)}/Ds=0$ (in $W_{\mathscr{C}}$) for every virtual (everywhere) isochoric motion (hence for all admissible values of $u_{\alpha/\beta}$ and $u_{\alpha/\beta\gamma}$ such that $u^{\alpha}{}_{/\alpha}=0=u^{\alpha}{}_{/\alpha\gamma}$).*

To prove the theorem we first assume $dl^{(i)}/Ds=0$ for given values of u^{α} and A_{α} and for all $u_{\alpha/\beta}$ and $u_{\alpha/\beta\gamma}$ such that

(a) $\quad u^{\alpha}{}_{/\alpha}=0\,,\qquad g^{\alpha\beta}u_{\alpha/\beta\gamma}=0\,.$

Let $\delta u_{\alpha/\beta}$ and $\delta u_{\alpha/\beta\gamma}$ denote the difference of two arbitrary solutions of (a). Since $\delta u_{\alpha/[\beta\gamma]}=0$ and $u^{\alpha}{}_{/\alpha}=u^{\alpha}{}_{/\overset{\perp}{\alpha}}$, $\delta u_{\alpha/\overset{\perp}{\beta}}$ and $(\delta u_{\alpha/\beta\gamma})^{\perp}$ are certainly included in quantities forming such a difference if in locally proper and natural co-ordinates we have

(b) $\quad \overset{\perp}{g}{}^{ab}\,\delta u_{a/b}=0\,,\qquad g^{ab}\,\delta u_{a/bc}=0\,,\qquad \overset{\perp}{\varepsilon}{}^{pbc}\,\delta u_{a/bc}=0\,.$

Then by (87.14)

(c) $$\delta\frac{dl^{(i)}}{Ds}=X^{(ab)}\,\delta u_{a/b}-m^{abc}\,\delta u_{a/bc}=0$$

for all solutions of (b). Then there are the Lagrangian multipliers p, μ^c, and $\lambda^a{}_r$ for which $(94.1)_2$ and

(d) $\quad m^{abc}=g^{ab}\mu^c+\lambda^a{}_r\overset{\perp}{\varepsilon}{}^{rbc}\qquad (g^{ab}=\delta^{ab})$

hold. By $(86.1)_2$ e.g. $m^{12c}=-m^{21c}$, i.e. $\lambda^1{}_r\overset{\perp}{\varepsilon}{}^{r2c}=-\lambda^2{}_r\overset{\perp}{\varepsilon}{}^{r1c}=\lambda^2{}_r\overset{\perp}{\varepsilon}{}^{1rc}$. This yields (for $c=3$) $\lambda^1{}_1=\lambda^2{}_2$ and (for $c=1$) $\lambda^1{}_3=0$. Now we easily see that $\lambda^r{}_s$ has the form $m\delta^r{}_s$. Hence $(d)_1$ becomes $m^{abc}=g^{ab}\mu^c+m\overset{\perp}{\varepsilon}{}^{abc}$ which by $(86.1)_2$ easily yields $(94.1)_1$. Hence by Theorem 94.1 $\mathscr{C}$ is a non-viscous fluid.

If $\mathscr{C}$ is a non-viscous fluid, then $dl^{(i)}/Ds=0$ obviously hold for every admissible solution $u_{\alpha/\beta}$, $u_{\alpha/\beta\gamma}$ of (a). q.e.d.

Let us remark that (94.1) *is equivalent to the condition that for every spatial elementary surface $d\sigma_{\alpha}$, the intrinsic torque $d\mathscr{M}_{\alpha}$ of contact forces through this surface*—cf. $(86.1)_1$—*and the resultant $X^{(\alpha\beta)}d\sigma_{\beta}$ of the symmetric stress should be normal to the surface.*

Indeed by $(86.1)_1$ and $(20.5)_2$, $(94.1)_1$ yields $d\mathcal{M}_\alpha = 2m\overset{\perp}{g}^{\alpha\beta} d\sigma_\beta$; hence $d\mathcal{M}_\alpha$ is parallel to $d\sigma_\alpha$. Now we conversely assume this parallelism for every $d\sigma_\beta$. Then by assuming also $g_{\alpha\beta} = \delta'_{\alpha\beta}$ and $d\sigma_r = 0$ for $\beta \neq r$, where r has been fixed arbitrarily, we easily deduce from $(86.6)_1$ that $\overset{\perp}{\varepsilon}_{\alpha\beta\gamma} m^{\beta\gamma r} = 2m_{(a)} \delta_a^{\ r}$ holds for some scalar $m_{(a)}$. Then, for an arbitrary $d\sigma_\beta$, we obtain $2m\,d\sigma_r = d\mathcal{M}_r = 2m_{(r)} d\sigma_r$, so that $m = m_{(1)} = m_{(2)} = m_{(3)}$. Now we easily see that there is a scalar m, independent of $d\sigma_\alpha$, for which we have $\overset{\perp}{\varepsilon}_{\alpha\beta\gamma} m^{\beta\gamma\tau} = 2m g_\alpha^{\ \tau}$. By $(20.5)_1$ and $(86.1)_2$ this yields

$$\text{(e)} \quad m^{\rho\sigma\tau} = \tfrac{1}{2} \overset{\perp}{\varepsilon}^{\rho\sigma\alpha} \overset{\perp}{\varepsilon}_{\alpha\beta\gamma} m^{\beta\gamma\tau} = \overset{\perp}{\varepsilon}^{\rho\sigma\alpha} m g_\alpha^{\ \tau},$$

hence $(94.1)_1$. The proof of the part of the italicized assertion above, that we have not proved concerns $X^{(\alpha\beta)}$ and is well known.

Since m—cf. $(94.1)_1$—does not affect conservation equations [§ 89] we can assume $m \equiv 0$. Then by $(94.1)_1$ and $(88.8)_1$, $X^{[\alpha\beta]} = 0$ for $E_{\alpha\beta} \equiv 0$. Then by (94.1) we conclude that a *non-viscous fluid possibly capable of couple stress behaves like an (ordinary) non-viscous fluid (*i.e. *incapable of couple stress)* and, in particular, for it the resultant dR_α of contact forces through an arbitrary elementary spatial surface is normal to the surface.

Let us now remark that for such a fluid the conservation equations (23.3) hold under the definition (24.3) of $\mathcal{U}^{\alpha\beta}$ ($\mathcal{M}_{\alpha\beta} \equiv 0 \equiv E_{\alpha\beta}$). Furthermore, since $k = k^* \mathcal{D}^{-1} = k^* a^{*1/2} (\det \|C_{LM}\|)^{-1/2}$—cf. $(58.3)_1$, $(56.9)_1$—and $(57.7)_2$ hold, for our non-viscous fluid $\mathcal{C}$ possibly capable of couple stress we can replace the arguments T, ε_{LM}, and C_{BLM} of w, η, and Y^{LM}—cf. conditions a) and b) in § 93—with T, k, and suitable parameters $p_1, \ldots, p_n$. It is now easy to see that the considerations made in §§ 28, 29 hold for $\mathcal{C}$, so that $\mathcal{C}$ behaves as a non-viscous fluid, as far as motion is concerned. More precisely we have—cf. $(28.8)_1$ and (28.10)

$$\mathcal{F} = \mathcal{F}(T, k, y^L), \qquad \eta = -\frac{\partial \mathcal{F}}{\partial T}, \qquad X^{(\alpha\beta)} = k^2 \frac{\partial \mathcal{F}}{\partial k} \overset{\perp}{g}^{\alpha\beta}. \tag{94.2}$$

By (89.4) the second Cauchy equation $(88.8)_1$ becomes for $\mathcal{C}$

$$X^{[\alpha\beta]} = (m_{/\gamma} + m A_\gamma) \overset{\perp}{\varepsilon}^{\alpha\beta\gamma} \quad \text{for} \quad E_{\alpha\beta} \equiv 0. \tag{94.3}$$

Equation (93.1) of energy balance was deduced from (23.1) under the definition (87.3) of $\mathcal{U}^{\alpha\beta}$. Now we assume $E_{\alpha\beta} = {}^{4+h}E_{\alpha\beta}$ for $h = 2$ or $h = 3$—cf. (40.18). Then by $(40.21)_2$ and $(43.9)_{2,3}$ we can turn (93.1) into the following extension of (43.8):

$$k \frac{Dw}{Ds} + \frac{d_h l^{(i)}}{Ds} = k \frac{dQ}{Ds} + k \frac{d_h \lambda}{Ds} \qquad \left(\frac{d_h l^{(i)}}{Ds} = \frac{d l^{(i)}}{Ds} \ \text{for} \ X^{\alpha\beta} = {}^{h+4}X^{\alpha\beta} \right) (h = 2, 3), \tag{94.4}$$

where ${}^{h+4}X_{\alpha\beta}$ is the determination of $X_{\alpha\beta}$ to be used together with the determination ${}^{h+4}E_{\alpha\beta}$ of $E_{\alpha\beta}$—cf. (43.1)—, where $d_h \lambda$ is defined by (40.16), and where $d_h l^{(i)}$ is $dl^{(i)}$—cf. (87.14)—for $X^{\alpha\beta} = {}^{h+4}X^{\alpha\beta}$.

From (94.4) and the second principle (25.1) we deduced (46.2) in the case $m^{\alpha\beta\gamma} \equiv 0$. This deduction is still holding in the general case—cf. $(94.4)_{2,3}$.

Now we assume $h=3$ with a view to dealing with solids. Furthermore we understand Y^{LM} as given by $(59.1)_{2,3}$ for $X^{\rho\sigma} = {}^7X^{\rho\sigma}$ $(Y^{LM} = {}^7Y^{LM})$. Then on the basis of $(58.3)_1$, $(92.5)_2$, $(94.4)_{2,3}$, $(40.16)_2$, and $(58.10)_1$ we can turn (46.2) for $n=3$ into the first of the relations

$$-k^*\eta DT - (Y^{(BR)} + 2\mathfrak{m}_*^{BLM} \overset{-1}{C}{}^{RS} C_{SLM}) D\varepsilon_{BR} + \mathfrak{m}_*^{BLM} DC_{[BL]M} \\ \geqslant k^*(D\mathscr{F} - E_L^* D\pi_*^L - H_L^* D\mu_*^L) \equiv k^*(D\bar{\mathscr{F}} + \pi_*^L DE_L^* + \mu_*^L DH_L^*). \tag{94.5}$$

The second is an identity (due to the definition $(46.4)_{1,2}$ of $\bar{\mathscr{F}}$.

We say that at P^* $\mathscr{C}$ is a *polarizable elastic body capable of couple stress if (at P^*)* $\bar{w}$, η, Y^{LM}, π_*^L, and μ_*^L are functions of ε_{LM}, C_{BLM}, E_L^*, H_L^*, and T, and if constraints are absent (i.e. all admissible values of $\varepsilon_{LM}, \ldots, T$ are physically compatible with all admissible values of $D\varepsilon_{LM}$, DC_{BLM}, DE_L^*, DH_L^*, and DT).

Let $\mathscr{C}$ be such a body at P^*. Then from (94.5) we can deduce in the usual way that—cf. $(93.6)_{1,2}$, (93.7), and $(72.4)_{3,4}$

$$\eta = -\frac{\partial\bar{\mathscr{F}}}{\partial T}, \quad \mathfrak{m}_*^{BLM} = k^* \frac{\partial\bar{\mathscr{F}}}{\partial C_{[BL]M}} + \lambda\varepsilon^{*BLM},$$
$$Y^{(AB)} = -k^* \frac{\partial\bar{\mathscr{F}}}{\partial\varepsilon_{AB}} - k^*\left(\frac{\partial\bar{\mathscr{F}}}{\partial C_{[AL]M}} \overset{-1}{C}{}^{BS} + \frac{\partial\bar{\mathscr{F}}}{\partial C_{[BL]M}} \overset{-1}{C}{}^{AS}\right) C_{SLM}, \tag{94.6}$$
$$\pi_*^L = -\frac{\partial\bar{\mathscr{F}}}{\partial E_L^*}, \quad \mu_*^L = -\frac{\partial\bar{\mathscr{F}}}{\partial H_L^*}, \quad \bar{\mathscr{F}} = \bar{\mathscr{F}}(T, \varepsilon_{LM}, C_{[BL]M}, E_L^*, H_L^*).$$

Using the Helmholtz postulate we can derive the equation $T = \partial\bar{w}/\partial\eta$, where $\bar{w}$ is defined by $(46.4)_{1,3}$, and the equations obtained from $(94.6)_{2\text{–}6}$ by the substitution of $\bar{w}$ for $\bar{\mathscr{F}}$.

We say that $\mathscr{C}$ is a *polarizable non-viscous fluid possibly capable of couple stress*, if it is elastic, polarizable, and capable of couple stress at all of its material points, and if for it $dl^{(i)}/Ds$ vanishes identically in every virtual isochoric motion.

Substantially the same reasoning as that in the first part of this section proves that (94.1) holds for the fluid above; furthermore the constitutive equations (46.7), (46.8), and (46.10) also hold for it, so that we have (46.9), which yields $\overset{+}{E}{}^{[\alpha\beta]} = 0$ for $E_{\alpha\beta} = {}^6E_{\alpha\beta}$. Then $(94.3)_1$ also hold in the present case. Since in addition we can make $m \equiv 0$ in that m does not influence the behaviour of materials, we can conclude that *the above fluid $\mathscr{C}$ behaves as an (ordinary) polarizable non-viscous fluid* [§ 47].

§ 95. Some Preliminary Variational Formulas Related to Second Order Lagrangian Kinematics and the Variation of Space-Time Metric

Following Bressan [1972b], in this section we prove the basic variational formula (95.19) which includes couple stress and in § 96 we shall apply it to state a variational principle for elasticity with couple stress.

From $(53.16)_2$ and $(53.7)_1$ we deduce

$$C_{BLM}=x^{\beta}{}_{B}\overset{\perp}{g}_{\beta\gamma}\alpha^{\gamma}{}_{L|M} \qquad (\alpha^{\gamma}{}_{L}=\overset{\perp}{g}{}^{\gamma}{}_{\rho}x^{\rho}{}_{L}). \tag{95.1}$$

Furthermore $(95.1)_2$, $(17.6)_1$, and $(52.8)_2$ yield

$$\alpha^{\gamma}{}_{L|M}=\overset{\perp}{g}{}^{\gamma}{}_{\rho}x^{\rho}{}_{L|M}+u^{\gamma}{}_{|M}u^{\dagger}_{L}+u^{\gamma}u_{\rho|M}x^{\rho}{}_{L} \qquad (u^{\dagger}_{L}\equiv u_{\rho}x^{\rho}{}_{L}), \tag{95.2}$$

so that by $(53.16)_1$

$$\alpha^{\gamma}{}_{LM}=\overset{\perp}{g}{}^{\gamma}{}_{\rho}x^{\rho}{}_{L|M} \quad \text{for} \quad u^{\dagger}_{L}=0 \tag{95.3}$$

and by $(95.1)_1$

$$C_{BLM}=x^{\beta}{}_{B}\overset{\perp}{g}_{\beta\rho}x^{\rho}{}_{L|M}+x^{\beta}{}_{B}u_{\beta|M}u^{\dagger}_{L}. \tag{95.4}$$

We want to calculate $\delta_g C_{BLM}$ [§ 69]. Therefore we first remark that by (53.6)

$$x^{\rho}{}_{L|M}=x^{\rho}{}_{L,M}+\begin{Bmatrix}\rho\\ \lambda\ \ \mu\end{Bmatrix}x^{\lambda}{}_{L}x^{\mu}{}_{M}-\begin{Bmatrix}H\\ L\ \ M\end{Bmatrix}^{*}x^{\rho}{}_{H}+u^{\dagger}_{M}\frac{D}{Ds}x^{\rho}{}_{L}, \tag{95.5}$$

and that by $(95.2)_2$ and (69.2b)

$$\delta_g u^{\dagger}_{L}=(x^{\alpha}{}_{L}u^{\beta}+\tfrac{1}{2}u^{\dagger}_{L}u^{\alpha}u^{\beta})\,\delta g_{\alpha\beta}. \tag{95.6}$$

Then by (95.5)

$$\delta_g x^{\rho}{}_{L|M}=x^{\lambda}{}_{L}x^{\mu}{}_{M}\,\delta_g\begin{Bmatrix}\rho\\ \lambda\ \ \mu\end{Bmatrix}+\frac{Dx^{\rho}{}_{L}}{Ds}x^{\mu}{}_{M}u^{\sigma}\delta g_{\mu\sigma} \quad \text{for} \quad u^{\dagger}_{L}=0. \tag{95.7}$$

Let $u^{\dagger}_{L}=0$ hold. Then we have $(54.2)_1$, i.e. the first of the equalities

$$x^{\rho}{}_{L}=\alpha^{\rho}{}_{L}, \qquad \overset{\perp}{g}{}^{\rho}{}_{\sigma}\frac{D}{Ds}x^{\sigma}{}_{L}=u^{\rho}{}_{|L} \qquad (u^{\dagger}_{L}=0). \tag{95.8}$$

The second follows from $(95.1)_2$, $(95.8)_3$, and $(57.6)_1$.

By (95.7) and (95.8)

$$\overset{\perp}{g}_{\beta\rho}\delta_g x^{\rho}{}_{L|M}=\alpha^{\lambda}{}_{L}\alpha^{\mu}{}_{M}\overset{\perp}{g}_{\beta\rho}\delta_g\begin{Bmatrix}\rho\\ \lambda\ \ \mu\end{Bmatrix}+u_{\beta|L}\alpha^{\mu}{}_{M}u^{\sigma}\delta g_{\mu\sigma} \qquad (u^{\dagger}_{L}=0), \tag{95.9}$$

so that by (95.4), (69.4), (95.6), (95.8), and convention $(91.1)_1$

$$\delta_g C_{BLM} = \alpha^\beta{}_B x^\rho{}_{L|M} \overset{+}{g}_\beta{}^\lambda \overset{+}{g}_\rho{}^\sigma \delta g_{\lambda\sigma} + g_{\beta\rho} \alpha^{\beta\lambda\mu}_{BLM} \delta_g \begin{Bmatrix} \rho \\ \lambda \ \mu \end{Bmatrix}$$

$$+ \alpha^\beta{}_B u_{\beta|L} \alpha^\mu{}_M u^\sigma \delta g_{\mu\sigma} + \alpha^\beta{}_B u_{\beta|M} \alpha^\lambda{}_L u^\sigma \delta g_{\lambda\sigma} \quad (u^\dagger_L = 0). \tag{95.10a}$$

By $(53.13)_1$ and (95.3) we can turn (95.10a) into the equality

$$\delta_g C_{BLM} = \alpha^\lambda{}_\beta \alpha^\sigma{}_{LM} \delta g_{\lambda\sigma} + \alpha^{\beta\lambda\mu}_{BLM} \left(g_{\beta\rho} \delta_g \begin{Bmatrix} \rho \\ \lambda \ \mu \end{Bmatrix} + u_{\beta/\lambda} u^\sigma \delta g_{\mu\sigma} + u_{\beta/\mu} u^\sigma \delta g_{\lambda\sigma} \right), \tag{95.10b}$$

which certainly holds for $u^\dagger_L \neq 0$ also.

Let us now remark that for (x) locally geodesic we have

$$g_{\beta\rho} \delta_g \begin{Bmatrix} \rho \\ \lambda \ \mu \end{Bmatrix} = \delta_g \{\lambda\mu, \beta\} - \begin{Bmatrix} \rho \\ \lambda \ \mu \end{Bmatrix} \delta g_{\beta\rho} = \tfrac{1}{2} [(\delta g_{\beta\mu})_{/\lambda} + (\delta g_{\lambda\beta})_{/\mu} - (\delta g_{\lambda\mu})_{/\beta}], \tag{95.10c}$$

which by $(86.1)_2$ yields

$$m^{\beta\lambda\mu} g_{\beta\rho} \delta_g \begin{Bmatrix} \rho \\ \lambda \ \mu \end{Bmatrix} = m^{\beta\lambda\mu} (\delta g_{\beta\mu})_{/\lambda}. \tag{95.11}$$

Since $\delta_g \begin{Bmatrix} \rho \\ \lambda \ \mu \end{Bmatrix}$ and $(\delta g_{\beta\mu})_{/\lambda}$ are tensors, (95.11) holds in every frame.

As was remarked below (93.7), the analogues of $(93.6)_{2,\dots,4}$ for $w = \hat{w}(\eta, \varepsilon_{LM}, C_{BLM}, y^L)$ hold, so that

$$\begin{cases} w = \hat{w}(\eta, \varepsilon_{LM}, C_{BLM}, y^L), & \mathfrak{m}^{BLM}_* = k^* \dfrac{\partial w}{\partial C_{[BL]M}} + \lambda \varepsilon^{*BLM}, \\ Y^{(LM)} = -k^* \dfrac{\partial w}{\partial \varepsilon_{LM}} - 2\mathfrak{m}^{(LAB}_* \overset{-1}{C}{}^{M)S} C_{SAB}, & \dfrac{\partial w}{\partial C_{(BL)M}} = 0. \end{cases} \tag{95.12}$$

By $(61.3)_2$ $\varepsilon^{*BLM} \delta_g C_{BLM} = 0$. Hence $(95.12)_2$, and $(92.1)_3$ and (95.10b) yield

$$\frac{k^*}{\mathscr{D}} \frac{\partial w}{\partial C_{[BL]M}} \delta_g C_{BLM} = \frac{1}{\mathscr{D}} \mathfrak{m}^{BLM}_* \delta_g C_{BLM} = \mathfrak{A}^{\lambda\sigma} \delta g_{\lambda\sigma}$$

$$+ m^{\beta\lambda\mu} \left(g_{\beta\rho} \delta_g \begin{Bmatrix} \rho \\ \lambda \ \mu \end{Bmatrix} + u_{\beta/\lambda} u^\sigma \delta g_{\mu\sigma} + u_{\beta/\mu} u^\sigma \delta g_{\lambda\sigma} \right) \tag{95.13}$$

respectively, where

$$\mathfrak{A}^{\rho\sigma} = \frac{1}{\mathscr{D}} \mathfrak{m}^{BLM}_* \alpha^\rho{}_B \alpha^\sigma{}_{LM}. \tag{95.14}$$

Now let us remark that $(53.16)_2$, $(56.16)_2$, and $(53.16)_1$ yield

$$\overset{-1}{C}{}^{MS} C_{SAB} \alpha^{\sigma}{}_{M} = \overset{-1}{C}{}^{MS} g_{\lambda\mu} \alpha^{\lambda}{}_{S} \alpha^{\mu}{}_{AB} \alpha^{\sigma}{}_{M} = \overset{\perp}{g}{}^{\lambda\sigma} g_{\lambda\mu} \alpha^{\mu}{}_{AB} = \alpha^{\sigma}{}_{AB}. \tag{95.15}$$

respectively. Then

$$\alpha^{(\rho}{}_{L} \alpha^{\sigma)}{}_{AB} = \overset{-1}{C}{}^{MS} C_{SAB} \alpha^{(\rho}{}_{L} \alpha^{\sigma)}{}_{M} = \overset{-1}{C}{}^{MS} C_{SAB} \alpha^{\rho}{}_{(L} \alpha^{\sigma}{}_{M)}. \tag{95.16}$$

This yields—cf. (95.14)

$$\mathscr{D}\, \mathfrak{A}^{(\rho\sigma)} = \mathfrak{m}_{*}^{LAB} \alpha^{(\rho}{}_{L} \alpha^{\sigma)}{}_{AB} = \mathfrak{m}_{*}^{(LAB} \overset{-1}{C}{}^{M)S} C_{SAB} \alpha^{\rho}{}_{L} \alpha^{\sigma}{}_{M}. \tag{95.17}$$

By $(57.7)_2$ $\delta_g C_{LM} = 2\delta_g \varepsilon_{LM}$, so that (69.5) yields the first of the equalities

$$\begin{aligned} 2\frac{k^*}{\mathscr{D}} \frac{\partial w}{\partial \varepsilon_{LM}} \delta_g \varepsilon_{LM} &= \frac{k^*}{\mathscr{D}} \frac{\partial w}{\partial \varepsilon_{LM}} \alpha^{\rho}{}_{L} \alpha^{\sigma}{}_{M} \delta g_{\rho\sigma} \\ &= -\left(\frac{1}{\mathscr{D}} Y^{(LM)} \alpha^{\rho}{}_{L} \alpha^{\sigma}{}_{M} + 2\mathfrak{A}^{(\rho\sigma)}\right) \delta g_{\rho\sigma} = -(X^{(\rho\sigma)} + 2\mathfrak{A}^{(\rho\sigma)}) \delta g_{\rho\sigma}. \end{aligned} \tag{95.18}$$

The second follows from $(95.12)_3$ and (95.17), while $(59.2)_{1,2}$ yield $(95.18)_3$. From $(69.9)_1$, $(58.3)_1$, and $(95.12)_{1,4}$ we deduce the first of the equalities

$$\begin{aligned} k\delta_g w_c &= \frac{k^*}{\mathscr{D}} \frac{\partial w}{\partial \varepsilon_{LM}} \delta_g \varepsilon_{LM} + \frac{k^*}{\mathscr{D}} \frac{\partial w}{\partial C_{[BL]M}} \delta_g C_{BLM} \\ &= -\tfrac{1}{2} X^{(\rho\sigma)} \delta g_{\rho\sigma} + m^{\beta\lambda\mu} g_{\beta\rho} \delta_g \left\{ {\rho \atop \lambda\ \mu} \right\} + 2m^{\beta(\lambda\mu)} u_{\beta/\lambda} u^{\sigma} \delta g_{\mu\sigma} \\ &= -\tfrac{1}{2} X^{(\rho\sigma)} \delta g_{\rho\sigma} + m^{\beta\lambda\mu} (\delta g_{\beta\mu})_{/\lambda} - v^{\mu} u^{\sigma} \delta g_{\mu\sigma}, \end{aligned} \tag{95.19}$$

while $(95.19)_2$ follows from (95.18) and (95.13). Lastly we deduce $(95.19)_3$ from (95.11), $(86.1)_2$, and $(87.1)_1$.

§ 96. A Variational Principle Involving Couple Stress and the Variation of Space-Time Metric

We regard the field $g_{\alpha\beta} = \hat{g}_{\alpha\beta}(x^{\rho})$ and the motion $\hat{x}^{\rho}$ of $\mathscr{C}$ in S_4—cf. $(52.2)_1$—as given. Let C_4 be a bounded 4-dimensional domain in S_4 where $\hat{g}_{\alpha\beta}$ and $\hat{x}^{\rho}$ are of class $C^{(2)}$, i.e. are twice continuously differentiable. We denote the boundary of C_4 by $\mathscr{F} C_4$ and regard its outward face as positive. Let $\delta g_{\alpha\beta}$ be a variation of

$g_{\alpha\beta}$ of class $C^{(2)}$ in C_4. Furthermore assume

$$\delta g_{\alpha\beta}=0=(\delta g_{\alpha\beta})_{,\gamma}, \quad \text{hence} \quad (\delta g_{\alpha\beta})_{/\gamma}=0, \quad \text{on } \mathscr{F} C_4. \tag{96.1}$$

By $(96.1)_1$ and some well known integral properties of the divergence,

$$\int_{C_4} m^{\beta\lambda\mu}(\delta g_{\beta\mu})_{/\lambda}\sqrt{-g}\,dx=-\int_{C_4} m^{\rho\lambda\sigma}{}_{/\lambda}\delta g_{\rho\sigma}\sqrt{-g}\,dx, \tag{96.2}$$

where, as well in § 69 and the remainder of the book, dx stands for $dx^0\,dx^1\,dx^2\,dx^3$. Then, by (95.19), $(69.9)_2$, and equality (69.10b), proof of which is unaffected by couple stresses, we obtain

$$\delta_g\int_{C_4}\rho\sqrt{-g}\,dx=\int_{C_4}[\sqrt{-g}\,k\,\delta_g w_c+w_c\delta_g(k\sqrt{-g})]\,dx=-\tfrac{1}{2}\int_{C_4}\mathscr{U}^{\rho\sigma}\delta g_{\rho\sigma}\sqrt{-g}\,dx, \tag{96.3}$$

where

$$\mathscr{U}^{\rho\sigma}=\rho u^\rho u^\sigma+X^{(\rho\sigma)}+2m^{(\rho\lambda\sigma)}{}_{/\lambda}+2v^{(\rho}u^{\sigma)}. \tag{96.4}$$

The definition (87.3) of the total energy tensor proposed by Bressan [1966a] reduces to equality (96.4) exactly, for an elastic body, $\mathscr{C}$, that is capable of couple stress and not of heat conduction, in the absence of electromagnetic phenomena. By (69.12) and (96.3) *in this case the validity in* C_4 *of the gravitation equations* $(23.1)_1$ *is equivalent to the following variational condition:*

$$\delta_g I=0 \quad \textit{where} \quad I=\int_{C_4}\left(R+\frac{16\pi h}{c^4}\rho\right)\sqrt{-g}\,dx \tag{96.5}$$

for every twice continuously differentiable variation $\delta g_{\alpha\beta}$ *of* $g_{\alpha\beta}$ *that fulfills condition* (96.1).

Gravitation equations $(23.1)_1$ imply conservation equations $(23.3)_1$. In §§87,88 we proved that under the definition (87.3) of $\mathscr{U}_{\alpha\beta}$ the classical limit ($c\to\infty$) of equations $(23.3)_1$ is equivalent to the Cauchy equations and the equation of energy balance in the classical theory of couple stress—cf. G. Grioli [1960], [1962, Chap. X], and Toupin [1962], [1964]. By the afore-mentioned equivalence, condition (96.5) is in accordance with the same classical theory of couple stress, so that in particular it is physically acceptable.[4)] Then it can be used as a variational principle valid for the body $\mathscr{C}$ in the absence of heat conduction, instead of gravitation equations.

Let us remark that the exact equivalence of (96.5) to the gravitation equations $(23.1)_1$, proved under the definition (87.3) of $\mathscr{U}_{\alpha\beta}$ (proposed by Bressan [1966a]) and in the present particular (reversible) case, could not be foreseen. By the criteria on which this proposal was based, (at least) a difference constituted by terms of the order c^{-n} with $n\geqslant 2$ could reasonably be expected. The natural character of the variational principle (96.5) valid in the present particular (reversible) case, shows that gravitation equations are quite satisfactory (also in the general case) under the definition (87.3) of $\mathscr{U}_{\alpha\beta}$.

§ 97. A Variational Principle Involving the Variation of World Lines in the Presence of Couple Stresses. On Constitutive Equations

Let the assumptions on $\mathscr{C}$ made in § 96 still hold; in particular we regard the field $\hat{g}_{\alpha\beta}$ and the motion $\mathscr{M}$ of $\mathscr{C}$, i.e. $\hat{x}^\rho$, as given. But now we consider the arbitrary variation $\delta^* x^\rho$ of $\mathscr{M}$—i.e. of world lines—spoken of in § 70, so that, in particular, the decomposition formula (70.6b) holds. Let $\hat{g}_{\alpha\beta}[\hat{x}^\rho]$ be in $C^{(2)}[C^{(3)}]$.

Theorem 97.1. *Let $\mathscr{C}$ be uncapable of heat conduction and let its specific internal energy w have the expression $\hat{w}(y^L, \varepsilon_{LM}, C_{[BL]M})$. Then (in the absence of electromagnetic phenomena) the validity in C_4 of the conservation equations* (23.1)$_1$ *for the expression* (96.4) *of $\mathscr{U}_{\alpha\beta}$ is equivalent to the following variational condition:*

$$\delta_{\mathscr{M}} \int_{C_4} \rho \sqrt{-g}\, dx = 0, \quad or \quad \delta_{\mathscr{M}} I = 0, \quad with \quad I = \int_{C_4} \left(R + \frac{16\pi h}{c^4} \rho \right) \sqrt{-g}\, dx \tag{97.1}$$

for every variation $\zeta_\rho d\lambda$ of $\mathscr{M}$ that is of class $C^{(3)}$ and fulfills the boundary conditions

$$\zeta_\rho = 0 = \zeta_{\rho,\sigma} \quad on \quad \mathscr{F} C_4 . \tag{97.2}$$

Indeed the assumptions on w and $\mathscr{U}^{\alpha\beta}$ made in Theorem 70.2(a), (b) hold for $n=2$. Moreover (96.3) was deduced under the definition (96.4) of $\mathscr{U}_{\alpha\beta}$ from the only boundary condition (96.1)$_1$. Hence (70.18) holds for all $\delta g_{\alpha\beta}$ that satisfy (70.19) for $n=2$. Then by Theorem 70.2 (c) the conservation equations (23.3)$_1$ hold in C_4 if and only if $\delta_{\mathscr{M}} A = 0$ for all $\delta\mathscr{M}$ that satisfy (70.21) for $n=2$, i.e. (97.2). By (70.17)$_4$ $\delta_{\mathscr{M}} A = 0$ if and only if either of equalities (97.1)$_{1,2}$ holds. Hence our thesis is true. q.e.d.

This equivalence theorem holds in particular in special relativity ($R=0$), where the two alternatives (97.1)$_{1,2}$ coincide. It is natural to consider (97.1) as a variational principle to be postulated in special relativity instead of conservation equations.

By the equivalence theorems proved in sections 96 and 97 we can assert the following, in general relativity, for elastic bodies uncapable of heat conduction:

(i) gravitation equations imply that I—cf. (97.1)$_3$—should be stationary with respect to both the variations of space time metric that are of class $C^{(2)}$ and fulfill (96.1) [§ 96], and the variations of world lines that are of class $C^{(3)}$ and fulfill (97.2);

(ii) the variational condition (97.1) follows from the variational principle (96.5).[5)]

Now remark that *we have*

$$2m^{\beta\rho\mu}{}_{/[\beta\mu]} = m^{\beta\mu\rho}{}_{/\beta\mu} = -m^{\sigma\lambda\rho}{}_{/\lambda\sigma} . \tag{97.3}$$

Indeed by $(86.1)_2$, (87.6), and the cyclic and skewsymmetry properties of the Riemann tensor

$$2m^{\beta\mu\rho}{}_{/\beta\mu}=m^{\beta\mu\lambda}R_{\beta\mu\lambda}{}^{\rho}=-(m^{\mu\lambda\beta}+m^{\lambda\beta\mu})R_{\beta\mu\lambda}{}^{\rho}=(m^{\beta\lambda\mu}+m^{\beta\lambda\mu})R_{\beta\mu\lambda}{}^{\rho}=4m^{\beta\rho\mu}{}_{/[\beta\mu]}\,.$$

Hence $(97.3)_1$ holds. We deduce $(97.3)_2$ from $(86.1)_2$. q.e.d.

Let us remark that the elastic body $\mathscr{C}$ being considered is characterized by the two functions $w=\hat{w}(y^L,\varepsilon_{LM},C_{BLM})$ and $\lambda=\hat{\lambda}(y^L,\varepsilon_{LM},C_{BLM})$ through (95.12). Thus, on the one hand, the determinations of $\mathscr{U}_{\alpha\beta}$ and the gravitation equations for $\mathscr{C}$ are characterized by the same functions. On the other hand the variational conditions (96.5) is characterized by $\hat{w}$ alone. Hence by the equivalence assertion involving (96.5) $\hat{\lambda}$ cannot influence gravitation equations. Therefore *we can describe the behaviour of* $\mathscr{C}$ *also by keeping* $\hat{w}$ *unaltered and by assuming* $\hat{\lambda}=0$. *Under this assumption*

$$\mathfrak{m}_*^{BLM}=\frac{\partial w}{\partial C_{BLM}},\qquad \mathfrak{m}_*^{B[LM]}=0=m^{\beta[\lambda\mu]},\qquad m^{[\beta\lambda\mu]}=0 \tag{97.4}$$

and

$$v^\alpha=2m^{\alpha\rho\sigma}u_{\rho/\sigma},\qquad m^{\beta\rho\mu}{}_{/\beta\mu}=0=m^{\sigma\lambda\rho}{}_{/\sigma\lambda}\,. \tag{97.5}$$

Indeed $(97.4)_1$ follows from $(95.12)_{2,4}$ and yields $(97.4)_2$ by (A 4.3) and $(61.3)_2$. Then $(97.4)_3$ holds by $(92.1)_3$ and yields $(97.4)_4$ by $(86.1)_2$. We deduce (97.5) from $(97.4)_4$ and $(87.1)_1$. By $(97.4)_4$ $m^{\beta\rho\mu}{}_{/[\beta\mu]}=0$; hence $(97.5)_3$ holds by (97.3). Thence $(97.5)_2$ follows by $(86.1)_2$.

§ 98. On General Materials of Order $n=2$ in the Adiabatic Case. Variations of $g_{\alpha\beta}$ and World Lines

Following Pitteri [1975a] we extend sections 96 and 97 on polar materials to general materials of order $n=2$, except that we use the reduction Theorem 70.2 to deal with variations of world lines.

In section 21 the specific internal energy w was defined in terms of gravitational mass. This determination of w, based on the principle of equivalence of mass and energy, does not require the knowledge of the expression of the work $dl^{(i)}$ of internal forces, in contrast to what is done in classical physics. This fact is useful in dealing with general elastic materials of order $n\geqslant 2$; more in particular it allows us to avoid the choice of an appropriate classical definition of this work and any relativization of its.

As well as in sections 96 and 97 we assume that the body $\mathscr{C}$ has a vanishing coefficient of thermal conduction, so that the specific entropy η can be supposed to be constant in the world tube $W_{\mathscr{C}}$ of $\mathscr{C}$ and in particular in its portion C_4. We now regard w as a (twice continuously differentiable) function of the form

$w=\hat{w}(y^L, \varepsilon_{LM}, C_{BLM})$. Let $\hat{g}_{\alpha\beta}$ and $\hat{x}^\rho$ be in $C^{(2)}$. We define the Lagrangian and Eulerian hyperstress tensors $\mathfrak{m}_*^{BLM}$ $[m^{\beta\lambda\mu}]$

$$\mathfrak{m}_*^{BLM} = k^* \frac{\partial w}{\partial C_{BLM}}, \qquad m^{\beta\lambda\mu} = \frac{1}{\mathscr{D}} \alpha^{\beta\lambda\mu}_{BLM} \mathfrak{m}_*^{BLM} \tag{98.1}$$

and the Lagrangian and Eulerian stress tensors Y^{LM} and $X^{\lambda\mu}$:

$$Y^{LM} = -k^* \frac{\partial w}{\partial \varepsilon_{LM}} - 2\mathfrak{m}_*^{LAB} \overset{-1}{C}{}^{MS} C_{SAB}, \qquad X^{\lambda\mu} = \frac{1}{\mathscr{D}} \alpha^\lambda{}_L \alpha^\mu{}_M Y^{LM} \tag{98.2}$$

where $(98.1)_2$ is based on (91.1) and coincides with $(92.1)_3$. Since by $(61.3)_2$ $C_{B[LM]}=0$, (98.1) yields—cf. (A 4.3) in Appendix A

$$\mathfrak{m}_*^{B[LM]} = 0, \qquad m^{\beta[\lambda\mu]} = 0. \tag{98.3}$$

We further set

$$v^\alpha = -m^{\lambda\alpha\mu} u_{\lambda/\mu}, \qquad \hat{v}^\alpha = -m^{(\lambda\alpha)\mu} u_{\lambda/\mu} \quad (\text{hence } u_\alpha v^\alpha = 0 = u_\alpha \hat{v}^\alpha) \tag{98.4}$$

and

$$\mathscr{U}^{\alpha\beta} = \rho u^\alpha u^\beta + X^{(\alpha\beta)} + m^{\alpha\lambda\beta}{}_{/\lambda} - m^{\lambda\alpha\beta}{}_{/\lambda} + m^{(\alpha\beta)\lambda}{}_{/\lambda} + 2v^{(\alpha} u^{\beta)}. \tag{98.5}$$

According to the remark at the end of section 97, the behaviour of any polar material can be based on constitutive equations of the form (95.12) with $\lambda \equiv 0$, and this implies (97.4) and (97.5), so that $(98.1)_1$, (98.3), and $(98.4)_1$ are theorems for these materials. The same holds for (98.2) by $(98.1)_1$ and $(95.12)_3$ as far as the symmetric part of Y^{LM} is concerned. By $(86.1)_2$, the expression (98.5) of $\mathscr{U}^{\alpha\beta}$ is the one used for polar materials, i.e. (87.3). Thus we have shown the acceptability of those among the definitions above that in connection with general materials of order 2 introduce some magnitudes already used for polar materials.

Let us remember that for polar materials $m^{\beta\lambda\mu}$ has a physical Eulerian definition. In addition the validity of conditions (a) to (c) in section 93, which characterize elastic materials, is independent of the reference configuration C^*. In connection with the form (87.3) of $\mathscr{U}^{\alpha\beta}$ we proved that a material is elastic if and only if it has constitutive equations of the form (93.6). Hence *if a material has such constitutive equations in connection with a choice of* C^*, *the same holds for every other choice of* C^*. Now it is outstanding the question whether or not the tensor $m^{\beta\lambda\mu}$ defined in terms of $\hat{w}$ by means of the non-Eulerian definition (98.1), which involves C^*, is independent of C^*. Pitteri [1975a, sect. 4] proved that the answer is affirmative.

Let us further remark that by $(86.1)_2$ and $(98.4)_2$, $\hat{v}^\alpha = 0$ *holds for polar materials.* In addition by (98.1) the tensor $m^{[\beta\lambda]\mu}$ is the Eulerian counterpart of $\partial w / \partial C_{[BL]M}$, so that on the basis of $(95.12)_2$ with $\lambda \equiv 0$, it is natural to call $m_*^{[BL]M}$ $[m^{[\beta\lambda]\mu}]$ *Lagrangian [Eulerian] couple stress tensor* also in the general case.

Theorem 98.1. *A continuous body, $\mathscr{C}$, that is uncapable of heat conduction and has the specific internal energy of the form $w=\hat{w}(y^L, \varepsilon_{LM}, C_{BLM})$ satisfies the variational condition (96.5) ($\delta_g I=0$) for every twice continuously differentiable solution $\delta g_{\alpha\beta}$ of (96.1) in C_4, if and only if the gravitation equations $(23.1)_1$ hold under the definition (98.5) of $\mathscr{U}^{\alpha\beta}$.*

Proof. By $(58.3)_1$ and $(98.1)_1$

$$k\delta_g w_c = \frac{k^*}{\mathscr{D}}\frac{\partial w}{\partial \varepsilon_{LM}}\delta_g \varepsilon_{LM} + \frac{k^*}{\mathscr{D}} m_*^{BLM}\delta_g C_{BLM} \qquad (w_c = w + c^2). \tag{98.6}$$

Now from (95.10c) we deduce, instead of (95.11),

$$m^{\beta\lambda\mu} g_{\beta\rho}\delta_g \begin{Bmatrix} \rho \\ \lambda\ \mu \end{Bmatrix} = (m^{[\beta\lambda]\mu} + \tfrac{1}{2} m^{(\beta\mu)\lambda})(\delta g_{\beta\mu})_{/\lambda}. \tag{98.7}$$

We consider the definition (95.14) of $\mathfrak{A}^{\rho\sigma}$ as holding again. Thence, by $(98.1)_1$ and (98.2), we deduce again $(95.13)_2$, (95.17), and (95.18). Substitution of $(95.13)_2$ and (95.18) into (98.6) yields the first of the following generalizations of equalities (95.19)—see $(98.3)_2$

$$\begin{aligned} k\delta_g w_c &= -\tfrac{1}{2} X^{(\rho\sigma)}\delta g_{\rho\sigma} + m^{\beta\lambda\mu} g_{\beta\rho}\delta_g \begin{Bmatrix} \rho \\ \lambda\ \mu \end{Bmatrix} + 2m^{\beta\lambda\mu} u_{\beta/\lambda} u^\sigma \delta g_{\mu\sigma} \\ &= -\tfrac{1}{2} X^{(\rho\sigma)}\delta g_{\rho\sigma} + (m^{[\beta\lambda]\mu} + \tfrac{1}{2} m^{(\beta\mu)\lambda})(\delta g_{\beta\mu})_{/\lambda} - 2v^\mu u^\sigma \delta g_{\mu\sigma}. \end{aligned} \tag{98.8}$$

The second follows from (98.7), $(98.3)_2$, and $(98.4)_1$.

By the divergence theorem the *only boundary conditions* $(96.1)_1$ imply the analogues for $m^{[\beta\lambda]\mu}$ and $m^{(\beta\lambda)\mu}$ of condition (96.2) on $m^{\beta\lambda\mu}$. Furthermore $\rho = kw_c$ and $w_c = w + c^2$—cf. (69.9). Hence by (69.10b), (98.8), and (98.5), we deduce the validity of (96.3) for the new determination (98.5) of $\mathscr{U}^{\alpha\beta}$, under the *only* boundary condition $(96.1)_1$. Then by (69.12) we see that also in the present case characterized by (98.5), the variational condition (96.5)—i.e. the vanishing of $\delta_g I$ for all twice continuously differentiable solutions of (96.1)—is equivalent to the validity of the Einstein gravitation equation $(23.1)_1$ in C_4. q.e.d.

Now we consider again the variation $\zeta_\rho d\lambda$ of the motion $\hat{x}^\rho$ of $\mathscr{C}$ introduced in section 97; and we assume that $\hat{g}_{\alpha\beta}$ and $\hat{x}^\rho$ are in $C^{(3)}$.

Theorem 98.2. *A continuous body $\mathscr{C}$, that is uncapable of heat conduction and has the specific internal energy of the form $w=\tilde{w}(y^L, \varepsilon_{LM}, C_{BLM})$ satisfies the variational condition $(97.1)_1$ or $(97.1)_2$ ($\delta_{\mathscr{M}} I=0$) for every variation $\zeta_\rho d\lambda$ of $\mathscr{M}$ that is of class $C^{(3)}$ and fulfills (97.2), if and only if the conservation equations $(23.3)_1$ hold in C_4 for the expression (98.5) of $\mathscr{U}^{\alpha\beta}$.*

Since (96.3) holds also in the present case for all variations $\delta g_{\alpha\beta}$ in $C^{(2)}$ that satisfy $(96.1)_1$ as was remarked above, the proof of Theorem 98.2 can be obtained substantially from the one of Theorem 97.1 by substituting C_{BLM} for $C_{[BL]M}$.

§ 99. Variational Principles for Elastic Materials of any Order $n \geqslant 1$, not Capable of Heat Conduction

In this section the results of Pitteri [1975b] are presented. On the whole we have shortened his proofs on the basis of Theorem 70.2—cf. fts. 6, 7 in Chapter 7.[6)]

Fix $n \geqslant 1$ and let the specific internal energy w of $\mathscr{C}$ have an expression of the form $w = \hat{w}(y^L, C_{BL_1}, \ldots, C_{BL_1 \ldots L_n})$. We also assume that $\hat{w}$ is in $C^{(2)}$, $\hat{g}_{\alpha\beta}$ is in $C^{(n)}$ and $\hat{x}^\rho$—cf. (52.2)—is in $C^{(n+1)}$.

Theorem 99.1. *If the double tensor $T^{\alpha\cdots}{}_{A\ldots}$ has a tensorial $\hat{t}$-independent expression in the representation $\hat{x}^\rho$ of the motion $\mathscr{M}$ of $\mathscr{C}$*—cf. section 70,

$$\delta_g T^{\alpha\cdots}{}_{A\cdots|M} = (\delta_g T^{\alpha\cdots}{}_{A\cdots})_{|M} + \alpha^\rho{}_M u^\sigma \frac{DT^{\alpha\cdots}{}_{A\cdots}}{Ds} \delta g_{\rho\sigma} + T^{\rho\cdots}{}_{A\ldots} \alpha^\sigma{}_M \delta_g \begin{Bmatrix} \alpha \\ \rho\ \sigma \end{Bmatrix} + \cdots. \tag{99.1}$$

Indeed fix the event point $\mathscr{E}$. We can choose the time parameter $t = \hat{t}(x)$ in such a way that $u^\dagger = 0$ at $\mathscr{E}$—cf. (52.8). Then by (53.6) and by (95.6) and (53.8) we respectively have at $\mathscr{E}$

$$\delta_g T^{\alpha\cdots}{}_{A\cdots|M} = \delta_g T^{\alpha\cdots}{}_{A\cdots;M} + \frac{DT^{\alpha\cdots}{}_{A\cdots}}{Ds} \delta_g u^\dagger_M, \qquad \delta_g u^\dagger_M = \alpha^\rho{}_M u^\sigma \delta g_{\rho\sigma}. \tag{99.2}$$

By writing e.g. $\delta_g T^{\alpha\cdots}{}_{A\cdots,\tau}$ for $\delta_g(T^{\alpha\cdots}{}_{A\cdots,\tau})$ we obviously have

$$\delta_g T_{\cdots}^{\cdots}{}_{,\tau} = (\delta_g T_{\cdots}^{\cdots})_{,\tau}, \quad \delta_g T_{\cdots}^{\cdots}{}_{,M} = (\delta_g T_{\cdots}^{\cdots})_{,M}, \qquad \delta_g x^\rho{}_M = 0 = \delta_g \begin{Bmatrix} L \\ M\ N \end{Bmatrix}, \tag{99.3}$$

so that by (A2.3), (A2.4), and (A2.6) in Appendix A

$$\begin{aligned} \delta_g T^{\alpha\cdots}{}_{A\cdots;M} &= \left(\delta_g T^{\alpha\cdots}{}_{A\cdots,\tau} + \begin{Bmatrix} \alpha \\ \rho\tau \end{Bmatrix} \delta_g T^{\rho\cdots}{}_{A\cdots} + T^{\rho\cdots}{}_{A\cdots} \delta_g \begin{Bmatrix} \alpha \\ \rho\tau \end{Bmatrix} + \cdots \right) x^\tau{}_M \\ &\quad + (\delta_g T^{\alpha\cdots}{}_{A'\cdots}) \begin{Bmatrix} A' \\ AM \end{Bmatrix} + \cdots = (\delta_g T^{\alpha\cdots}{}_{A\cdots})_{;M}. \end{aligned} \tag{99.4}$$

This and (99.2) yield equality (99.1), whose validity is independent of the choice of $\hat{t}$. q.e.d.

By $(53.7)_1$ and $(53.16)_{1,2}$, from $(99.3)_3$ and from (69.4) and (99.1) we respectively deduce

$$\begin{aligned} \delta_g C_{BL_1\ldots L_n} &= x^\rho{}_B (\delta_g \overset{+}{g}_{\rho\sigma}) \alpha^\sigma{}_{L_1\ldots L_{n-1}|L_n} + \alpha_{\sigma B} \delta_g \alpha^\sigma{}_{L_1\ldots L_{n-1}|L_n} \\ &= \alpha^\rho{}_B \alpha^\sigma{}_{L_1\ldots L_n} \delta g_{\rho\sigma} + \alpha_{\rho B} \left[\frac{D\alpha^\rho{}_{L_1\ldots L_n}}{Ds} \alpha^\sigma{}_{L_n} u^\tau \delta g_{\sigma\tau} \right. \\ &\quad \left. + (\delta_g \alpha^\rho{}_{L_1\ldots L_{n-1}})_{|L_n} + \alpha^\gamma{}_{L_1\ldots L_{n-1}} \alpha^\tau{}_{L_n} \delta_g \begin{Bmatrix} \rho \\ \gamma\tau \end{Bmatrix} \right]. \end{aligned} \tag{99.5}$$

Since $2\begin{Bmatrix}\rho\\ \gamma\tau\end{Bmatrix}=g^{\rho\sigma}\{g_{\gamma\sigma,\tau}+g_{\sigma\tau,\gamma}-g_{\gamma\tau,\sigma}\}$, by using locally geodesic co-ordinates we easily see that

$$\delta_g\begin{Bmatrix}\rho\\ \gamma\tau\end{Bmatrix}=B_{\gamma\tau}{}^{\rho\alpha\beta\sigma}(\delta g_{\alpha\beta})_{/\sigma}, \tag{99.6}$$

where

$$2B_{\gamma\tau}{}^{\rho\alpha\beta\sigma}=g^{\rho\alpha}g^{\beta}{}_{(\gamma}g^{\sigma}{}_{\tau)}+g^{\rho\beta}g^{\alpha}{}_{(\gamma}g^{\sigma}{}_{\tau)}-g^{\rho\sigma}g^{\alpha}{}_{(\gamma}g^{\beta}{}_{\tau)}\,. \tag{99.7}$$

By the tensor character of both sides of (99.6) this equality holds in all frames.

Theorem 99.2. *For every* $n\geqslant 1$

$$\delta_g\alpha^{\rho}{}_{L_1\ldots L_n}=K^{\rho\beta_1\beta_2}_{L_1\ldots L_n}\delta g_{\beta_1\beta_2}+\sum_{i=3}^{n+1}K^{\rho\beta_1\ldots\beta_i}_{L_1\ldots L_n}(\delta g_{\beta_1\beta_2})_{/\beta_3\ldots\beta_i}, \tag{99.8}$$

where the $\hat{t}$-independent double tensors $K^{\rho\beta_1\ldots\beta_i}_{L_1\ldots L_n}$ $(2\leqslant i\leqslant n+1=2,3,\ldots)$ are defined recursively (with respect to n and i) by (99.9) *to* (99.13) *below*—cf. (99.7)

$$K^{\rho\beta_1\beta_2}_{L_1}=u^{\rho}u^{\beta_1}\alpha^{\beta_2}{}_{L_1}, \tag{99.9}$$

$$K^{\rho\beta_1\beta_2}_{L_1\ldots L_n}=u^{\rho}u^{\beta_1}\alpha^{\beta_2}{}_{L_1\ldots L_n}+\left(\frac{D}{Ds}\alpha^{\rho}{}_{L_1\ldots L_{n-1}}\right)\alpha^{\beta_1}{}_{L_n}u^{\beta_2}+\overset{\perp}{g}{}^{\rho}{}_{\sigma}K^{\sigma\beta_1\beta_2}_{L_1\ldots L_{n-1}|L_n}\quad (n>1) \tag{99.10}$$

$$\begin{aligned}K^{\rho\beta_1\beta_2\beta_3}_{L_1\ldots L_n}=\overset{\perp}{g}{}^{\rho}{}_{\sigma}[&K^{\sigma\beta_1\beta_2\beta_3}_{L_1\ldots L_{n-1}|L_n}+K^{\sigma\beta_1\beta_2}_{L_1\ldots L_{n-1}}\alpha^{\beta_3}{}_{L_n}\\ &+\alpha^{\gamma}{}_{L_1\ldots L_{n-1}}\alpha^{\tau}{}_{L_n}B_{\gamma\tau}{}^{\sigma\beta_1\beta_2\beta_3}]\quad (n>1)\,,\end{aligned} \tag{99.11}$$

$$K^{\rho\beta_1\ldots\beta_i}_{L_1\ldots L_n}=\overset{\perp}{g}{}^{\rho}{}_{\sigma}(K^{\sigma\beta_1\ldots\beta_i}_{L_1\ldots L_{n-1}|L_n}+K^{\sigma\beta_1\ldots\beta_{i-1}}_{L_1\ldots L_{n-1}}\alpha^{\beta_i}{}_{L_n})\quad (i=3,\ldots n;\ n>1)\,, \tag{99.12}$$

$$K^{\rho\beta_1\ldots\beta_{n+1}}_{L_1\ldots L_n}=\overset{\perp}{g}{}^{\rho}{}_{\sigma}K^{\sigma\beta_1\ldots\beta_n}_{L_1\ldots L_{n-1}}\alpha^{\beta_{n+1}}{}_{L_n}\quad (n>1)\,. \tag{99.13}$$

Indeed, since $\delta_g g^{\rho}{}_{\sigma}=0$, by (69.2a) and (69.2b)

$$\delta_g\overset{\perp}{g}{}^{\rho}{}_{\sigma}=\delta_g(u^{\rho}u_{\sigma})=u^{\rho}u_{\sigma}u^{\alpha}u^{\beta}\delta g_{\alpha\beta}+u^{\rho}u^{\beta}\delta g_{\sigma\beta}=u^{\rho}u^{\beta}\overset{\perp}{g}{}_{\sigma}{}^{\alpha}\delta g_{\alpha\beta}\,. \tag{99.14}$$

Hence $(99.3)_3$ and $(53.7)_1$ yield

$$\delta_g\alpha^{\rho}{}_{L_1}=x^{\sigma}{}_{L_1}\delta_g\overset{\perp}{g}{}^{\rho}{}_{\sigma}=u^{\rho}u^{\beta}\overset{\perp}{g}{}_{\sigma}{}^{\alpha}x^{\sigma}{}_{L_1}\delta g_{\alpha\beta}=u^{\rho}u^{\beta_1}\alpha^{\beta_2}{}_{L_1}\delta g_{\beta_1\beta_2}\,. \tag{99.15}$$

Thus the theorem holds for $n=1$. Now fix $n>1$ and assume that (99.8) holds for $n-1$. Furthermore by $(\delta g_{\beta_1\beta_2})_{/\beta_3\ldots\beta_i}$ for $i=2$ let us mean $\delta g_{\beta_1\beta_2}$; then by $(53.13)_1$

$$(\delta_g\alpha^{\sigma}{}_{L_1\ldots L_{n-1}})_{|L_n}=\sum_{i=2}^{n}[K^{\sigma\beta_1\ldots\beta_i}_{L_1\ldots L_{n-1}|L_n}(\delta g_{\beta_1\beta_2})_{/\beta_3\ldots\beta_i}+K^{\sigma\beta_1\ldots\beta_i}_{L_1\ldots L_{n-1}}\alpha^{\beta_{i+1}}_{L_n}(\delta g_{\beta_1\beta_2})_{/\beta_3\ldots\beta_{i+1}}]. \tag{99.16}$$

From $(53.16)_1$ and from (99.14), (99.1), and (99.6) we obtain

$$\delta_g \alpha^\rho{}_{L_1\ldots L_n} = (\delta_g \overset{\perp}{g}{}^\rho{}_\sigma)\alpha^\sigma{}_{L_1\ldots L_{n-1}|L_n} + \overset{\perp}{g}{}^\rho{}_\sigma \delta_g \alpha^\sigma{}_{L_1\ldots L_{n-1}|L_n} = \overset{\perp}{g}{}^\rho{}_\sigma (\delta_g \alpha^\sigma{}_{L_1\ldots L_{n-1}})_{|L_n}$$
$$+\left(u^\rho u^{\beta_1} \overset{\perp}{g}{}_\sigma{}^{\beta_2} \alpha^\sigma{}_{L_1\ldots L_{n-1}|L_n} + \alpha^{\beta_1}{}_{L_n} u^{\beta_2} \frac{D}{Ds} \alpha^\rho{}_{L_1\ldots L_{n-1}}\right) \delta g_{\beta_1\beta_2}$$
$$+\overset{\perp}{g}{}^\rho{}_\sigma \alpha^\gamma{}_{L_1\ldots L_{n-1}} \alpha^\tau{}_{L_n} B_{\gamma\tau}{}^{\sigma\beta_1\beta_2\beta_3} (\delta g_{\beta_1\beta_2})_{/\beta_3}\,. \tag{99.17}$$

By the definitions (99.10) to (99.13), from $(53.16)_1$, (99.16), and (99.17) we deduce the validity of (99.8) for n. We conclude that (99.8) holds for every $n \geq 1$. q.e.d.

Corollary. *We have*

$$\delta_g C_{BL_1\ldots L_n} = \sum_{i=2}^{n+1} H^{\beta_1\ldots\beta_i}_{L_1\ldots L_n} (\delta g_{\beta_1\beta_2})_{/\beta_3\ldots\beta_i}\,, \tag{99.18}$$

where

$$H^{\beta_1\beta_2}_{BL_1\ldots L_n} = \alpha^\rho{}_B K^{\rho\beta_1\beta_2}_{L_1\ldots L_n} + \alpha^{\beta_1}{}_B \alpha^{\beta_2}{}_{L_1\ldots L_n}\,; \quad H^{\beta_1\ldots\beta_i}_{BL_1\ldots L_n} = \alpha_{\rho B} K^{\rho\beta_1\ldots\beta_i}_{L_1\ldots L_n} \quad (i=3,\ldots,n+1). \tag{99.19}$$

Indeed by $(53.16)_2$ $C_{BL_1\ldots L_n} = \overset{\perp}{g}_{\rho\sigma} \alpha^\rho{}_B \alpha^\sigma{}_{L_1\ldots L_n}$, so that by (69.4) and (99.15)

$$\delta_g C_{BL_1\ldots L_n} = \alpha^\rho{}_B \alpha^\sigma{}_{L_1\ldots L_n} \delta g_{\rho\sigma} + \overset{\perp}{g}_{\rho\sigma} \alpha^\rho{}_B \delta_g \alpha^\sigma{}_{L_1\ldots L_n}\,.$$

Thence (99.18) follows by (99.8) and (99.19). q.e.d.

Theorem 99.3. *Let $\mathscr{C}$ be uncapable of heat conduction and let its specific internal energy have the form $w = \hat{w}(y^L, C_{BL_1}, \ldots, C_{BL_1\ldots L_n})$. Moreover let $\mathscr{U}^{\alpha\beta}$ be the symmetric tensor $\mathscr{U}^{\alpha\beta}_{[n]}$ that is defined recursively by*—cf. (99.19)

$$\mathscr{U}^{\rho\sigma}_{[1]} = \rho u^\rho u^\sigma - k \frac{\partial \tilde{w}}{\partial C_{BL_1}} \alpha^\rho{}_B \alpha^\sigma{}_{L_1}\,; \quad \text{for} \quad m>1$$
$$\mathscr{U}^{\rho\sigma}_{[m]} = \mathscr{U}^{\rho\sigma}_{[m-1]} - 2k \frac{\partial \tilde{w}}{\partial C_{BL_1\ldots L_m}} H^{(\rho\sigma)}_{BL_1\ldots L_m} - 2\sum_{i=3}^{m+1} (-1)^i \left(k \frac{\partial \tilde{w}}{\partial C_{BL_1\ldots L_i}} H^{(\rho\sigma)\beta_3\ldots\beta_i}_{BL_1\ldots L_i}\right). \tag{99.20}$$

Then

$$2\delta_g \int_{C_4} \rho \sqrt{-g}\, dx = -\int_{C_4} \mathscr{U}^{\alpha\beta} \delta g_{\alpha\beta} \sqrt{-g}\, dx \tag{99.21a}$$

for all variations $\delta g_{\alpha\beta}$ that are in $C^{(n+1)}$ and satisfy the condition

$$(\delta g_{\alpha\beta})_{,\lambda_1\ldots\lambda_i} = 0 \quad (i=0,\ldots,n-2) \quad \text{on} \quad \mathscr{F} C_4\,. \tag{99.22}$$

Indeed in (59.21a) we can replace δ_g with $\delta_g^{[n]}$, where—cf. (69.9)

$$\delta_g^{[m]} \int_{C_4} \rho \sqrt{-g}\, dx = \int_{C_4} \left[w_c \delta_g (k\sqrt{-g}) + k\sqrt{-g} \sum_{i=1}^{m} \frac{\partial w_c}{\partial C_{BL_1\ldots L_i}} \delta_g C_{BL_1\ldots L_i} \right] dx\,. \tag{99.23}$$

Furthermore from (69.10b) we obtain, for $m=1$,

$$2\delta_g^{[m]}\int_{C_4}\rho\sqrt{-g}\,dx=-\int_{C_4}\mathscr{U}^{\alpha\beta}_{[m]}\,\delta g_{\alpha\beta}\sqrt{-g}\,dx\,. \tag{99.21b}$$

Now assume that (99.21b) holds for $m-1$ ($\geqslant 1$). From (99.23) and (99.18) we deduce the first two among the equalities

$$\begin{aligned}(\delta_g^{[m]}-\delta_g^{[m-1]})\int_{C_4}\rho\sqrt{-g}\,dx&=\int_{C_4}\frac{\partial\hat{w}}{\partial C_{BL_1\ldots L_m}}(\delta_g C_{BL_1\ldots L_m})k\sqrt{-g}\,dx\\ &=\sum_{i=2}^{m+1}\int_{C_4}k\frac{\partial\hat{w}}{\partial C_{BL_1\ldots L_m}}H^{\beta_1\ldots\beta_i}_{BL_1\ldots L_m}(\delta g_{\beta_1\beta_2})_{/\beta_3\ldots\beta_i}\sqrt{-g}\,dx\\ &=\sum_{i=2}^{m+1}(-1)^i\int_{C_4}\left(k\frac{\partial\tilde{w}}{\partial C_{BL_1\ldots L_m}}\right)_{/\beta_3\ldots\beta_i}H^{\beta_1\ldots\beta_i}_{BL_1\ldots L_m}\,\delta g_{\beta_1\beta_2}\sqrt{-g}\,dx\end{aligned} \tag{99.24}$$

respectively, while Greens' lemma and (99.22) yield $(99.24)_3$. Then, by $(99.20)_2$, (99.21b) holds for m. We conclude that (99.21b) holds for every $m\geqslant 1$ and in particular for $m=n$. q.e.d.

Theorem 99.4. *Under the assumptions made in Theorem 99.3 the gravitation equations* $(23.1)_1$ *hold in* C_4 *for* $\mathscr{U}^{\alpha\beta}=\mathscr{U}^{\alpha\beta}_{[n]}$—cf. (99.20)—*if and only if*

$$\delta_g I=0 \quad \textit{where} \quad I=\int_{C_4}\left(R+\frac{16\pi h}{c^4}\rho\right)\sqrt{-g}\,dx \tag{99.25}$$

for all variations $\delta g_{\alpha\beta}$ *that are in* $C^{(2)}$ *and* $C^{(n-1)}$, *and that satisfy* $(99.22)_1$ *for* $0\leqslant i\leqslant \max(1,n-2)$.

Indeed for the variations above (69.12) and (99.21a) hold, so that $(99.25)_1$ is equivalent to $\int_{C_4}(A^{\alpha\beta}+8\pi hc^{-4}\mathscr{U}^{\alpha\beta})\delta g_{\alpha\beta}\sqrt{-g}\,dx=0$. This implies the theorem. q.e.d.

Theorem 99.5. *Under the assumptions made in Theorem 99.3 and under the definitions (99.20) and* $(99.25)_2$ *the conservation equations* $\mathscr{U}^{\alpha\beta}{}_{/\beta}=0$ *hold in* C_4 *for* $\mathscr{U}^{\alpha\beta}=\mathscr{U}^{\alpha\beta}_{[n]}$ *if and only if* $\delta_{\mathscr{M}}I=0$ *for all variations* $\zeta^\rho\,d\lambda$—cf. (70.3)—*of the motion* $\mathscr{M}$ *of* $\mathscr{C}$, *that are of class* $C^{(n)}$ *in* C_4 *and satisfy the boundary conditions*

$$\zeta^\rho{}_{,\lambda_1\ldots\lambda_i}=0 \quad (0\leqslant i\leqslant n-1) \quad \textit{on} \quad \mathscr{F}C_4\,. \tag{99.26}$$

Indeed the assumptions on w and $\mathscr{U}^{\alpha\beta}$ made in Theorem 70.1 are fulfilled. In addition Theorem 99.3 holds. Then Theorem 70.2 (c) implies the theorem. q.e.d.

Footnotes to Chapter 10

[1] Kröner [1959/60] emphasized the necessity of considering $d\mathcal{M}_\alpha \neq 0$ under certain physical circumstances.

[2] Incidentally Cosserats theory of generalized media [1907] was revived in 1958 by Günter and by Ericksen and Truesdell, who developed a theory in classical physics more general than the one of polar materials.

[3] Cf. Truesdell and Toupin [1960, (241.4), p. 609] and note that these authors use the opposite of the classical analogues of our tensors $X^{\alpha\beta}$, $m^{\alpha\beta\gamma}$, and $c q^\alpha$.

[4] We have to remember that Berdicewski [1966] refers to materials more general than those considered in the present chapter. He postulates the vanishing of the variation of I for certain variations of $g_{\alpha\beta}$ ($\delta'_g I = 0$) and for certain variations of the motion $\mathcal{M}$ of $\mathscr{C}$ ($\delta'_{\mathcal{M}} I = 0$). Then he deduces that certain two local conditions (generalized gravitation and conservation equations) are equivalent to the conditions $\delta'_g I = 0$ and $\delta'_{\mathcal{M}} I = 0$ respectively and involve two tensors of total energy $\mathscr{U}^{(1)}_{\alpha\beta}$ and $\mathscr{U}^{(2)}_{\alpha\beta}$ respectively. Berdicewski writes that he states no general criterium to know whether these tensors coincide. However he proves that they certainly do in certain interesting cases. In the theory presented in this book the analogues of $\mathscr{U}^{(1)}_{\alpha\beta}$ and $\mathscr{U}^{(2)}_{\alpha\beta}$ coincide—cf. Theorem 70.2. Pitteri [1975a, b] proved that they do so also in connection with general materials of order $n \geqslant 2$—cf. §§ 97—99 and in particular footnote 5 in Chapter 10.

Let us add that Berdicewski seems not to be interested in checking carefully the accord, or possibly the exact equivalence, between the afore-mentioned local equations such as conservation equations $(23.3)_1$ [their classical limits] and some already stated relativistic [classical] theory on the same subject. Probably this occurs because actually the experimental tests of the latter (classical) theories are unsatisfactory, as far as so general materials are concerned.

The topic above is dealt with by Bressan [1972b] as far as couple stress are concerned. In particular he proved the accordance of (96.5) with the classical theory of couple stress by Grioli and Toupin.

[5] As was hinted at in footnote 4 in Chapter 10, Berdicewski [1966] postulates both variational conditions $\delta'_g I = 0$ and $\delta_{\mathcal{M}} I = 0$—cf. (96.5), $(97.1)_2$. In correspondence with them he considers two different total energy tensors, say $\mathscr{U}^{(1)}_{\alpha\beta}$ and $\mathscr{U}^{(2)}_{\alpha\beta}$, having the roles of our $\mathscr{U}_{\alpha\beta}$ in gravitation equations $(23.1)_1$ and conservation equations $(23.3)_1$ respectively. He writes that, generally, only the first is symmetric and that they are different. He adds that $\mathscr{U}^{(1)}_{\alpha\beta}$ and $\mathscr{U}^{(2)}_{\alpha\beta}$ coincide in particular cases, e.g. for non-viscous fluids or ordinary elastic bodies.

Berdicewski's results may induce people to think that in general $\mathscr{U}^{(1)}_{\alpha\beta} \neq \mathscr{U}^{(2)}_{\alpha\beta}$ (and that the equations $\mathscr{U}^{(1)/\beta}_{\alpha\beta} = 0$ and $\mathscr{U}^{(2)/\beta}_{\alpha\beta} = 0$ should be mutually independent) also in the case of couple stress. This is disproved by Bressan [1972b]—cf. assertion (ii) above, which implies $\mathscr{U}^{(1)}_{\alpha\beta} \equiv \mathscr{U}^{(2)}_{\alpha\beta}$ in the case of couple stress. Incidentally this result is related with Bressan's systematic use of Lagrangian spatial derivatives.

[6] The general theory developed in § 99 can be applied to the case of couple stresses [the case of general elastic materials of order $n=2$] dealt with in §§ 96, 97 [§ 98]. However one of the main aims of §§ 96—97 was e.g. the relation between the gravitation equations arising from the variational principle $\delta_g I = 0$ and those already stated in the case of couple stresses [§ 88]. Similarly a main aim of § 98 was the connection between the results presented there and their analogues for couple stresses stated in §§ 96—97. In connection with theses facts the use of the results achieved in § 99 did not seem useful to shorten the treatments made in sections 96—98, to that these sections were preferred to be kept independent of the present one.

Appendix A

Double Tensors

§ A1. Definition of Double Tensors Related to Two Topological Spaces

Let I be a non-empty set and K a class of bijections—i.e. one-to-one functions—of I onto open subsets of the euclidean space E_n of the n-tuples of real numbers. We call the ordered couple $\mathscr{R}=(I,K)$ a *topological space of dimension n and order* ω ($\omega=0,1,\ldots$) if there exists a bijection $\phi\in K$ such that K is the set of the bijections ψ of I onto open subsets of E_n for which; (i) the bijection $\phi\psi^{-1}$ (of the counterdomain $\mathscr{C}(\psi)$ of ψ onto the one $\mathscr{C}(\phi)$ of ϕ) is ω times continuously differentiable and (ii) in case $\omega\geqslant 1$ the Jacobian of $\phi\psi^{-1}$ is non-zero everywhere.

If $\phi\in K$ we may call ϕ a *regular co-ordinate system*—or briefly a regular frame—of $\mathscr{R}$. Co-ordinate systems which are not elements of K may also be useful, e.g. the one of spherical co-ordinates in case $\mathscr{R}$ is the ordinary 3-dimensional space.

If $P\in I$—so that P is called a *point* of $\mathscr{R}$—, $\phi\in K$, and $(x^1,\ldots,x^n)=\phi(P)$, then we can write $x^i=\phi^i(P)$.

Now let $\mathscr{R}'=(I',K')$ be another topological space, of dimension n' and order ω'. Let $P\in I$ and $P'\in I'$ hold, and let

$$\{T_{\rho_1\ldots\rho_a}{}^{\sigma_1\ldots\sigma_b}{}_{R_1\ldots R_c}{}^{S_1\ldots S_d}\}=\mathbf{T}(\phi,\phi') \tag{A 1.1}$$

by a system of $n^{a+b}n'^{c+d}$ real numbers defined for $\phi\in K$ and $\phi'\in K'$. We also assume that for every $\bar\phi\in K$ and $\bar\phi'\in K'$

$$\bar T_{\alpha_1\ldots}{}^{\beta_1\ldots}{}_{A_1\ldots}{}^{B_1\ldots}=\frac{\mathscr{D}^p\mathscr{D}'^q}{|\mathscr{D}^p\mathscr{D}'^q|}T_{\rho_1\ldots}{}^{\sigma_1\ldots}{}_{R_1\ldots}{}^{S_1\ldots}\frac{\partial x^{\rho_1}}{\partial\bar x^{\alpha_1}}\cdots\frac{\partial\bar x^{\beta_1}}{\partial x^{\sigma_1}}\cdots\frac{\partial y^{R_1}}{\partial\bar y^{A_1}}\cdots\frac{\partial\bar y^{B_1}}{\partial y^{S_1}}\cdots \tag{A 1.2}$$

where $\{\bar T_{\alpha_1\ldots}{}^{\beta_1\ldots}{}_{A_1\ldots}{}^{B_1\ldots}\}=\mathbf{T}(\bar\phi,\bar\phi')$, where the partial derivative $\partial\bar x^\beta/\partial x^\sigma$ of the function $\bar x^\beta=\bar\phi^\beta[\phi^{-1}(x^1,\ldots,x^n)]$ is evaluated at the point $\phi(P)$ of E_n and the analogue holds for $\partial x^\rho/\partial\bar x^\alpha$, $\partial\bar y^R/\partial y^A$, and $\partial\bar y^B/\partial y^S$, and where $p,q\in\{0,1\}$ and $\mathscr{D}[\mathscr{D}']$ is the jacobian of the transformation $\bar\phi\phi^{-1}$ of $\mathscr{C}(\phi)$ onto $\mathscr{C}(\bar\phi)$

$[\bar{\phi}'\phi'^{-1}$ of $\mathscr{C}(\phi')$ onto $\mathscr{C}(\phi^{-1})]$. Then we call **T** a *double tensor of covariant order* (a,c) *and contravariant order* (b,d). We also say that **T** is attached to the point P of $\mathscr{R}$ with its first $a+b$ indices and to the point P' of $\mathscr{R}'$ with its last $c+d$ indices. (The lower indices are covariant, the upper indices contravariant).

Furthermore for $p=q=0$ we say that the double tensor is *absolute* and for $p=1$ and $q=0$, for $p=0$ and $q=1$, and for $p=q=1$ we say that it is *axial* with respect to $\mathscr{R}$, to $\mathscr{R}'$, and to both $\mathscr{R}$ and $\mathscr{R}'$ respectively.

The scalars $T_{\rho_1\ldots}{}^{\sigma_1\ldots}{}_{R_1\ldots}{}^{S_1\ldots}$—see (A1.1)—are called the components of **T** in the frames ϕ and ϕ'.

The system $T_{\rho_1\ldots}{}^{\sigma_1\ldots}{}_{R_1\ldots}{}^{S_1\ldots}$ of $n^{a+b}n'^{c+d}$ scalars depending on ϕ and ϕ' is a double tensor (in $\mathscr{R}$ and $\mathscr{R}'$, attached to P and P'), if and only if for fixed $\rho_1,\ldots,\sigma_1,\ldots$ $[R_1,\ldots,S_1,\ldots]$ the system $T_{\rho_1\ldots}{}^{\sigma_1\ldots}{}_{R_1\ldots}{}^{S_1\ldots}$ of n'^{c+d} $[n^{a+b}]$ scalars is a tensor in $\mathscr{R}'[\mathscr{R}]$ attached to $P'[P]$. A scalar is a double tensor of zero covariant and contravariant orders.

By multiplying e.g. the components of an absolute tensor $\mathbf{T}_{(1)}$ by those of another, $\mathbf{T}_{(2)}$, we obtain a double tensor, $\mathbf{T}_{(3)}$, with $a_3=a_1+a_2,\ldots,d_3=d_1+d_2$ where (a_i,c_i) is the covariant order of $\mathbf{T}_{(i)}$ and (b_i,d_i) the contravariant one $(i=1,2,3)$.

Similarly, if we have two double tensors of the same orders, and we add their components with the same indices, we obtain a double tensor of the same orders.

§ A2. Partial Covariant Derivative and Total Covariant Derivative Based on a Mapping

Let the double tensor **T** attached to the points P of $\mathscr{R}$ and P' of $\mathscr{R}'$ be defined for $P[P']$ belonging to the open region R of $\mathscr{R}$ $[R'$ of $\mathscr{R}']$. Then the components of **T** in the frames ϕ of $\mathscr{R}$ and ϕ' of $\mathscr{R}'$ are regarded as functions of the co-ordinates x^ρ of P in ϕ and those y^L of P' in ϕ'. In this case we speak of a *double tensor field*.

Let $g_{\rho\sigma}$ be (the components of) a symmetric tensor, **g**, in $\mathscr{R}$ defined over the whole $\mathscr{R}$. Suppose that the order ω of $\mathscr{R}=(I,K)$ is greater than or equal to 2 and that $g_{\alpha\beta}$ is a twice continuously differentiable function of x^ρ. We call the triple $S=(I,K,\mathbf{g})$ a *metric space*, and **g** the *metric tensor* of S. It is used, in particular, to raise and lower indices, and to construct Christoffel symbols:

$$2\{\alpha\beta,\gamma\}=g_{\gamma\beta,\alpha}+g_{\alpha\gamma,\beta}-g_{\alpha\beta,\gamma} \qquad \left(f_{,\alpha}=\frac{\partial f}{\partial x^\alpha}\right), \tag{A2.1}$$

$$\begin{Bmatrix}\gamma\\ \alpha\beta\end{Bmatrix}=\{\alpha\beta,\rho\}\,g^{\rho\gamma} \qquad (g_{\alpha\rho}g^{\rho\beta}=\delta_\alpha{}^\beta)\,. \tag{A2.2}$$

Let $S'=\{I',K',\mathbf{g}'\}$ be another metric space. We denote by $\{AB,C\}'$ and $\begin{Bmatrix}C\\ AB\end{Bmatrix}'$ the Christoffel symbols constructed by means of g'_{AB}.

Let $T_{\rho_1\ldots}{}^{\sigma_1\ldots}{}_{R_1\ldots}{}^{S_1\ldots}$ be a double tensor field in S and S'. Now we define its partial covariant defivatives in S and S' respectively:

$$T_{\rho_1\ldots}{}^{\sigma_1\ldots}{}_{R_1\ldots}{}^{S_1\ldots}{}_{/\alpha}=\frac{\partial T_{\ldots}^{\ldots}}{\partial x^\alpha}-\left\{\begin{matrix}\beta\\ \rho_1\,\alpha\end{matrix}\right\}T_{\beta\rho_2\ldots}{}^{\ldots}-\cdots+\left\{\begin{matrix}\sigma_1\\ \alpha\,\beta\end{matrix}\right\}T_{\rho_1\ldots}{}^{\beta\sigma_2\ldots}{}_{R_1\ldots}{}^{S_1\ldots}+\ldots,\tag{A2.3}$$

$$\begin{aligned}T_{\rho_1\ldots}{}^{\sigma_1\ldots}{}_{R_1\ldots}{}^{S_1\ldots}{}_{/L}=\frac{\partial T_{\ldots}^{\ldots}}{\partial y^L}&-\left\{\begin{matrix}B\\ R_1\,L\end{matrix}\right\}'T_{\rho_1\ldots}{}^{\sigma_1\ldots}{}_{BR_2\ldots}{}^{S_1\ldots}-\ldots\\&+\left\{\begin{matrix}S_1\\ L\,B\end{matrix}\right\}'T_{\rho_1\ldots}{}^{\sigma_1\ldots}{}_{R_1\ldots}{}^{BS_2\ldots}+\cdots.\end{aligned}\tag{A2.4}$$

For fixed $R_1,\ldots,S_1,\ldots$ $T_{\rho_1\ldots}{}^{\sigma_1\ldots}{}_{R_1\ldots}{}^{S_1\ldots}$ is a tensor in S. Furthermore, by writing (A1.2) for $\phi=\bar{\phi}$ and by taking the partial covariant derivatives in S of the two sides, we obtain

$$\begin{aligned}\bar{T}_{\rho_1\ldots}{}^{\sigma_1\ldots}{}_{A_1\ldots}{}^{B_1\ldots}=T_{\alpha_1\ldots}{}^{\beta_1\ldots}{}_{R_1\ldots}{}^{S_1\ldots}{}_{/\alpha}&\frac{\mathscr{D}'^q}{|\mathscr{D}'^q|}\frac{\partial y^{R_1}}{\partial\bar{y}^{A_1}}\cdots\frac{\partial\bar{y}^{B_1}}{\partial y^{S_1}}\cdots\\+T_{\alpha_1\ldots}{}^{\beta_1\ldots}{}_{R_1\ldots}{}^{S_1\ldots}&\left(\frac{\mathscr{D}'^q}{|\mathscr{D}'^q|}\frac{\partial y^{R_1}}{\partial\bar{y}^{A_1}}\cdots\frac{\partial\bar{y}^{B_1}}{\partial y^{S_1}}\cdots\right)_{/\alpha}.\end{aligned}$$

Since $\partial y^R/\partial\bar{y}^A$, $\partial\bar{y}^B/\partial y^S$, and $\mathscr{D}'$ are independent of x^ρ and have no Greek indices, by (A2.3) the last covariant derivative above vanishes. Thus we see that for fixed $\rho_1,\ldots,\sigma_1,\ldots,\alpha$, $T_{\rho_1\ldots}{}^{\sigma_1\ldots}{}_{R_1\ldots}{}^{S_1\ldots}{}_{/\alpha}$ is a tensor in S'. Hence the partial covariant derivative (A2.3) is a double tensor. The same holds for (A2.4).

For fixed $R_1,\ldots,S_1,\ldots$, $T_{\rho_1\ldots}{}^{\sigma_1\ldots}{}_{R_1\ldots}{}^{S_1\ldots}{}_{/\alpha}$ is a tensor in S. Furthermore, by applying (A1.2) for $\phi=\bar{\phi}$ and by taking the partial covariant derivatives in S of the two sides, we easily see—on the basis of the independence of $\partial y^R/\partial\bar{y}^L$ and $\partial\bar{y}^B/\partial y^S$ of x^α—that for fixed $\rho_1,\ldots,\sigma_1,\ldots,\alpha$, $T_{\rho_1\ldots}{}^{\sigma_1\ldots}{}_{R_1\ldots}{}^{S_1\ldots}{}_{/\alpha}$ is a tensor in S'. Hence the partial covariant derivative (A2.3) is a double tensor. The same holds for (A2.4).

Let $P=P(P')$ be a continuously differentiable mapping of S' into S. It can also be written as follows: $x^\rho=x^\rho(y^L)$. Then $x^\rho{}_L=\partial x^\rho/\partial y^L$ is a double tensor field defined over S'. Indeed

$$\bar{x}^\rho{}_L=\frac{\partial\bar{x}^\rho}{\partial\bar{y}^L}=\frac{\partial\bar{x}^\rho}{\partial x^\sigma}x^\sigma{}_{;M}\frac{\partial y^M}{\partial\bar{y}^L}\qquad\left(x^\sigma{}_{;M}=\frac{\partial x^\sigma}{\partial y^M}=x^\sigma{}_M\right).\tag{A2.5}$$

We set

$$T_{\rho_1\ldots}{}^{\sigma_1\ldots}{}_{R_1\ldots}{}^{S_1\ldots}{}_{;L}=T_{\ldots}^{\ldots}{}_{/\alpha}x^\alpha{}_{;L}+T_{\ldots}^{\ldots}{}_{/L},\tag{A2.6}$$

so that $T_{\ldots}^{\ldots}{}_{;L}$ is a double tensor and we call it *total covariant derivative* of the double tensor $\mathbf{T}(P,P')$ based on the mapping $P=P(P')$.

The operation "$_{;L}$" defined by (A 2.6) obviously has the following properties—cf. Ericksen [1960, p. 811]:

a) *If* $T_{\dots}^{\dots}$ *is of the special form* $T_{\rho_1\dots}{}^{\sigma_1\dots}$ [$T_{R_1\dots}{}^{S_1\dots}$], *then* $T_{\dots}^{\dots}{}_{;L}$ *reduces to* $T_{\dots}^{\dots}{}_{/\alpha}x^{\alpha}{}_{L}$ [*to* $T_{\dots}^{\dots}{}_{/L}$].

b) *If we first apply the operation "$_{;L}$" to the double tensor field* $\mathbf{T}(P,P')$ *and then we replace* P *by* $P(P')$, *we obtain the same result as if we apply the operation "$_{;L}$" to* $\mathbf{T}[P(P'),P']$.

Furthermore we obviously have

$$g_{\alpha\beta;A}=g_{\alpha}{}^{\beta}{}_{;A}=g^{\alpha\beta}{}_{;A}=0=g'_{LM;A}=g'_{L}{}^{M}{}_{;A}=g'^{LM}{}_{;A} \tag{A 2.7}$$

so that the operation "$_{;A}$" commutes with the one of raising or lowering indices. The usual derivation rule of products holds for "$_{;A}$": e.g.

$$(T_{\alpha\beta}{}^{LM}\,\theta^{\beta}{}_{M})_{;A}=T_{\alpha\beta}{}^{LM}{}_{;A}\,\theta^{\beta}{}_{M}+T_{\alpha\beta}{}^{LM}\,\theta^{\beta}{}_{M;A}\,. \tag{A 2.8}$$

Let us remark that by (A 2.6) and $(52.4)_3$

$$x^{\rho}{}_{A;B}=x^{\lambda}{}_{A}\begin{Bmatrix}\rho\\ \lambda\,\mu\end{Bmatrix}x^{\mu}{}_{B}+x^{\rho}{}_{/AB}=x^{\rho}{}_{B;A}\,. \tag{A 2.9}$$

§ A3. On Differentiation of Double Tensors, Functions of Double Tensors. Case of Arguments Fulfilling Typical Regular Conditions

Let the absolute double tensor $T_{\alpha\dots}{}^{A\dots}$ (of the kind considered in § A2) be a function $T_{\alpha\dots}{}^{A\dots}(H^{\lambda\dots}{}_{L\dots})$ of the double tensor $H^{\lambda\dots}{}_{L\dots}$ and possibly of other arguments. Then—cf. (A 1.2)—we have

$$\bar{T}_{\beta\dots}{}^{B\dots}(\bar{H}^{\mu\dots}{}_{M\dots},\dots)=\frac{\partial x^{\alpha}}{\partial\bar{x}^{\beta}}\cdots\frac{\partial\bar{y}^{B}}{\partial y^{A}}\cdots T_{\alpha\dots}{}^{A\dots}\left(\bar{H}^{\mu\dots}{}_{M\dots}\frac{\partial x^{\lambda}}{\partial\bar{x}^{\mu}}\cdots\frac{\partial\bar{y}^{M}}{\partial y^{L}}\cdots,\dots\right) \tag{A3.1}$$

whence

$$\frac{\partial\bar{T}_{\beta\dots}{}^{B\dots}}{\partial\bar{H}^{\mu\dots}{}_{M\dots}}=\frac{\partial x^{\alpha}}{\partial\bar{x}^{\beta}}\cdots\frac{\partial\bar{y}^{\beta}}{\partial y^{A}}\cdots\frac{\partial T_{\alpha\dots}{}^{A\dots}}{\partial H^{\lambda\dots}{}_{L\dots}}\frac{\partial x^{\lambda}}{\partial\bar{x}^{\mu}}\cdots\frac{\partial\bar{y}^{M}}{\partial y^{L}}\cdots. \tag{A 3.2}$$

Thence we easily see that $\partial T_{\alpha\dots}{}^{A\dots}/\partial H^{\lambda\dots}{}_{L\dots}$ is a double tensor having the covariant [contravariant] indices of $T_{\alpha\dots}{}^{A\dots}$ as covariant [contravariant] indices and the covariant [contravariant] indices of $H^{\lambda\dots}{}_{L\dots}$ as contravariant [covariant] indices.

Incidentally we can prove that the left-hand side of (A 3.2) is a double tensor by applying the quotient rule of the differential $dT_{\alpha\dots}{}^{A\dots}=(\partial T_{\alpha\dots}{}^{A\dots}/\partial H^{\lambda\dots}{}_{L\dots})dH^{\lambda\dots}{}_{L\dots}$.

In order to fix ideas we now identify S_n with space-time S_4, so that $g_{\rho\sigma}dx^\rho dx^\sigma$ has the signature $+2$, and $S'_{n'}$ with the material 3-dimensional space S_3^* whose metric $a_{LM}^* dy^L dy^M$ is strictly positive-definite. From

$$T_{\cdots}^{\cdots}=T_{\cdots}^{\cdots}(H^{\lambda_1\cdots}{}_{L_1\cdots}), \quad g^{\lambda_1\alpha}H_\alpha{}^{\lambda_a\cdots}{}_{L_1\cdots}=H^{\lambda_1\cdots}{}_{L_1\cdots}=a_{L_1A}^*H^{\lambda_1\cdots A}{}_{L_2\cdots} \tag{A 3.3}$$

we deduce

$$\frac{\partial T_{\cdots}^{\cdots}}{\partial H_\alpha{}^{\lambda_2\cdots}{}_{L_1\cdots}}=g^{\alpha\lambda_1}\frac{\partial T_{\cdots}^{\cdots}}{\partial H^{\lambda_1\lambda_2\cdots}{}_{L_1\cdots}}, \qquad \frac{\partial T_{\cdots}^{\cdots}}{\partial H^{\lambda_1\cdots A}{}_{L_2\cdots}}=a_{L_1A}^*\frac{\partial T_{\cdots}^{\cdots}}{\partial H^{\lambda_1\cdots}{}_{L_1L_2\cdots}}. \tag{A 3.4}$$

Incidentally (A 3.4) can also be deduced by considering $dT_{\cdots}^{\cdots}$ as a linear function of each of the systems of differentials $dH^{\lambda_1\cdots}{}_{L_1\cdots}$, $dH_\alpha{}^{\lambda_2\cdots}{}_{L_1\cdots}$ and $dH^{\lambda_1\cdots A}{}_{L_2\cdots}$.

The case when the function above, say $\hat{T}_{\cdots}^{\cdots}(\;)$, is defined only on a possibly non-linear manifold $\mathcal{M}$ in a pseudo-Euclidean space is dealt with in this section from a rather general point of view. More in particular, it is remarked that, on the one hand, since the derivative $\partial\hat{T}_{\cdots}^{\cdots}/\partial H_{\cdots}^{\cdots}$ of that function—defined by condition (A 3.7) below—is not uniquely determined, it is natural to identify it with the *normally vanishing derivative* introduced below (A 3.9 b). This derivative is meaningful in all cases with the exception of a few ones characterized by Theorem A 3.1. On the other hand, it is natural to express $\partial\hat{T}_{\cdots}^{\cdots}/\partial H_{\cdots}^{\cdots}$ as the gradient of a suitable extension $\tilde{T}_{\cdots}^{\cdots}$ of the field $\hat{T}_{\cdots}^{\cdots}$ to any neighborhood $\mathcal{N}_{\mathcal{M}}$ of $\mathcal{M}$—cf. Theorem A 3.2. In section A 4 several examples of practical interest are considered. It is shown that in them the extension $\tilde{T}_{\cdots}^{\cdots}$ of $\hat{T}_{\cdots}^{\cdots}$ that is usually taken into account is *normally constant*, i.e. its gradient is the normally vanishing derivative—see below Theorem A 3.2.

The values of the above variable $H^{\lambda_1\cdots\lambda_r}{}_{L_1\cdots L_s}$ $(=H^{\lambda\cdots}{}_{L\cdots}=H_{\cdots}^{\cdots})$ are double tensors attached to S_4 at x^α and to S_3^* at y^L. They constitute a pseudo-Euclidean space $S^{(N)}$ of dimension $N=4^r\cdot 3^s$, where the norm of $H_{\cdots}^{\cdots}$ is identified with $H^{\lambda\cdots}{}_{L\cdots}H_{\lambda\cdots}{}^{L\cdots}$.

We also assume that the double tensor $T_{\cdots}^{\cdots}=T_{\alpha\cdots}{}^{A\cdots}$ is a function $\hat{T}_{\cdots}^{\cdots}(H_{\cdots}^{\cdots})$ of $H_{\cdots}^{\cdots}$ defined on the $(N-p)$-dimensional continuously differentiable manifold $\mathcal{M}$ represented by

$$f_{(i)}(H^{\lambda\cdots}{}_{L\cdots})=0 \qquad (i=1,\ldots,p') \tag{A 3.5a}$$

with $p'\geqslant p$. Of course the system (A 3.5a) is supposed to be invariant under changes of frame. This usually occurs in that the p' $f_{(i)}(H_{\cdots}^{\cdots})$ can be divided into sets each of which is the set of the components of a double tensor of any non-negative rank. As is apparent in case system (A 3.5a) is $H^{[\lambda\mu]}=0$ $(p'=16,\ p=6)$ [§ A4], the equality $p'=p$—hence the independence of equations (A 3.5)—may be uncompatible with the tensor form of $f_{(i)}(H^{\lambda\cdots}{}_{L\cdots})$.

The tangent space $S_{\mathbf{H}}^{(N-p)}$ at the point $\mathbf{H}$ of $\mathcal{M}$ has dimension $N-p$ and is represented by the equations

$$\frac{\partial f_{(i)}}{\partial H^{\lambda \cdots}{}_{L \ldots}} dH^{\lambda \cdots}{}_{L \ldots} = 0 \qquad (i = 1, \ldots, p') \tag{A 3.6}$$

in $dH^{\lambda \cdots}{}_{L \ldots}$. Let $S^c_{\mathbf{H}}$ be a p-space complementary to $S^{(N-p)}_{\mathbf{H}}$ in $S^{(N)}$ (so that $S^{(N-p)}_{\mathbf{H}} \cap S^c_{\mathbf{H}}$ has dimension zero) and a continuous function of $\mathbf{H}$ defined on $\mathscr{M}$.

Suppose that $H_{\ldots}^{\ldots}$ fulfills (A 3.5), i.e. $\mathbf{H} \in \mathscr{M}$, and that the function $\hat{T}_{\ldots}^{\ldots}$ of domain $\mathscr{M}$ is *differentiable* at $\mathbf{H}$. We mean that in connection with two particular frames (x) and (y) for S_4 and S^*_3, there is a linear function $(\partial \hat{T}_{\ldots}^{\ldots} / \partial H^{\lambda \cdots}{}_{L \ldots}) dH^{\lambda \cdots}{}_{L \ldots}$ of $dH^{\lambda \cdots}{}_{L \ldots}$ such that

$$\frac{\partial \hat{T}_{\ldots}^{\ldots}}{\partial H^{\lambda \cdots}{}_{L \ldots}} dH^{\lambda \cdots}{}_{L \ldots} = \hat{T}_{\ldots}^{\ldots}(H^{\lambda \cdots}{}_{L \ldots}) + dH^{\lambda \cdots}{}_{L \ldots}) - \hat{T}_{\ldots}^{\ldots}(H^{\lambda \cdots}{}_{L \ldots}) + \textcircled{2} \quad \text{for} \quad d\mathbf{H} \in S^{(N-p)}_{\mathbf{H}} \tag{A 3.7}$$

(i.e. (A 3.7) holds for every solution $dH^{\lambda \cdots}{}_{L \ldots}$ of (A 3.6) at $H_{\ldots}^{\ldots}$) where

$$\lim_{M \to 0} \frac{\textcircled{2}}{M} = 0 \quad \text{with} \quad M = (\delta_{\lambda\mu} \ldots a^{*LM} \ldots H^{\lambda \cdots}{}_{L \ldots} H^{\mu \cdots}{}_{M \ldots})^{1/2} \geqslant 0 \,. \tag{A 3.8}$$

Remark that M is not invariant under changes of the frame (x) by the presence of $\delta_{\lambda\mu}$ in $(\text{A } 3.8)_2$. Its replacement with $g_{\lambda\mu}$ would realize this invariance but would render $(\text{A } 3.8)_1$ false.

Let us define the coefficient $\partial \hat{T}_{\ldots}^{\ldots} / \partial H^{\lambda \cdots}{}_{L \ldots}$ in arbitrary frames according to the law (A 1.2) for double tensors, regarding the indices $\lambda \ldots$ [$L \ldots$] as covariant [contravariant]. Then in spite of the afore-mentioned lack of (metrical) invariance the topological property (A 3.7) is invariant under changes of the frames (x) and (y), and under lifting and lowering indices.

The restriction on $S^c_{\mathbf{H}}$ of the differential $d\hat{T}_{\ldots}^{\ldots} = (\partial \hat{T}_{\ldots}^{\ldots} / \partial H^{\lambda \cdots}{}_{L \ldots}) dH^{\lambda \cdots}{}_{L \ldots}$ can be chosen arbitrarily. We chose it identically zero, which is equivalent to requiring

$$\beta_{(i)}{}^{\lambda \cdots}{}_{L \ldots} \partial \hat{T}_{\ldots}^{\ldots} / \partial H^{\lambda \cdots}{}_{L \ldots} = 0 \qquad (i = 1, \ldots, p) \,, \tag{A 3.9 a}$$

where the p vectors $\beta_{(i)}{}^{\lambda \cdots}{}_{L \ldots}$ in $S^{(N)}$ constitute a basis of $S^c_{\mathbf{H}}$. Conditions (A 3.7) and (A 3.9a) define uniquely the double tensor $\partial \hat{T}_{\ldots}^{\ldots} / \partial H^{\lambda \cdots}{}_{L \ldots}$. We call it *derivative of the field $\hat{T}_{\ldots}^{\ldots}$ (on $\mathscr{M}$) with respect to $H^{\lambda \cdots}{}_{L \ldots}$, vanishing along $S^c_{\mathbf{H}}$*.

From (A 3.7) and (A 3.9a) we see that *the derivative $\partial \hat{T}_{\ldots}^{\ldots} / \partial H^{\lambda \cdots}{}_{L \ldots}$ above vanishes on $\mathscr{M}$ if and only if $\hat{T}_{\ldots}^{\ldots}$ is constant on $\mathscr{M}$*. This derivative depends on $S^c_{\mathbf{H}}$ effectively except if it vanishes for a choice of $S^c_{\mathbf{H}}$ ($p > 0$). It is natural to try and render it determined once for all by a suitable choice of $S^c_{\mathbf{H}}$ and in particular by identifying $S^c_{\mathbf{H}}$ with the p-space $S^{\perp}_{\mathbf{H}}$ orthogonal to $S^{(N-p)}_{\mathbf{H}}$ at $\mathbf{H}$, so that the vanishing condition (A 3.9 a) can be turned into

$$\frac{\partial f_{(i)}}{\partial H_{\lambda \ldots}{}^{L \ldots}} \frac{\partial \hat{T}_{\ldots}^{\ldots}}{\partial H^{\lambda \cdots}{}_{L \ldots}} = 0 \qquad (i = 1, \ldots, p') \qquad \left(\frac{\partial f_{(i)}}{\partial H_{\lambda \ldots}{}^{L \ldots}} = g^{\lambda\mu} \ldots a^*_{LM} \ldots \frac{\partial f_{(i)}}{\partial H^{\mu \cdots}{}_{M \ldots}} \right). \tag{A 3.9 b}$$

If this holds, we shall call $\partial\hat{T}_{\dots}^{\dots}/\partial H^{\dots}_{\dots}$ the *normally vanishing derivative* of $\hat{T}_{\dots}^{\dots}$. This derivative is meaningful in case $S_{\mathbf{H}}^{\perp}$ and $S_{\mathbf{H}}^{(N-p)}$ are complementary. That this hypothesis is non-trivial (in pseudo-Euclidean spaces) and essential can be shown by identifying (A 3.5 a) with

$$f(H^\lambda)\equiv w_\lambda H^\lambda=0 \quad \text{where} \quad w_\lambda w^\lambda=0, \quad w_\lambda \neq 0. \tag{A 3.5b}$$

Indeed by (A 3.5 b) $\mathcal{M}$ is a hyperplane and, for $\mathbf{H}\in\mathcal{M}$, $S_{\mathbf{H}}^{(N-p)}=\mathcal{M}$; furthermore $f(H^\lambda+\xi w^\lambda)=0$ for every real ξ, so that $S_{\mathbf{H}}^{\perp}$ contains a straight line through $\mathbf{H}$, that is parallel with w^λ and belongs to $S_{\mathbf{H}}^{(N-p)}$. There is a vector b^λ such that $b^\lambda w_\lambda=1$. Set, under $(\text{A } 3.5\text{b})_2$,

$$\hat{T}(H^\lambda)=b_\lambda H^\lambda, \quad \text{hence} \quad \hat{T}(\xi w^\lambda)=\xi.$$

Then by (A 3.7)

$$1=\frac{d}{d\xi}\hat{T}(\xi w^\lambda)=\frac{\partial\hat{T}}{\partial H^\lambda}w^\lambda,$$

which is incompatible with the analogue of (A 3.9 b) for $\hat{T}$: $w^\lambda\partial\hat{T}/\partial H^\lambda=0$.

Now, denoting the characteristic of any matrix M by κ_M, we prove, in the general case, the following theorem (which is not essential in section A 4).

Theorem A 3.1. *The spaces* $S_{\mathbf{H}}^{(N-p)}$ *and* $S_{\mathbf{H}}^{\perp}$ *are complementary if and only if*

$$\kappa_{M'}=p \quad \textit{where} \quad M'=\|\gamma_{ij}\| \quad \textit{with} \quad \gamma_{ij}=\frac{\partial f_{(i)}}{\partial H^{\lambda\dots}{}_{L\dots}}\frac{\partial f_{(j)}}{\partial H_{\lambda\dots}{}^{L\dots}} \quad (i,j=1,\dots,p'). \tag{A 3.10}$$

Indeed fix the frames (x) and (y) and assume $\kappa_M=p$ $[\kappa_{M'}=\kappa_M\neq p]$ where $M=\|\gamma_{ij}\|$ $(i,j=1,\dots,p)$. Then any vector of $S_{\mathbf{H}}^{\perp}$,

$$dH^{\lambda\dots}{}_{L\dots}=\sum_{j=1}^{p}c_j\partial f_{(j)}/\partial H_{\lambda\dots}{}^{L\dots},$$

belongs to $S_{\mathbf{H}}^{(N-p)}$ if and only if it fulfills (A 3.6) for $i=1,\dots,p$, which is equivalent to $\gamma_{i1}c_1+\dots+\gamma_{ip}c_p=0$ $(i=1,\dots,p)$. This implies [does not imply] $c_1=\dots=c_p=0$, i.e. the spaces $S_{\mathbf{H}}^{(N-p)}$ and $S_{\mathbf{H}}^{\perp}$ are [are not] complementary. q.e.d.

Let us add that for $\mathbf{H}\in\mathcal{M}$ we can define a p-dimensional neighborhood $\mathcal{N}_{\mathbf{H}}$ of $\mathbf{H}$ in $S_{\mathbf{H}}^{c}$ in such a way, that the intersection of $\mathcal{N}_{\mathbf{H}}$ and $\mathcal{N}_{\mathbf{H}'}$ is empty for $\mathbf{H},\mathbf{H}'\in\mathcal{M}$ and $\mathbf{H}\neq\mathbf{H}'$. Then we can prolong $\hat{T}_{\dots}^{\dots}$ into a function $\tilde{T}_{\dots}^{\dots}$ defined on the N-dimensional region $\mathcal{N}_{\mathcal{M}}$ described by $\mathcal{N}_{\mathbf{H}}$ when $\mathbf{H}$ describes $\mathcal{M}$, by requiring $\tilde{T}_{\dots}^{\dots}$ to be a linear function of $\mathbf{H}$ in every $\mathcal{N}_{\mathbf{H}}$; it is easy to do this in such a way that

$$\frac{\partial\tilde{T}_{\dots}^{\dots}}{\partial H^{\lambda\dots}{}_{L\dots}}dH^{\lambda\dots}{}_{L\dots}=\frac{\partial\hat{T}_{\dots}^{\dots}}{\partial H^{\lambda\dots}{}_{L\dots}}dH^{\lambda\dots}{}_{L\dots} \quad \text{on } \mathcal{M} \text{ for } dH^{\lambda\dots}{}_{L\dots}\in S_{\mathbf{H}}^{(N-p)} \tag{A 3.11}$$

where the right-hand side is defined by (A 3.7).

Although some parts of the theorem below are obvious, its actual version has been written fully to sum up the main facts about the derivative being considered.

Theorem A 3.2. *Suppose that* (a) $H^{\lambda\ldots}{}_{L\ldots}=\phi^{\lambda\ldots}{}_{L\ldots}(H'^{\mu\ldots}{}_{M\ldots})$ *is a continuously differentiable mapping of a neighborhood* $\mathcal{N}_{\mathcal{M}}$ *of* $\mathcal{M}$ *onto* $\mathcal{M}$, *whose restriction on* $\mathcal{M}$ *is the identity,*

(b) $\hat{T}_{\ldots}^{\ldots}(H^{\ldots}{}_{\ldots})$ *is a continuously differentiable double tensor field on* $\mathcal{M}$, *and*

(c) $\mathbf{H}\in\mathcal{M}$ *and* $S^{\phi}_{\mathbf{H}}$ *is the p-space tangent at* $\mathbf{H}$ *to the p-dimensional manifold* $\sigma_{\mathbf{H}}$:

$$\phi^{\mu\ldots}{}_{M\ldots}(H'^{\lambda\ldots}{}_{L\ldots})=H^{\mu\ldots}{}_{M\ldots}=\text{const}. \tag{A 3.12}$$

Consequently,

(α) *if* $\partial\hat{T}_{\ldots}^{\ldots}/\partial H_{\ldots}^{\ldots}$ *is the derivative of* $\hat{T}_{\ldots}^{\ldots}$ *vanishing along* $S^{c}_{\mathbf{H}}=S^{\phi}_{\mathbf{H}}$, *so that* (A 3.9 a) *holds, then*

$$\frac{\partial\tilde{T}_{\ldots}^{\ldots}}{\partial H^{\lambda\ldots}{}_{L\ldots}}=\frac{\partial\hat{T}_{\ldots}^{\ldots}}{\partial H^{\lambda\ldots}{}_{L\ldots}}\quad\text{on }\mathcal{M},$$
$$\text{where}\quad \tilde{T}_{\ldots}^{\ldots}(H'^{\lambda\ldots}{}_{L\ldots})=\hat{T}_{\ldots}^{\ldots}[\phi^{\mu\ldots}{}_{M\ldots}(H'^{\mu\ldots}{}_{M\ldots})]\quad\text{for}\quad \mathbf{H}'\in\mathcal{N}_{\mathcal{M}}, \tag{A 3.13}$$

(β) *if* $S^{\phi}_{\mathbf{H}}\subset S^{\perp}_{\mathbf{H}}$, *then* $S^{\phi}_{\mathbf{H}}=S^{\perp}_{\mathbf{H}}$ *and*

$$\frac{\partial f_{(i)}}{\partial H_{\lambda\ldots}{}^{L\ldots}}\,\frac{\partial\tilde{T}_{\ldots}^{\ldots}}{\partial H^{\lambda\ldots}{}_{L\ldots}}=0\quad\left[f_{(i)}\left(\frac{\partial\tilde{T}_{\ldots}^{\ldots}}{\partial H'^{\mu\ldots}{}_{M\ldots}}\right)=0\text{ for }f_{(i)}\text{ linear}\right]\quad(i=1,\ldots,p'), \tag{A 3.14}$$

(γ) $\partial\tilde{T}_{\ldots}^{\ldots}/\partial H_{\ldots}^{\ldots}$ *is the normally vanishing derivative of the field* $\tilde{T}_{\ldots}^{\ldots}$ *if and only if* (A 3.14) *holds, and*

(δ) $S^{\phi}_{\mathbf{H}}\subset S^{\perp}_{\mathbf{H}}$ *(hence* $S^{\phi}_{\mathbf{H}}=S^{\perp}_{\mathbf{H}}$*) if and only if*

$$\frac{\partial f_{(i)}}{\partial H_{\lambda\ldots}{}^{L\ldots}}\,\frac{\partial\phi^{\mu\ldots}{}_{M\ldots}}{\partial H'^{\lambda\ldots}{}_{L\ldots}}=0\quad(i=1,\ldots,p')\quad(\mathbf{H}'=\mathbf{H}\in\mathcal{M}). \tag{A 3.15}$$

E.g. $\mathcal{M}$ can be represented by (A 4.1) below and (A 3.12) can be $H'^{(\mu\nu)}=H^{\mu\nu}=\text{const}$.

In case $(\text{A 3.13})_2$ and (A 3.15) hold, $\tilde{T}_{\ldots}^{\ldots}$ can be called a *normally constant extension* of $\hat{T}_{\ldots}^{\ldots}$ (on the neighborhood $\mathcal{N}_{\mathcal{M}}$ of $\mathcal{M}$). Let us remark, first, that by theses (γ) and (δ) *its gradient coincides with the normally vanishing derivative of* $\hat{T}_{\ldots}^{\ldots}$ *(on* $\mathcal{M}$*)*. Second, remark that *if* $\mathcal{M}$ *is linear, then the normally vanishing derivative* $\partial\tilde{T}_{\ldots}^{\ldots}/\partial H_{\lambda\ldots}{}^{L\ldots}$ satisfies the *same equations (representing* $\mathcal{M}$*) as* $\mathbf{H}^{\lambda\ldots}{}_{L\ldots}$—cf. $(\text{A 3.14})_2$. Let us call $T_{\ldots}^{\ldots}$ *the (linear) normally constant extension* of $\hat{T}_{\ldots}^{\ldots}$ in case $\phi^{\mu\ldots}{}_{M\ldots}(H'^{\lambda\ldots}{}_{L\ldots})$ is linear.

Proof of Theorem A 3.2. Let $\beta_{(i)}{}^{\lambda\ldots}{}_{L\ldots}$ $(i=1,\ldots,p)$ be p linearly independent tensors belonging to $S^{\phi}_{\mathbf{H}}$. Furthermore, by the definitions of $\sigma_{\mathbf{H}}$ and $S^{\phi}_{\mathbf{H}}$, for $\mathbf{H}'=\mathbf{H}\in\mathcal{M}$, and for fixed $\mu\ldots$ and $M\ldots$, $\partial\phi^{\mu\ldots}{}_{M\ldots}/\partial H'^{\lambda\ldots}{}_{L\ldots}$ is orthogonal to $S^{\phi}_{\mathbf{H}}$, so that

$$\beta_{(i)}{}^{\lambda\ldots}{}_{L\ldots}\frac{\partial\phi^{\mu\ldots}{}_{M\ldots}}{\partial H'^{\lambda\ldots}{}_{L\ldots}}=0 \quad (\mathbf{H}'=\mathbf{H}\in\mathcal{M}), \tag{A 3.16}$$

hence by the definition $(\text{A}\,3.13)_2$, on the one hand,

$$\beta_{(i)}{}^{\lambda\ldots}{}_{L\ldots}\frac{\partial\tilde{T}_{\ldots}^{\;\ldots}}{\partial H'^{\lambda\ldots}{}_{L\ldots}}=\beta_{(i)}{}^{\lambda\ldots}{}_{L\ldots}\frac{\partial\hat{T}_{\ldots}^{\;\ldots}}{\partial H^{\mu\ldots}{}_{M\ldots}}\frac{\partial\phi^{\mu\ldots}{}_{M\ldots}}{\partial H'^{\lambda\ldots}{}_{L\ldots}}=0 \quad (\mathbf{H}'=\mathbf{H}\in\mathcal{M}). \tag{A 3.17}$$

In addition $\partial\tilde{T}_{\ldots}^{\;\ldots}/\partial H_{\ldots}^{\;\ldots}$ obviously fulfills condition (A 3.7) in $\partial\hat{T}_{\ldots}^{\;\ldots}/\partial H_{\ldots}^{\;\ldots}$. On the other hand, conditions (A 3.7) and (A 3.9a) determine $\partial\hat{T}_{\ldots}^{\;\ldots}/\partial H_{\ldots}^{\;\ldots}$. Hence $(\text{A}\,3.13)_1$ holds. Thus we have proved thesis (α).

Likewise, since (A 3.7) and (A 3.9b) determine the normally vanishing derivative $\partial\hat{T}_{\ldots}^{\;\ldots}/\partial H_{\ldots}^{\;\ldots}$ and $\partial\tilde{T}_{\ldots}^{\;\ldots}/\partial H_{\ldots}^{\;\ldots}$ fulfills (A 3.7), these derivatives coincide if and only if (A 3.14) holds. Thus thesis (γ) has been proved.

In order to prove thesis (β) let $S_{\mathbf{H}}^{\phi}=S_{\mathbf{H}}^{\perp}$ hold. Hence $S_{\mathbf{H}}^{\phi}=S_{\mathbf{H}}^{\perp}$, because $S_{\mathbf{H}}^{\phi}$ and $S_{\mathbf{H}}^{\perp}$ have the same dimension p. Then, setting $S_{\mathbf{H}}^{c}=S_{\mathbf{H}}^{\phi}$, (A 3.13) holds by thesis (α); furthermore $S_{\mathbf{H}}^{c}=S_{\mathbf{H}}^{\perp}$, whence (A 3.9b). Hence we have (A 3.14). Thus thesis (β) holds. Thesis (δ) is now obvious. q.e.d.

Recall that in the exceptional case (A 3.5b) $M'=\|\gamma_{ij}\|=\|\gamma_{11}\|$, where $\gamma_{11}=w_\lambda w^\lambda=0$—cf. $(\text{A}\,3.10)_3$—i.e. $\kappa_{M'}=0$. Then, by Theorem A 3.1, for $\mathbf{H}\in\mathcal{M}$ the spaces $S_{\mathbf{H}}^{(N-p)}$ and $S_{\mathbf{H}}^{\perp}$ are not complementary.

Common practice agrees with the following

Convention. If the double tensor field $\hat{T}_{\ldots}^{\;\ldots}(H_{\ldots}^{\;\ldots})$ is defined on a manifold $\mathcal{M}$ such that at every point $H_{\ldots}^{\;\ldots}\in\mathcal{M}$ the spaces $S_{\mathbf{H}}^{(N-p)}$ and $S_{\mathbf{H}}^{\perp}$ are complementary, then $\partial\hat{T}_{\ldots}^{\;\ldots}/\partial H_{\ldots}^{\;\ldots}$ is identified with the normally vanishing derivative of $\hat{T}_{\ldots}^{\;\ldots}$ and is simply called *the derivative of* $\hat{T}_{\ldots}^{\;\ldots}$.

§ A 4. Partial Derivative of any Double Tensor $\hat{T}_{\ldots}^{\;\ldots}(H_{\ldots}^{\;\ldots})$ Defined Only for Values of $H_{\ldots}^{\;\ldots}$ that are Symmetric, Skewsymmetric, Spatial, or Subject to Other Particular Conditions

Let $T_{\ldots}^{\;\ldots}(H_{\ldots}^{\;\ldots})$ be a double tensor defined on the manifold $\mathcal{M}$ introduced in section A 3. We now consider four examples corresponding to various choices of $\mathcal{M}$. Our first example is the case when the argument $H_{\rho\sigma}{}^{\ldots}$ is symmetric with respect to ρ and σ, so that the equations (A 3.5) representing $\mathcal{M}$ are

$$f_{\alpha\beta}(H_{\rho\sigma}{}^{\ldots})\equiv H_{[\alpha\beta]}{}^{\ldots}\equiv g_{[\alpha}{}^{\rho}g_{\beta]}{}^{\sigma}H_{\rho\sigma}{}^{\ldots}=0\,. \tag{A 4.1}$$

Let us set, for any $H_{\rho\sigma}{}^{\ldots}$ (and by interchanging the roles of $H_{\ldots}$ and $H'_{\ldots}$ in (A 3.12) and $(\text{A}\,3.13)_2$)

$$\tilde{T}_{\ldots}^{\;\ldots}(H_{\alpha\beta}{}^{\ldots})=\hat{T}_{\ldots}^{\;\ldots}(H'_{\alpha\beta}{}^{\ldots}) \quad \text{where} \quad H'_{\alpha\beta}{}^{\ldots}=g_\alpha{}^{(\rho}g_\beta{}^{\sigma)}H_{\rho\sigma}{}^{\ldots}=\varphi(H_{\rho\sigma})\,. \tag{A 4.2}$$

Hence $\tilde{T}\ldots^{\cdots}(H_{\alpha\beta}{}^{\cdots})=\tilde{T}(H'_{\alpha\beta}{}^{\cdots})$, so that

$$\frac{\partial\tilde{T}\ldots^{\cdots}}{\partial H_{\rho\sigma}{}^{\cdots}}=g_\alpha{}^{(\rho}g_\beta{}^{\sigma)}\frac{\partial\tilde{T}\ldots^{\cdots}}{\partial H'_{\alpha\beta}{}^{\cdots}},\quad\text{i.e.}\quad\frac{\partial\tilde{T}\ldots^{\cdots}}{\partial H_{[\alpha\beta]}{}^{\cdots}}=0\quad\text{for}\quad H_{\alpha\beta}{}^{\cdots}=H_{(\alpha\beta)}{}^{\cdots}. \tag{A4.3}$$

This symmetry property of $\partial\tilde{T}\ldots^{\cdots}/\partial H_{\alpha\beta}{}^{\cdots}$ and definition (A 4.1) imply (A 3.14)$_{1,2}$, so that $\tilde{T}\ldots^{\cdots}$ is the normally constant extension of $\hat{T}\ldots^{\cdots}$—see the remarks below Theorem A 3.2.

By interchanging $_{(\,)}$ with $_{[\,]}$ in the preceding reasoning we see that, in case $\hat{T}\ldots^{\cdots}(H_{\rho\sigma}{}^{\cdots})$ is defined only for $H_{(\rho\sigma)}{}^{\cdots}=0$, we obtain

$$\frac{\partial\tilde{T}\ldots^{\cdots}(H\ldots^{\cdots})}{\partial H_{(\rho\sigma)}{}^{\cdots}}=0\quad\text{for}\quad\tilde{T}\ldots^{\cdots}(H_{\rho\sigma}{}^{\cdots})=\hat{T}\ldots^{\cdots}(H_{[\rho\sigma]}{}^{\cdots}). \tag{A4.4}$$

Thus *the (normally vanishing) derivative of a double tensor field with respect to an argument having any symmetry or skewsymmetry property has the same property.* Incidentally the second remark below Theorem A 3.2 is a generalization of this fact.

As our third example let u^α be any time-like unit vector ($u^\alpha u_\alpha=-1$), e.g. the 4-velocity, and assume that the equations of $\mathscr{M}$ include (A 4.5)$_2$ below:

$$f(H_\alpha)\equiv u^\alpha H_\alpha=0. \tag{A4.5}$$

Let us set ($H'_\alpha=\phi_\alpha(H_\rho)=\overset{\perp}{g}_\alpha{}^\rho H_\rho$ and)

$$\tilde{T}\ldots^{\cdots}(H_\rho)=\hat{T}\ldots^{\cdots}(\overset{\perp}{g}_\alpha{}^\rho H_\rho),\quad\text{hence}\quad\tilde{T}(H_\alpha)=\tilde{T}(H_\alpha+\xi u_\alpha)(\xi\text{ real}). \tag{A4.6}$$

As a consequence

$$\begin{aligned}0&=\frac{d}{d\xi}\tilde{T}(H_\alpha+\xi u_\alpha)=\frac{\partial\tilde{T}\ldots^{\cdots}(H_\alpha+\xi u_\alpha)}{\partial H_\rho}u_\rho;\\ &\text{hence}\quad\frac{\partial\tilde{T}\ldots^{\cdots}(H_\alpha)}{\partial H_\rho}u^\rho=0\quad(\mathbf{H}\in\mathscr{M}).\end{aligned} \tag{A4.7}$$

The comparison of (A 4.5) with (A 4.7)$_3$ shows that (A 3.15) holds, besides (A 3.13)$_2$, so that $\partial\tilde{T}\ldots^{\cdots}/\partial H_\alpha$ is the normally vanishing derivative of $\hat{T}\ldots^{\cdots}$.

As our fourth example let $\mathscr{M}$ be represented, in given frames (x) and (y), by

$$\begin{aligned}&f_{LM}(R^\rho{}_A)\equiv\overset{\perp}{g}_{\alpha\beta}R^\alpha{}_LR^\beta{}_M-a^*_{LM}=0,\\ &f_L(R^\rho{}_A)\equiv u_\alpha R^\alpha{}_L=0,\qquad\det\|R^r{}_A\|\geqslant0.\end{aligned} \tag{A4.8}$$

We consider the neighborhood $\mathscr{N}_{\mathscr{M}}$ of $\mathscr{M}$ formed by the double tensors $H^\rho{}_A$ with $\det\|H^r{}_A\|>0$, and we set

$$R^\alpha{}_L=\phi^\alpha{}_L(H^\rho{}_\mu)\quad\text{with}\quad\overset{\perp}{g}{}^\rho{}_\alpha H^\alpha{}_\mu=R^\rho{}_L\mathscr{D}^L{}_\mu,\qquad\mathscr{D}_L{}^A\mathscr{D}_{AM}=H^\rho{}_LH^\sigma{}_M\overset{\perp}{g}_{\rho\sigma} \tag{A4.9}$$

where (A 4.9)$_2$ is the polar decomposition of $\overset{\perp}{H}{}^{\rho}{}_{M}$, so that the tensor $\mathscr{D}_L{}^A$ is (strictly) positive definite—cf. (54.5). With this in mind we set, under conditions (A 4.8),

$$\tilde{T}_{\dots}^{\dots}(R^{\alpha}{}_L \mathscr{D}^{LM}) = \hat{T}_{\dots}^{\dots}(R^{\alpha}{}_L), \quad \text{where} \quad \mathscr{D}_{[LM]} = 0 \quad (\mathscr{D}_{LM} \text{ positive definite}). \tag{A 4.10}$$

Conditions (A 4.10)$_{2,3}$ are fulfilled by the definition

$$\mathscr{D}^{LM} = a^*{}_{(A}{}^{L} a^*{}_{B)}{}^{M} K^{AB}, \tag{A 4.11}$$

when K_{AB} is in a suitable neighborhood $\mathscr{N}$ of a^*_{AB}. Then

$$0 = \frac{\partial}{\partial K^{LM}} \tilde{T}_{\dots}^{\dots}(H^{\alpha B}) = R^{\alpha}{}_A a^{*(A}{}_L a^{*B)}{}_M \frac{\partial \tilde{T}_{\dots}^{\dots}(H^{\alpha B})}{\partial H^{\alpha B}} \quad \text{where} \quad H^{\alpha B} = R^{\alpha}{}_L \mathscr{D}^{LM}. \tag{A 4.12}$$

For $\mathscr{D}_{LM} = a^*_{LM}\,(\in \mathscr{N})$, by (A 4.12)$_{1,2}$ we certainly have the second of the relations

$$\begin{aligned} \frac{\partial f_{LM}}{\partial R_{\alpha M}} \frac{\partial \tilde{T}_{\dots}^{\dots}}{\partial H^{\alpha M}} &\equiv 2R^{\alpha}{}_{(L} \frac{\partial \tilde{T}_{\dots}^{\dots}}{\partial H^{\alpha}{}_{M)}} = 0\,, \\ \frac{\partial f_L}{\partial R_{\alpha\rho}} \frac{\partial \tilde{T}_{\dots}^{\dots}}{\partial H^{\alpha\rho}} &\equiv u^{\alpha} \frac{\partial \tilde{T}_{\dots}^{\dots}}{\partial H^{\alpha\rho}} = 0 \quad \text{for} \quad R^{\alpha}{}_L = H^{\alpha}{}_L\,, \end{aligned} \tag{A 4.13}$$

while (A 4.13)$_{3,4}$ follow from (A 4.8)$_{3,4}$ after our third example.

From (A 4.8)$_1$ we easily deduce (A 4.13)$_1$. By (A 4.13) (and Theorem A 3 (γ)) $\partial \tilde{T}_{\dots}^{\dots}/\partial R^{\alpha}{}_L$ is the normally vanishing derivative of the field $\hat{T}_{\dots}^{\dots}$.

Appendix B

Two Uniqueness Properties of $E_{\alpha\beta}$

The first of the theorems to be proved in this section concerns the expression of the electromagnetic energy tensor $E_{\alpha\beta}$ in the empty part $R_4^{(0)}$ of space time—cf. footnote 3 in Chapter 4. The second—taken from Bressan [1967b]—concerns the determinations of $E_{\alpha\beta}$ within matter.

In $R_4^{(0)}$ we have $f_{\alpha\beta}=F_{\alpha\beta}$ and it is natural to consider $E_{\alpha\beta}$ as a universal function of $F_{\alpha\beta}$, $f_{\alpha\beta}$, $g_{\alpha\beta}$, and u_α—cf. condition (a) in § 39, and $(39.1)_1$. Hence $E_{\alpha\beta}$ has the same expression in $R_4^{(0)}$ and in non-polarizable materials. Then, in order to prove that under natural assumptions, $E_{\alpha\beta}={}^3E_{\alpha\beta}$—cf. (37.6)—must hold in $R_4^{(0)}$ [Theorem B1 below] let us remark that by (34.3) and the Ohm law (34.4) —where the existence of anisotropic electric conductors is implicitly assumed— we have

$$J^\alpha=\bar{\sigma}^{\alpha\beta}E_\beta+\bar{\rho}u^\alpha. \tag{B1}$$

Then the following assumption appears true:

a) *An arbitrary value* $F_{\alpha\beta}$ *with* $E_\alpha=F_{\alpha\beta}u^\beta\neq 0$ *is physically compatible with any one* $J^{(\lambda)}$ *of four linearly independent values* $J_\alpha^{(0)},\dots,J_\alpha^{(3)}$ *for the density* J_α *of true electric 4-current*[1], *with the local condition* $u_{\alpha/\beta}=0$, *with zero polarizations, and with arbitrary values of* $E_{\alpha\beta/\gamma}$ *which fulfill the conditions (Maxwell equations)*

$$\varepsilon^{\alpha\beta\gamma\delta}F_{\beta\gamma/\delta}=0\,,\qquad g^{\gamma\delta}F_{\beta\gamma/\delta}=J_\beta^{(\lambda)}. \tag{B2}$$

Theorem B1. *Let the electromagnetic energy tensor* $E_{\alpha\beta}$ *be admissible and simple, i.e. let it fulfill conditions* (a) *to* (d) *in* § 39, *so that, first, we have*

$$E_{\alpha\beta}=\phi_{\alpha\beta}(F_{\rho\sigma},f_{\rho\sigma},g_{\rho\sigma},u_\rho)\quad\text{with}\quad \phi_{\alpha\beta}(0,0,g_{\rho\sigma},u_\rho)=0\,, \tag{B3}$$

second, for every non-polarizable material and along every physically possible process we have—cf. (37.6)

$$E_{[\alpha\beta]}=0\,,\qquad E_{\alpha\beta}{}^{/\beta}={}^3E_{\alpha\beta}{}^{/\beta}\quad\text{for}\quad F_{\rho\sigma}\equiv f_{\rho\sigma}\quad\text{and}\quad u_{\rho/\sigma}=0\,, \tag{B4}$$

[1] Of course, given $F_{\alpha\beta}$, hence E_α, some values of J_α are not physically compatible with E_α, because e.g. $E^\alpha J_\alpha\geqslant 0$ must hold.

and lastly $E_{\alpha\beta}$ is a spatially isotropic function of $F_{\rho\sigma}$ (and $f_{\rho\sigma}$) [Definition 38.1] *which is twice continuously differentiable. Then*

$$T_{\alpha\beta} \equiv E_{\alpha\beta} - {}^3E_{\alpha\beta} = 0 \quad for \quad f_{\alpha\beta} = F_{\alpha\beta}\,. \tag{B5}$$

Proof. By $(\mathrm{B}3)_{1,2}$ and definitions $(\mathrm{B}5)_1$ and (37.6) $T_{\alpha\beta}$ has, for $F_{\rho\sigma} = f_{\rho\sigma}$, an expression of the form

$$T_{\alpha\beta} = \psi_{\alpha\beta}(F_{\rho\sigma}, g_{\rho\sigma}, u_\rho) \quad \text{with} \quad \psi_{\alpha\beta}(0, g_{\rho\sigma}, u_\rho) = 0 \quad (\text{for } F_{\rho\sigma} = f_{\rho\sigma})\,. \tag{B6}$$

By $(\mathrm{B}5)_1$, $(\mathrm{B}6)_1$, and (37.6) again, (B4) implies

$$T_{[\alpha\beta]} = 0\,, \qquad T^{\alpha\beta}{}_{/\beta} = \frac{\partial \psi^{\alpha\beta}}{\partial F_{\rho\sigma}} F_{\rho\sigma/\beta} = 0 \quad \text{for} \quad F_{\rho\sigma} \equiv f_{\rho\sigma} \quad \text{and} \quad u_{\rho/\sigma} = 0\,. \tag{B7}$$

Let us consider an arbitrary value of $F_{\alpha\beta}(= -F_{\beta\alpha})$ with $E_\alpha = F_{\alpha\beta}u^\beta \neq 0$. Consequently $(\mathrm{B}7)_3$ must hold for every non polarizable material and every process, at every event point where $u_{\alpha/\beta} = 0$ holds. Hence, for every one $J_\alpha^{(\lambda)}$ of the linearly independent 4-vectors $J_\alpha^{(0)}, \dots, J_\alpha^{(3)}$ mentioned in the existence assumption above, $(\mathrm{B}7)_3$ must hold for every tensor $F_{\rho\sigma/\beta}(= -F_{\sigma\rho/\beta})$ which fulfills the Maxwell equations (B2).

If $(\mathrm{B}7)_3$ holds for some values of $F_{\rho\sigma/\beta}$, then it holds for every linear combination of them. Furthermore, if we consider the values of $F_{\rho\sigma/\beta}$, each of which fulfills equations (B2) for one of the values $0 \dots 3$ of λ, and we consider all (finite) linear combinations of these values, then we obtain the values of $F_{\rho\sigma/\beta}$ which only fulfill the first Maxwell equation $(\mathrm{B}2)_1$. Hence $(\mathrm{B}7)_3$ must hold for every solution $F_{\rho\sigma/\beta}$ $(= -F_{\sigma\rho/\beta})$ of $(\mathrm{B}2)_1$. Then there are 16 scalars $\lambda^\alpha{}_\tau$ for which we have

$$\frac{\partial \psi^{\alpha\beta}}{\partial F_{\rho\sigma}} = \lambda^\alpha{}_\tau \varepsilon^{\tau\beta\rho\sigma}\,, \quad \text{whence} \quad -6\lambda^\alpha{}_\eta = \varepsilon_{\eta\beta\rho\sigma} \frac{\partial \psi^{\alpha\beta}}{\partial F_{\rho\sigma}}\,. \tag{B8}$$

Since the second partial derivatives of $\psi^{\alpha\beta}$ with respect to $F_{\rho\sigma}$ and $F_{\beta\gamma}$, and with respect to $F_{\beta\gamma}$ and $F_{\rho\sigma}$ are equal by their continuity, from $(\mathrm{B}8)_1$ we deduce

$$\frac{\partial \lambda^\alpha{}_\tau}{\partial F_{\beta\gamma}} \varepsilon^{\tau\beta\rho\sigma} = \frac{\partial \lambda^\alpha{}_\tau}{\partial F_{\rho\sigma}} \varepsilon^{\tau\beta\beta\gamma} = 0 \quad (\beta \text{ unsummed})\,. \tag{B9}$$

Now let us fix the indices $\alpha, \eta, \beta, \gamma$ with $\eta \neq \beta$ arbitrarily. Furthermore let $\eta, \beta, \rho, \sigma$ be a permutation of the 4-tuple $0, \dots, 3$. Then the left-hand side of $(\mathrm{B}9)_1$ becomes $\pm\sqrt{-g}\,\partial\lambda^\alpha{}_\eta/\partial F_{\beta\gamma}$, so that (B9) implies that $\lambda^\alpha{}_\eta$ is independent of $F_{\beta\gamma}$ in case $\eta \neq \beta$.

Furthermore $\lambda^\alpha{}_\eta$ certainly does not depend on $F_{\eta\eta}$, and for $\eta \neq \gamma$ we have —cf. § A4— $\partial\lambda^\alpha{}_\eta/\partial F_{\eta\gamma} = -\partial\lambda^\alpha{}_\eta/\partial F_{\gamma\eta}$.

We conclude that $\lambda^\alpha{}_\eta$ is independent of $F_{\beta\gamma}$ $(\beta, \gamma = 0, \dots, 3)$.

Since $E_{\alpha\beta}$—cf. $(\mathrm{B}3)_1$—is a spatially isotropic function of $F_{\rho\sigma}$ [Definition 38.1], by (37.6) and $(\mathrm{B}5)_1$ the same holds for $T_{\alpha\beta}$, i.e. $\psi_{\alpha\beta}$. Then, by Theorem 38.4, $\partial\psi^{\alpha\beta}/\partial F_{\rho\sigma}$ and—cf. $(\mathrm{B}8)_2$—$\lambda^\alpha{}_\eta$ also are spatially isotropic functions of $F_{\rho\sigma}$.

Furthermore $\lambda^\alpha{}_\eta$ is independent of $F_{\rho\sigma}$. Then the tensor $\lambda^\alpha{}_\eta$ is spatially isotropic, so that it has the form $(38.3)_1$:

$$\lambda^\alpha{}_\eta = \mathscr{H}\overset{\perp}{g}{}^\alpha{}_\eta + \mathscr{K} u^\alpha u_\eta = \mathscr{H} g^\alpha{}_\eta + (\mathscr{H}+\mathscr{K}) u^\alpha u_\eta . \tag{B10}$$

Then $(B8)_1$, $(B6)_2$, and $(20.4)_1$ imply

$$T^{\alpha\beta} = \psi^{\alpha\beta}(F_{\rho\sigma}, g_{\rho\sigma}, u_\rho) = \mathscr{H}\varepsilon^{\alpha\beta\rho\sigma} F_{\rho\sigma} + (\mathscr{H}+\mathscr{K}) u^\alpha \overset{\perp}{\varepsilon}{}^{\beta\rho\sigma} F_{\rho\sigma} . \tag{B11}$$

This and the symmetry assertion $(B7)_1$ imply $T^{\alpha\beta}=0$. Thus $(B5)_2$ has been proved for every $F_{\alpha\beta}$ with $E_\alpha = F_{\alpha\beta} u^\beta \neq 0$ and for $f_{\alpha\beta} = F_{\alpha\beta}$. Hence, by the continuity of $\psi_{\alpha\beta}(F_{\rho\sigma}, g_{\rho\sigma}, u_\rho)$ we can assert (B5). q.e.d.

With a view to proving a second uniqueness theorem on $E_{\alpha\beta}$ we take the possibility of anisotropic electric conductivities into account again, in order to justify the following existence assumption which is rather weak and certainly acceptable.

b) A typical *(physically possible) value of the couple* $F_{\alpha\beta}$, $f_{\alpha\beta}$ *is physically compatible with any one* $J_\alpha^{(\lambda)}$ *of four suitable linearly independent values* $J_\alpha^{(0)}, \ldots, J_\alpha^{(3)}$ *of* J_α, *with arbitrary solutions* $F_{\alpha\beta/\gamma}$, $f_{\alpha\beta/\gamma}$ *of the Maxwell equations* (35.1) *and with arbitrary values of* $u_{\alpha/\beta}$.

Theorem B2. *Let* $\Pi^{(e)}$ *be a given function of* $F_{\alpha\beta}$, $f_{\alpha\beta}$, $F_{\alpha\beta/\gamma}$, *and* $f_{\alpha\beta/\gamma}$—*so that by* (35.1) *and* (35.2) $\Pi^{(e)}$ *may depend on* J_α *and* J'_α. *Then there is at most one simple and admissible tensor of electromagnetic energy* $E_{\alpha\beta}$ [Definition 39.1] *for which* $(36.2)_2$—*i.e.* $c u_\alpha E^{\alpha\beta}{}_{/\beta} = \Pi^{(e)}$—*holds for every matter element* $d\mathscr{C}$ *and every process* $\mathscr{P}$ *physically possible for* $d\mathscr{C}$.

Proof. Let $\theta_{\alpha\beta}$ be the difference of two simple and admissible electromagnetic energy tensors of the kind being considered. Then—cf. Definition 39.1—$\theta_{\alpha\beta}$ is a function $\theta_{\alpha\beta}(F_{\rho\sigma}, f_{\rho\sigma}, g_{\rho\sigma}, u_\rho)$ of the explicitly written arguments, such that the first of the equalities

$$\theta_{\alpha\beta}(F_{\rho\sigma}, F_{\rho\sigma}, g_{\rho\sigma}, u_\rho) = 0 , \qquad u^\alpha \theta_{\alpha\beta}{}^{/\beta} = 0 \tag{B12}$$

is a mathematical identity and the second holds for every matter element along every physically possible process.

By (22.9) $(B12)_2$ becomes, in locally natural and proper co-ordinates,

$$-u^\alpha \theta_{\alpha\beta}{}^{/\beta} = k \frac{D}{Ds}\frac{\theta}{k} + \theta''^r{}_{/r} + (\theta'_r + \theta''_r) A^r + (\overset{\perp}{\theta}{}^{rs} + \theta \overset{\perp}{g}{}^{rs}) u_{r/s} = 0 . \tag{B13}$$

We momentarily assume $u_{\alpha/\beta}=0$, which yields $A_\alpha = 0 = \dot{k}$—cf. $(21.3)_3$. Then by (B13) we have

$$\frac{\partial \theta_0{}^\beta}{\partial F_{\rho\sigma}} F_{\rho\sigma/\beta} + \frac{\partial \theta_0{}^\beta}{\partial f_{\rho\sigma}} f_{\rho\sigma/\beta} \equiv -u^\alpha \theta_{\alpha\beta}{}^{/\beta} = \frac{D\theta}{Ds} + \theta''^r{}_{/r} = 0 . \tag{B14}$$

Given $F_{\rho\sigma}$, $f_{\rho\sigma}$, and the solution $F_{\rho\sigma/\beta}$ of $(35.1)_2$, by the existence assumption (b) there are four linearly independent values $J_\alpha^{(0)}, \ldots, J_\alpha^{(3)}$ of J_α such that for $\lambda=0, \ldots, 3$ (B14) holds for every solution $f_{\rho\sigma/\beta}\,(=-f_{\sigma\rho})$ of $(35.1)_2$ with $J_\alpha=J_\alpha^{(\lambda)}$. Then (B14) holds for every $f_{\rho\sigma/\beta}$, so that $\partial\theta_0{}^\beta/\partial f_{\rho\sigma}=0$; hence by (B12) the relation

$$\theta_{\alpha\beta}(F_{\rho\sigma}, f_{\rho\sigma}, g_{\rho\sigma}, u_\rho)=0 \tag{B15}$$

holds identically for $\alpha=0$—cf. $(17.7)_{1,3}$. Then by convention (17.13) we have $\theta\equiv\theta''_r\equiv 0$.

By the identities above and (B13), $\theta'_r A^r + \overset{\perp}{\theta}{}^{rs} u_{r/s}=0$ holds for every A^r and $u_{r/s}$. In addition θ'_r and $\overset{\perp}{\theta}{}^{rs}$ are independent of A^r and $u_{r/s}$. As a consequence $\theta'_r\equiv\overset{\perp}{\theta}{}^{rs}\equiv 0$. Then (B15) holds also for $\alpha=1,2,3$.

By the definition of $\theta_{\alpha\beta}$ (B15) implies the thesis. q.e.d.

Appendix C

On the Divergence of Spatial Vectors in Space-Time

In the first part of this appendix we show that from the kinematic point of view the space-time divergence $\mathrm{q}^{\alpha}{}_{/\alpha}$ and not the spatial divergence $\mathrm{q}^{\alpha}{}_{/\overset{\perp}{\alpha}}$ of the spatial vector q^{α} is the relativistic analogue of the classical (spatial) divergence $\bar{\mathrm{q}}^{r}{}_{/r}$. In the second part we strive to show that in the case of heat conduction Eckart's relativistic expression (24.5) of the heat kq_{ass} absorbed by heat conduction per unit time and proper volume, is better than the "kinematical" relativization $-cq^{\alpha}{}_{/\alpha}$ of the classical version $-\bar{q}^{r}{}_{/r}$ of kq_{ass} $[cq^{r}=\bar{q}^{r}]$ as far as the first principle is concerned.[1)]

The motion of the body $\mathscr{C}$ is characterized by the time-like unitvector field u^{α}. We now consider a field γ^{α} of the same kind, and we denote by $\mathfrak{F}$ the ideal fluid whose motion is characterized by this field. We also consider the scalar field ψ which expresses the density with respect to $\mathfrak{F}$ of any physical magnitude $\mathfrak{M}$.

Let us represent the motion of a point of $\mathfrak{F}$ by the functions $x^{\alpha}=x^{\alpha}(s)$, where s is the arc length, so that in locally proper and natural co-ordinates we have

$$\gamma^{\alpha}=\frac{Dx^{\alpha}}{Ds}=\frac{V^{\alpha}}{\sqrt{1-\mathscr{V}^{2}}} \quad \text{where} \quad V^{\alpha}=\frac{Dx^{\alpha}}{Dx^{0}} \quad \text{and} \quad \mathscr{V}^{2}=\delta_{rs}V^{r}V^{s}. \tag{C1}$$

The proper density ψ_u of $\mathfrak{M}$ (we mean the one with respect to $\mathscr{C}$) is characterized by the condition $\psi_u dV_{\mathscr{C}}=\psi dV_{\mathrm{F}}$, where $dV_{\mathrm{F}}\,[dV_{\mathscr{C}}]$ is the volume of an element, $d\mathfrak{F}$, of $\mathfrak{F}$ measured by an observer stationary with respect to $\mathfrak{F}$ $[\mathscr{C}]$. Then $dV_{\mathscr{C}}=\sqrt{1-\mathscr{V}^{2}}\,dV_{\mathrm{F}}$, so that we have

$$\psi_u=\frac{\psi}{\sqrt{1-\mathscr{V}^{2}}}=\psi\gamma^{0}=-\psi\gamma^{0}u_{0}\,; \quad \text{hence} \quad \psi_u=-\psi\gamma^{\beta}u_{\beta}. \tag{C2}$$

We set

$$\mathrm{q}_{\alpha}=\psi\overset{\perp}{\gamma}_{\alpha}, \quad \text{hence} \quad \mathrm{q}_{\alpha}u^{\alpha}=0. \tag{C3}$$

From $(\mathrm{C3})_1$, $(\mathrm{C2})_4$, and $(17.6)_1$ we deduce

$$\psi\gamma^{\alpha}=\mathrm{q}^{\alpha}-\psi\gamma_{\beta}u^{\beta}u^{\alpha}=\mathrm{q}^{\alpha}+\psi_u u^{\alpha}. \tag{C4}$$

1) This appendix is based on Bressan [1970].

Now we assume that the magnitude $\mathfrak{M}$ is stationary with respect to $\mathfrak{F}$, so that the conservation equation

$$(\psi\gamma^\alpha)_{/\alpha}=0 \quad [\gamma^\alpha\gamma_\alpha=-1] \tag{C5}$$

holds. This occurs e.g. in case $\mathfrak{F}$ is a gas being diffusing in $\mathscr{C}$ and $\mathfrak{M}$ is its conventional mass [§ 21]. Then $q^\alpha=\psi\overset{+}{\gamma}{}^\alpha$ is its flux vector.

Of course $\mathfrak{M}$ can be identified with the rest mass of this fluid provided the internal energy of the fluid is constant; however $\mathfrak{M}$ cannot be identified with the mass of $\mathfrak{F}$ relative to an observer who is stationary with respect to $\mathscr{C}$.

By (C4) and (21.5) for $k\tau_{\dots}^{\dots}=q^\alpha$, equation (C5) is equivalent to

$$-q^\alpha{}_{/\alpha}=(\psi_u u^\alpha)_{/\alpha}\equiv k\frac{D}{Ds}\frac{\psi_u}{k} \quad \left(k=k^*\frac{dC^*}{dC} \text{—cf. (21.2)}\right). \tag{C6}$$

Since $D(k\,dC)=D(k^*dC^*)=0$—cf. $(\mathrm{C}6)_3$—and $(\mathrm{C}6)_{1,2}$ hold, we have

$$-q^\alpha{}_{/\alpha}dC=k\,dC\frac{D}{Ds}\frac{\psi_u}{k}=\frac{D}{Ds}(\psi_u dC). \tag{C7}$$

Then the quantity $-q^\alpha{}_{/\alpha}dC\,Ds$ (and not $-q^\alpha{}_{/\overset{+}{\alpha}}dC\,Ds$) equals the increment $D(\psi_u dC)$ in Ds of the part $d\mathfrak{M}$ of $\mathfrak{M}$ which is in $d\mathscr{C}$. We conclude that *if the physical magnitude $\mathfrak{M}$ is stationary with respect to $\mathfrak{F}$*—cf. (C5)—, *then* $-q^\alpha{}_{/\alpha}dC\,Ds$ —cf. $(\mathrm{C}3)_1$—*can be interpreted as the part of $\mathfrak{M}$ that "enters" $d\mathscr{C}$ in Ds algebraically.*

Let us remark that from $(\mathrm{C}2)_1$ and from $(\mathrm{C}3)_2$, (C4), and $(\mathrm{C}5)_2$, respectively, we deduce

$$\psi\psi_u\geqslant 0, \quad q^\alpha q_\alpha-\psi_u{}^2=-\psi^2\leqslant 0. \tag{C8}$$

Now let us give the (regular) spatial field q^α in the world tube $W_{\mathscr{C}}$ of $\mathscr{C}$ arbitrarily, and let T_4 be a bounded domain formed with ∞^3 vector lines. We can fix the space-like cross-section σ_3 of T_4 and the values of ψ_u on σ_3 in such a way that the corresponding solution $\psi_u=\psi_u(x)$ of the differential equation (C6) in ψ_u makes the left hand side of $(\mathrm{C}8)_2$ negative in T_4, so that $q^\alpha+\psi_u u^\alpha$ is time-like.

Then $(\mathrm{C}8)_{1,2}$ determine ψ $(\neq 0)$ and (C4) determines the (time-like) unit vector γ^α of $q^\alpha+\psi_u u^\alpha$.

Since $(\mathrm{C}8)_1$, $(\mathrm{C}3)_2$, and (C4) yield $\gamma^\alpha u_\alpha<0$, γ^α is pointing towards future as well as u^α.

We conclude that *the above arbitrary spatial field q^α can be interpreted in T_4 as the flux vector of the conventional mass of a gas diffusing within $\mathscr{C}$.*

Now let the above field q^α represent the flux of the magnitude $\mathfrak{M}$ with respect to $\mathscr{C}$. At first let there be no sources for $\mathfrak{M}$. Then it is natural to consider $\mathfrak{M}$ as stationary with respect to a suitable fluid, $\mathfrak{F}$, whose motion with respect to $\mathscr{C}$ is parallel with q^α $(\gamma_{/\overset{+}{\alpha}}\parallel q_\alpha)$.

Furthermore, when the flux q^α of $\mathfrak{M}$ with respect to $\mathscr{C}$ is given, the space-time distribution of $\mathfrak{M}$ compatible with this flux is determined up to the addition of a distribution of $\mathfrak{M}$ which is stationary with respect to $\mathscr{C}$. Hence the density of $\mathfrak{M}$ (with respect to $\mathfrak{F}$) can be assumed to be positive in T_4. Then we can represent the flux of $\mathfrak{M}$ in $\mathscr{C}$ (i.e. $\mathfrak{F}$ and the distribution of $\mathfrak{M}$ on it) by means of the above gas, provided the proper density of the gas is assumed to be equal ψ.

The foregoing considerations of relativistic kinematics allow us to consider $-\mathfrak{q}^\alpha{}_{/\alpha} dC$ $(q^\alpha u_\alpha = 0)$, and not $-\mathfrak{q}^\alpha{}_{/\overset{*}{\alpha}} dC$, as the relativistic kinematical analogue of the classical expression $-\mathfrak{q}^r{}_{/r}$ for the quantity of $\mathfrak{M}$ entering $d\mathscr{C}$ per unit time (in case $\mathfrak{q}^r$ is the flux vector of $\mathfrak{M}$).

The same considerations directly concern a case where $\mathfrak{M}$ has no sources. However our conclusion is compatible with the general case and cannot be accepted only in the special case.

Now let $\mathfrak{M}$ have sources (and the above flux vector $\mathfrak{q}^\alpha$). We can associate with the field $\mathfrak{q}^\alpha$ the above gas again. Then the time derivative (with respect to $\mathscr{C}$) of the conventional mass of the gas in $d\mathscr{C}$ equals the contribution of the flux $\mathfrak{q}^\alpha$ of $\mathfrak{M}$ to the analogue derivative of the quantity of $\mathfrak{M}$ present in $d\mathscr{C}$. The remaining part of the latter time derivative is due to the sources of $\mathfrak{M}$ in $d\mathscr{C}$.

If $-\mathfrak{q}^r{}_{/r}$ is a term in a classical physical equation, then the foregoing considerations serve to interpret $-\mathfrak{q}^\alpha{}_{/\alpha}$ in case $-\mathfrak{q}^\alpha{}_{/\alpha}$ is a term in the corresponding relativistic equation; however those considerations do not at all compel us to use $-\mathfrak{q}^\alpha{}_{/\alpha}$ in the latter equation as the relativization of the term $-\mathfrak{q}^r{}_{/r}$ in the former.

For instance, the classical expression for the heat $k q_{\text{ass}}$, absorbed by heat conduction per unit (proper) volume and time, is $-\bar{q}^r{}_{/r}$ and the corresponding relativistic expression is $-c(q^\alpha{}_{/\alpha} + q^\alpha A_\alpha)$—cf. (24.5) and $(25.3)_3$ i.e. $\bar{q}^r = c q^r$.

We know the meaning of $-c q^\alpha{}_{/\alpha}$. Eckart [1940, p. 921] pointed out that the very small term $c q^\alpha A_\alpha$ can be interpreted on the basis of the mass-energy equivalence principle as the work done by a flow of heat through accelerated matter. We shall try to justify the presence of the term $c q^\alpha A_\alpha$ in the first principle from the physical point of view.

On the one hand we consider a mass point P in classical physics. Let m be its mass, v_r its velocity with respect to the fluid $\mathfrak{F}'$, and let α^r be the (dragging) acceleration of $\mathfrak{F}'$ with respect to locally non-rotating and freely falling spaces. Then *the power relative to $\mathfrak{F}'$ of the dragging force acting on P is $(-m\alpha^r)v_r = -(m v_r)\alpha^r$, so that it depends on m and v_r only through the momentum $m v_r$ of P.*

In addition let us consider the acceleration of P with respect to $\mathfrak{F}'$ and the velocity gradient of $\mathfrak{F}'$ with respect to (inertial spaces or) locally freely falling and non rotating frames, as given. Hence $m\alpha^r$ is that part of the total force on P evaluated with respect to the above spaces, which is due to the acceleration of $\mathfrak{F}'$. *If this part is exerted by $\mathfrak{F}'$, then $\mathfrak{F}'$ must spend for this the power $m\alpha^r v_r$ which equals the opposite $m v^r \alpha_r$ of the above (dragging) work.*

On the other hand, we know that if $\mathfrak{f}^r$ is the flux vector of a fluid—so that $\mathfrak{f}^r d\sigma_r$ equals the mass of the fluid which crosses the arbitrary elementary oriented surface $d\sigma_r$ toward its positive face per unit time—then $\mathfrak{f}^r$ equals the momentum μv^r of the fluid per unit volume.

Let us remark that by the mass-energy equivalence principle we can associate to the heat flow in $d\mathscr{C}$, represented by cq^α, a mass flow with respect to $d\mathscr{C}$, whose momentum per proper volume is $\mu v^r = c^{-1} q^r$.

We now consider, in general relativity, the motion of $d\mathscr{C}$ with respect to a locally natural and proper frame, (x), which is an analogue of the classical frames which are locally freely falling and non-rotating. Furthermore we consider the heat flux q^α with respect to $\mathfrak{F}\ (=\mathscr{C})$. Then $\alpha_r = c^2 A_r$ is the dragging acceleration and by the above considerations $-\mu v^r \alpha_r = -c q^\alpha A_\alpha$ is the dragging power acting on the flowing mass associated with the heat flux q^α.

Furthermore the field cq^α—which in ordinary cases can be thought of as a function of $\theta_\alpha = T_{/\frac{1}{\alpha}} + T A_\alpha$—cf. $(25.2)_2$—, determines the above heat flow with respect to $\mathscr{C}$. The local intrinsic acceleration of $d\mathscr{C}$ contributes to determine the heat flow in $d\mathscr{C}$ with respect to a locally natural and proper frame. By the above considerations $d\mathscr{C}$ must spend for this contribution the opposite of the dragging power $-\mu v^r \alpha_r$ per unit proper volume, i.e. $c q^\alpha A_\alpha$. This contributes to decrease the internal energy of $d\mathscr{C}$ in accordance with the presence of the term $-c q^\alpha A_\alpha$ in the right-hand side of Eckart's version of the first principle—cf. (24.5), $(24.6)_1$.

The thermodynamic tensor $Q_{\alpha\beta}$—cf. $(24.3)_2$—is symmetric. Its symmetry is required only in general relativity by gravitation equations. In special relativity $Q_{\alpha\beta}$ may be replaced by its term $u_\alpha q_\beta$; this replacement is equivalent to crossing out from $(24.3)_2$ the term $q_\alpha u_\beta$, whose contribution to the first principle is the very little term $q^\alpha A_\alpha$—cf. (24.5), $(24.6)_1$.

The foregoing considerations based on the mass-energy equivalence principle aim at showing that the thermodynamic tensor $Q_{\alpha\beta}$ is more satisfactory than $u_\alpha q_\beta$ also in special relativity.

Appendix D

On the Lie Derivatives $\mathscr{L}_u$, D^c, D_c and the Lagrangian Representation $\overset{\perp}{T}_{\dots}^{\dots} \to \overset{*}{T}_{\dots}^{\dots}$. Application to Linear Elasticity

We present the Lie derivative $\mathscr{L}_u$ with respect to the 4-velocity u^α of matter (which is defined on $W_{\mathscr{C}}$ and fulfills the condition $u^\alpha u_\alpha = -1$), as a slight modification of D^c and D_c—cf. (73.4), (73.5). In harmony with this we also generalize definitions $(73.1)_{1,2}$ and some relations, e.g. (73.8).

We present this and some applications of it to elasticity, in order to make it easier to compare the present relativistic theory of materials with papers of other authors, on the same subject.

We define $\mathscr{L}_u T_{\rho_1\dots\rho_n}{}^{\sigma_1\dots\sigma_p}{}_{A\dots}{}^{B\dots}$ by

$$\mathscr{L}_u T_{\rho_1\dots A\dots}{}^{\sigma_1\dots B\dots} = \frac{D}{Ds} T_{\dots}^{\dots} + u^\gamma{}_{/\rho_1} T_{\gamma\rho_2\dots}{}^{\dots} + \dots - u^{\sigma_1}{}_{/\gamma} T_{\rho_1\dots}{}^{\gamma\sigma_2\dots} - \dots, \tag{D1}$$

so that latin (material) indices have the role of parameters.

By (73.5) and (73.4)

$$(\mathscr{L}_u T^{\rho_1\dots\rho_n})^\perp = \frac{D^c}{Ds} T^{\rho_1\dots\rho_n}, \qquad (\mathscr{L}_u T_{\rho_1\dots\rho_n})^\perp = \frac{D_c}{Ds} T_{\rho_1\dots\rho_n}, \tag{D2}$$

and, unlike D^c and D_c, the Lie derivative does not commute with the operation of raising indices, but it commutes with contraction and fulfills the rule for the derivation of the product, e.g.

$$\mathscr{L}_u(T_{\alpha\rho}{}^\beta{}_A{}^H V^{\rho B}{}_H) = (\mathscr{L}_u T_{\alpha\rho}{}^\beta{}_A{}^H) V^{\rho B}{}_H + T_{\alpha\rho}{}^\beta{}_A{}^H \mathscr{L}_u V^{\rho B}{}_H . \tag{D3}$$

Given the double tensor $T_{\rho_1\dots\rho_n A\dots}{}^{\sigma_1\dots\sigma_p B\dots}$, let us generalize the definitions $(73.1)_{1,2}$ into

$$\overset{*}{T}_{L_1\dots L_n A\dots}{}^{M_1\dots M_p B\dots} \alpha^{\sigma_1}{}_{M_1}\dots\alpha^{\sigma_p}{}_{M_p} = \alpha^{\rho_1}{}_{L_1}\dots\alpha^{\rho_n}{}_{L_n} \overset{\perp}{T}_{\rho_1\dots\rho_n A\dots}{}^{\sigma_1\dots\sigma_p B\dots}. \tag{D4}$$

The analogue of (D2) holds for $T_*^{\dots}$, $T^*_{\dots}$, and $\overset{*}{T}_{\dots}^{\dots}$:

$$\overset{*}{T}_{\dots} = T^*_{\dots}, \qquad \overset{*}{T}{}^{\dots} = T_*^{\dots}. \tag{D5}$$

Furthermore

$$\overset{*}{T}{}^{L_1}{}_{L_2\dots}{}^{\dots} = \overset{-1}{C}{}^{L_1H} \overset{*}{T}_{HL_2\dots}{}^{\dots}, \qquad \overset{*}{T}_{L_1L_2\dots}{}^{\dots} = C_{L_1H} \overset{*}{T}{}^H{}_{L_2\dots}{}^{\dots}. \tag{D6}$$

It is easy to write the analogue of (73.8) for $\overset{*}{T}\dots^{\dots}$ and $\mathscr{L}_u$:

$$\left(\frac{D}{Ds}\overset{*}{T}_{L_1\dots L_n A\dots}{}^{M_1\dots M_p B\dots}\right)\alpha^{\sigma_1}{}_{M_1}\dots\alpha^{\sigma_p}{}_{M_p}=\alpha^{\rho_1}{}_{L_1}\dots\alpha^{\rho_n}{}_{L_n}(\mathscr{L}_u\overset{+}{T}_{\rho_1\dots\rho_n A\dots}{}^{\sigma_1\dots\sigma_p B\dots})^{\perp}. \tag{D7}$$

By (53.15) and (D4) for $n-2=0=p$

$$(\overset{+*}{g}_{LM}=)\overset{*}{T}_{LM}=C_{LM} \quad \text{for} \quad T_{\rho\sigma}=\overset{+}{g}_{\rho\sigma}, \tag{D8}$$

so that by $(53.7)_1$ and (D7)[1)]

$$\frac{DC_{LM}}{Ds}=\alpha^{\rho}{}_{L}\alpha^{\sigma}{}_{M}\mathscr{L}_u\overset{+}{g}_{\rho\sigma}. \tag{D9}$$

In locally co-moving co-ordinates $(x^r\equiv y^r)\,\alpha^{\rho}{}_{L}=\delta^{\rho}{}_{L}$—cf. $(62.4)_{2,3}$. Hence by (D8), (D9), and $(57.6)_{2,3}$

$$\mathscr{L}_u\overset{+}{g}_{rs}=\frac{DC_{rs}}{Ds}=2u_{(r/s)} \quad (x^r\equiv\delta^r{}_L y^L). \tag{D10}$$

Let us assume $Dt\equiv Dx^0\equiv Ds$, hence $u^{\rho}=\delta^{\rho}{}_0$. Then

$$u^{\rho}{}_{/\sigma}=u^{\rho}{}_{,\sigma}+\begin{Bmatrix}\rho\\ \gamma\sigma\end{Bmatrix}u^{\gamma}=\begin{Bmatrix}\rho\\ 0\sigma\end{Bmatrix}, \quad \frac{D}{Ds}T\dots^{\dots}=T\dots^{\dots}{}_{/0} \quad (u^{\rho}\equiv\delta^{\rho}{}_0); \tag{D11}$$

furthermore $T\dots^{\dots}$ is a function of x^{ρ}, so that

$$\frac{DT_{\rho_1\dots A\dots}{}^{\sigma_1\dots B\dots}}{Ds}=T\dots^{\dots}{}_{/0}=T\dots^{\dots}{}_{,0}-\begin{Bmatrix}\gamma\\ 0\sigma\end{Bmatrix}T_{\gamma\rho_2\dots}{}^{\dots}-\dots+\begin{Bmatrix}\sigma_1\\ \gamma 0\end{Bmatrix}T_{\rho_1\dots}{}^{\gamma\sigma_2\dots}+\dots. \tag{D12}$$

Then $(D11)_{1,2}$ and (D1) imply

$$\mathscr{L}_u T_{\rho_1\dots A\dots}{}^{\sigma_1\dots B\dots}=T_{\rho_1\dots A\dots}{}^{\sigma_1\dots B\dots}{}_{,0} \quad (u^{\rho}\equiv\delta^{\rho}{}_0). \tag{D13}$$

Let us consider the following choice of the internal energy function $\tilde{w}$ in $(63.4)_3$:

$$\tilde{w}(\eta,\varepsilon_{AB})=\tfrac{1}{2}\overset{*}{\gamma}{}^{LMAB}\varepsilon_{LM}\varepsilon_{AB}, \quad \overset{*}{\gamma}{}^{LMAB}=\overset{*}{\gamma}{}^{ABLM}=\overset{*}{\gamma}{}^{BALM}, \tag{D14}$$

where $\overset{*}{\gamma}{}^{LMAB}$ is a function of the specific entropy η. Then from $(63.4)_2$ and from $(58.3)_1$ and $(59.2)_{1,2}$, respectively, we deduce

$$Y^{LM}=k^*\overset{*}{\gamma}{}^{LMAB}\varepsilon_{AB}, \quad X^{\rho\sigma}=\chi^{\rho\sigma AB}\varepsilon_{AB} \text{ where } \chi^{\rho\sigma AB}=k\alpha^{\rho}{}_{L}\alpha^{\sigma}{}_{M}\overset{*}{\gamma}{}^{LMAB}. \tag{D15}$$

1) Bennoun [1964a, b] uses $\overset{+}{g}_{\alpha\beta}$ and $\mathscr{L}_u\overset{+}{g}_{\alpha\beta}$ as arguments of constitutive functions.

We consider an adiabatic process, so that (besides $\overset{*}{k}$) the coefficients $\overset{*}{\gamma}{}^{LMAB}$ are constant along world-lines. Then by (D7) and (D15)$_3$

$$\left(\mathscr{L}_u \frac{\chi^{\rho\sigma AB}}{k}\right)^{\perp} = 0 . \tag{D16}$$

In Rainer [1963] ((D14) is disregarded and) (D15)$_2$ holds together with

$$\mathscr{L}_u \chi_{\rho\sigma}{}^{AB} = 0, \quad \text{hence} \quad \frac{D}{Ds} \overset{*}{\chi}_{RS}{}^{AB} = 0 . \tag{D17}$$

By (59.2)$_{1,2}$ and by (D6)$_1$ and (D15)$_2$, respectively, we have—disregarding (D17)

$$Y^{LM} = \frac{1}{\mathscr{D}} \overset{*}{X}{}^{LM} = \frac{1}{\mathscr{D}} \overset{-1}{C}{}^{RL} \overset{-1}{C}{}^{SM} \overset{*}{\chi}_{RS}{}^{AB} \varepsilon_{AB} \text{—cf. (62.7).} \tag{D18}$$

This and (D17)$_2$ express the Lagrangian law for the elastic material considered in Rayner's purely mechanical theory. As is proved in Bressan [1964a], (D18) and (D17)$_2$ are not compatible with the existence of a strain energy (unless $Y^{LM} \equiv 0$), hence (D14) cannot hold in Rayner's theory for any nontrivial choice of $\hat{w}$.

References

Aero, E. L., Kuvshinskii, E. V. [1960]: Fundamental equation of the theory of elastic media with rotationally interacting particles (in Russian). Fizika Tverdogo Tela **2**, 1399. English Transl. Soviet Physics Solid State **2**, 1272—1821 (1961).

Alfvén, H. [1944]: On the existence of electromagnetic hydrodynamic waves. Ark. f. mat. astr. o. fys. 29 B, N 2.

Alts, T., Müller, I. [1972]: Relativistic thermodynamics of simple heat conducting fluids. Arch. Rational. Mech. Anal. **48**, 245.

Anderson, J. [1973]: Principles of relativity physics. New York: Academic Press. 484 pp.

Bennoun, J. F. [1964a]: Sur les équations d'un milieu matériel en relativité générale. C. R. Acad. Sci. Paris **258**, 94.

[1964b]: Sur les représentations hydrodynamique et thermodynamique des milieux élastiques en relativité générale. C. R. Acad. Sci. Paris **259**, 3705.

[1965]: Etude des milieux continus élastiques et thermodynamiques en relativité générale. Ann. Inst. H. Poincaré (A) **3**, 41.

Benvenuti, P. [1960]: Formulazione relativa delle equazioni dell'elettromagnetismo in Relativita generale. Ann. Scuola Normale Sup. Pisa (3) **14**, 171.

Berdicewski, V. L. [1966]: The constructions of models of continuous systems by means of a variational principle. Prikladnaia mathematica i mechanica (Applied mathematics and mechanics) **30** Moscow, 510.

Bogy, D., Naghdi, D. [1970]: On heat conduction and wave propagation in rigid solids. Journal Math. Phys. **11**, 917.

Bopp, F., see Sommerfeld, A.

Born, M. [1909]: Die Theorie des starren Elektrons in der Kinematic des Relativitätsprinzips. Ann. Phys. **30**, 1.

[1911]: Elastizitätstheorie und Relativitätstheorie. Phys. Z. **12**, 69.

Bragg, L. [1965]: On relativistic worldlines and motions and on non-sentient response. Arch. Rational Mech. Anal. **18**, 127.

[1970]: Relativistically dynamic elastic dielectrics. J. Mathematical Phys. **11**, 318.

Bressan, A. [1962]: Metodo di assiomatizzazione in senso stretto della meccanica classica. Applicazione di esso ad alcuni problemi di assiomatizzazione non ancora completamente risolti. Rend. Sem. Mat. Univ. di Padova **32**, 55.

[1963a]: Sui sistemi continui nel caso asimmetrico. Ann. Mat. Pura Appl. (IV) **62**, 169.

[1963b]: Cinematica dei sistemi continui in relatività generale. Ann. Mat. Pura Appl. **62**, 99.

[1963c]: Termo-magneto-fluido-dinamica in relatività generale. Gas perfetti. Riv. Mat. Univ. Parma **4**, 57.

[1963d]: Onde ordinarie di discontinuità nei mezzi elastici con deformazioni finite in relatività generale. Riv. Mat. Univ. Parma **4**, 23.

[1964a]: Termodinamica e magneto-visco-elasticità con deformazioni finite in relatività generale. Rend. Sem. Mat. Univ. Padova **34**, 1.

[1964b]: Una teoria di relatività includente, oltre all'elettromagnetismo e alla termodinamica, le equazioni costitutive dei materiali ereditari. Sistemazione assiomatica. Rend. Sem. Mat. Univ. Padova **34**, 74.

[1965]: Qualche proprietà di unicità del tensore energetico del campo elettromagnetico. Rend. Circ. Mat. di Palermo (II) **14**, 147.
[1966a]: Coppie di contatto in relatività. Ann. Scuola Normale Sup. Pisa (III) **20**, 63.
[1966b]: Elasticità relativistica con coppie di contatto. Ricerche di Mat. **15**, 169.
[1966c]: Sul teorema di Poynting e sul tensore energetico del campo elettromagnetico. Rend. Circ. Mat. di Palermo (II) **15**, 133.
[1966d]: Sui fluidi capaci di elettromagnetostrizione dai punti di vista classico e relativistico. Ann. Mat. Pura Appl. (IV) **74**, 317.
[1966e]: Elasticità relativistica con elettro-magneto-strizione. Ann. Mat. Pura Appl. (IV) **74**, 383.
[1967a]: On relativistic thermodynamics. Nuovo Cimento, Suppl. (X) **48**, 201.
[1967b]: Ancora sul teorema di Poynting. Rend. Accad. Naz. Lincei (VIII) **42**, 491.
[1968]: Sistemi polari in relatività. Symposia Mathematica (Ist. Naz. Mat. Roma) **1**, 289.
[1969]: On the influence of gravity on elasticity. Meccanica, N. 3, **4**, 195.
[1970]: Sul significato cinematico della divergenza spazio-temporale dei vettori spaziali rappresentanti un flusso nel cronotopo relativistico. Rend. Acc. Naz. Lincei (VIII) **49**, 57.
[1972a]: Sui conduttori ideali in Relatività. Ann. Mat. Pura Appl. (IV) **94**, 177.
[1972b]: Principi variazionali relativistici e coppie di contatto. Ann. Mat. Pura Appl. (IV) **94**, 201.
[1972c]: On the principles of material frame indifference and local equivalence. Meccanica **7**, 3.
[1972d]: A general interpreted modal calculus. New Haven-London, Yale: University Press. 327 pp.
[1972e]: Le equazioni gravitazionali di Einstein dedotte rigorosamente postulando qualche versione relativistica dell'equazione di Poisson. Parte 1. Forma di un sotituto relativistico dell'equazione di Poisson. Rend. Acc. Naz. Lincei (VIII), **53**, 118. Parte 2. Deduzione delle equazioni gravitazionali mediante un sostituto relativistico dell'equazione di Poisson giustificabile in modo diretto e naturale. Rend. Acc. Naz. Lincei (VIII), **53**, 200.
[1974a]: On the usefulness of modal logic in the axiomatization of physics, PSA 1972 (Proceeding of 1972 meeting of the Philosophy of Science Association). Dortrecht-Boston: D. Reidel Publ. co. p. 285.
[1974b]: Comments on Suppes' paper: The essential but implicit role of modal conepts in science. PSA 1972. Dortrecht-Boston: Reidel, p. 315.
[1974c]: On some relativistic version of the principle of material indifference in relativity. Lettere al Nuovo Cimento, p. 134.
Brevik, I. [1970a]: Electromagnetic energy-momentum tensor within material media. Physics Letters **31** A, p. 50.
[1970b]: Electromagnetic energy-momentum tensor. Part 1. Minkowski's tensor. Mat. Phys. Medd. Dan. Vid. Selsk. **37**, n° 11. Part 2. Discussion of various tensor forms. Mat. Phys. Medd. Dan. Vid. Selsk. **37**, n° 13.
Bridgman, P. W. [1927]: The logic of modern physics. New York: The Macmillan Comp.
[1949]: Reflections of a physicist. Philosophical library, New York.
Bruhat, Y. (Fourer) [1959]: Fluides chargés de conductivité infinie. C. R. Acad. Sci. Paris **248**, 2558.
[1960]: Fluides relativistes de conductibilité infinie. Astronautica Acta **6**, 354.
[1966]: Etude des Equations des fluides chargés relativistes inductifs et conducteurs. Commun. math. Phys. **3**, 334.
Burks, A. W. [1951]: The logic of causal propositions Mind **60**, 363.
Carstiou, F. [1963]: Relativistic Fluid Mechanics and Magnetohydrodynamics. New York, London, 125.
Cartan, E. [1922]: Sur les équations de la gravitation d'Einstein. J. Math. Pures Appl. **1**, 141 (Oeuvres III, 1, 549).
Carter, B., Quintana, H. [1972]: Foundations of general relativistic high pressure elasticity theory. Proc. R. Lon. Soc. A. **331**, 57.
[1975]: Stationary elastic rotational deformation of relativistic neutron star model (preprint).
Cattaneo, C. [1948]: Sulla conduzione del calore. Atti Sem. Mat. Fis. Univ. Modena **3**, 3.
[1958]: General Relativity: relative standard mass, momentum, energy and gravitational field in a general system of reference. Nuovo Cimento **10**, 318.
[1959]: Proiezioni naturali e derivazione trasversa in una varietà Riemanniana a metrica iperbolica normale. Ann. Mat. Pura Appl. **48**, 361.
[1960]: Introduzione alla teoria einsteiniana della gravitazione. Veschi Roma.

[1963]: Principi di conservazione e teoremi di Gauss in relatività generali. Rend. Mat., Ediz. Cremonese. Roma **21**, 373.

[1969]: Sur une conjecture concernant le tenseur d'énergie et impulsion d'un fluide relativiste non visquex. Colloques international de la recherche scientifique. N 184, 213.

[1971a]: Essai d'une théorie relativiste de l'elasticité. C. R. Acad. Sci. Paris **272** (A), 1421—1424.

[1971b]: Equations relativistes des petits mouvements adiabatiques d'un corps élastique. C. R. Acad. Sci. Paris **273** (A), 533.

[1973]: Dinamica relativistica di un mezzo elastico in regime adiabatico. Bollettino UMI (4), **8**. Suppl., 49.

Coleman, B. D., Mizel, V. J. [1963]: Thermodynamics and departures from Fourier's law of heat conduction. Arch. Rational Mech. Anal. **13**, 245.

Coleman, B. D., Gurtin, M. E. [1967]: Equipresence and constitutive equations for rigid heat conductors. Z. Angew. Math. Phys. **18**, 199.

Coleman, B. D., Noll, W. [1959]: On the thermostatic of continuous media. Arch. Rat. Mech. Anal. **4**, 97.

Cosserat, E., F. [1907]: Sur la mécanique générale. C. R. Acad. Sci. Paris **145**, 1139—1142.

Cresswell, M. J., see Hughes, G. E.

Dällenbach [1919]: Die allgemein kovarianten Grundgleichungen des elektromagnetischen Feldes im Innern ponderabler Materie vom Standpunkt der Elektronentheorie. Ann. Physik **58**, 523.

Dantzig, D. van [1934a]: Electromagnetism independent of metrical geometry I—IV. Proc. Akad. Wet. Amsterd. **37**, 521, 526, 643, 825.

[1934b]: The fundamental equations of electromagnetism independent of metrical geometry. Proc. Cambridge Phil. Soc. **30**, 421.

[1937a]: Über das Verhältnis von Geometric und Physik. C. r. Congr. In. Math. Oslo (1936) **2**, 225.

[1937b]: Some possibilites of the future development of the notions of space and time. Erkenntnis **7**, 142.

[1939]: On relativistic thermodynamics. Proc. Kond. Ned. Akad. v. Wetensch. **42**, 601.

[1940]: On the thermodynamics of perfectly perfect fluids. Proc. Kond. Ned. Akad. v. Wetensch. Amsterdam **43**, 387.

Daucourt, G. [1963]: Matter and gravitational shock waves in general relativity. Arch. Rat. Mech. and Analysis **3**, 55.

Dehnen, H. [1966]: Über Allgemein-relativistische Dynamik. Wiss. Z. Univ. Jena, Math.-nat. R. **15**, 15.

Dirac, P. A. M. [1938]: A new basis for cosmology. Proc. Roy. Soc., London A **165**, 199.

Dixon, R. C., Eringen, A. C. [1965]: A dynamical theory of polar elastic dielectrics. Int. J. Engng. Sci. **3**, 359.

Donder, T. De [1921]: La gravifique Einsteinienne. Paris: Gauthier-Villars.

[1931]: Théorie invariantive de l'élasticité à déformations finies. Bull. de l'Acad. Roy. Belg. Cl. des Sci. (Octobre).

Donder, T. De, Dupont, Y. [1932—1933]: Théorie relativiste de l'élasticité et de l'electromagnetostriction. Acad. Roy. Belg. Bull. Cl. Sci. **18**, 680, 782, 899; **19**, 370.

[1936—1937]: Théorie nouvelle de la dynamique des systèmes continus. Acad. Roy. Belg. Bull. Cl. Sci. **22**, 907, 992, 1378; **23**, 17, 102.

Eckart, C. [1940a]: The thermodynamics of irreversible processes. I The simple fluid. Phys. Rev. **58**, 267.

[1940b]: The thermodynamics of irreversible processes. III Relativistic theory of the simple fluid. Phys. Rev. **58**, 919.

Edelen, D. G. B. [1963]: On the foundations of relativistic energy mechanics. Nuovo Cimento **30**, 290.

[1964a]: Material momentum-energy tensors and the calculus of variations. Proc. Nat. Acad. Sci. U.S.A. **51**, 367.

[1964b]: Deformations and momentum-energy complexes. Arch. Rational Mech. Anal. **16**, 316.

Eglit, M. E. [1965]: A generalisation of the model of an ideal compressible fluid (Russian). PMM **29**, N 2, 351.

Ehlers, J. [1961]: Beiträge zur relativistischen Mechanik kontinuierlicher Medien. Abh. d. Math.-nat. K. d. Akad. d. Wiss. u. d. Lit. Nr. **11**, 793.

Einstein, A. [1905]: Zur Elektrodynamik bewegter Körper. Ann. Physik Lpz. (4) **17**, 891.

[1916]: Die Grundlage der allgemeinen Relativitätstheorie. Ann. Phys. (4) **49**, 769.

[1955]: The Meaning of Relativity. Princeton 5th ed.

Ericksen J. L. [1960]: Tensor fields (Appendix). Handbuch der Physik, Band III/1, 794—858. Berlin-Göttingen-Heidelberg: Springer.

Eringen, A. C., see Dixon, R. C.

Eringen, A. C., see Grot, A.

Eringen, A. C. [1970]: On a theory of general relativistic thermodynamics and viscous fluids. Proceeding of the conference on "A critical Review of the foundations of Relativistic and Classical Thermodynamics". Baltimore: Mono Book Corp. p. 483.

Finzi, B. [1931]: Meccanica relativistica ereditaria. Atti Soc. It. Progr. Sci. **20**, 8.

Finzi, B., Pastori, M. [1961]: Calcolo tensoriale e applicazioni. 2ª ed. Bologna: Zanichelli.

Fock, V. A. [1939a]: On the motion of finite masses in the general theory of relativity. Z. Eksper. Teoret. Fiz. **9**, 375.

[1939b]: Sur le mouvement des masses finies d'aprés la théorie de gravitation einsteinienne. Acad. Sci. USSR J. Phys. **1**, 81.

[1950]: Some applications of the ideas of Lobachewsky's non-Euclidean geometry to physics. Symposium by A. T. Kotelnikov and V. A. Fock. Some applications of Lobachewsky' ideas to Mechanics and Physics. Gostekhizdat.

[1964]: The theory of space time and gravitation. 2nd ed. (transl. by N. Kemmer) Pergamon Press.

Friedrichs, K. [1928]: Eine invariante Formulierung des Newtonschen Gravitationsgesetzes und des Grenzüberganges vom Einsteinschen zum Newtonschen Gesetz. Math. Ann. **98**, 566.

Grioli, A. C. [1971]: Una derivata interessante la teoria relativistica dei materiali non semplici. Rend. Sem. Mat. Univ. Padova **45**, 183.

[1972]: Su alcune esperienze ideali riguardanti il principio di indifferenza materiale. Rend. Sem. Mat. Univ. Padova. **47**, 139.

[1973]: Sulla derivata di Cattaneo e la derivata lagrangiana spaziale. Rend. Acc. Naz. Lincei (VIII) **55**, 450.

Grioli, G. [1960]: Elasticità asimmetrica. Ann. Mat. Pura Appl. (IV) **50**, 389.

[1962]: Mathematical theory of elastic equilibrium. Ergebnisse der angewandten Mathematik, Nr. 7. Berlin-Göttingen-Heidelberg: Springer.

Groot, S. R., Mazur, P. [1962]: Non-equilibrium thermodynamics. Amsterdam: North Holland Publishing Company.

Groot De, S. R., see Kluitenberg, G. A.

Grot, R. A., Eringen, A. C. [1966]: Relativistic continuum mechanics. Int. J. Engng. Sci. **4**, 611.

Gurtin, M. E. [1965a]: Thermodynamics and possibility of long range interaction in rigid heat conductors. Arch. Rational Mech. Anal. **18**, 335.

[1965b]: Thermodynamics and the possibility of spatial interaction in elastic materials. Arch. Rational Mech. Anal. **19**, 339.

Gurtin, M. E., see Coleman, B. D.

Gyorgyi, G. [1954]: Die Bewegung des Energiemittelpunktes und des Energie-Impuls-Tensors des elektromagnetischen Feldes in Dielektrika. Acta Phys. Hung. **4**, 121.

Herglotz, G. [1911]: Über die Mechanik des deformierbaren Körpers vom Standpunkte der Relativitätstheorie. Ann. Phys. (4) **36**, 493—533.

Hönl, H., Dehnen, H. [1966]: Allgemein-relativistische Dynamik und Machsches Prinzip. Z. Phys. **191**, 313.

Hughes, G. E., Cresswell, M. J. [1968]: An introduction to modal logic. London: Methuen and Co. Ltd., 388 pp.

Hughes, W. F. [1961]: Relativistic magnetohydrodynamics and irreversible thermodynamics. Proc. of the Cambridge Philosophical Society **57**, 878.

Jankiewicz, C. [1962]: A proof of uniqueness of the energy momentum tensor of the electromagnetic field in Riemannian space (in Russian). Bull. int. Acad. pol. Sci. Serie Sci. Math. Astr. Phys. **10**, 299 & 403.

Kleene, S. C. [1952]: Introduction to metamathematics. Amsterdam: North Holland Publishing Co.

Kluitenberg, G. A., De Groot, S. R., Mazur, P. [1953]: Relativistic thermodynamics of irreversible processes. I (heat conduction, diffusion, viscous flow and chemical reaction formal part). Physica **19**, 689. II (heat conduction and diffusion physical part) Physica **19**, 1079.

Kluitenberg, G.A., Mazur, P. [1955]: thermodynamics of irreversible processes. IV (Systems with polarization and magnetization in an electromagnetic field). Physica **21**, 148.

Kneissler, L. [1949]: Die Maxwellsche Theorie in veränderter Formulierung. Wien: Springer.
Kottler, F. [1922]: Maxwellsche Gleichungen und Metrik. Sitzgsber. Akad. Wiss. Wien (IIa) **131**, 119—144.
Kraniŝ, M. [1966]: Relativistic hydrodynamics with irreversible thermodynamics without the paradox of infinite velocity of heat conduction. Il nuovo Cimento **62** B, N 1, 51.
Kretschmann, E. [1917]: Über den physikalischen Sinn der Relativitätspostulate, A. Einsteins neue und seine ursprüngliche Relativitätstheorie. Ann. Physik **55**, 241.
Kröner, E. [1959/60]: Allgemeine Kontinuumstheorie der Versetzungen und Eigenspannungen. Arch. Rational Mech. Anal. **4**, 273.
Kuvshinskii, E. V., see Aero, E. C.
Lamla, E. [1912]: Über die Hydrodynamik des Relativitätsprinzip. Dissertation. Berlin.
Lanczos, C. [1923]: Ein vereinfachendes Koordinatensystem für die Einsteinschen Gravitationsgleichungen. Physikalische Zeitschrift **23**, 537.
Landau, L. D., Lifschitz, E. M. [1959]: Fluid Mechanics (trans. Sykes, J. B., Reid, W. H.) London.
Levi Civitá, T. [1917]: Sulla espressione analitica spettante al tensore gravitazionale nella teoria di Einstein. Rend. Acc. Lincei **26**, 381.
[1928]: Fondamenti di meccanica relativistica. Bologna: Zanichelli.
Lianis, G. [1966]: The formulation of constitutive equations in continuum relativistic physics. Purdue University, School of Aeronautics, Astronautics and Engineering Sciences. Preprint.
[1973]: The general form of constitutive equations in Relativistic physics. Il Nuovo Cimento **14** B, 57.
[1974]: Relativistic Thermodynamics of viscoelastic dielectrics. Arch. Rational Mech. Anal. **55**, 300.
Lianis, G., see Ramirez, G. A.
Liu, I. S., Müller, I. [1972]: On the thermodynamics of fluids in electromagnetic fields. Arch. Rational Mech. Anal. **46**, 149.
Lichnerowicz, A. [1955]: Théories relativistes de la gravitation et de l'électromagnétisme. Paris: Masson.
[1965a]: Théorémes d'existence et d'unicité pour un fluide thermodynamique relativiste. C. R. Acad. Sci. Paris **260**, 3291.
[1965b]: Etude mathématique des équations de la magnétohydrodynamique relativiste. C. R. Acad. Sci. Paris **260**, 4449.
[1967]: Relativistic hydrodynamics and magnetohydrodynamics. New York, Amsterdam: W. A. Benjamin, Inc.
Lüst, R. [1959]: Über die Ausbreitung von Wellen in einem Plasma. Fortschr. d. Physik **7**, 503.
Maugin, G. [1971]: Magnetized deformable media in general relativity. Ann. Inst. H. Poincaré (A) **15**, 275.
[1972]: Sur une possible définition du principe d'indifférence materielle en relativité générale. C. R. Acad. Sci. Paris (A) **275**, 319.
Mazur, P., see Kluitenberg, G. A.
Mizel, V. J., see Coleman, B. D.
Möller, C. [1952]: The theory of relativity. Oxford: Claredon Press.
Müller, I. [1967]: Zum Paradoxon der Wärmeleitungstheorie. Z. Physik **198**, 329.
[1969]: Toward relativistic thermodynamics. Arch. Rational Mech. Anal. **34**, 259.
[1970]: A possible experiment on the principle of objectivity. (Preprint.)
[1972]: On the frame dependence of stress and heat flux. Arch. Rat. Mech. Anal. **45**, 241.
Müller, I., see Alts, T. and Liu, I. S.
Naghdi, P., see Bogy, D.
Noll, W. [1955]: On the continuity of the solid and fluid states. J. Rational Mech. Anal. **4**, 13.
[1958]: A mathematical theory of the mechanical behaviour of continuous media. Arch. Rational Mech. Anal. **2**, 197.
Noll, W., see Truesdell, C. and Coleman B. D.
Painlevé, P. [1922]: Les axiomes de la mechanique. Gauthier Villars, Paris.
Pastori, M., see Finzi, B.
Pham Mau Quan [1955]: Sur une théorie relativiste des fluides thermodynamiques. Ann. Mat. Pura Appl. **38**, 121.

[1956]: Etude électromagnétique et thermodynamique d'un fluide relativiste chargé. J. Rational Mech. Anal. **5**, 473.

Pichon, G. [1965]: Les ondes élastiques dans les milieux chargés. C. R. Acad. Sci. Paris **260**, 3299.

Pitteri, M. [1973]: On the foundations of electromagnetism for polarizable bodies. Il Nuovo Cimento **18**, 144.

[1975a]: Two variational principles of second order materials in general relativity. Ann. Mat. Pura e Appl. (IV) **106**, 315.

[1975b]: Elastic materials of any order $n \geqslant 1$ from the variational point of view in general relativity. Being printed on Ann. Mat. Pura e Appl. (IV).

Pratelli, A. M. [1961]: Casi estremi nella dinamica relativistica di fluidi elettricamente conduttori. Missili **3**, 1.

[1965]: Discontinuità e ipersuperfici caratteristiche in magnetofluido dinamica. Ann. Mat. Pura Appl. (IV) **69**, 41.

Rancoita, G. M. [1959]: Forze in un corpo polarizzato e magnetizzato. Suppl. Nuovo Cimento **11**, 183.

Ramirez, G. A., Lianis, G. [1968]: Relativistic kinematics of deformable continua. I Acta Mechanica **6**, 326.

Rayner, C. B. [1963]: Elasticity in general relativity. Proc. Roy. Soc. London A **272**, 44.

Saini, G. L. [1961a]: Singular hypersurfaces of order one in relativistic magneto-fluid dynamics. Proc. Roy. Soc. London A **260**, 61.

[1961b]: Theory of schoc-dynamics. J. Math. Mech. **10**, 887.

Schmutzer, G. [1964]: Zu den Grundlagen der allgemein-relativistischen Kontinuumsmechanik und Thermodynamik. Ann. Physik **14**, 56.

Schöpf, H. G. [1964a]: Allgemeinrelativistische Prinzipien der Kontinuumsmechanik. Ann. Physik **12**, 337.

[1964b]: Bewegte anisotrope Dielektrika in relativistischer Sicht. Ann. Physik **13**, 41.

[1964c]: Die Lagrangeschen Koordinaten als Feldvariable in der allgemeinrelativistischen Kontinuumsmechanik. Ann. Physik **14**, 121.

[1965a]: Elastische Stoßwellen im Rahmen der allgemeinen Relativitätstheorie. Ann. Physik **15**, 348.

[1965b]: Grundbegriffe der allgemeinrelativistischen Magnetohydrodynamik für ideale Leiter. Ann. Physik **16**, 114.

[1967]: Fourdimensional covariant kinematics of continuum media I. Abhandlungen der deutschen Akademie der Wissenschaften zu Berlin, Klasse für Math., Physik und Technik. Heft 3.

Sedov, L. [1965a]: Mathematical methods for the establishment of new models in continuum physics. Usp. Mat. Nauk. **20**, 121 (Russian).

[1965b]: On the energy-momentum tensor and macroscopic interaction in gravitational fields and material mediums. Dokl. Akad. Nauk. **164**, 519 (Russian).

Schild, A., see Synge, J. L.

Signorini, A. [1954]: Meccanica Razionale. 2nd. ed. Vol. II. Roma: Perella.

Söderholm, L. [1970]: A principle of objectivity for relativistic continuum mechanics. Arch. Rat. Mech. Anal. **39**, 89.

[1976]: The principle of material frame indifference and material equations of gases. Int. Engng. Sci. **14**, 523.

Sommerfeld, A., Bopp, F. [1950]: Zum Problem der Maxwellschen Spannung. Annalen der Physik **8**, 41.

Stratton, J. A. [1941]: Electromagnetic theory. McGraw-Hill Book Co.

Stückelberg, E. C. G. [1962]: Relativistic thermodynamics. III (Velocity of elastic waves and related problems). Helvetica Physica Acta **35**, 568.

Stückelberg, E. C. G., Wanders, G. [1953]: Thermodynamique en Relativité générale. Helvetica Physica Acta **26**, 307.

Suppes, P. [1972]: The essential but implicit role of modal concepts in science. PSA 1972, Dortrecht-Boston: Reidel, p. 305.

Synge, J. L. [1959]: A theory of elasticity in general relativity. Math. Z. **72**, 82.

[1960]: Relativity: the general theory. Amsterdam: North Holland.

[1965]: Relativity: the special theory. 2nd ed. Amsterdam: North Holland.

Synge, J. L., Schild, A. [1954]: Tensor Calculus. Univ. of Toronto Press, Toronto.

Taub, A. H. [1948]: Relativistic Rankine Hugoniot equations. Phys. Rev. (2) **74**, 328.
[1954]: General relativistic variational principle for perfect fluids. Phys. Rev. **94**, 1468.
[1963]: Hydrodynamics and general relativity. Fund. Topics in rel. fluid mechanics and magnetohydrodynamics. (Prod. Symp.) New York, p. 21.
Tolman, R. C. [1930]: On the weight of heat and thermal equilibrium in general relativity. Phys. Rev. **35**, 904.
[1949]: Relativity, Thermodynamics and Cosmology. Oxford: Claredon Press.
Tolman, R. C., Ehrenfest [1930]: Temperature equilibrium in a static gravitational field. Phys. Rev. **36**, 1791.
Toupin, R. A. [1956]: The elastic dielectric. J. Rational Mech. Anal. **5**, 849.
[1957/58]: World invariant kinematics. Arch. Rational Mech. Anal. **1**, 181.
[1962]: Elastic materials with couple stresses. Arch. Rational Mech. Anal. **11**, 385.
[1964]: Theories of elasticity with couple stresses. Arch. Rational Mech. Anal. **17**, 85—112.
Toupin, R. A., see Truesdell, C.
Truesdell, C. [1961]: General and exact theory of waves in finite elastic strain. Arch. Rational Mech. Anal. **8**, 263.
[1976]: Correction of two errors in the kinetic theory of gases which have been used to cast unfounded doubt upon the principle of material frame indifference. Meccanica, 11, p. 196.
Truesdell, C., Toupin, R. A. [1960]: The classical field theories. Handbuch der Physik, Band III/1, pp. 226—793. Berlin-Göttingen-Heidelberg: Springer.
Truesdell, C., Noll, W. [1965]: The non-linear field theories of mechanics. Handbuch der Physik, Band III/3. Berlin-Heidelberg-New York: Springer.
Valcovici, V. [1946]: Sur une interprétation du tourbillon et sur la rotation des directions principales de la déformation. Mathematica (Timisoara) **22**, 57.
van Fraassen, B. C. [1972]: Bressan and Suppes on modality. PSA 1972, Dortrecht-Boston: Reidel, p. 323.
Vernotte, P. [1958]: Les paradoxes de la théorie continue de l'équation de la chaleur. C. R. Acad. Sci. Paris **246**, 3154.
Volterra, V. [1909]: Sulle equazioni integro-differenziali della teoria dell'elasticità. Rend. Lincei (5) **18**, 295. Opere **3**, 288.
[1912]: Sur les équations integro-différentielles et leurs applications. Acta Math. **35**, 295.

Index

Abraham's tensor 99
absolute space 123
— — denied by Einstein's theory of relativity 1
absolute temperature 63
acoustic axis 174
— —, generalized 202
axiomatic foundations of our theory 35—39, 216—218, 223 footn. 1

Black body radiation 119
Born rigidity 192
— Theorem 195
Bridgman, P. W., concerning Newton's space-time frame 5

Cattaneo, C., space-time divergence 44
—, standard time 45
—, heat conduction 70, 72
—, pressure as energy density 84
Cauchy, A. L., equation for continous media 61
— law of motion 231
chronometry, chronometric tensor 37
Clausius Duhem inequality 65, 66
couple stress 226

deformation, pure 145
derivative, absolute — of double tensors 142
— commutation formulas for Lagranian spatial and total covariant derivatives 234, 235
—, covariant partial and total 260, 261
— Lie derivatives, application to linear elasticity 277
—, normally vanishing 264
—, total (time orthogonal) geodesic 206
— — with respect to a double tensor, that may fulfil some regular conditions 261
dielectric tensors 92, 93
dilation coefficient for line elements 149
— surface elements 150,151
— volume elements 149
divergence, of tensors 43, 58, 158
— of spatial vectors 273
double tensor, related to two topological spaces 258
dual $*F_{\alpha\beta}$ of tensor $F_{\alpha\beta}$, introduced to extend Maxwell equations to the case of deforming, rotating and accelerated media 93

Eckart, C., about heat conduction in general and special relativity 26
Einstein equations, complying with Lorentz transformations and including electromagnetic field 1
— in general relativity 25
— neglecting heat conduction 29 footn. 17
— including electromagnetic and thermal phenomena influencing gravitation 29 footn. 18
— gravitation equations in special and general relativity 60
elastic materials, visco-elastic 167
— —, capable of couple stress or polar 238, 240
— —, polarizable 189
— —, polarizable viscous and/or non viscous fluids 122—127
— —, non viscous and/or possibly viscous fluids 78, 79, 80
electric charge, proper density 91
— current 91
— 4-current 91
— field 91
— induction 91
— polarizations 91
electromagnetic, Poynting's theorem of electromagnetic energy 112
— tensor 92
— —, 1st 92
— —, 2nd 92
ellipticity's strong conditions 175
energy, basic theorems on the tensor 109, 111, 112
—, electromagnetic, per unit proper volume 113
—, specific internal, (gravitational mass, variing with) 54
—, tensor of electromagnetic 99, 101, 110, 111
— — see Poynting 112, 113

equation, conservation 60, 61
—, constitutive 167, 189, 190
—, continuity 54
— —, proofs 52, 153
— 1st Cauchy equations of continous media 60, 61, 63, 80, 128, 129, 198
— 2nd Cauchy equations 60, 62, 97, 231
—, Einstein gravitation 60
—, Fourier 64, 65
— —, discussion 69
Maxwell equations 93, 95, 96
— — of energy balance, 1st principle 61, 63, 116, 239
equivalence, physical, of frames 16
—, — of general frames in general relativity 16, 17
Eringen, A. C., and Grot point of view on heat conduction complying with Bressan's 68, 90
Eringen and Dixon about polarizable material 90
event point, primitive notions 35, 216, 223 footn. 1
— —, privileged absolute concept 17
— material point 216

Fermi, E., transport of spatial vectors 147
Fermi triad, (three vectors undergoing Fermi transport) 147
fluids, characterization of non viscous 78—79
—, non viscous 77—80
—, polarizable non viscous 121
—, polarizable viscous 125
—, viscous 77
flux, convected 56
— — applicated to linear elasticity 277
—, co-rotational 57
—, heat flux vector 62
frames, classical locally natural 13, 15, 39, 47
—, co-moving 45
—, Euclidian 47
—, locally Euclidian 48, 49
—, locally geodesic 39
—, locally proper 13, 41, 45, 47, 49
—, relations between classical and relativistic locally natural 46
Fock's conjecture about frames in relativity 19, 23, 77, 123
Fock's critical notes to Einstein's failing to understand their own theories 23
force, contact 7, 226, 227
—, density of ponderomotive 4-force 97, 109
—, elimination of forces at a distance 7
—, ponderomotive force and 4-force 97, 103
free energy, specific 77
Friedrich, world invariance in classical dynamics 8, 28 footn. 10

gas, perfect 80, 81
geodesic 13
Grioli, G., theories about polar materials 226

heat conduction, 9, 10, 61, 75, 85, 87, 119, 120, 273
— —, coefficient (or tensor) of heat conductivity 64, 69
— —, paradox of heat propagation 69
historical hints, at relativistic theories of fluids 87
— — at the development of relativity 1, 2, 25, 26
— — at theories of elastic and more general materials in relativity (Lagranian point of view) 137, 164
— — at theories of elastic waves 168, 169
— — at theories of materials with memory and fading memory in relativity 203
— — at theories of piezo-elasticity and magneto-elastic waves 188
— — at theories of polar materials 226
— — concerning tensors of electromagnetic energy (Minkowski's tensor) 108
— — in the absence or presence of electromagnetic phenomena 118
history, in materials with memory, intrinsic kinematic history up to the instant t of order n... 205
— rotationally and/or translationally equivalent thermo-electro-magneto-kinematic-magneto histories 207
—, Lagranian thermo-electrokinematic 209, 212
— thermo-electromagnetokinematic history of order n... 207

ideal conductor 128, 196
— fluid 45
image, material 152
— of a spatial tensor 190—192
— of a spatial vector 152
indices 36, 139
inertial space 3
intrinsic acceleration 40
invariance of conductors 196
— of physical equations 16
invariance world in classical physics 8
isotropy, spatial 75
— functions 106
— — and tensors 102

Levi Civita, T., on space metric 11, 22
Lorentz frame 2
— transformation 1

magnetic fields 91
— inductions 91

magnetic polarizations 91
mass equivalence principle of energy and mass 6
—, generalized inertial mass quadric 202
—, inertial mass quadric 174
—, invariant mass density 55
—, proper density of conventional mass 55
—, proper gravitational mass 54
—, proper reference or conventional mass 54
material, elastic 164
— — image 152
— — metric 139
—, piezo-elastic or polarizable elastic materials 188
— point 164
— polar material, i.e. capable of couple stress 226
— with memory 203
Maxwell equations 61, 63, 93, 95, 96, 116
metric material 139
—, space time metric, according to Synge 36
—, spatial 44
Minkowski, metric 3
— tensor 99, 100, 109
Müller, J. 73—75

necessary in physically possible worlds 219, 224
Noll, W., material frame indifference 29, footn. 2, 23, 104, 203, 223

Ohm law 91
operational character of physical concepts 4

particle 10, 29 footn. 16
—, test 10, 29 footn. 16
Piola-Kirchhoff, stress tensors 155
physically possible process 219, 224 footn. 16
polar decompositions of the spatial position gradient 145
polarizable material and fluid 164, 121
polarization quadric 174
— —, generalized 202
ponderomotive force, work of ponderomotive forces 97
Poynting's vector, theorem of possibly deforming media 112, 113
principal axis and triad of strain 146, 177
— directions of tensions 177
— material axis of tension 177
— wave 177, 178
principle, 1st, of thermodynamics 61, 63, 116, 239
—, 2nd 64, 121
— — combined with the equation of energy balance 77, 79, 121, 125, 165, 188, 189
—, equivalence, of mass and energy 6
—, local equivalence 14
— of material frame indifference, or objectivity principle in full form 209
— — in the 1st and 2nd reduced form 209
—, physical, of relativity in the sense of Fock 16, 17
—, varionational, in the absence of couple stress 182, 186
— — presence of couple stress 248, 252
process, world (thermo-electromagnetokinetic, admissible, actual ...) 220, 221
proper length 4
— mass 6, 54
— time 3
— volume 52
— space, momentarily 3

reference configuration 139
— metric (also material metric) 139
— physical state 139
— representation of a configuration 218
relativity, what we mean by general 23
rigid 192
rotation 145

Schöpf, G. 164, 179, 202 footn. 3
Sedov, L. 26, 109, 119, 226
space, absolute 1
spatial index of a tensor 42
— metric 44
— projector 41
— section (cross) 38
— rotation 102
specific internal energy 54
— physical magnitude 55
standard time and velocity (according to C. Cattaneo) 45, 171
stationary tensor 194
Stefan Boltzman laws 119
strain tensor 146
— — principal axis or triad 146, 177
stress tensor, eulerian 62
— —, 1st and 2nd Piola-Kirchhoff tensors 155
Synge 35—37, 137

Taub equations for shock waves 85
— principles for ideal fluids 179
thermometer, pocket- 120
tensor, isotropic, spatially 103
—, natural decomposition of 42, 43
—, non-working part of the couple stress tensor 233
—, spatial 42
—, spatial part of 42, 43
—, spatial Ricci tensor 51
—, stationary 194
—, temporal and mixed parts of tensors 43
thermal conduction inequality 65—67

thermodynamic tensor 62, 73, 85
time parameter 139
— orthogonal parameter 141
Tolman, R. C., pocket temperature in relativity 120
torque, intrinsic, of forces 226
Toupin, R. A. theories 28 footn. 9—11

Uniqueness, properties of the tensor of electromagnetic energy 105, 269

variation of space-time metric 179, 247, 253
— of world lines 182, 249, 253
varionational principles 182, 186, 248, 252
velocity, 4-velocity 40
— angular and deformation velocities 56

waves in elastic or piezo-elastic bodies 172, 200
— in kinematic treatment 81, 168
— in non viscous fluids 81, 131
— in perfect gases 84
— in the isotropic case 176
work of contact forces 156, 228
— — — per unit reference proper volume 156, 238
— of ponderomotive forces per unit proper time and actual proper volume 97
— of the electromagnetic field under microscopical conditions on matter per unit conventional mass 110, 111, 155
world line 28, 216
— — invariance 8, 16
— — process, corresponding to a set of constitutive functionals 220, 221
— — tube 40

Volume 14: Bell
The Physics of Large Deformation of Crystalline Solids
166 figures. X, 253 pages. 1968

Volume 15: Buchholz
The Confluent Hypergeometric Function. With Special Emphasis on Its Applications
XVIII, 238 pages. 1969

Volume 16: Slepian
Mathematical Foundations of Network Analysis
XI, 195 pages. 1968

Volume 17: Gavalas
Nonlinear Differential Equations of Chemically Reacting Systems
10 figures. IX, 107 pages. 1968

Volume 18: Marti
Introduction to the Theory of Bases
XII, 149 pages. 1969

Volume 19: Knops, Payne
Uniqueness Theorems in Linear Elasticity
IX, 130 pages. 1971

Volume 20: Edelen, Wilson
Relativity and the Question of Discretization in Astronomy
34 figures. XII, 186 pages. 1970

Volume 21: McBride
Obtaining Generating Functions
VIII, 100 pages. 1971

Volume 22: Day
The Thermodynamics of Simple Materials with Fading Memory
8 figures. X, 134 pages. 1972

Volume 23: Stetter
Analysis of Discretization Methods for Ordinary Differential Equations
12 figures. XVI, 388 pages. 1973

Volume 24: Strieder, Aris
Variational Methods Applied to Problems of Diffusion and Reaction
12 figures. IX, 109 pages. 1973

Volume 25: Bohl
Monotonie: Lösbarkeit und Numerik bei Operatorgleichungen
9 Abbildungen. IX, 255 Seiten. 1974

Volume 26: Romanov
Integral Geometry and Inverse Problems for Hyperbolic Equations
21 figures. VI, 152 pages. 1974

Volume 27: Joseph
Stability of Fluid Motions I
57 figures. XIII, 282 pages. 1976

Volume 28: Joseph
Stability of Fluid Motions II
39 figures. XIV, 274 pages. 1976

Springer-Verlag
Berlin
Heidelberg
New York